T0328615

Satellite and Terrestrial Radio Positioning Techniques

Satellite and Terrestrial Radio Positioning Techniques

A Signal Processing Perspective

Edited by

Davide Dardari

Emanuela Falletti

Marco Luise

AMSTERDAM • BOSTON • HEIDELBERG • LONDON
NEW YORK • OXFORD • PARIS • SAN DIEGO
SAN FRANCISCO • SINGAPORE • SYDNEY • TOKYO

Academic Press is an imprint of Elsevier

Academic Press is an imprint of Elsevier
The Boulevard, Langford Lane, Kidlington, Oxford, OX5 1GB, UK
225 Wyman Street, Waltham, MA 02451, USA

First edition 2012

British Library Cataloguing in Publication Data
A catalogue record for this book is available from the British Library.

Library of Congress Cataloging-in-Publication Data
A catalog record for this book is available from the Library of Congress.

ISBN: 978-0-08-101596-4

For information on all Academic Press publications
visit our web site at *www.elsevierdirect.com*

Printed and bound in the UK

11 12 13 14 15   10 9 8 7 6 5 4 3 2 1

Contents

Preface

Reliable and accurate positioning and navigation is critical for a diverse set of emerging applications calling for advanced signal-processing techniques. This book provides an overview of some of the most recent research results in the field of signal processing for positioning and navigation, addressing many challenging open problems.

The book stems from the European Network of Excellence in Wireless Communications NEWCOM++, in which I was privileged to be involved as both an external observer and a contributor. The Network of Excellence is an initiative of the European Commission, which gives an opportunity to excellent researchers across the continent to build new levels of collaboration. Within the framework of this initiative, there has been an activity focused on the development of signal-processing techniques to provide high-accuracy location awareness.

This book considers many different aspects and facets of positioning and navigation techniques. It begins with "classical" technologies for positioning in satellite systems (e.g., GPS and Galileo) and in terrestrial cellular networks. The reader will also find new topics including the *ultimate bounds* on the accuracy of positioning systems determined by noise and interference; the description and performance of some new techniques such as *direct positioning* that aim at making GPS work with very weak received radio signals (e.g., indoors); as well as the techniques to optimally *combine* the measurements coming from radio signals and from different sensors like inertial platforms (e.g., gyroscopes). The new field of *cooperative* positioning is also discussed, wherein many nodes exchange signals and information to increase the accuracy of their positions, and finally the exciting field of super-accurate indoor ranging with ultra-wide bandwidth (UWB) radio signals is thoroughly addressed.

The combination of theory and experimentation in the NEWCOM++ project has led to practical results that the readers can find in the last part of the book. As an example of the direct application of the research forefront to real-world problems, fusion techniques for integration of multiple sensor measurements based on experimental data are explored. I hope this book can serve as a reference for anyone who is interested in the field of positioning and navigation.

Moe Z. Win
Associate Professor
Massachusetts Institute of Technology

Foreword

Many of the readers of this book may have had the occasion to get acquainted with the adventures of Harry Potter in the best-selling works by J.K. Rowling. If so, they will have noticed that young Harry has got something that is called the "Marauder's Map": a piece of parchment that shows every inch of the magical school of Hogwarts, as well as the ever-changing, real-time location of Harry's friends and foes. *Wow*, if it is in Harry Potter's book, it must be something *magic*, the layman wonders. But, the readers of this book know better: it is not magic, but *technology*. In the cold language of engineers, the Marauder's map is a *geographic information system* (GIS) with a dedicated *positioning* plug-in that tracks real-time, a set of authorized users, and show their locations upon a the map on a display. The GIS is something that anyone can have on his/her smartphone at a small cost. But, something that heavily relies on a number of different techniques ranging from radio transmission to geometric computation, from data mining to Kalman filtering, and all of them deriving from the common, unifying umbrella of *signal processing*, that represents the common background of the many positioning appliances that are now widespread in developed countries, like the GPS car navigators. Such ubiquitous positioning devices, in cars or in smartphones, are the basis for a number of innovative context-aware services that are nowadays already available. For example, looking for a pharmacy in a chaotic big city is no longer like treasures hunting, but we are only at the beginning: in the coming years, we will see the advent of *high-definition* situation-aware applications, based on the availability of positioning information with submeter accuracy, and required to operate even in harsh propagation environments such as inside buildings. The number of newly offered services is only limited by phantasy, and is expected to grow exponentially, together with the corresponding market revenues.

However, the path towards this goal is still challenging. Some of the current positioning technologies were primarily designed for different applications (e.g., managing a communication network), and are not optimized for providing accurate and ever-available location information. In addition, none of the positioning technologies currently available or under development ensures service coverage in different heterogeneous environments (e.g., outdoor, indoor, at sea, and on the road), and high-definition positioning accuracy. In conclusion, the integration of different positioning technologies is the pivotal aspect for future seamless positioning systems, and the key to ignite a new era of ubiquitous location-awareness.

So far, most books related to positioning address the topic focusing on a specific system, for example, satellite-based or terrestrial, or are single-technology oriented (GPS or RF Tags just to mention a few). However, the mechanism with which the different positioning systems derive information about the user location share, in many cases, the same fundamental approach. In addition, the design of future seamless positioning systems cannot leave aside a global knowledge of different technologies if their efficient integration has to be pursued.

With this in mind, we tried to provide in this book a broad overview of satellite and terrestrial positioning and navigation technologies under the common denominator of *signal processing*. We are convinced that every positioning problem can be ultimately cast into the issue of designing a signal processor (to be specific, a *parameter estimator*) which provides the most accurate user's location, starting from a set of noisy position-dependent measurements collected through signal exchanges between the wireless devices involved. Our aim was not to simply give a mere description of the various current positioning standards or technologies. Rather, we intended to introduce and illustrate the theoretical foundation that lies behind them, and to describe a few advanced practical solutions to the positioning issue, strengthened by case studies based on experimental data.

This book takes advantage of the contribution of several experts participating to the European Network of Excellence NEWCOM++, of which it represents one of the main outcomes. Most of the material has been originated from a bunch of enthusiastic young researchers working in a cooperative environment. The readers may have noticed that this is an edited book, with many contributors. Although, it may be difficult to coordinate and homogenize the work of so many researchers (and we hope we succeeded in this goal), this is a case where "diversity" shines. The different approaches to the general issue of positioning coming from different institutions and research "schools" will be apparent to the readers – we do hope that such diversity (that in our opinion is the added-value of the book) will contribute widening his/her perspective on the subject.

This book is intended for PhD students and researchers who aim at creating a solid scientific background about positioning and navigation. It is also intended for engineers who need to design positioning systems and want to understand the basic principles underlying their performance. Even if less importance is given to an exhaustive description of available literature, the table of contents is also designed to provide a book useful for the beginners.

For a brief survey of the basic theory of positioning and navigation, the first three chapters may be read, whereas more advanced concepts and techniques are provided in the successive chapters.

Specifically, Chapter 1 introduces the concept of radio positioning and states the mathematical problem of determining the position of a mobile device in a certain reference frame, using measurements extracted from the propagation of radio waves between certain reference points and the mobile device. It presents a classification of the wireless positioning systems based, on one hand, the kind of information (or measurement) they extract from the propagating signal and on the other hand, the kind of network infrastructure established among the devices involved in the localization process. Then, it goes through an introductory description of the main positioning systems examined in the book, namely satellite systems, their terrestrial augmentation and assistance systems, terrestrial network-based systems (e.g., cellular networks, wireless LANs, wireless sensor networks, and ad-hoc networks).

Finally, an overview of the fundamental mathematical methodologies suited to resolve the radio positioning problem in the above-cited contexts is given, in tight association with the signal processing approaches able to implement them in a technological context.

Chapter 2 presents an overview of the satellite-based positioning systems, with particular emphasis on the American GPS, the forthcoming European Galileo and the modernized Russian GLONASS, which provide almost global coverage of the Earth Global Navigation Satellite Systems (GNSSs).

First, the "space segment" of such systems, in terms of transmitted signal formats and occupied bands is described. Then, the architecture of a typical satellite navigation receiver is discussed in detail, as it has several peculiar requirements and features with respect to a communication-oriented transceiver. A discussion of the main sources of error in the position estimate is then presented. The last part of the chapter is devoted to present the so-called "augmentation systems", a category of mostly terrestrial network-based systems aimed at providing support to the GNSS receiver to improve the accuracy or the availability of its position estimate. Examples of such systems are: differential GPS, EGNOS, network RTK, and assisted GNSS.

The fundamental technologies and signal processing approaches to estimate the position of a mobile device using terrestrial networks-based radio communication systems are addressed in Chapter 3. The potential position-related information that can be extracted from a propagating signal is reviewed, namely: received signal strength (RSS), time-of-arrival (TOA), time-difference-of-arrival (TDOA), and angle-of-arrival (AOA).

Then the fundamental techniques to derive the position information from a collection of such measurements are explained, according to the classification in geometric techniques (either deterministic or statistical) and mapping (or fingerprinting) techniques. The most common sources of error affecting the above-mentioned processes are then analyzed.

The chapter continues presenting the positioning approaches typically adopted in different network technologies (i.e., cellular networks, wireless LANs, and wireless sensor networks), addressing the underlying signal format, the most suited kind of measurement and the associated positioning and navigation algorithms. Particular attention is devoted to the ultra-wideband technology, as the most promising signal format to implement high performance terrestrial positioning.

Several factors impact in practice on the achievable accuracy of wireless positioning systems. However, theoretical bounds can be set in order to determine the best accuracy, one may expect in certain conditions as well as to obtain useful benchmarks when assessing the performance of practical schemes. Chapter 4 is dedicated to the presentation of several such bounds, mostly derived from the Cramér-Rao bound (CRB) framework. Theoretical performance bounds related to the ranging estimation via time-of-arrival from UWB signals are derived and discussed, also taking into account the critical conditions such as the multipath propagation. Also, the improved Ziv-Zakai bound family is introduced as a tighter benchmark in the case of dense scattering, where the CRB falls in the ambiguity region.

Then, novel results are presented, related to the derivation of performance limits for innovative positioning approaches, such as direct position estimation (DPE) in GNSS, cooperative terrestrial localization, and a recent analysis on the interference-prone systems, such as multicarrier systems.

Chapter 5 presents a collection of the latest research results in the field of wireless positioning, carried out within the NEWCOM++ Network of Excellence. It shows a necessarily-partial panorama of the "hottest topics" in advanced wireless positioning, within the applicative and technological framework drawn in the previous chapters.

The focus is first oriented to the recent advances in UWB positioning algorithms, considering a frequency-domain approach for TOA estimation, a joint TOA/AOA estimation algorithm, the impairment due to interference, and the mitigation of the nonline-of-sight bias effect. Then, an application of MIMO systems for positioning is discussed. Non-conventional geometrical solutions for positioning are represented by the bounded-error distributed estimation and the projection onto convex sets (POCS) approach. POCS is then revisited in the context of cooperative positioning, together with a cooperative least-squares approach and a distributed algorithm based on belief propagation. Finally, the cognitive positioning concept is introduced as a feature of cognitive radio terminals. After deriving the expected performance bound, optimum signal design for positioning purposes is addressed and positioning approaches are discussed.

Chapter 6 is devoted to present the several signal processing strategies to combine together, in a seamless estimation process, position-related measurements coming from different technologies and/or systems (e.g., TOA and TDOA measurements in terrestrial networks, TOA and RSS measurements, or even satellite and terrestrial systems, or satellite and inertial navigation systems). This approach, generally indicated as "hybridization", promises to provide better accuracy with respect to its stand-alone counterparts, or better availability thanks to the diversity of the employed technologies. For example, hybridization between satellite and inertial systems is expected to compensate the respective fragilities of the two systems, namely: the relatively high error variance of the former and the drift of the latter.

The mathematical framework where hybridization is developed is Bayesian filtering. The generic structure is reviewed and the well-known Kalman filter and its variants are inserted in the framework, with examples of applications to positioning problems. Then the particle filter approach is explained, with its most used variants.

Examples of hybrid localization algorithms are then shown, starting from an hybrid terrestrial architecture, then passing to the architectures that blend GNSS and inertial measurements, using either the Kalman filter approach or the direct position estimation approach. Finally, an example of hybrid localization based on GNSS and peer-to-peer terrestrial signaling is presented.

Chapter 7, the final part of this book, is dedicated to some case studies. Real-world application examples of positioning and navigation systems, which are the results of experimental activities performed by the researchers involved in the NEWCOM++ Network of Excellence, are reported.

Acknowledgements

The authors would like to thank Sergio Benedetto, the Scientific Director of the NEWCOM++ Network of Excellence, for his unique capability of leading and managing this large network during these years. They would also like to explicitly acknowledge the support and cooperation of the Project Officers of the European Commission, Peter Stuckmann and Andy Houghton, that who facilitated the development of the research activities of NEWCOM++. The writing of this book would not have been possible without the contribution of all partners involved in the NEWCOM++ "Localization and Positioning" work package which the authors M. Luise and D. Dardari had the honor to lead. The authors Special specially thanks go to Carles Fernández-Prades, Sinan Gezici, Monica Nicoli, and Erik G. Ström, for their invaluable contribution to the structure and organization of the book.

Acronyms and Abbreviations

ACGN	additive colored Gaussian noise
ACK	acknowledge
ACRB	average CRB
ADC	analog-to-digital converter
AEKF	adaptive extended Kalman filter
AFL	anchor-free localization
AGNSS	assisted GNSS
AGPS	assisted GPS
AltBOC	alternate binary offset carrier
AN	anchor node
AOA	angle of arrival
AOD	angle of departure
AP	access point
API	application programming interface
ARNS	aeronautical radio navigation services
ARS	accelerated random search
A-S	anti-spoofing
AS	azimuth spread
ASIC	application-specific integrated circuit
AWGN	additive white Gaussian noise
BCH	Bose–Chaudhuri–Hocquenghem
BCRB	Bayesian CRB
BIM	Bayesian information matrix
BLAS	basic linear algebra subprograms
BLUE	best linear unbiased estimator
BOC	binary offset carrier
BP	belief propagation
BPF	band-pass filter
bps	bits per second
BPSK	binary phase shift keying
BPZF	band-pass zonal filter
BS	base station
BSC	binary symmetric channel
BTB	Bellini–Tartara bound
BTS	base transceiver station
C/A	coarse/acquisition
C/NAV	commercial/navigation
C/N_0	carrier-to-noise density ratio
CAP	contention access period
CBOC	composite binary offset carrier
CC	central cluster

CCK	complementary code keying
CDF	cumulative density function
CDM	circular disc monopole
CDMA	code division multiple access
CE-POCS	orthogonal projection onto circular and elliptical convex sets
CFP	contention free period
CH	cluster head
CIR	channel impulse response
CKF	cubature Kalman filter
CL	civil-long
CM	civil-moderate
CNLS	constrained NLS
CNSS	compass navigation satellite system
Coop-OA	cooperative OA
Coop-POCS	cooperative POCS
COTS	commercial off-the-shelf
CP	cognitive positioning
CPICH	common pilot channel
CPM	continuous-phase-modulated
C-POCS	orthogonal projection onto circular convex set
CPR	channel pulse response
CPS	cognitive positioning system
cps	chips per second
CPU	central processing unit
CR	cognitive radio
CRB	Cramér–Rao lower bound
CRC	cyclic redundancy check
CRPF	cost-reference particle filter
CS	control segment/commercial service
CSI	channel state information
CSS	chirp spread spectrum
CTS	clear-to-send
CW	continuous wave
DAA	detect and avoid
DAB	digital audio broadcasting
DCM	direction cosine matrix
DE	differential evolution
DEPE	delay estimation through phase estimation
DFE	digital front-end
DFT	discrete Fourier transform
DGPS	differential GPS
DIFS	DCF interframe spacing
DL	down-link
DLL	delay-locked loop

DMLL	distributed maximum log-likelihood
DOA	direction of arrival
DoD	Department of Defense
DP	direct path
DPCH	dedicated physical channel
DPE	direct position estimation
DS	delay spread
DSP	digital signal processor
DSSS	direct sequence spread spectrum
DVB	digital video broadcasting
dwMDS	distributed weighted multidimensional scaling
EB	energy-based
ECEF	Earth-centered, Earth-fixed
ED	energy detector
EEPROM	electrically erasable programmable read-only memory
EGNOS	European geostationary navigation overlay system
EIRP	effective isotropic radiated power
EKF	extended Kalman filter
EKFBT	extended Kalman filter with bias tracking
E-L	early-minus-late
EPE	Ekahau positioning engine
E-POCS	orthogonal projection onto elliptical set
ERQ	enhanced robust quad
ESA	European Space Agency
EU	European Union
F/NAV	freely accessible navigation
FB-MCM	filter-bank multicarrier modulation
FCC	Federal Communications Commission
FDMA	frequency division multiple access
FEC	forward error correction
FFD	full function device
FFT	fast Fourier transform
FHSS	frequency hopping spread spectrum
FIM	Fisher information matrix
FLL	frequency-locked loop
FMT	filtered multitone
FOC	full operational capability
FPGA	field-programmable gate array
FPK	Flächen-Korrektur-Parameter (area correction parameters)
GAGAN	GPS-aided GEO augmented navigation
GANSS	Galileo/additional navigation satellite systems
GDOP	geometric dilution of precision
GEO	geostationary
GFSK	Gaussian-shaped binary frequency shift keying

GIOVE	Galileo in-orbit validation element
GIS	geographical information system
GLONASS	global orbiting navigation satellite system
GNSS	global navigation satellite system
GPIB	general purpose interface bus
GPRS	general packet radio service
GPS	global positioning system
GS	geodetic system
GSM	global system for mobile communications
GST	Galileo system time
GUI	graphical user interface
IIDL	hardware description language
HDLA	high-definition location awareness
HDSA	high-definition situation aware
hdwMDS	hybrid dwMDS
HEO	highly inclined elliptical orbits
HMM	hidden Markov model
HOW	handover word
HPOCS	hybrid POCS
HW	hardware
I	in-phase
i.i.d.	independent, and identically distributed
I/NAV	integrity/navigation
IBERT	integrated bit error ratio tester
IC	integrated circuit
ICD	interface control document
ICT	information and communication technologies
IE	informative element
IF	intermediated frequency
IGSO	inclined geosynchronous orbit
ILS	instrument landing system
IMU	inertial measurement unit
INR	interference-to-noise power ratio
INS	inertial navigation system
IODC	issue of data clock
IODE	issue of data ephemeris
IP	intellectual property
IR	impulse radio
IRNSS	regional navigation satellite system
IR-UWB	impulse radio UWB
ISM	industrial scientific medical
ISO/IEC	International Organization for Standardization / International Electrotechnical Commission
ISRO	Indian Space Research Organization

IST	information society technologies
ITU	International Telecommunication Union
ITS	intelligent transportation system
IVP	inertial virtual platform
JBSF	jump back and search forward
KF	Kalman filter
KNN	k-nearest-neighbor
LAAS	local area augmentation system
LAMBDA	least-squares ambiguity decorrelation adjustment
LAN	local area network
LAPACK	linear algebra package
LBS	location-based service
LCS	location services
LDC	low duty cycle
LDPC	low-density parity check
LEO	localization error outage
LIFO	last-in first-out
LLC	logical link control
LLR	log-likelihood ratio
LNA	low noise amplifier
LOB	line of bearing
LOS	line of sight
LRT	likelihood ratio test
LS	least-squares
LSB	least significant bit
LTE	long-term evolution
LVDS	low-voltage differential signaling
MAC	medium access control
MAP	maximum a posteriori
MAI	multiple access interference
MBOC	multiplexed binary offset carrier
MB-UWB	multiband UWB
MC	multicarrier
MCAR	multiple carrier ambiguity resolution
MCRB	modified CRB
MEO	medium earth orbit
MEMS	electromechanical systems
MF	matched filter
MGF	moment generating function
MHT	multiple-hypotheses testing
MIMO	multiple-input multiple-output
MISO	multiple-input single-output
ML	maximum likelihood
MLE	maximum likelihood estimator

MMSE	minimum mean square error
MOM	method of moments
MP	multipath
MPC	multipath component
MPEE	multipath error envelope
MRC	maximal ratio combining
MS	mobile station
MSAS	multifunctional satellite augmentation system
MSB	most significant bit
MSE	mean square error
MSEE	mean square estimation error
MSK	minimum-shift-keying
MST	minimum spanning tree
MTSAT	multifunctional transport satellite
MUI	multiuser interference
MV	minimum variance
N/A	not available
NAV	navigation
NAVSTAR	navigation system for timing and ranging
NB	narrowband
NBI	narrowband interference
NCO	numerically controlled oscillator
NDIS	network driver interface specification
NED	north-east-down
NFR	near-field ranging
NLOS	non-line of sight
NLS	nonlinear least squares
NMEA	National Marine Electronics Association
NMV	normalized minimum variance
NN	neural network
NOLA	nonoverlapping assumption
NPE	Navizon positioning engine
NQRT	new quad robustness test
NRE	nonrecurring expenditures
NRZ	nonreturn to zero
NSI5	nonstandard I5
NSQ5	nonstandard Q5
NTP	network time protocol
OA	outer approximation
OCS	operational control segment
OEM	original equipment manufacturer
OFDM	orthogonal frequency division multiplexing
OMA	open mobile alliance
OMUX	output multiplexer

OOB	out of band
OQPSK	offset quadrature phase-shift keying
OQRT	original quad robustness test
ORQ	original robust quad
OS	open service
OTD	observed time difference
OTDOA	observed TDOA
P2P	peer-to-peer
PAM	pulse amplitude modulation
PAN	personal area network
PC	personal computer
PDA	personal digital assistant
pdf	probability density function
PDP	power delay profile
PF	particle filter
PHR	physical header
PHY	physical layer
PLL	phase-locked loop
PN	pseudonoise
PND	personal navigation device
POC	payload operation center
POCS	projections onto convex sets
POR	projection onto rings
PPM	pulse position modulation
ppm	parts per million
PPS	precise position service
PR	pseudorandom
PRN	pseudorandom noise
PRS	public regulated service
PRT	partial robustness test
PSD	power spectral density
PSDP	power spatial delay profile
PSDU	physical service data unit
PSK	phase shift keying
PVT	position, velocity, and time
PW	pulse width
pTOA	pseudo time of arrival
PV	position–velocity
Q	quadrature phase
QPSK	quadrature phase shift keying
QZSS	quasi-zenith satellite system
RDMV	root derivative minimum variance
RDSS	radio determination satellite service
RF	radio frequency

RFD	reduced function device
RFID	radio frequency identification
RIMS	ranging and integrity monitoring stations
RLE	robust location estimation
RMS	root mean square
RMSE	root mean square error
RMV	root minimum variance
RN	reference node
RNSS	regional navigation satellite system
ROA	rate of arrival
ROC	receiver operational characteristic
ROM	read-only memory
RQ	robust quadrilateral
RRC	root raised cosine/radio resource control
RRLP	radius resource location protocol
RSS	received signal strength
RT	robust trilateration
RTCM	radio technical commission for maritime services
RTK	real-time kinematic
RTLS	real-time locating system
RTS	ready to send
RTT	round-trip time
RV	random variable
RX	receiver
SA	selective availability
SAR	search and rescue
SAW	surface acoustic wave
SBAS	satellite-based augmentation system
SBS	serial backward search
SBSMC	serial backward search for multiple clusters
SCKF	square-root cubature Kalman filter
SCPC	single channel per carrier
SDR	software defined radio
SDS	symmetric double sided
SET	SUPL enabled terminal
SFD	start-of-frame delimiter
SHR	synchronization header
SIFS	short interframe spacing
SIMO	single-input multiple-output
SIR	sequential importance resampling
SIS	signal-in-space
SISO	single-input single-output
SLP	SUPL location platform
SMA	subminiature version A

SMC	sequential Monte Carlo
SMR	signal-to-multipath ratio
SNIR	signal-to-noise-plus-interference ratio
SNR	signal-to-noise ratio
SoL	safety of life
SPKF	sigma-point Kalman filter
sps	symbols per second
SPS	standard position service
SQKF	square-root quadrature Kalman filter
SRN	secondary reference node
SRS	same-rate service
SS	spread spectrum
SS-CPM	spread spectrum continuous-phase-modulated
SS-GenMSK	spread-spectrum generalized-minimum-shift-keying
ST	simple thresholding
SUPL	secure user-plane location
SV	satellite vehicle
SVD	singular value decomposition
SW	software
SYNCH	synchronization preamble
TCAR	three carrier ambiguity resolution
TDE	time delay estimation
TDOA	time difference of arrival
TH	time hopping
TH-PPM	time-hopping pulse position modulation
TI	trilateration intersection
TLM	telemetry
TLS	total least squares
TLS-ESPRIT	total least-squares estimation of signal parameters via rotational invariance techniques
TMBOC	time-multiplexed binary offset carrier
TNR	threshold-to-noise ratio
TOA	time of arrival
TOF	time of flight
TOW	time of week
TRS	two-rate service
TTFF	time-to-first-fix
TW-TOA	two-way TOA
TX	transmitter
UE	user equipment
UERE	user equivalent range error
UKF	unscented Kalman filter
UL	uplink
ULA	uniform linear array

ULP	user location protocol
UMTS	universal mobile telecommunications system
UN	unknown node
URE	user range error
U.S.	United States
US	user segment
UT	user terminal
UTC	coordinated universal time
UTM	universal transverse Mercator
UTRA	UMTS terrestrial radio access
UWB	ultra-wide bandwidth
VANET	vehicular ad hoc network
VHDL	VHSIC hardware description language
VHSIC	very high speed integrated circuit
VNA	vector network analyzer
VRS	virtual reference station
WAAS	wide area augmentation system
WADGPS	wide area differential GPS
WARN	wide area reference network
WB	wideband
WBI	wideband interference
WCDMA	wideband code division multiple access
WE	wireless extensions
WED	wall extra delay
WGS84	world geodetic system
WiMAX	worldwide interoperability for microwave access
WLAN	wireless local area network
WLS	weighted least squares
WMAN	wireless metropolitan area network
WPAN	wireless personal area network
WRAPI	wireless research application programming interface
WRR	pulse width to average multipath component rate of arrival ratio
wrt	with respect to
WSN	wireless sensor network
WT	wireless tools
WWB	Weiss–Weinstein bound
ZZB	Ziv–Zakai lower bound

Introduction

Davide Dardari, Emanuela Falletti, Francesco Sottile

1.1 THE GENERAL ISSUE OF WIRELESS POSITION LOCATION

1.1.1 Context and Applications

Locating is a process used to determine the location of one position relative to other defined positions, and it has been a fundamental need of human beings ever since they came into existence. In fact, in the pretechnological era, several tools based on observation of stars were developed to deal with this issue.

In the technological era, it is possible to localize persons and objects in real time by exploiting radio transmissions (in the following denoted as wireless transmissions). In this context, the global positioning system (GPS) is for sure the most popular example of satellite-based positioning system, which makes it possible for people with ground receivers to pinpoint their geographic location [24].

Nowadays, position awareness is becoming a fundamental issue for new location-based services (LBSs) and applications. Specifically, wireless positioning systems have attracted considerable interest for many years [1, 7, 12–14, 16, 22, 23, 26, 28, 29, 33, 35, 40].

One of the leading applications of positioning techniques is transportation in general, and intelligent transportation systems (ITSs) in particular, including accident management, traffic routing, roadside assistance, and cargo tracking [17], which span the mass utilization of the well-known GPS. Safety is one of the main motivations for civilian mobile position location, whose implementation is mandatory for the emergency calls originated by dialing 112 (in Europe) or 911 numbers (in the U.S.A.) [18, 21]. Furthermore, LBSs are nowadays attracting more and more interest and investments, since they pave the way for completely new market strategies and opportunities, based on mobile local advertising, personnel tracking, navigation assistance, and position-dependent billing [23, 28]. A pictorial representation of a context-aware service management architecture is shown in Fig. 1.1.

In the coming years, we will see the emergence of *high-definition situation-aware* (HDSA) applications capable of operating in harsh propagation environments, where GPS typically fails, such as inside buildings and in caves. Such applications require positioning systems with submeter accuracy [14]. Reliable localization in such conditions is a key enabler for a diverse set of applications, including logistics, security tracking (the localization of authorized persons in high-security areas), medical services (the monitoring of patients), search and rescue operations (communications with fire fighters or natural disaster victims), control of home appliances, automotive safety, and military systems. It is expected that the global revenues coming from real-time locating systems (RTLSs) technology will amount to more than six billion Euros in 2017 [6].

Satellite and Terrestrial Radio Positioning Techniques. DOI: 10.1016/B978-0-12-382084-6.00001-5

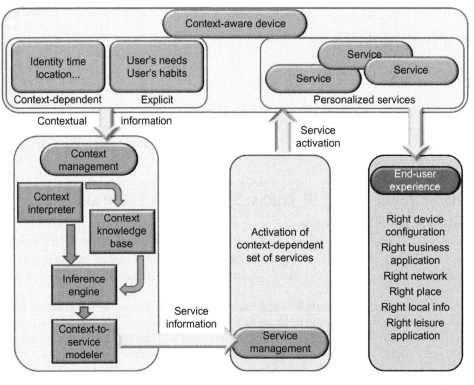

FIGURE 1.1

Concept of context-aware service management architecture.

As will be clear during the reading of this book, none of the current and under-study positioning technologies alone is able to ensure service coverage in different heterogeneous environments (e.g., outdoor, indoor) while offering high-definition positioning accuracy. The integration of different positioning technologies appears to be key to seamless future RTLSs, which will ignite a new era of ubiquitous location awareness.

1.1.2 Classification of Wireless Positioning Systems

The primary characteristic of wireless *position location* is that it implies the presence of an "active" terminal, whose position has to be determined. This situation is fundamentally different from *radio-location*, which usually refers to finding a "passive" distant object that by no means participates in the location procedure; for example, radars implement a radiolocation procedure. For this reason, radio-location is often related to military and surveillance systems. On the contrary, an "active" terminal performing position location is supposed to actively participate in determining its own position, taking appropriate measurements and receiving/exchanging wireless information with some reference station(s). The position information is generally used by the terminal itself, but can also be forwarded to some kind of control station responsible for the activities of the terminal. Position location refers therefore to a large family of systems, procedures, and algorithms, born in the military field but

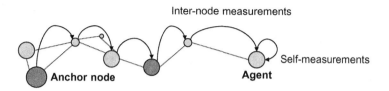

FIGURE 1.2

General positioning network.

recently expanded in a countless set of civil applications. In this book, the terms "position location," "positioning," and "localization" are interchangeable.

A fundamental difference exists between *position location* and *(radio)navigation*. Indeed, navigation refers to "the theory and practice of planning, recording, and controlling the course and position of a vehicle, especially a ship or aircraft."[1] This means that navigation systems are able not only to determine the punctual position of the terminal but also to *track its trajectory* after the first position fix. In navigation, trajectory tracking is more than a mere sequence of independent location estimates, since it often involves the estimation of tri-axial velocity and possibly acceleration.

Wireless positioning systems have a number of reference wireless nodes (*anchor* nodes) at fixed and precisely known locations in a coordinate reference frame and one or more mobile nodes to be located (often referred to as *agent*, *target* or *mobile user*) (see Fig. 1.2). The terminology is not universal, but it depends on the technology behind: In cellular-based positioning systems the term *base station* (BS) is used to refer to radio frequency (RF) devices with known coordinates, while *mobile station* (MS) is used to refer to RF devices with unknown coordinates, sometimes also indicated as *user terminal* (UT) or *user equipment* (UE). In the context of wireless sensor networks (WSNs), the RF devices are usually indicated as *nodes*, being an *anchor* node with known coordinates and an *agent* node with unknown coordinates.

Positioning typically occurs in two main steps: First, specific measurements are performed between nodes and, second, these measurements are processed to determine the position of agent nodes. A typical example of measured data is the distance between the nodes involved. This measurement is referred to as *ranging*. On the basis of the type of measurements carried out between nodes and the network configuration, wireless positioning systems can be classified according to different criteria, as explained in the following sections.

1.1.2.1 Classification Based on Available Measurements

Every signal or physical measurable quantity that conveys position-dependent information can be, in principle, exploited to estimate the position of the agent node. Depending on the node's hardware capabilities, different kinds of measurements are available based, for example, on RF, inertial devices (e.g., acceleration), infrared, and ultrasound. In particular, when radio signals are considered, useful position-dependent information can be derived by analyzing signal characteristics such as received signal strength (RSS), time of arrival (TOA), and angle of arrival (AOA), or just from the knowledge that two or more nodes are in radio visibility (connected). In Table 1.1 a classification of exploitable

---

[1]From the American Heritage® dictionary of the English language.

Table 1.1 Classification of Positioning Systems Based on Available Measurements

Measured Quantity	Positioning Scheme	Characteristic Aspect
Angle of arrival (AOA)	Angle based	Characterizes the direction of propagation Usually antenna arrays are required
Received signal strength (RSS)	Range based Fingerprinting Interferometric	Measurement of the received power
Time of arrival (TOA)	Range based	Measurement of the signal propagation delay
Time difference of arrival (TDOA)	Range difference based	Measurement of signals propagation delay difference
Near-field ranging (NFR)	Range based	Relates the distance to the angle between the electric and magnetic fields in near-field conditions
Radio visibility	Proximity range-free	Connectivity
Acceleration and angular velocity	Inertial	Provides linear and angular displacement
Earth magnetic field	Magnetic	Provides orientation information

position-dependent measurements is reported. The following is a brief overview, while further details are given in Chapter 3.

Angle-of-Arrival (AOA) Measurements

Angle-based techniques estimate the position of an agent by measuring the AOA of signals arriving at the measuring station. The signal source is located on the straight line formed by the measurement station and the estimated AOA (also called *line of bearing* (LOB)). When multiple independent AOA measurements are simultaneously available, the intersection of two LOBs gives the (2D) estimated position. With perfect measurements, the *positioning problem* to be solved in this case is the intersection of a number of straight lines in the 3D space. In practice, noise, finite AOA estimation resolution, and multipath propagation force the use of more than two angles. The measurement station, equipped with an antenna array that allows AOA estimation, can be either the terminal to be located (in this case, it measures the AOAs of signals from different anchor nodes) or the anchor nodes themselves (in this case, they sense the signal transmitted by the agent, estimating its AOA).

Received Signal Strength (RSS) Measurements

Power-Based Ranging

The simplest measurement, practically always available in every wireless device, is the received signal power or RSS. Based on the consideration that in general the further away the node, the weaker the received signal, it is possible to obtain an estimate of the distance between two nodes (*ranging*) by measuring the RSS. Theoretical and empirical models are used to translate the difference (in dB) between the transmitted signal strength (assumed known) and the received signal strength into a range estimate. RSS ranging does not require time synchronization between nodes. Unfortunately, signal propagation issues such as refraction, reflection, shadowing, and multipath cause the attenuation to correlate poorly with distance, resulting in inaccurate and imprecise distance estimates.

Fingerprinting

Fingerprinting, also referred to as *mapping* or *scene analysis*, is a method of mapping the measured data (e.g., RSS) to a known grid point in the environment represented by a data fingerprint. The data fingerprint is generated by the environment site-survey process during the off-line system calibration phase. During on-line system location, the measured data are matched to the existing fingerprints. Typical drawbacks of this method include variation of the fingerprint due to changes in geometry, for example simple closing of doors.

Interferometric

The technique relies on a pair of nodes transmitting sinusoids at slightly different frequencies. The envelope of the received composite signal, after band-pass filtering, varies slowly over time. The phase offset of this envelope can be estimated through RSS measurements and contains information about the difference in distance of the nodes involved. By making multiple measurements in a network with at least eight nodes, it is possible to reconstruct the relative location of the nodes in a 3D frame [27].

Time-of-Arrival (TOA) Measurements

Time-Based Ranging

Considering that the electromagnetic waves travel at the speed of light, that is, $c \simeq 3 \cdot 10^8$ m/s, the distance d between a pair of nodes can be obtained from the measurement of the propagation delay or time of flight (TOF) $\tau = d/c$, through the estimation of the signal (TOA). As is shown in Chapter 3, when wide bandwidth signals are employed and accurate time measurements are available, time-based ranging can provide high-accuracy positioning capabilities. However, time synchronization and measurement errors represent the main issues when designing time-based ranging techniques.

Time-Sum-of-Arrival systems measure the relative sum of ranges between the agent and the anchor nodes and define a position location problem as the intersection of three or more ellipsoids with foci at two anchors.

Time-Difference-of-Arrival (TDOA) systems measure the difference in range between transmitter–receiver pairs. A TDOA measure defines a hyperboloid of constant range-difference, with the anchors at the foci.

Connectivity

The simplest way to obtain useful measurements for positioning is *proximity*, where the mere connectivity information (yes/no) is used to estimate node position. The location information is provided as a proximity to the closest known anchor (*landmark*). The key advantage of this technique is that it does not require any dedicated hardware and time synchronization among nodes since the connection information is available in every wireless device. However, the kind of position-dependent information obtainable using such a kind of approach may be unsatisfactory.

Near-Field Ranging (NFR)

NFR adopts low frequencies (typically around 1 MHz) and consequently long wavelengths (around 300 m) [32]. The key idea of this method is to exploit the deterministic relationship that exists between the angle formed by electric and magnetic fields of the received signal and the distance between the transmitter and the receiver. This low-frequency approach to location provides greater obstacle penetration, better multipath resistance, and sometimes more accurate location solutions because of the extra information present in near-field as opposed to classical far-field higher frequency approaches. The main drawbacks of this technology are the large antennas required and the scarce energy efficiency.

Self-Measurements

Besides the exploitation of measurements of radio signal characteristics exchanged between nodes (*internode measurements*), a single node could also take advantage in determining its own position of local measurements (*self-measurements*) using on-board sensors such as inertial measurement units (IMUs). The recent progress of the low-cost electromechanical systems (MEMS) market has made IMUs very popular. An IMU may typically contain an accelerometer and a gyroscope. The accelerometer measures the acceleration of the device on which it is attached (rotational speed), in addition to the earth's gravity, whereas the gyroscope measures the angular rate of the device. These measurements do not provide the device position directly as they enable only the tracking of device displacements. Several strategies, usually based on the integration of measured data, can be adopted to derive the device's position. However, The ranging estimates can be obtained, for instance, through this integration phase induces position and orientation drifts due to measurement errors. This is the main limitation of inertial sensors to solve the positioning problem over long intervals of time. To mitigate these drifts, inertial devices can be coupled with a magnetometer to use the earth's magnetic field as a reference. As is explained in Chapter 6, the greatest advantage of adopting IMUs comes from their combination with some wireless positioning technique by means of data fusion signal processing algorithms.

1.1.2.2 Classification Based on Network Configuration

The network configuration and the set of available measurements affect the signal processing strategy (localization algorithm) to be used to solve the positioning problem.

Consider, for example, the classical problem of determining the position (x,y) of an agent by using ranging estimates d_i between the agent node and a set of N anchor nodes placed at known coordinates (x_i, y_i), with $i = 1, 2, \ldots, N$. The ranging estimates can be obtained, for instance, through TOA, RSS, or NFR measurements. Assuming for simplicity perfect distance estimates, the position of the agent can be found by means of simple geometric considerations. In fact, the ith anchor defines (in a 2D scenario) a circle centered in (x_i, y_i) with radius d_i (see Fig. 1.3). The point of intersection of the circles corresponds to the position of the agent. In a two-dimensional space, at least three anchor nodes are required.

Unfortunately, in the presence of distance estimation errors, the circles in general do not intersect in a unique position, thus making the localization problem more challenging, as addressed in detail in Chapters 2 and 3.

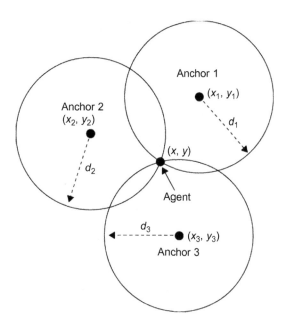

FIGURE 1.3

Example of geometric positioning.

Depending on the application constraints, only a small fraction of nodes might be aware of their positions (anchor nodes) being equipped with GPS receivers or deployed in known positions. The other nodes with unknown positions (agents) must estimate their positions by interacting with the anchor nodes. When a direct interaction with a sufficient number of anchor nodes is possible, *single-hop* localization algorithms can be adopted. On the contrary, cooperation between nodes is required to propagate, in a *multihop* and cooperative fashion, the anchor node position information to those nodes that cannot establish a direct interaction with anchor nodes.

In certain scenarios none of the nodes is aware of its absolute position (*anchor-free* scenario). An *absolute location* is the exact spot where the node resides, described within a shared reference frame for all located nodes. If the reference frame is the earth, the most used geodetic system (GS) is the world geodetic system (WGS84). However, in many applications the knowledge of absolute coordinates is not necessary (e.g., ad hoc battlefield and rescue systems). In these cases, only relative coordinates are estimated (sometimes called *virtual coordinates*) and ad hoc positioning algorithms have to be designed.

Positioning can be *terminal-centered,* when the agent performs distance measurements from the anchor nodes on the basis of radio signals transmitted by the anchor nodes, and carries out the calculations needed to determine its own position; or *network-centered,* when the signal transmitted by the agent is used by the anchor nodes (connected in a network) to compute the agent position, in which case the position information is then sent back to the agent.

A summary of this classification is presented in Table 1.2. Other possible classifications are based on the wireless technology adopted, such as *cellular* versus *sensor network* and *satellite* versus

Table 1.2 Classification Based on Network Configuration

Network Configuration	Characteristic Aspect
Anchor-based	Some nodes know their locations
Multihop	The distance from anchors can be obtained indirectly by means of intermediate nodes
Single-hop	The distance from anchors can be obtained by direct interaction
Anchor-free	None of the nodes knows its position. Only relative coordinates (virtual coordinates) can be found
Range-free	Only connectivity information is used
Terminal-centered	Specialized electronics within the mobile handset to determine its own location
Network-centered	Specialized location equipment within the network to determine mobile terminal's location

terrestrial systems, or on the coverage area, such as *indoor* versus *outdoor*. This kind of categorization is addressed in more detail in the dedicated Section 1.2.

1.1.3 Performance Metrics

The requirements of location-aware networks and technologies are driven by applications. Since the measurements used to estimate the agent's position are affected by some uncertainty (e.g., noise), the agent's position estimate will also be characterized by errors.

The *position estimation error* is given by the Euclidean distance between the estimated position $\hat{\mathbf{x}}$ and the true position $\mathbf{x}$ as

$$e(\mathbf{x}) = \|\hat{\mathbf{x}} - \mathbf{x}\|_2. \tag{1.1}$$

A local performance metric is the root mean square error (RMSE) of position estimates

$$\mathsf{RMSE} = \sqrt{\mathbb{E}\left\{e^2(\mathbf{x})\right\}}, \tag{1.2}$$

where $\mathbb{E}\{\cdot\}$ indicates statistical expectation over all (random) sources of error. The RMSE is often referred to as *accuracy* as it is a measure of the statistical deviation of the position estimate from the real position. A high accuracy corresponds to low RMSEs.

Precision describes the statistical deviation from a mean position, in particular the variance or the standard deviation of the (potentially biased) estimate. A high precision is represented as a low variance or standard deviation. For unbiased estimates, accuracy and precision coincide.

Other representations of accuracy and precision include (temporal/spatial) ratios of confidence, that is, being lower than some threshold for a certain percentage of time or of measurements. This representation can be seen as an *outage probability*,[2] with the definition of *outage event* as the event of

---

[2]The outage probability is a well-known concept for the performance evaluation of wireless communication systems; the similarity with the application to location-aware networks is in evaluating the probability that the quality of service will fall below a given target [40].

the error exceeding the error threshold e_{th}:

$$p_{out} = \mathcal{P}\{e(\mathbf{x}) > e_{th}\}, \tag{1.3}$$

where $\mathcal{P}\{X\}$ indicates the probability of the event X and e_{th} is the threshold (i.e., the maximum allowable) position estimation error, and the probability is evaluated over the ensemble of all possible spatial positions and time instants [39]. When evaluated over the localization area, the localization error outage (LEO) can be seen as a global performance index. An equivalent index often adopted in the literature is the cumulative density function (CDF) $F_e(e)$ of the position estimation error, which is given by the equation

$$F_e(e_{th}) = \mathcal{P}\{e(\mathbf{x}) \le e_{th}\} = 1 - p_{out}. \tag{1.4}$$

Other performance indexes are the *robustness* of the algorithm to some impairments, such as lack of radio visibility, and the *coverage*, the area where nodes can be localized. In particular, aspects related to the *localization update rate* (i.e., the number of times the position estimate is (re)calculated per second) are important in navigation systems (navigation of pedestrians and navigation of vehicles typically require different localization update rates) and intersect algorithm complexity and node cost.

1.2 POSITIONING AND NAVIGATION SYSTEMS

The positioning and navigation systems analyzed in this book are those for which there exists, or it is expected, a widespread personal use and for which the scientific and technological research has a prominent role in these years. Recalling the classifications discussed earlier, we now describe a *technological* discrimination between satellite and terrestrial positioning systems. A pictorial view of the main positioning technologies currently available and their level of coverage and accuracy is depicted in Fig. 1.4.

Satellite positioning systems rely on a constellation of artificial satellites rotating in well-known orbits and continuously transmitting signals used by the mobile terminals to perform ranging measurements. They are inherently *navigation* systems, while most recent terrestrial systems are intended for

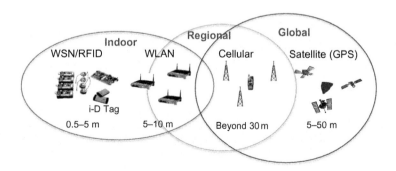

FIGURE 1.4

An illustration of the main positioning technologies, as well as their qualitative level of coverage and accuracy.

positioning only. The well-known global positioning system (GPS) is nowadays the primary global navigation satellite system (GNSS).

Terrestrial positioning systems rely on a network of ground-located reference stations. In the past, several terrestrial systems for maritime and avionic navigation were used: Decca, LORAN-C, TACAN, and VOR/DME just to mention a few [2, 25]. They are characterized by very specialized fields of application and high costs of installation and maintenance. In the long term, some of them will be superseded by GNSS. This generation of terrestrial navigation systems is beyond the scope of this book.

On the other hand, recent terrestrial position location systems were born as a sort of by-product of current wireless communications systems. One of the main differences between current satellite and terrestrial positioning systems is the fundamental purpose for which the signal traveling from the transmitter to the receiver has been designed: in the satellite case, the purpose is truly localization, whereas in the terrestrial case localization is often ancillary with respect to data communication. For this reason, technological challenges and scopes are different in the two cases. This is also the primary reason why satellite navigation technology is seen often as a sort of field different from telecommunications, perhaps closer to geographical sciences and earth observation.

Because of the variety of terrestrial wireless systems and modulation formats, many different approaches have been proposed so far to enable positioning in personal handsets and portable devices. These include terminal-centered and network-centered procedures for cellular networks, for which early proposals were studied more than fifteen years ago, with procedures tailored to the modulations and protocols for WLANs, wireless metropolitan area networks (WMANs), and WSNs. For example, new dedicated RTLS, based on radio frequency identification (RFID) or on promising transmission technologies such as ultra-wide bandwidth (UWB), have been recently introduced in the market as illustrated in Section 1.2.3. In Table 1.3 the main characteristics of a few relevant existing systems are summarized. A list of positioning systems using other technologies can be found in Ref. [16].

Nonetheless, it has to be recognized that nowadays a strong convergence path lies ahead, through the integration of navigation and communications devices, applications, and services (NAV/COM systems and services). A frontier of wireless positioning is the hybridization between satellite and terrestrial systems toward the concept of seamless positioning, whose main example is the assisted GPS service, which uses a terrestrial cellular network to improve GPS receiver performance.

1.2.1 Satellite-Based Systems

The navigation world has just witnessed an important milestone: the advent and full operability of a number of different satellite navigation systems aiming at competing with and complementing the GPS authority. Europe is urging the deployment of its Galileo global satellite system, Russia is radically modernizing its global orbiting navigation satellite system (GLONASS), and Japan and India have their own regional systems under development, while China is converting its initial regional Beidou system into a global one. The United States itself is investing significant resources for GPS modernization. The advent of this new panorama in the sky has fostered worldwide research in the field of satellite navigation and is going to deeply change the market of navigation receivers as well as the consumers' perspective, with new applications, new services, and increased availability. However, the undisputed lead among satellite-based systems belongs nowadays to GPS.

Table 1.3 Comparison of Existing Positioning Systems

Technology	Measurement Technique	Accuracy	Pros	Cons
GPS	TDOA	10–20 m	Earth scale coverage	Expensive infrastructure, only outdoor
Galileo	TDOA	1–5 m	Earth scale coverage	Expensive infrastructure, only outdoor
A-GNSS	TDOA	< 5 m	Country coverage	Scarce indoor accuracy
Cellular	E-OTD / OTDOA	50–500 m	Country coverage	Requires synchronized base stations
Cellular	CellID	Cell size	Country coverage	Scarce accuracy
WLAN	RSS–fingerprinting	1–5 m	Indoor coverage, low cost	Database required for fingerprinting, low accuracy
WSN (ZigBee)	RSS	1–10 m	Indoor coverage, low power consumption, low cost	Low accuracy
WSN (UWB)	TOA/ TDOA/ AOA	0.1–1 m	Indoor coverage, high accuracy	Short range (<20 m), problems in NLOS
RFID	Proximity	Connectivity range	Indoor coverage, low power consumption, low cost	Low accuracy, one tag per location
NFR	E.M. near-field characteristics	1–5 m	Indoor coverage, low cost	Low frequency, large antennas
INS	Acceleration/ angular rate/ earth magnetic field	1–5% of the traveled distance/ angle	Works everywhere	Position/orientation drift, magnetic disturbance in indoor

GPS is a satellite-based radio navigation system used to compute precise time and three-dimensional position anywhere on the earth. An illustration is provided in Fig. 1.5. GPS position solutions are accomplished by obtaining signal TOA measurements, or *pseudoranges*, from a minimum of four GPS satellites. These raw pseudoranges are the measured distances along the line of sight (LOS) of the signals broadcast by each of the N_{sat} satellites. The pseudorange ρ_k, for each satellite k, is

$$\rho_k = \sqrt{(x_k - x_u)^2 + (y_k - y_u)^2 + (z_k - z_u)^2} + c \cdot \Delta b_u, \quad k = 1, 2 \ldots, N_{sat},$$

where x_k, y_k, and z_k are the Earth-centered, Earth-fixed (ECEF) coordinates for satellite k, and c is the speed of light. The subscript u represents the user and x_u, y_u, and z_u are the user's ECEF coordinates. The user clock bias, Δb_u, is the offset between the receiver time reference and the (unknown) GPS time.

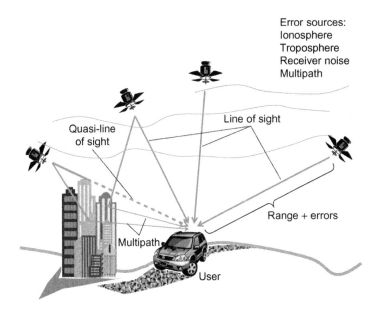

FIGURE 1.5

Basic GPS architecture.

As is detailed in Chapter 2, a set of (at least) four such equations is linearized and iteratively solved for the user position and clock bias using a least-squares (LS) computation [24]. The user's clock bias is a time-varying term that affects all pseudoranges and is caused by the following factors:

- Local oscillator drift and bias
- Satellite payload filter (analog and digital) propagation delays
- Antenna and receiver propagation/processing delays.

In principle, highly accurate position solutions may be obtained by solving the system of equations mentioned earlier. However, in general, there are several primary error sources to GPS. Two of these include unknown atmospheric errors, or delays, introduced by the ionosphere and troposphere. These effects cause the LOS signal to actually arrive later than predicted by the pseudorange equation. Multipath propagation is another primary pseudorange error source. Multipath signals are (usually undesired) signal reflections from the ground or other nearby obstacles. As opposed to the atmospheric effects, which directly affect the LOS signal TOA, multipath causes the GPS receiver to make erroneous measurements of the TOA of the signal.

The principles summarized earlier for the GPS are the basis to understand the architecture of all the satellite navigation systems currently in development, either global or regional. A more detailed discussion is available in Chapter 2.

1.2.2 Augmentation Systems and Assisted GNSS

GNSS augmentation systems were born to continuously provide robust and safe navigation especially when high precision or enhanced coverage or availability is required. *Accuracy, availability, integrity,* and *continuity* are the key performance of any GNSS, so that procedures and external aids to improve them have been developed under the label of the *augmentation systems* [30].

Augmentation systems attempt to correct for many of the dominant error sources in GNSS. It is basically accomplished by placing a reference station at a precisely known location in the vicinity of a user, or where high-accuracy navigation is required. The reference station measures the ranges to each of the satellites in view, demodulates the navigation message, and depending on the type of parameter, computes several types of corrections to be applied by the user's receiver in order to improve its performance. Then the station broadcasts its corrections to local users via a data link, so that position accuracies of a few centimeters are obtained. Augmentation works only against common mode, spatially correlated errors such as the ionosphere and troposphere delays. Multipath-induced errors, as well as interference-induced ones, are not common to the reference station and the user; therefore they cannot be recovered by means of any augmentation systems.

The main augmentation systems currently available are differential GPS (DGPS), satellite-based augmentation systems (SBASs), real-time kinematic (RTK) systems, and assisted GNSS (AGNSS) [37]. It is interesting to note that while DGPS, SBAS, and RTK require the deployment of a specific terrestrial network of reference stations and specific communication protocols, the AGNSS approach essentially exploits the network architecture of existing cellular communication systems, with specifically added features. For this reason AGNSS is a very promising technology, since it inherently implements the concept of NAV/COM integration.

1.2.3 Terrestrial Network-Based Systems

The trend toward personal use of navigation systems associated with LBSs requires that positioning devices be able to seamlessly work under various, variable, and critical conditions, such as inside warehouses, multistoreyed buildings, underground stores and parking, and indoor commercial and office campuses. Examples of applications are location detection of products stored in a warehouse, location detection of medical personnel or equipment in a hospital, location detection of firemen in a building on fire, detecting the location of police dogs trained to find explosives in a building, and finding tagged maintenance tools and equipment scattered all over a plant [14]. Unfortunately, GNSS indoor reception is dramatically impaired by strong attenuation due to walls and slabs and by the multipath effect. Therefore, indoor environments open challenging issues for GNSS signal processing and receiver design, to which new modulations (such as those foreseen for Galileo) and new navigation approaches (mainly, assisted GNSS services) try to give solution.

When there is an indoor receiver, signal reception is characterized by a strongly attenuated direct component and several reflected or scattered multipath components. The attenuation affecting the direct path can range from 10 to 25 dB, depending on the nature of the concrete, thus reducing the carrier power the receiver has to deal with from about -160 dBW to even -190 dBW; however, the nominal sensitivity in signal acquisition of current commercial receivers is around -178 dBW. Furthermore, indoor multipath and scattering effects become far more harmful. In such conditions, the use of basic GPS receivers is really questionable and substantially different approaches have to be adopted.

Nowadays much research is focused on the use of terrestrial wireless technology as a means of developing positioning and navigation systems that work where satellite systems fail (indoor environments, urban areas). New LBSs require a certain level of location accuracy to be met by the positioning systems, in spite of all the propagation problems typical of wireless communication, such as channel fading, low signal-to-noise ratio (SNR), multiuser interference, and multipath conditions.

Pioneering work on indoor positioning dates back to more than 10 years ago, but a lot of work is still going on to refine and get past those pioneering ideas, both in academia and industry [3, 15, 16]. Several wireless technologies have been studied for indoor positioning. Their distinguishing elements are:

- The positioning algorithm, which may use various types of measurement of the signal, such as TOA, AOA, and RSS.
- The physical layer of the network infrastructure used to communicate with the user's terminal.

One of the most promising technologies for indoor positioning and communications appears to be UWB [31, 33].

In this book the term *terrestrial network-based positioning and navigation systems* refers to those location systems that use wireless technologies entirely deployed on the ground. The most used wireless technologies of this kind are cellular networks, wireless local area networks (WLANs), wireless systems based on UWB, radio frequency identification (RFID) technology, and wireless sensor networks (WSNs).

Terrestrial network-based positioning systems can also be referred to as *local* or *short-range* systems, because their coverage area is restricted to the region where they are deployed. Thus, they differ from GNSS, whose coverage is global.

1.2.3.1 *Positioning in Cellular Networks*

Cellular networks rely on a set of base stations (BSs), with a coverage radius up to about tens of kilometers each. Nowadays they are widely deployed in all developed countries.

The most widespread positioning technology in cellular networks is based on TDOA [22]. For instance, GSM location is based on the existing *observed time difference* (OTD). OTD evaluates the time difference between signals traveling from two different BSs to an MS. At least three visible BSs are needed to estimate the MS position, obtained by intersecting hyperbolic lines having foci at the BSs' positions. The final location estimation accuracies in GSM-based location systems using OTD ranges from 50 to 500 m.

The signal parameter estimation method used in UMTS networks is the observed TDOA (OTDOA), which is based on the TDOA approach. Anyway, the accuracy of cellular-based positioning is quite modest, for this reason recent location estimation algorithms try to exploit any available information about the environment (e.g., fading conditions, Doppler frequency, and network topology) to attain higher accuracy through data fusion methods. Positioning in cellular systems is treated in Section 3.2.1.

1.2.3.2 *Positioning in Wireless Local Area Networks (WLANs)*

WLAN indoor locations are deployed in much smaller areas than in cellular networks. They are widely used both in private and public bodies such as company campuses, universities, corporations, airports, museums, and shopping malls. Outdoor WLANs' deployment can be seen only in small zones of large cities as hot spots. WLAN-based positioning solutions rely mostly on signal strength evaluation. Since received signal strength (RSS) measurement is part of the normal operating mode of a wireless transceiver, no other ad hoc hardware infrastructure is required. As is described in Section 3.3, the most used WLAN positioning techniques exploit fingerprinting methods [3].

1.2.3.3 *Positioning in Radio Frequency Identification (RFID)*

RFID technology has attracted an enormous interest worldwide, since the earliest pioneering ideas dating back to 1948. A number of applications can now be found in several fields such as logistics, automotive, surveillance, automation systems, and in general real-time object identification [9]. An RFID system consists of tags applied to objects and their readers. The reader interrogates the tags via a wireless link to obtain the data stored on them. When tag cost, size, and power consumption requirements become particularly stringent, passive or semipassive tag solutions are taken into consideration. Communication with passive tags usually relies on backscatter modulation, and the tag's control logic and memory circuits obtain the necessary power to operate from the RF signal sent by the reader.

Recent developments indicate a trend to hybridize active RFID and RTLS technologies [19]. Some RFID vendors are adopting or adapting RTLS concepts to provide additional functionalities for their products. Several systems rely on proximity-based positioning algorithms, which, in general, are not very accurate for many applications. The standard ISO/IEC 24730-2 [20] has been introduced in 2006 to fill the gap between the RFID and RTLS technologies. Some research efforts are also going on to merge RFID and UWB technologies toward extremely low-cost RTLS [5]. Positioning algorithms adopted in RFID-based RTLS are usually the same as those adopted in WLANs and WSNs.

1.2.3.4 *Positioning in WSNs*

A WSN in its simplest form can be defined as a network of (low-size and low-complex) devices denoted as nodes that can sense the environment and communicate the information gathered from the monitored field through wireless links. The data are forwarded, possibly via multiple hops relaying to a local sink (a controller or monitor) or to other networks through a gateway (as shown in Fig. 1.6.) [38]. The number of applications where WSNs are used today, or has been envisioned for the future, is quite large. Apart from the "core" applications regarding general monitoring of environment or processes, WSNs are or will be used also for applications in traffic safety, medicine, agriculture, logistics, and disaster relief, just to name a few.

In many (not to say all) WSN applications, a sensor reading is not of much use unless it is accompanied by the position at which the data were gathered. The positioning problem in WSNs can vary widely in character from network to network, and from application to application. The appropriateness of the

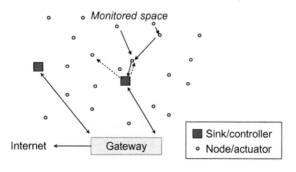

FIGURE 1.6

Example of a WSN.

Table 1.4 Factors Influencing the Choice of a WSN Positioning Algorithm

WSN Aspect	Important Algorithm Property
Available node hardware	Algorithm complexity
Available fixed infrastructure	Distributed vs. centralized algorithms
Expected network size (number of nodes, connectivity)	Algorithm scalability
Available input data	Positioning method, for example trilateration
Network application area	Positioning algorithm accuracy

approach to positioning for sensor nodes depends on the available hardware, measured data, infrastructure, and on the application requirements. In some cases, a fixed infrastructure can be installed throughout the network deployment area in order to aid the positioning of mobile sensor nodes. This infrastructure may include reference nodes at known location (anchors), or central processing stations with extended resources in terms of computational power and/or energy supply. The expected size of the network, that is, the node density and the coverage area of the network, also plays an important role in the design process. Some WSNs that have been envisioned in the literature involve thousands of sensor nodes densely spread out over very large areas. In such large networks, it is of paramount importance that the complexity of the positioning algorithm is not a rapidly increasing function of the number of nodes and/or connectivity level of the network; that is, *algorithm scalability* is often an important factor to consider. We list in Table 1.4 some of the principal WSN aspects, in terms of positioning algorithm design and of the algorithm properties they influence. Positioning in WSNs is addressed in Section 3.4.

1.2.3.5 *The Ultra-Wide Band (UWB) Technology*

UWB is promising for high-definition indoor positioning, as it can achieve very accurate short distance estimation. UWB is also a viable technology for short-range wireless indoor communication with a number of attractive potential features: high-rate transmission, low complexity, low cost, and low power consumption [4, 13]. This technology has generated considerable and increasing interest by many manufacturers in the United States since February 2002, when the Federal Communications Commission (FCC) opened up 7.5 GHz of spectrum (from 3.1 to 10.6 GHz) for use by UWB devices [8].

The traditional design approach for a UWB communication system uses baseband narrow time-domain pulses of very short duration, typically of the order of a nanosecond, thereby spreading the energy of the radio signal quite uniformly over a wide frequency band ranging from extremely low frequencies to a few gigahertz. This method is usually called *impulse radio UWB* (IR-UWB). A great advantage of the short pulse modulation is the possibility to estimate the TOA with a fine resolution, which translates in ranging estimation with an accuracy of less than one meter.

In March 2004 a technical group called Task Group TG4a was established under the IEEE 802.15 standardization framework. Its mission was to define an alternative physical layer (IEEE 802.15.4a), based on the UWB characteristics, for the IEEE 802.15.4 standard, the most used by WSNs. The two design goals of low cost and low power are achieved by a new PHY layer based on UWB through simple demodulation schemes, low bit rates, and low transmitted power. Low power consumption is also achieved through low duty cycle operations. The first commercial IEEE 802.15.4a compliant chip-set has been announced to be delivered in late 2011. In parallel, several companies proposing proprietary UWB products for RTLS are deeply involved in the development of the new IEEE 802.15.4f standard, which is devoted to specify a solution to precise indoor positioning with extremely low cost and low consumption tags.

1.3 APPLICATION OF SIGNAL PROCESSING TECHNIQUES TO POSITIONING AND NAVIGATION PROBLEMS

As shown in the examples reported in Section 1.1.2, in the absence of measurement errors, most positioning problems can be afforded following a *deterministic geometric* approach, where the location of the MS is directly determined from the position-related parameters extracted from the received signal through geometric relationships (e.g., intersections of circles and hyperboloids). In practice, measurements are subject to errors, and hence, such approaches may be useless. Then the positioning problem has to be addressed within a more general estimation theory framework.

Without loss of generality, the positioning problem can be stated as follows (see Fig. 1.7): Consider a generic scenario populated by a number of wireless nodes, where a subset of them are located in unknown positions (MSs). Let $\mathbf{x}$ be the set of MSs' position. We want to find an estimate $\hat{\mathbf{x}}$ of the MSs' position starting from a set of available measurements $\mathbf{r}$ (observations). This set may include measurements from BSs as well as inter-MS or self-measurements. Measurements can be either TOA, AOA, RSS, or even heterogeneous combinations of them. The main issue is to design the estimator $\hat{\mathbf{x}} = \hat{\mathbf{x}}(\mathbf{r})$ that minimizes some performance metric such as the RMSE. It is generally preferable to derive an estimator that provides an *unbiased* and *minimum variance* estimate.

Statistical geometry provides a theoretical framework that helps to solve the positioning problem even in the presence of measurement errors. Statistical techniques are based on a probabilistic

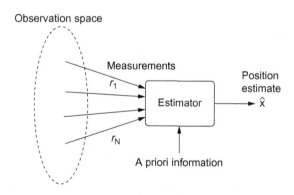

FIGURE 1.7

The position estimation problem.

description of the observations **r**. It is assumed that the elements of **r** are (realizations of) random variables (RVs) whose joint probability density function (pdf) depends on the position of the MS. An additive noise model is usually employed, meaning that the estimated position-related parameters are given by the sum of a deterministic term depending on the MS location plus a random noise term. Statistical techniques are classified as *parametric* or *nonparametric*, depending on whether a probabilistic description of the observation set **r** is available or not, respectively. Now we provide a brief overview of the main approaches typically followed in the estimation theory to solve the positioning problem. For further details, the reader is referred to classical estimation theory books [36].

1.3.1 Parametric Statistical Techniques

Parametric methods assume complete or partial statistical knowledge of the position-related para meters.

1.3.1.1 *Bayesian Estimators*

When a priori statistical characterization of the measurements **r** as well as of the parameter to estimate (the MSs' position **x**) is available, Bayesian techniques can be adopted. The unbiased minimum mean square error (MMSE) estimator is given by

$$\hat{\mathbf{x}}_{\text{MMSE}} = \int \mathbf{x} p(\mathbf{x}|\mathbf{r}) \, d\mathbf{x}, \tag{1.5}$$

where $p(\mathbf{x}|\mathbf{r})$ is the posterior conditional pdf of **x**.

Following another criterion, the maximum a posteriori probability (MAP) estimator is defined as

$$\hat{\mathbf{x}}_{\text{MAP}} = \arg\max_{\mathbf{x}} p(\mathbf{x}|\mathbf{r}), \tag{1.6}$$

which is equivalent to the MMSE estimator when the RVs **x** and **r** are jointly Gaussian.

1.3.1.2 *Maximum Likelihood Estimator*

When no a priori statistical characterization of MSs' positions is available, the minimum variance unbiased estimator does not always exist or, when it does, no straightforward procedures are available to find it. A popular, but in general suboptimum, estimator is the maximum likelihood (ML) estimator

$$\hat{\mathbf{x}}_{\text{ML}} = \arg\max_{\mathbf{x}} p(\mathbf{r}|\mathbf{x}), \tag{1.7}$$

where $p(\mathbf{r}|\mathbf{x})$ is the conditional pdf of the measurements conditioned on MSs' positions. The popularity of the ML estimator comes from the fact that it is asymptotically efficient; that is, for small measurement errors it tends to be a minimum variance unbiased estimator. Indeed, when an efficient estimator exists, the ML estimator will produce it.

The variance of any unbiased estimator is lower bounded by the Cramér–Rao lower bound (CRB), which is usually adopted as a performance benchmark for newly designed estimators. The fundamental limits are addressed in Chapter 4.

1.3.2 **Nonparametric Statistical Techniques**

1.3.2.1 *Least Squares (LS) Estimator*

When the (a priori) statistical characterization of measurements is not available, the standard approach is given by the LS method. Assume that the measurements can be expressed as follows:

$$\mathbf{r} = \mathbf{h}(\mathbf{x}) + \mathbf{n}. \tag{1.8}$$

where $\mathbf{h}(\mathbf{x})$ denotes the relation between the measurements and the MSs' position and $\mathbf{n}$ is the measurement noise. The LS estimate is the position $\hat{\mathbf{x}}_{LS}$ that minimizes the sum of the squared measurement errors as follows:

$$\hat{\mathbf{x}}_{LS} = \arg\min_{\mathbf{x}} (\mathbf{x} - \mathbf{h}(\mathbf{x}))^T (\mathbf{x} - \mathbf{h}(\mathbf{x})). \tag{1.9}$$

Since no probabilistic assumptions are made about the measurements, minimizing the LS error does not in general translate into minimizing the estimation error, and hence the LS is not optimal in general. Note that if the measurement error $\mathbf{n}$ is Gaussian distributed, the LS and ML estimators become equivalent.

1.3.3 **Nongeometric Techniques**

A completely different approach is followed by nongeometric techniques, where the measurements are not used to construct geometrical relationships, but rather used to obtain a sort of "signature" of each location of interest. An example of nongeometric technique is the *fingerprinting* (or *mapping*) positioning methods. Fingerprinting techniques are described in more detail in Section 3.1.2.3.

1.3.4 **Advanced Signal Processing Tools**

Even though the estimation theory has been a well-established one for several decades, the design of good estimators is still an active field of research because of the presence of possibly non-Gaussian impairments (e.g., multipath propagation, non-line-of-sight (NLOS) channel conditions, lack of time synchronization), different network configurations (centralized, distributed, cooperative, cognitive), and constraints such as computational complexity and energy efficiency. Therefore, several advanced signal processing tools have been developed. For example, when the set $\mathbf{r}$ involves inter-MS measurements (*cooperative localization*), the direct solution of the MMSE, ML, LS, or maximum a posteriori (MAP) estimators becomes unaffordable from the complexity point of view. In addition, measurements $\mathbf{r}$ may contain observations taken in different time instants in environments where MSs are continuously moving. In this case, tracking algorithms following a Bayesian approach can be used to estimate the position more accurately by exploiting the temporal correlation among successive observations. As is shown in Chapter 6, the Bayesian framework also allows an efficient integration of different positioning technologies (e.g., satellite and terrestrial or radio and inertial) as well as of MS mobility models into a single navigation system through *data fusion*.

1.3.4.1 *Bayesian Filtering*

In *Bayesian* filtering [11] the localization problem is modeled as a dynamic system where the vector state $\mathbf{x}_n$, at discrete time n, represents the coordinates of the MS. In particular, at time n the a posteriori

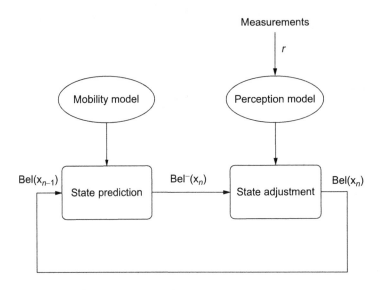

FIGURE 1.8

A schematic representation of Bayesian filtering for position tracking.

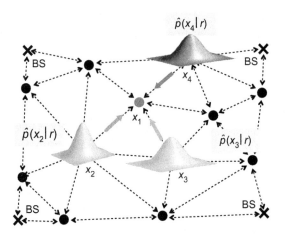

FIGURE 1.9

Positioning through belief propagation.

pdf $\text{Bel}(\mathbf{x}_n)$ of the state $\mathbf{x}_n$, called *belief*, is evaluated in two steps (see Fig. 1.8). In the first step, the belief function is updated according to the *mobility model* $p(\mathbf{x}_n|\mathbf{x}_{n-1})$, which represents the dynamic model for the system yielding $\text{Bel}(\mathbf{x}_n)$. The mobility model gives the description of the state variation $\mathbf{x}_{n-1} \to \mathbf{x}_n$, that is, the statistical description of MS movements. In the second step, the belief function is further updated to $Bel(\mathbf{x}_n)$ to account for the statistical information $p(\mathbf{r}_n|\mathbf{x}_n)$ on the position at time n starting from the measurement vector $\mathbf{r}_n$ collected at time n. This is the *perception model* and operates as an updater for the system state. Through the belief function, it is possible to identify the most likely state (MS position) at time n among all possible states.

As the implementation of Bayesian filters can be complex, several suboptimal approaches have been developed in the literature and are illustrated in Chapter 6.

1.3.4.2 *Belief Propagation*

Belief propagation techniques and their reduced complexity implementations, such as those based on factor graphs, represent powerful signal processing tools to solve the positioning problem in cooperative scenarios [10, 34, 40]. As shown in Fig. 1.9, a positioning network can be represented as an undirected graph where vertices are nodes with associated locations $\mathbf{x}_k$ and prior pdf, and edges (branches) interconnect nodes and allow the exchange of measurements with likelihood $p(\mathbf{r}_{n,k}|\mathbf{x}_k, \mathbf{x}_n)$. At each iteration, node k obtains an approximated a posteriori pdf (*belief*) $\hat{p}(\mathbf{x}_k|\mathbf{r})$ about its own position. Neighbors use their own beliefs and measurements to compute a belief about the kth node's position and send it to node k (message). The message exchange between nodes continues until the convergence of the algorithm is reached. Further details on positioning based on belief propagation are given in Chapter 5, where, in addition, a case study example based on this approach is provided (see Section 5.4.5.2).

References

[1] A. Sayed, A. Tarighat, N. Khajehnouri, Network-based wireless location, IEEE Signal Process. Mag. 22 (4) (2005) 24–40.

[2] S. Appleyard, R. Linford, P. Yarwood, Marine Electronic Navigation, second ed., Routledge & Kegan Paul, 1988.

[3] P. Bahl, V.N. Padmanabhan, RADAR: An in-building RF-based user location and tracking system, in: Proceedings of the IEEE Infocom 2000, Tel Aviv, Israel, 2000.

[4] D. Dardari, A. Conti, U. Ferner, A. Giorgetti, M.Z. Win, Ranging with ultrawide bandwidth signals in multipath environments, Proc. IEEE, Special Issue on UWB Technology & Emerging Applications, 97 (2) (2009) 404–426.

[5] D. Dardari, R. D'Errico, C. Roblin, A. Sibille, M.Z. Win, Ultrawide bandwidth RFID: The next generation? Proc. IEEE, Special Issue on RFID – A Unique Radio Innovation for the 21st Century, 98 (9) (2010) 1570–1582.

[6] R. Das, P. Harrop, RFID forecast, players and opportunities 2007–2017. http://www.idtechex.com, 2007.

[7] F. Gustafsson, F. Gunnarsson, Mobile positioning using wireless networks, IEEE Signal Process. Mag. 22 (4) (2005) 41–53.

[8] Federal Communications Commission, Revision of part 15 of the commission's rules regarding ultrawideband transmission systems, first report and order (ET Docket 98–153), Adopted Feb. 14, 2002, Released Apr. 22, 2002.

[9] K. Finkenzeller, RFID Handbook: Fundamentals and Applications in Contactless Smart Cards and Identification, second ed., John Wiley & Sons, 2004.

[10] D. Fontanella, M. Nicoli, L. Vandendorpe, Bayesian localization in sensor networks: Distributed algorithm and fundamental limits, in: 2010 IEEE International Conference on Communications (ICC), 2010, pp. 1–5.

[11] D. Fox, J. Hightower, L. Liao, D. Schulz, G. Borriello, Bayesian filtering for local estimation, IEEE Pervasive Comput. 2 (3) (2003).

[12] G. Sun, J. Chen, W. Guo, K.J. Ray Liu, Signal processing techniques in network-aided positioning, IEEE Signal Process. Mag. 22 (4) (2005) 12–23.

[13] S. Gezici, Z. Tian, G.B. Giannakis, H. Kobayashi, A. Molisch, H.V. Poor, et al., Localization via ultra-wideband radio, IEEE Signal Process. Mag. 22 (4) (2005) 70–84.

[14] H. Liu, H. Darabi, P. Banerjee, J. Liu, Survey of wireless indoor positioning techniques and systems, IEEE Trans. Syst. Man Cybern. C Appl. Rev. 37 (6) (2007) 1067–1080.

[15] A. Harter, A. Hopper, P. Steggles, A. Ward, P. Webster, The anatomy of a context-aware application, Wireless Networks 8 (2002) 187–197.

[16] J. Hightower, G. Borriello, Location systems for ubiquitous computing Computer 34 (8) (2001) 57–66.

[17] http://ec.europa.eu/information_society/activities/intelligentcar/icar/index_en.htm, 2008.

[18] http://ec.europa.eu/information_society/activities/intelligentcar/technologies/tech_07/index_en.htm, 2008.

[19] ISO/IEC 19762-5:2008. Information technology – automatic identification and data capture (AIDC) techniques – part 5: Locating systems.

[20] ISO/IEC 24730-2. Information technology – real-time locating systems (RTLS) – part 2: 2.4 GHz air interface protocol.

[21] J.H. Reed, K.J. Krizman, B.D. Woerner, T.S. Rappaport, An overview of the challenges and progress in meeting the E-911 requirement for location services, IEEE Commun. Mag. 36 (4) (1998) 30–37.

[22] J.J. Caffery Jr., G.L. Stuber, Overview of radiolocation in CDMA cellular systems, IEEE Commun. Mag. (1998) 38–45.

[23] K. Gratsias, E. Frentzos, V. Delis, Y. Theodoridis, Towards a taxonomy of location based services, in: Web and Wireless Geographical Information Systems, Lecture Notes in Computer Science, Springer, Berlin/Heidelberg, 2005.

[24] E.D. Kaplan, Understanding GPS: Principles and Applications, second ed., Artech House, Norwood, MA, 2006.

[25] M. Kayton, W. Fried, Avionics Navigation Systems. A Wiley-Interscience publication. J. Wiley, 1997.

[26] M. Vossiek, L. Wiebking, P. Gulden, J. Wieghardt, C. Hoffmann, P. Heide, Wireless local positioning, IEEE Microwave Mag. 4 (4) (2003) 77–86.

[27] M. Maroti, P. Völgyesi, S. Dora, B. Kusy, A. Nadas, A. Ledeczi, et al., Radio interferometric geolocation. in: Proceedings of the ACM Conference on Embedded Networked Sensor Systems, San Diego, 2005.

[28] P. Bellavista, A. Kupper, S. Helal, Location-based services: Back to the future. IEEE Pervasive Comput. 7 (2) (2008) 85–89.

[29] N. Patwari, J.N. Ash, S. Kyperountas, A.O. Hero, R.L. Moses, N.S. Correal, Locating the nodes: Cooperative localization in wireless sensor networks. IEEE Signal Process. Mag. 22 (4) (2005) 54–69.

[30] R.E. Phelts, Multicorrelator Techniques for Robust Mitigation of Threats to GPS Signal Quality, PhD thesis, Stanford University, 2001.

[31] Z. Sahinoglu, S. Gezici, I. Guvenc, Ultra-Wideband Positioning Systems: Theoretical Limits, Ranging Algorithms, and Protocols, Cambridge University Press, 2008.

[32] H.G. Schantz, A real-time location system using near-field electromagnetic ranging, in: 2007 IEEE Antennas and Propagation Society International Symposium, Honolulu, HI, 2007.

[33] Y. Shen, M. Win, Fundamental limits of wideband localization – part I: A general framework, IEEE Trans. Inf. Theory 56 (10) (2010) 4956–4980.

[34] Y. Shen, H. Wymeersch, M. Win, Fundamental limits of wideband localization – part II: Cooperative networks. Inf. Theory IEEE Trans. 56 (10) (2010) 4981–5000.

[35] T.S. Rappaport, J.H. Reed, B.D. Woerner, Position location using wireless communications on highways of the future, IEEE Commun. Mag. 34 (10) (1996) 33–41.

[36] H.L.V. Trees, Detection, Estimation, and Modulation Theory: Part I, second ed., John Wiley & Sons, Inc., New York, 2001.

[37] F. Van Diggelen, A-GPS: Assisted GPS, GNSS, and SBAS, GNSS Technology and Applications Series, Artech House, 2009.

[38] R. Verdone, D. Dardari, G. Mazzini, A. Conti, Wireless Sensor and Actuator Networks: Technologies, Analysis and Design, Elsevier, 2008.

[39] M.Z. Win, A. Conti, S. Mazuelas, Y. Shen, W.M. Gifford, D. Dardari, Network localization and navigation via cooperation, IEEE Commun. Mag. (2011).

[40] H. Wymeersch, J. Lien, M. Win, Cooperative localization in wireless networks, Proc. IEEE 97 (2) (2009) 427–450.

Satellite-Based Navigation Systems

2

Giacomo Bacci, Emanuela Falletti, Carles Fernández-Prades,
Marco Luise, Davide Margaria, Francesca Zanier

2.1 GLOBAL NAVIGATION SATELLITE SYSTEMS (GNSSs)

By far the most well-known and widespread system currently used for localization of a radio-receiving terminal is the global positioning system (GPS), invented and designed in the 1970s and fully operational since the 1990s. Global navigation satellite systems (GNSSs) are considered so strategic in the many applications of modern life (leisure, defense and civil protection, agriculture, and transportation, just to cite a few) that a number of independent GNSSs were and are being designed and deployed after the GPS. In this chapter, the main characteristics of the GNSSs available today and foreseen in the near future are presented: the US GPS, the European Galileo, the Russian GLONASS, and the Chinese Compass (aka BeiDou 2). Consistent with the aim of the book, the analysis is focused mostly on the format and characteristics of the transmitted signal-in-space (SIS), that is, modulation, carrier frequency, bandwidth, and associated services.

2.1.1 Global Positioning System (GPS)

The NAVSTAR-GPS (NAVigation System for Timing And Ranging–Global Positioning System) project was officially launched in 1973 by the US Department of Defense (DoD) to design and deploy a positioning service with global coverage and continuous-time availability. The GPS was originally developed for authorized (military) use only, and subsequently was released for civil users in 1983. For a detailed history of the GPS project, the interested reader may consult Ref. [38].

Any GNSS system, in particular the GPS, is structured into three *segments*: the satellite constellation (i.e., the *space segment* (SS)), the ground control/monitoring network (the operational control segment (OCS)), and the collection of all user receiver devices (the user segment (US)). The OCS tracks and maintains the spacecrafts, monitors satellite health and signal integrity, and maintains the orbital configuration of the space segment. In addition, the OCS monitors and corrects the satellite clock, and updates the constellation *ephemeris*,[1] as well as other fundamental parameters. The US is typically a specific user receiver equipment that processes the GPS signals to determine the user's position, velocity, and time (PVT).

---

[1] The ephemeris are a set of parameters that allow precise calculation of the coordinates of each satellite in the sky. They are fundamental to perform positioning of the user receiver.

At the time of writing (2011), the GPS constellation consisted of 31 satellites positioned on six earth-centered orbital planes with five to six satellites on each plane. The current constellation is composed of satellites from Block IIA, launched between 1990 and 1997, during which the system was declared fully operational, Block IIR (1997–2004), and Block IIR-M (2004–present). The first next-generation satellite in Block IIF was launched in May 2010, with more to follow, while Block III satellites are planned to be employed for a post-2010 deployment. The nominal orbital period of a GPS is one-half of a sidereal day (approximately 11 h 58 min). The orbits are nearly circular and equally spaced about the equator at a 60° separation with a nominal inclination of 55° relative to the equator, whereas the orbital radius is approximately 26,600 km (20,200 km average height above ground). The GPS constellation provides a 24-h global user navigation and time determination capability.

From its very inception, the GPS was meant to provide two *services*, corresponding to two different signal formats: the standard position service (SPS) and the precise position service (PPS). The SPS is designed for the civil community, whereas the PPS is slated for the United States authorized military and selected government agency users. Further details are provided in the next subsections.

Frequency Plan

The so-called GPS *frequency plan* is shown in Fig. 2.1. The plan contains the allocated bandwidth and the relevant names for the different GPS bands. The current "interface" between the GPS space segment and GPS receiver (the navigation US) consists of two RF links at different carrier frequencies: L1, 1575.42 MHz, and L2, 1227.60 MHz. An additional frequency, namely L5, operating at 1176.45 MHz, is reserved for future services, as detailed in Section 2.1.1.3. The distribution of services on such frequencies is summarized in Fig. 2.2. All of these frequencies are coherent with the fundamental GPS frequency of exactly $f_0 = 10.23$ MHz. The three aforementioned frequencies can in fact be related to the fundamental clock frequency as $f_{L1} = 1575.42$ MHz $= 154 \cdot f_0, f_{L2} = 1227.60$ MHz $= 120 \cdot f_0$, and $f_{L5} = 1176.45$ MHz $= 115 \cdot f_0$. The value $f_0 = 10.23$ MHz is a nominal frequency as it appears to an observer on the ground. To compensate for relativistic effects due to satellite motion, the actual fundamental GPS frequency of oscillators aboard the spacecrafts is set before launch to $f_0^{(act)} = 10.22999999543$ MHz, so that a GPS receiver on the ground would receive a clock signal at exactly the desired frequency f_0. However, the oscillation frequency of the carriers L1, L2, and L5

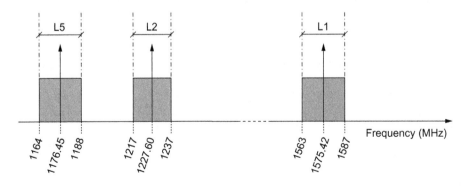

FIGURE 2.1

GPS frequency plan.

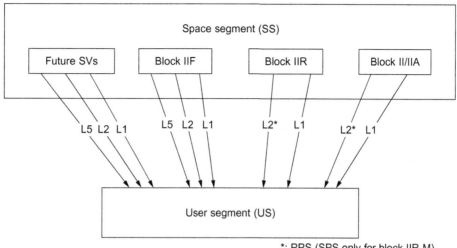

FIGURE 2.2

GPS minimum space segment to user segment interfaces [5, 6].

as received by the GPS user terminal changes in time owing to *Doppler shift* caused by the relative motion of transmitter and receiver. Such a frequency shift is bounded in the range ±5 kHz, at least for terrestrial users, and has to be compensated for by the receiver.

The main relevant parameters of the GPS signals are reported in Table 2.1, further information on the GPS signal format being presented in the next subsections. As shown in Fig. 2.2, the L1 channel provides both the SPS and the PPS, whereas the L2 channel currently accommodates the PPS only. The satellites in block IIR-M, IIF, and in subsequent blocks will also provide two additional services: The L2 civil-moderate (L2 CM) code and the L2 civil-long (L2 CL) code (see also Section 2.1.1.1). Such additional features belong to the *modernization program*, which also includes a new navigation (NAV) signal for civil use, L5 [5] (see Section 2.1.1.3 for further details). The GPS L1 frequency is also the carrier frequency for the *satellite-based augmentation system* (SBAS) signals [41] (see Section 2.3.2 for more information).

Multiple Access

All GPS satellites share the same frequency bands, making use of the *code division multiple access* (CDMA) technique. The transmitted signal on each subband is a low-rate binary phase shift keying (BPSK) digital signal containing the so-called *navigation* (NAV) data that are used by the receiver to perform ranging. Digital data are "covered" via modulo-2 addition with a higher-rate pseudorandom noise (PRN) binary ranging code running at the *chip rate*. Each satellite vehicle (SV) is assigned a unique PRN code, and all of the PRN sequences are nearly uncorrelated with one another, so that all SV signals can be separated and detected by just *correlating* the received signal with a properly shifted local replica of the satellite-specific codes. Three different PRN ranging codes are transmitted by GPS SV [6]: (1) the precision (P) code, which is the principal NAV ranging code; (2) the Y code, used

Table 2.1 Main Characteristics of GPS Signals

Frequency Band	E1/L1	L2	E5/L5
Signal name	GPS L1	GPS L2	GPS L5
Carrier frequency (MHz)	1575.42	1227.60	1176.45
Multiple-access technique	CDMA	CDMA	CDMA
Signal bandwidth (MHz)	20.46	20.46	24
Services	PPS[a], SPS	PPS[a], SPS[b]	SPS
PRN code rate (Mcps)	1.023 (C/A), 10.23 (P)	0.5115 (C/A)[b], 10.23 (P)	10.23
PRN code length (c)	1023 (C/A), $2.35 \cdot 10^{14}$ (P)	10230 (C/A)[c], $2.35 \cdot 10^{14}$ (P)	10230
NAV message data (bps)	50	50[c]	25
NAV message symbols (sps)	50	50	50
Modulation	BPSK	BPSK	BPSK

[a] Authorized users only.
[b] Only Block IIR-M SVs and future SVs.
[c] Length of the L2 CM code (see Section 2.1.1.2 for further details).

in place of the P code whenever the anti-spoofing (A-S)[2] mode of operation is activated; and (3) the coarse/acquisition (C/A) code, which is used primarily for acquisition of the P (or Y) code and is also the only one used in civil/commercial applications (see also Section 2.1.1.1).

2.1.1.1 *GPS L1*

The two codes on the L1 carrier are the P code and the C/A code. The P code has a chip rate of 10.23 Mcps (with a chip time roughly equal to 97.8 ns), so that the main lobe of the spectrum spans an RF (null-to-null) bandwidth of 20.46 MHz. The particular binary ranging sequence (code) is generated by the combination of two PRN codes with the same chip rate. One of the two component codes is periodic with a repetition length of 15,345,000 chips (corresponding to a repetition period of exactly 1.5 s); the other one has a length 15,345,037 chips. The two numbers, 15,345,000 and 15,345,037, are relative primes, so that the code length generated by the combination (chip-by-chip XOR) of these two codes is $15,345,000 \times 15,345,037 = 2.35469592765E + 14$ chips, or 23,017,555.5 seconds, which is slightly longer than 38 weeks. However, the actual length of the P code of every satellite is smaller: each satellite uses a *different* one-week-long segment of the long P code, with a one-week repetition period. The 38-week-long segment is in fact subdivided into 37 different sections, and each segment is assigned to a different SV. There are currently a total of 32 satellite identification numbers (different P-code segments), with five more (33–37) reserved for other uses such as ground transmission. To perform acquisition on the signal, the absolute time of the week must be known very accurately, and this information can be found in the C/A code signal.

The C/A code chip rate is 1.023 Mcps (chip time of roughly 977.5 ns). Therefore, the null-to-null bandwidth of the main lobe of the spectrum is 2.046 MHz. To also accommodate the P code, the transmitting bandwidth of GPS satellites across the L1 carrier is approximately 20 MHz, so that the

---

[2] The A-S mode is a protection level introduced by the control system to guard against fake transmissions of satellite signals (also called *spoofing*) by encrypting the P code to form the Y code.

C/A code spectrum bears its own main lobe as well as several sidelobes. The repetition period of the C/A code is much shorter than that of the P code, namely 1023 chips. With a chip rate of 1.023 Mcps, the C/A code repetition time is exactly 1 ms. This means that in good conditions of signal-to-noise ratio and with a small carrier Doppler shift, the GPS receiver can find the start of the C/A code via a trial-and-error procedure in an observation time as small as 1 ms. In practice, the observation (*acquisition*) time has to be longer to compensate for noise and Doppler. In order to find the beginning of a C/A code in the received signal only a very limited data record is needed such as 1 ms. As with the P code, each satellite is assigned a different PRN code belonging to the same set of Gold codes [13] with length $1023 = 2^{10} - 1$.

At the time of writing (2011), the L1 frequency carried the C/A and P(Y) signals, while the L2 frequency contained only the P(Y) signal (see Section 2.1.1.2 for further details). The C/A and P(Y) signals for the *i*th satellite on the L1 carrier are modulated on the in-phase and quadrature components of the bandpass signal, respectively:

$$s_{L1,i}(t) = A \cdot P_i(t) \cdot D_i(t) \cos(2\pi f_{L1} t + \phi_1)$$
$$+ \sqrt{2} A \cdot C_i(t) \cdot D_i(t) \sin(2\pi f_{L1} t + \phi_1), \tag{2.1}$$

where $s_{L1,i}(t)$ is the band-pass signal at the L1 frequency; A is the amplitude of the P code; $P_i(t) = \pm 1$ and $C_i(t) = \pm 1$ are the binary P code and C/A code, respectively; $D_i(t) = \pm 1$ are the NAV data; f_{L1} is the L1 frequency; and ϕ_1 is the initial phase. The P(Y), C/A, and the carrier frequencies are all *coherent* (phase locked); that is, they all bear a fixed frequency and phase relation to one another. All baseband digital signals (code and/or data) bear a simple nonreturn to zero (NRZ) format with (ideally) rectangular pulses.

The bit rate (which is also equal to the symbol rate, since GPS L1 does not contain any channel coding) is 50 Hz. Each data bit is 20 ms long, and contains 20 replicas of the C/A code. As can be seen in Fig. 2.3, the message structure is articulated across 25 frames. Every frame is made up of five subframes, each subframe consisting of 10 words, and each word is finally 30 bits long. The most significant bit (MSB) of all words is transmitted first. Considering a bit rate of 50 bps, the complete GPS message lasts 12.5 min. Further information on the contents and format of the subframes (stability of the GPS universal clock, ephemeris, ionospheric corrections, GPS/UTC time conversion, *almanacs*[3] is found in Ref. [6].

2.1.1.2 *GPS L2*

The L2 signal only contains the P (or Y) code, and data modulation can be switched on or off upon ground command:

$$s_{L2,i}(t) = \frac{A}{\sqrt{2}} \cdot P_i(t) \cdot D_i(t) \cos(2\pi f_{L2} t + \phi_2), \tag{2.2}$$

where $s_{L2,i}(t)$ is the band-pass signal at frequency L2; A is the amplitude of the P code onto the GPS L1 channel; $P_i(t) = \pm 1$ represents the phase of the P code; $D_i(t) = \pm 1$ is the data signal ($D_i(t) = 1$ if data modulation is off); f_{L2} is the L2 frequency; and ϕ_2 is the initial phase. Again, the format is NRZ

---

[3]The almanac is a collection of data about the orbits of all satellites in the system that is transmitted by all satellites, and allows coarse determination of the position of the satellites in the sky. It is refreshed on a periodic basis (minutes) and allows the receiver to compute, for instance, the satellite in visibility to perform initial acquisition.

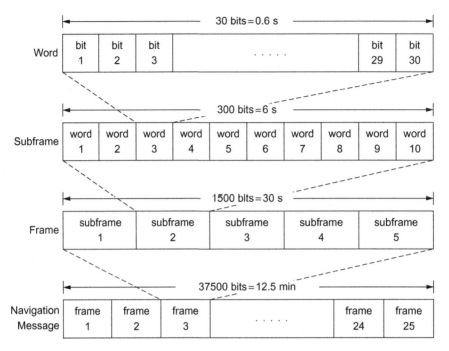

FIGURE 2.3

GPS NAV data structure.

for all baseband digital signals. The current NAV message on the L2 channel is the same as on the L1 channel (see Section 2.1.1.1). However, from SVs in block IIR-M on, the actual NAV data signal can have different formats: the signal always runs at 50 sps, but it may have channel coding to give better protection against noise. With no coding, the bit rate is equal to the symbol rate (as for L1), while with channel coding, the bit rate is 25 bps, and a *convolutional code* with rate 1/2 (i.e., with the addition of one protection (redundant) bit every NAV data bit). The details of the actual format of $D_i(t)$ are beyond the scope of the book and can be found in Ref. [6].

2.1.1.3 *GPS L5 and Modernized GPS*
From the launch of the first GPS satellite in 1978 until the end of 2004 there were three navigation signals on two frequencies. Since the launch of the first Block IIR-M satellite (Sept. 26th, 2005), the SPS is also available on the L2 channel (as described in Section 2.1.1.2). In the near future, the number of navigation signals will increase from three to seven and the number of frequencies from two to three (see Sections 2.1.1.1–2.1.1.3). In addition, the new signals will have substantially better characteristics, including a pilot carrier, much longer codes, the use of forward error correction, and a more flexible message structure with much better resolution.

For instance, the L5 carrier bears the so-called safety of life service. The basic structure of the L5 signal is somewhat similar to the L1 signal: a carrier that is BPSK modulated with PRN ranging codes

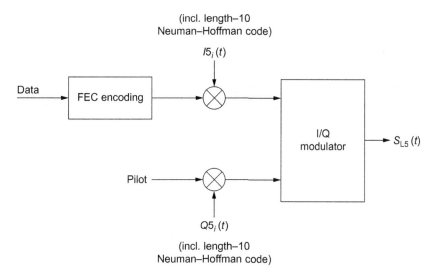

FIGURE 2.4

GPS L5 signal format.

and a superimposed data stream, as is shown in Fig. 2.4. But L5 will have two distinct equi-power subsignals: one with and one without a superimposed navigation data message, the latter acting as a data-less *pilot* signal. The first subsignal lies on the in-phase component of the bandpass L5 signal, and has the so-called binary I5 code and a 100 symbol/s data message (50 bps NAV message with rate 1/2 channel coding); the Q-component has the Q5 QPSK (quadri-phase) PRN code, and it has no NAV data. The pilot signal with the Q5 code is used to perform signal acquisition and tracking, a function that is somewhat improved by the absence of data modulation. The relevant data expression is not presented here—a similar arrangement is followed by the Galileo signals that is explored in more detail in the sections to follow. The PRN L5 codes $I5_i(t)$ and $Q5_i(t)$ for the ith SV are independent (but time synchronized) and both have a repetition time of 1 ms at a chip rate of 10.23 Mbps (repetition length 10230 chips) [6].

New and modern civil signals will be on L2 and on the new L5 frequency. The current GPS modernization plan, however, leaves the L1 frequency with the only the legacy C/A signal for civil applications. The next level of modernization will include a new signal for the L1 frequency, named L1C. With the addition of L1C, all the three GPS frequencies would then provide a modernized civil signal, completing the GPS modernization process. In the future, even with new and modern L2 and L5 signals, L1 is expected to remain the most important civil frequency. This is primarily because it is less affected by ionospheric refraction error than L2 or L5. (L1 has only 61% of the L2 error and 56% of the L5 error.)

In order to allow *interoperability* between GPS and Galileo systems, the U.S.A. and the European Union (EU) commenced and recently completed negotiations about the compatibility of Galileo L1 signals with both military and civil GPS signals [1]. As part of these negotiations, the U.S. Department of State proposed that the United States would implement a new signal on L1 with binary offset carrier

(BOC) modulation if Europe would do the same on Galileo [7]. In the next section, we revise the formats of the European Galileo GNSS, including such novel modulation formats.

2.1.2 **Galileo**

Galileo is the European GNSS providing a global positioning service under civilian control. It has to be interoperable with GPS and GLONASS, the American and Russian GNSSs, respectively. The Galileo program was officially initiated on May 26, 2003 by the EU and the European Space Agency (ESA). The following are some details on Galileo, more than what was reported for GPS, since it is relatively difficult to find such information in previous textbooks.

The fully deployed Galileo system consists of 30 medium earth orbit (MEO) satellites (27 operational plus three spares) positioned in three circular MEO planes at a nominal average orbit semimajor axis of $29,601.297$ km, and at an inclination of the orbital planes of $56°$ with reference to the equatorial plane. At the time of writing (2011), the Galileo project was in the validation phase. The two *Galileo In-Orbit Validation Element* (GIOVE)-A and GIOVE-B satellites were placed into orbit in 2008 to provide experimental results for the GPS–Galileo interoperable signal using the multiplexed binary offset carrier (MBOC) modulation in accordance with the agreement [1] drawn up by the EU and the United States [15].

Five different services are expected from Galileo:

- An *open service* (OS) providing all information such as positioning, navigation, and timing services, free of charge, for mass market navigation applications, interoperable with other GNSSs, and competitive with the GPS standard positioning services.
- A *safety of life* (SoL) compliant with the needs of safety critical users such as civil aviation, maritime, and rail domain. The safety of life (SoL) includes high integrity and authentication capability, although the activation of these possibilities will depend on the user communities. The SoL service includes service guarantees also.
- A *commercial service* (CS), aimed at generating commercial revenues through the provision of added value over the OS, i.e., dissemination of encrypted navigation-related data, ranging and timing for professional use, service guarantees, high integrity level, accurate timing services, high data rate broadcasting, provision of ionospheric delay modes, local differential correction signals and controlled access.
- A *public regulated service* (PRS), for application devoted to European and member states, in particular for critical applications and activities of strategic importance (defense/civil protection). It makes use of a robust signal and is controlled by member states. This service provides service guarantees, high integrity, full range of value-added features and an access controlled by encryption.
- A *search and rescue* (SAR) service providing assistance to the COSPAS-SARSAT[4] system by detecting emergency beacons and forwarding return link messages to the emergency beacons. It is a service for SAR applications providing near-real-time reception of distress messages and precise location of alerts [12].

---

[4]COSPAS-SARSAT is a satellite system designed to provide distress alert and location data to assist SAR operations, using spacecraft and ground facilities to detect and locate the signals of distress beacons operating at 406 MHz or 121.5 MHz. The goal of this system is to support all organizations in the world with responsibility for SAR operations, whether at sea, in the air, or on land.

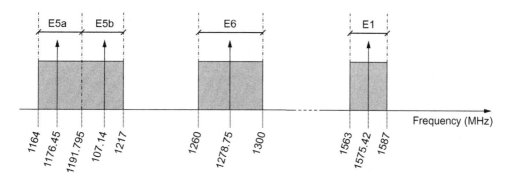

FIGURE 2.5

Galileo frequency plan.

Frequency Plan

Figure 2.5 shows the frequency plan employed by the Galileo system. Three independent signals are permanently transmitted by all Galileo satellites [16]: E5, E6, and E1. The E5 link is further split into two RF links denoted by E5a and E5b. The frequency bands were selected in the allocated spectrum for radio navigation satellite services and in addition to that, the E5a, E5b, and L1 bands are included in the allocated spectrum for aeronautical radio navigation services (ARNS) used by civil-aviation users for dedicated safety-critical applications. All Galileo transmitting satellites share the same frequency bands, making use of the CDMA technique. Spread spectrum signals are transmitted including different ranging codes per I/Q signal component, per bandpass signal, per carrier frequency, and per Galileo satellite.

The E1 signal is an open-access signal transmitted in the L1 band and encompassing, much like the GPS L5 signal, a data channel E1-B and a pilot (or data-less) channel E1-C. It has unencrypted ranging codes and navigation data, which are accessible to all users. The E1-B data stream also contains unencrypted integrity messages (that is, information about the possible failures of unavailability of satellites) and encrypted commercial data. The E1 signal supports the OS, the CS, and the SoL. The E1-A signal channel contains a restricted-access signal for PRS. The governmental restricted access is provided by an encryption algorithm to encode both ranging codes and navigation data.

E6 is a commercial-access signal transmitted in the E6 band that includes a data channel E6-B and a pilot channel E6-C. Its ranging codes and data are encrypted. The E6 signal is a dedicated signal for supporting the CS. Similar to E1-A, the E6-A signal channel contains a restricted-access signal for PRS, also providing antispoofing and other security features.

E5a is an open-access signal transmitted in the E5 band that includes a data channel and a pilot channel. The E5a signal has unencrypted ranging codes and navigation data, which are accessible by all users; it transmits the basic data to support navigation and timing functions, and also supports the OS. E5b is an open-access signal transmitted in the E5 band that includes a data channel and a pilot channel. It has unencrypted ranging codes and navigation data, accessible to all users. The E5b data stream also contains unencrypted integrity messages and encrypted commercial data, and supports OS, CS, and SoL.

Table 2.2 Main Characteristics of Galileo Signals

Frequency Band	E1/L1	E6	E5/L5	
Signal name	Galileo E1	Galileo E6	Galileo E5a	Galileo E5b
Carrier frequency (MHz)	1575.42	1278.75	1176.45	1207.14
Multiple-access technique	CDMA	CDMA	CDMA	CDMA
Signal bandwidth (MHz)	24.552	40.92	51.15[a]	
Services	OS, CS, SoL, PRS[b]	CS, PRS[b]	OS	OS, CS, SoL
PRN code rate (Mcps)	1.023	5.115	10.23	10.23
PRN code length (c)	4092 (p), 25 (s)[c]	n/a	10230 (p) 20 (s)[d]	10230 (p) 4 (s)[d]
NAV msg data (bps)	125	n/a	25	125
NAV msg symbols (sps)	250	1000	50	250
Modulation	CBOC	BPSK	AltBOC(15,10)[a]	

[a] The E5a and E5b signals are modulated onto a single E5 carrier using a technique known as AltBOC(15,10) (see Section 2.1.2.3 for further details). The resulting bandwidth is 51.15 MHz.
[b] Authorized users only.
[c] Pilot (data-less) channel only.
[d] The length of the secondary code for the pilot (data-less) channel is 100.

Table 2.2 summarizes the main parameters of the Galileo signals, including multiple-access techniques and signal modulations, which will be further detailed in the following subsections.

Multiple Access

Galileo uses the CDMA technique to perform multiple access by different satellites to the same subband. The Galileo ranging codes are built from so-called primary (p) and secondary (s) codes, by using a tiered code construction. The primary spreading codes can be either *linear feedback shift register* pseudonoise sequences or *optimized* pseudonoise sequences. Secondary code sequences are used to modify successive repetitions of a primary code. The chips of the secondary code are used to control the polarity of each epoch of the primary code sequence (logical sum) [16].

2.1.2.1 *Galileo E1*

The E1 signal is transmitted over a bandwidth of 20.46 MHz on the same carrier frequency used by the GPS L1 signal (1575.42 MHz). As anticipated in Section 2.1.2, it is composed of a restricted-access signal, transmitted on the data channel E1-A, an open-access signal (E1-B), and a pilot signal (E1-C). In the following sections, channels E1-B and E1-C are focused, since details about E1-A are not public. Figure 2.6 outlines the modulation scheme adopted for the E1 signal, namely the composite binary offset carrier (CBOC) modulation. The E1 CBOC signal components are generated as follows:

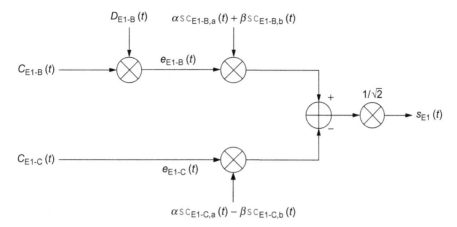

FIGURE 2.6

Modulation scheme for the E1 CBOC signal.

- The data signal $e_{\text{E1-B}}(t)$ contains the integrity navigation (I/NAV) message $D_{\text{E1-B}}(t)$ and is spread with the ranging code $C_{\text{E1-B}}(t)$. On top of this, it is further modulated by the square-wave subcarriers $s_{C_{\text{E1-B,a}}}(t)$ and $s_{C_{\text{E1-B,b}}}(t)$.
- The pilot signal $e_{\text{E1-C}}(t)$ contains only the ranging code $C_{\text{E1-C}}(t)$ (including its secondary code), and then it is modulated with the square-wave subcarriers $s_{C_{\text{E1-C,a}}}(t)$ and $s_{C_{\text{E1-C,b}}}(t)$.

As shown in Table 2.2, the ranging code chip rate is 1.023 Mcps for both the E1-B and the E1-C channels. The length of both primary codes is 4092, and the E1-C channel also uses a secondary code with a length of 25. The square-wave subcarriers of CBOC in Fig. 2.6 are

$$s_{C_{\text{E1-X,y}}}(t) = \text{sgn}\left[\sin\left(2\pi R_{\text{S,E1-X,y}} \cdot t\right)\right], \tag{2.3}$$

where $X \in \{B,C\}$, $y \in \{a,b\}$, $R_{\text{S,E1-X,a}} = 1.023$ MHz, and $R_{\text{S,E1-X,b}} = 6R_{\text{S,E1-X,a}} = 6.138$ MHz. The parameters α and β are chosen in such a way that the combined power of the $s_{C_{\text{E1-B,b}}}$ and $s_{C_{\text{E1-C,b}}}$ subcarrier components equals 1/11 of the total power of $e_{\text{E1-B}}$ plus $e_{\text{E1-C}}$, before the application of any band limitation. This yields $\alpha = \sqrt{10/11}$ and $\beta = \sqrt{1/11}$. Because of this value of the coefficients, the CBOC modulation for the E1 signal is often referred to as CBOC(1, 6, 1/11). Both the pilot and the data components are modulated onto the same carriers, with a power sharing of 50%.

The rationale of CBOC is simple to understand: displace a considerable fraction of the total signal power away from the center of the alllocated band [20]. This can be seen in Fig. 2.7, which reports the power spectral density (PSD) of the signal as a function of the normalized offset from the carrier frequency. From the analysis of the CRB for time delay estimation (4.3), we know that this improves the accuracy of positioning. The spectrum for the BOC(1,1) in Fig. 2.7 refers to a simple modulation format, which can be obtained from Fig. 2.6 by letting $\beta = 0$.

The data transmitted on the E1 carrier come from the so-called I/NAV navigation message (Fig. 2.8). The complete 1-*second* message is a sequence of frames, organized in 24 subframes, each

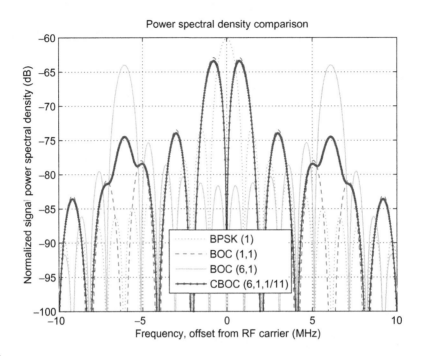

FIGURE 2.7

Comparison between the power spectral densities of BPSK(1), BOC(1,1), BOC(6,1) and CBOC(6,1,1/11) modulation formats.

subframe being composed of 15 nominal pages (one even page followed by one odd page). The symbol page starts with a page synchronization field, which contains a fixed synchronization pattern (whose length depends on the page type), followed by a page symbol field, which is the result of the channel encoding process. The net bit rate of the I/NAV data stream is 125 bps prior to channel encoding with a convolutional code with rate 1/2 and a constraint length of 7 [16], thus providing a symbol rate equal to 250 sps.

2.1.2.2 *Galileo E6*

The E6 signal occupies a bandwidth of 40.92 MHz across a carrier placed at 1278.75 MHz. It is a dedicated signal for CS and PRS. As with E1, only the E6-B and E6-C channels are dealt with, since E6-A is dedicated to authorized (PRS) users only.

Figure 2.9 outlines the modulation scheme adopted for the E6 signal. The two components are as follows:

- The signal $e_{E6-B}(t)$ contains the commercial navigation (C/NAV) message type data stream $D_{E6-B}(t)$, spread by the ranging code $C_{E6-B}(t)$.
- The signal $e_{E6-C}(t)$ is the pilot channel and contains only the ranging code $C_{E6-C}(t)$.

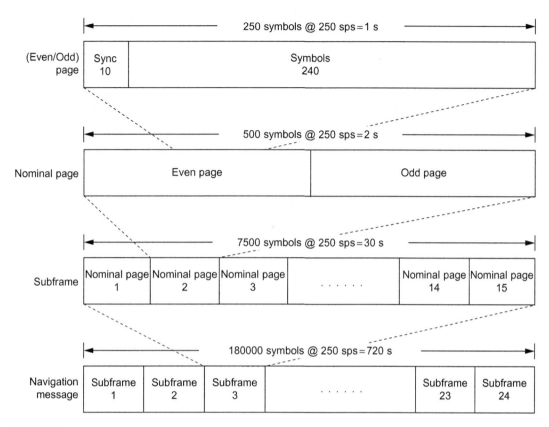

FIGURE 2.8

Galileo I/NAV data structure.

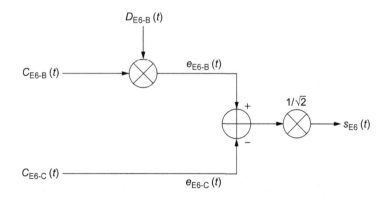

FIGURE 2.9

Modulation scheme for the E6 signal.

As shown in Table 2.2, the chip rate on E6 is 5.115 Mcps (as). The length of the codes is currently not available in the public releases of the Galileo interface control document (ICD) [16]. The data message has a symbol rate equal to 1000 sps. Further details about the data message are currently not available.

2.1.2.3 Galileo E5

The E5 signal nominally occupies a total bandwidth of 51.15 MHz as a result of the combination of two navigation signals, namely E5a and E5b. E5a is an open-access signal transmitted in the E5 band that includes a data channel and a pilot channel. The E5a signal has unencrypted ranging codes and navigation data, which are accessible by all users (OS). E5b supports OS, CS, and SoL with a mixture of open and encrypted data/codes. The structure of the E5 signal is such that the two components E5a and E5b can be concurrently modulated onto a single E5 carrier using a technique known as alternate BOC (AltBOC). The two signals can be processed independently by the user receiver, or equivalently, the E5 signal can be processed as a single wide-bandwidth signal with an appropriate implementation. Figure 2.10 describes the E5 modulation scheme. It is seen that the complete signal contains the following components (with a 50% power split between the respective data and pilot channels):

- The signal $e_{E5a\text{-}I}(t)$, which contains the navigation message F/NAV $D_{E5a\text{-}I}(t)$, spread with the unencrypted ranging code $C_{E5a\text{-}I}(t)$.
- The pilot signal $e_{E5a\text{-}Q}(t)$ containing only the unencrypted ranging code $C_{E5a\text{-}Q}(t)$.
- The signal $e_{E5b\text{-}I}(t)$, which contains the navigation message I/NAV $D_{E5b\text{-}I}(t)$, spread with the unencrypted ranging code $C_{E5b\text{-}I}(t)$.
- The pilot signal $e_{E5b\text{-}Q}(t)$ containing only the unencrypted ranging code $C_{E5b\text{-}Q}(t)$.

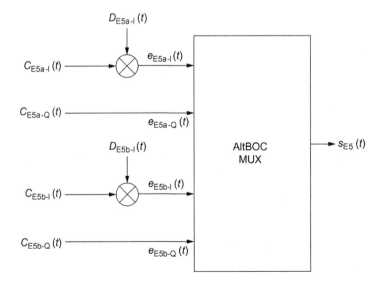

FIGURE 2.10

Modulation scheme for the E5 AltBOC signal.

The chip rate of the ranging codes for E5a and E5b is 10.23 Mcps $= 10 \times 1.023$ Mcps, and the square-wave subcarrier frequency of the AltBOC is $R_{S,E5} = 15.345$ MHz $= 15 \times 1.023$ MHz, according to the scheme shown in Fig. 2.10. In view of these parameters, the AltBOC modulation used for Galileo E5 is often referred to as AltBOC(15,10). As reported in Table 2.2, the ranging codes for E5a and E5b comprise a primary code, with a length of 10,230 chips and a duration of 1 ms, and a secondary code with a length of 100, for the pilot channel, and with lengths of 20 and 4, for signal E5a-I and E5b-I, respectively. As already mentioned in Section 2.1.2, the primary spreading codes can be either linear feedback shift-register-based pseudorandom sequences or optimized pseudonoise sequences [16].

The AltBOC modulation depicted in Fig. 2.10 deserves more attention. The resulting signal $s_{E5}(t)$ produced by the AltBOC is

$$s_{E5}(t) = \frac{\left(e_{E5a\text{-}I}(t) + j \cdot e_{E5a\text{-}Q}(t)\right)}{2\sqrt{2}} \cdot \left[s_{CE5\text{-}S}(t) - j \cdot s_{CE5\text{-}S}\left(t - T_{S,E5}/4\right)\right]$$

$$+ \frac{\left(e_{E5b\text{-}I}(t) + j \cdot e_{E5b\text{-}Q}(t)\right)}{2\sqrt{2}} \cdot \left[s_{CE5\text{-}S}(t) + j \cdot s_{CE5\text{-}S}\left(t - T_{S,E5}/4\right)\right]$$

$$+ \frac{\left(\overline{e}_{E5a\text{-}I}(t) + j \cdot \overline{e}_{E5a\text{-}Q}(t)\right)}{2\sqrt{2}} \cdot \left[s_{CE5\text{-}P}(t) - j \cdot s_{CE5\text{-}P}\left(t - T_{S,E5}/4\right)\right]$$

$$+ \frac{\left(\overline{e}_{E5b\text{-}I}(t) + j \cdot \overline{e}_{E5b\text{-}Q}(t)\right)}{2\sqrt{2}} \cdot \left[s_{CE5\text{-}P}(t) + j \cdot s_{CE5\text{-}P}\left(t - T_{S,E5}/4\right)\right], \qquad (2.4)$$

where $T_{S,E5} = 1/R_{S,E5}$ and the dashed components $\overline{e}_{E5a\text{-}I}(t)$, $\overline{e}_{E5a\text{-}Q}(t)$, $\overline{e}_{E5b\text{-}I}(t)$, and $\overline{e}_{E5b\text{-}Q}(t)$ are defined as

$$\overline{e}_{E5a\text{-}I}(t) = e_{E5a\text{-}Q}(t) \cdot e_{E5b\text{-}I}(t) \cdot e_{E5b\text{-}Q}(t)$$

$$\overline{e}_{E5a\text{-}Q}(t) = e_{E5a\text{-}I}(t) \cdot e_{E5b\text{-}I}(t) \cdot e_{E5b\text{-}Q}(t)$$

$$\overline{e}_{E5b\text{-}I}(t) = e_{E5a\text{-}I}(t) \cdot e_{E5a\text{-}Q}(t) \cdot e_{E5b\text{-}Q}(t)$$

$$\overline{e}_{E5b\text{-}Q}(t) = e_{E5a\text{-}I}(t) \cdot e_{E5a\text{-}Q}(t) \cdot e_{E5b\text{-}I}(t). \qquad (2.5)$$

The signals $s_{CE5\text{-}S}(t)$ and $s_{CE5\text{-}P}(t)$ represent the digital, four-level subcarrier functions for the *single* signals and the *product* signals, respectively:

$$s_{CE5\text{-}S}(t) = \sum_{i=-\infty}^{+\infty} AS_{|i|_8} \, \text{rect}\left(\frac{t - iT_{S,E5}/8}{T_{S,E5}/8}\right)$$

$$s_{CE5\text{-}P}(t) = \sum_{i=-\infty}^{+\infty} AP_{|i|_8} \, \text{rect}\left(\frac{t - iT_{S,E5}/8}{T_{S,E5}/8}\right), \qquad (2.6)$$

where $\text{rect}(t/T)$ denotes the rectangular pulse with duration T, $|\cdot|_X$ denotes the modulo-X operator, and the coefficients AS_i and AP_i are chosen according to Table 2.3. Figure 2.11 shows one period of the subcarrier functions $s_{CE5\text{-}S}(t)$ (darker line) and $s_{CE5\text{-}P}(t)$ (lighter line).

The reader will notice that this formulation of the AltBOC modulation is quite cumbersome and hinders its real rationale. The technique is the result of a number of concurrent requirements: (1) obtaining a constant-envelope signal that multiplexes the four E5 signals; (2) conveniently displacing

Table 2.3 AltBOC Subcarrier Coefficients. $|i|_8$ Indicates the Modulo-8 Operator

$|i|_8$	0	1	2	3	4	5	6	7
$2AS_i$	$\sqrt{2}+1$	1	-1	$-\sqrt{2}-1$	$-\sqrt{2}-1$	-1	1	$\sqrt{2}+1$
$2AP_i$	$-\sqrt{2}+1$	1	-1	$\sqrt{2}-1$	$\sqrt{2}-1$	-1	1	$-\sqrt{2}+1$

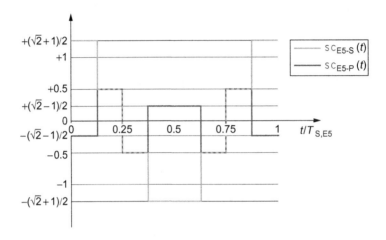

FIGURE 2.11

One period of the sideband subcarriers for AltBOC modulation.

the baseband spectrum so as the signal power is concentrated mostly around the subcarrier frequencies to minimize the CRB; (3) being able to independently demodulate the two components in the E5a-E5 subchannels. In particular, the motivation of the values in Table 2.3 can be easily understood by looking at the I-Q diagram of the modulated signal (Fig. 2.12) [16], which shows the 8PSK nature of the resulting signal, with a constant envelope as required:

$$s_{E5}(t) = \exp\left\{ j\frac{\pi}{4}k(t) \right\}, \quad k(t) \in \{1,\dots,8\}. \tag{2.7}$$

The navigation message transmitted onto the E5 carrier includes both the I/NAV message (on E5b-I) and the F/NAV message type (on E5a-I). The framing of F/NAV is presented in Fig. 2.13. Every frame is made up of 12 subframes, each subframe consisting of 5 pages; each page is 10 seconds long. F/NAV data include clock, ephemeredes, and satellite almanacs [16]. As shown in Table 2.2, the bit rate for the E5a-I signal is 25 bps, which results in a symbol rate of 50 sps after rate-1/2 convolutional encoding, whereas for E5b-I the bit rate is 125 bps, for a gross symbol rate of 250 sps.

2.1.3 GLONASS

GLONASS ("GLObal'naya NAvigatsionnaya Sputnikovaya Sistema," global navigation satellite system) is a GNSS developed by the former Soviet Union as a counterpart of the United States' GPS. Initially, GLONASS was targeted to the USSR Army needs: targeting and navigation for ballistic missiles with world coverage. The setting up of the system started in 1976 to reach full deployment in

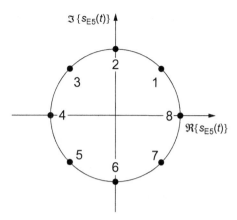

FIGURE 2.12

8-PSK phase-state diagram of E5 AltBOC signal.

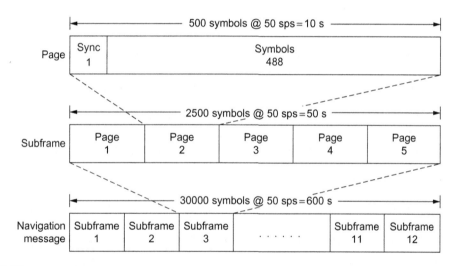

FIGURE 2.13

Galileo F/NAV data structure.

1995. At that time, the constellation encompassed 24 satellites and transmitted on the L1 band using frequency division multiple access (FDMA) (different frequency bands for different satellites). In the following years, the lack of funding, due to collapse of the Russian economy, deeply harmed system efficiency, and in 1999 GLONASS became officially a dual-use system,[5] by a Presidential decree [18] and, at the beginning of the new century, the GLONASS reconstruction was boosted by the Russian extra-gain due to the oil and gas export. Now the GLONASS system is undergoing modernization,

---

[5]That is, a system intended for both military and civil applications.

with the civil side managed by the Russian Space Agency. In 2008, only 12 satellites were in full operational status, but new satellites, six in 2007 and six in 2008, were launched. Since 2004, Russia has been involving India in the GLONASS system upgrade, but the details of the Indian involvement are not public. According to the Russian government program, the full GLONASS constellation with 24 healthy satellites should have been operational in 2009 [44] but the system should reach the GPS/Galileo performance only in 2011 [43].

2.1.3.1 *System Characteristics*
GLONASS Ground Segment
Like GPS and Galileo, GLONASS is based on several ground stations that monitor and control the satellites and their signals. The GPS and Galileo stations are spread on the globe, to guarantee a better control of the constellation and its signals. On the contrary, for political and military reasons, the GLONASS ground segment is completely located inside the former Soviet Union territory, since the Soviet Union never had territories or trusted allies abroad. This geographic concentration is a weakness of GLONASS because it reduces the precision and the efficiency of the whole system.

GLONASS Space Segment
The full GLONASS space segment consists of 21 operational satellites plus 3 spare satellites located on three different orbital planes spaced by 120°. The satellites within the same orbit plane are separated from one another by 45°. The orbits are circular with an inclination angle of 64.8°, and the orbital height above ground level is about 19,140 km (i.e., slightly lower than the GPS satellites). The orbit period is 11 h and 16 min. The whole constellation has a revolution period of 8 days. The 24 spacecraft constellation is currently under modernization, with the progressive replacement of old-generation satellites with new-generation ones. The GLONASS constellation will ultimately be expanded to 30 satellites, although this will require an upgrade to the control segment, which currently can handle only 24 spacecraft [17].

2.1.3.2 *GLONASS Navigation Signals*
Contrary to all the other GNSSs in current use or design, GLONASS uses FDMA to separate the different satellites. The GLONASS satellites broadcast their signals on *different frequencies* in the L1 and L2 bands (see the frequency plan in Fig. 2.14) but they all use the *same ranging code*. The carrier frequency of the nth channel is identified by a frequency offset $n \times 0.5625$ MHz, $n = \pm1, \pm2, \ldots$ from

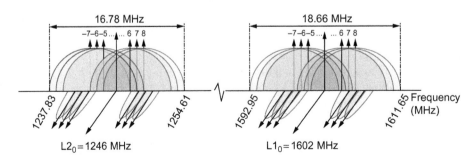

FIGURE 2.14

GLONASS L1 and L2 signals [42].

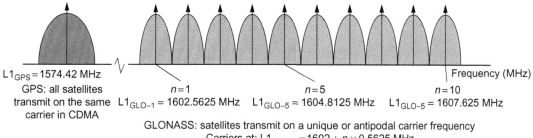

FIGURE 2.15

Comparison of GPS and GLONASS L1 frequencies [25].

Table 2.4 GLONASS Signal Parameters

Parameter	GLONASS	GPS
Number of satellites	21 + 3 spares	21 + 3 spares
Number of orbital planes	3	6
Orbital plane inclination (deg)	64.8	55
Orbital radius (km)	25,510	26,560
Fundamental clock frequency (MHz)	5	10.23
Signal separation technique	FDMA	CDMA
L1 frequency (MHz)	1598.0625–1609.3125	1575.42
L2 frequency (MHz)	1242.9375–251.6875	1227.6
C/A chip rate (Mchip/s)	0.511	1.023
P chip rate (Mchip/s)	5.11	10.23
C/A code length (chips)	511	1023
P code length (chips)	$5.11 \cdot 10^6$	$6.187104 \cdot 10^{12}$
Superframe length (bits – minutes)	7500 – 2.5	37,500 – 12.5
Frame length (bits – seconds)	100 – 2	30 – 0.6

the central L1 carrier, $f_{L1,0} = 1602$ MHz, as shown in Fig. 2.15. The signal timing and frequency are derived from a cesium atomic clock operating at the fundamental universal frequency of 5 MHz.

Using the same nomenclature as for GPS, a C/A code is transmitted by all GLONASS satellites on the L1 band (on L2 by the new GLONASS-M satellites only), and a P code is present on both L1/L2 bands. Because of the orbital arrangement described earlier, GLONASS has antipodal satellites separated by 180° in latitude: each couple of satellites can transmit on the same frequency since the two will never both appear in view at the same time. Table 2.4 summarizes the main parameters of the GLONASS signals in comparison with GPS, including multiple-access techniques and signal modulation. Figure 2.15 compares the GPS and GLONASS L1 frequencies, whereas Fig. 2.16 shows the structure of the GLONASS navigation message as a part of navigation radio signal, which includes ephemerides, information on satellites clocks, time marks, and almanac.

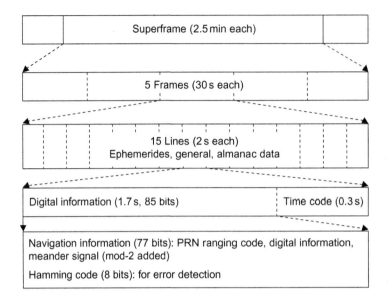

FIGURE 2.16

GLONASS navigation message: the carrier signal is BPSK modulated with the code sequence and (XOR) the navigation data.

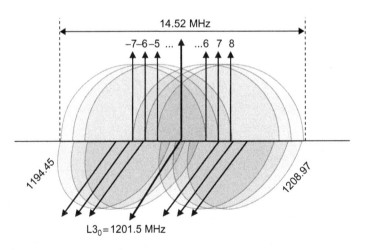

FIGURE 2.17

The GLONASS L3 frequency plan [42].

The second generation of GLONASS satellites provides a number of new features (new design and improved lifetime). In particular, the GLONASS-M satellites, launched in recent years, broadcast a new FDMA civil signal on L2. The next generation of GLONASS-K satellites will provide a third

FDMA civil signal on L3, as sketched in Fig. 2.17, and some CDMA GPS- and Galileo-compatible signals. The first launch was scheduled in 2009, while the complete constellation with all GLONASS-K or later generation satellites should be in place in the 2017–2020 time frame [17].

2.1.4 Compass/BeiDou and Regional GNSSs

Compass is the incoming Chinese GNSS. China started the development of an autonomous GNSS in the 1960s, but only during the 1980s did the project actually gain momentum. In 1994, China approved a new satellite system for navigation purposes on the basis of the radio determination satellite service (RDSS), a technology from GPS [31]. The first Chinese system was named BeiDou, from the Chinese name of the Northern Star, the brightest star of the *Ursa Minor* constellation. BeiDou was born like a regional dual system (both military an civil), to provide navigation and timing to China and surrounding areas.

The evolution of the BeiDou system is called BeiDou-2 or, more commonly, *Compass*. The first satellite of the Compass navigation satellite system (CNSS), which is a MEO satellite, was launched in April 2007. It started covering China and parts of neighboring countries in 2008, and will develop into a global constellation step by step.

The first-generation BeiDou used an "active" bi-directional radio-positioning technique based on TDOA information passed among geostationary (GEO) satellites acting as transceivers, a ground station, and the user equipment. Positioning with the RDSS technology needs a central ground station, at least two geosynchronous satellites, and a large number of *benchmarks*, that is, reference transceivers positioned in fixed and scattered ground sites, used to calibrate the navigation signals. As shown in Fig. 2.18, RDSS operates through appropriate bi-directional communications between satellite transponders and user terminals [9]. The payload operation center (POC) sends out a navigation or *polling* signal through one of the BeiDou satellites. The user terminals reply to this signal via both satellites, and the total double-hop time from the POC through the satellite then down to the user and back is recorded. On the basis of this time-lapse measurement, the known locations of the two satellites, and an estimate of the user altitude, the user's location can be determined by the POC. After this computation, the POC sends again the positioning information to the interested user. Since it is the POC that evaluates the position for all subscribers to the system, BeiDou can be easily used for fleet management and communication. Nonetheless, this system design required a complicated two-way radio terminal, and had a limited capacity and coverage (restricted to East Asia).

Compass is not to be considered as an extension of BeiDou-1, since it is based on a more conventional passive technology. In particular, it uses MEO satellites and CDMA-based navigation messages similar to GPS [49].

The first constellation of BeiDou used four GEO satellites, but upon completion the space segment of CNSS will consist of 5 GEO and 30 MEO satellites. Compass is expected to be a full GNSS and will broadcast CDMA signals in the same frequency bands shared by the European and American systems. The Compass space segment is peculiar, because it will be made up of two different kinds of satellites, GEO and MEO, two different coverages, regional and global, and two different kinds of ranging signal processing, RDSS and CDMA, since it will still support the old BeiDou-1 system.

The BeiDou RDSS transponders operate using the L-band for the uplink (1610–1625 MHz) and the S-band for the downlink (2438.5–2500 MHz). The services are and will be available and operative in mainland China only. Compass will use both the BeiDou RDSS system for the GEO satellites

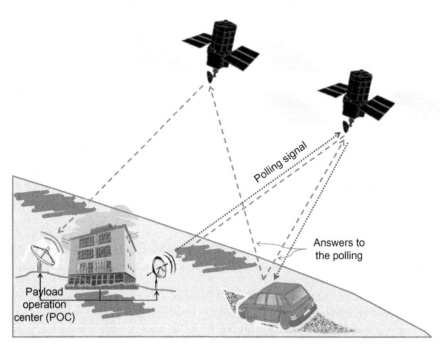

FIGURE 2.18

Scheme of a RDSS positioning system like BeiDou.

and CDMA for the MEOs. As with GPS and Galileo, two fundamental services will be provided by the MEO satellites. One is the *open service*, which is designed to provide users with positioning accuracy within 10 meters, velocity accuracy within 0.2 meter per second, and timing accuracy within 50 nanoseconds. The other is the *authorized service*, which should offer more accurate and reliable positioning, velocity, timing, communication services, and integrity information for authorized users. The frequency bands foreseen for the CNSS are 1195.14–1219.14 MHz, 1256.52–1280.52 MHz, 1559.05–1563.15 MHz, and 1587.69–1591.79 MHz [10].

More regional GNSSs are being designed nowadays. The two main ones are the *regional navigation satellite system (IRNSS)* and the Japanese *quasi-zenith satellite system (QZSS)*. Such systems as are still in an early development stage are not discussed here in further detail. Further details can be found in Refs. [26], [30], and [29].

2.2 GNSS RECEIVERS

User terminals including the radio receiver are a fundamental part of a navigation system—it is the receiver that ultimately performs the function of a user's position computation and applies the so-called navigation algorithms. Different from the space segment, which remains basically a niche market dominated by a few strong manufacturers and directed by the standardization bodies, the user segment has been a "conquest territory" for many manufacturers, according to their own interest in different market segments. Receiver design has been an arena of intense and competitive research and development, leading to the commercialization of receivers with ever improved performance, reduced time-to-first-fix, reduced probability of losing satellite tracking, reduced size, and reduced cost and consumption. Many of these outstanding results were obtained after clever use of *signal processing* techniques, so that the technical and scientific community has elaborated a well-established theory and practice of the navigation receivers, which are summarized in the following section. In particular, this section gives an overview of the fundamental architecture of a GNSS receiver, discussing the main functionalities and the theoretical principles at their basis. After a general description of the overall architecture and of the RF front-end, the three main functions of *signal acquisition*, *signal tracking*, and *navigation processing* are addressed and analyzed, followed by a general discussion of the main sources of error.

2.2.1 Overall Architecture

The function of a GNSS receiver is to determine the user position (and/or time, velocity) on the basis of the signals coming from the satellites. This is usually accomplished by a combination of suitable hardware (HW) and software (SW) techniques, or, perhaps more properly, *analog* and *digital* processing techniques. In the course of their evolution, GNSS receivers have seen a progressive shift of the analog-to-digital conversion as close as possible to the receiving antenna, so that modern receivers are mostly digital, with the analog section being limited to RF signal filtering, amplification, and down-conversion. All of the subsequent processing is thus performed through digital signal processor (DSP) techniques after signal digitization in the analog-to-digital converter (ADC).

Independent of the adopted analog/digital partitioning, the basic functional structure of a GNSS receiver is shown in Fig. 2.19.

Six different main subsystems can be identified:

- Antenna
- Radio frequency chain (RF chain)
- Analog-to-digital converter (ADC)
- Signal processing
- Navigation processing
- Receiver carrier generator or local oscillator (LO).

The signals transmitted from the satellites are received by an L-band *antenna*, which is right-hand circularly polarized (RHCP) and has nearly hemispherical (i.e., above the local horizon) coverage.

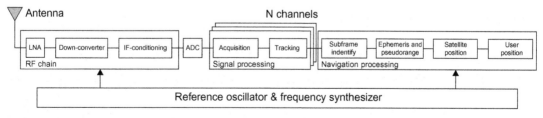

FIGURE 2.19

General GNSS receiver architecture.

Typically, the antenna is followed by a passive band-pass prefilter to minimize out-of-band RF interference.

The output of the antenna is the input of the *RF chain*, wherein the incoming signal is amplified via a low-noise (*pre*)amplifier (LNA), and is down-converted to a lower intermediate frequency (IF) f_{IF}. In a *superheterodyne* receiver, this is accomplished through mixing the received signal at the carrier frequency f_c with a local RF oscillation at the frequency $(f_0 + f_{IF})$. The high frequency $(2f_0 + f_{IF})$ arising in the mixing is filtered by a band-pass filter, the so-called precorrelation filter. The bandwidth of such a filter must be large enough to allow the desired signal to pass unaltered, in spite of possible Doppler shifts and drifts in the reference oscillator. On the other hand, the bandwidth must also be kept as small as possible, to reduce noise and interference at the RF-chain output.

After the RF chain, an ADC is used to digitize the signal prior to performing *signal processing* on the resulting high-data-rate digital stream. For each satellite that is being processed, an independent *channel* is needed to accomplish signal acquisition, tracking, and signal/navigation processing. One of the fundamental properties of the receiver is the number of parallel channels that they concurrently can process. Modern receivers have up to a few tens of parallel channels. In each channel, the incoming signal has to be "associated" with the proper transmitting satellite to be processed. Thus, the IF digitized signal received from a certain satellite at the discrete time instant n and associated with one of the receiver's channels can be written as

$$y_{IF}[n] = \sqrt{2P_r} c_{IF}(nT_c - \tau) D(nT_c - \tau) \cos\left(2\pi(f_{IF} + f_D)nT_c + \phi_{IF}\right) + v[n], \tag{2.8}$$

where P_r is the received signal power at RF, $c_{IF}(nT_c - \tau)$ is the spreading (ranging) code associated with the satellite, filtered through the front-end and possibly modulated by a subcarrier (e.g., in the BOC(1, 1)), $D(nT_c - \tau)$ is the navigation data bit, T_c is the code chip duration, τ is the code delay that measures the distance of the receiver from the satellite, f_D is the Doppler frequency shift, and ϕ_{IF} is the IF carrier phase. The additive noise term $v[n]$ also includes, besides the thermal noise at the antenna, interference effects from both external sources and other satellites with nonzero code cross-correlation. In the expression seen earlier, the unknown parameters of specific interest for positioning are the code delay τ, the Doppler frequency f_D, and the navigation data $D(nT_c - \tau)$.

The process of assigning a channel to a specific satellite is the *acquisition phase*, during which a coarse estimation of carrier frequency and ranging code delay of the satellite is performed. Once acquisition is accomplished, the received signal is *correlated* with a local replica of the reference ranging code of the satellite (where "correlation" stands for "multiplication by the local code and accumulation over a certain integration time").

The output of such correlation on a certain time window is the input of the *tracking loops* for precise estimation of the code timing and the carrier frequency/phase. Once the tracking loops are in the *lock mode*, the navigation message can be decoded, and the time delay of the ranging code is measured.

The parameters produced by the signal processing block (NAV data and time delay of each satellite) represent the input of the *navigation processing* functions. During navigation, the time delay measurements are translated into values of the distances from the user terminal to the received satellites (the so-called pseudoranges), and the NAV message is decoded to find *ephemeris*, and from those, to compute the satellite positions in the sky. From the satellite positions and the pseudoranges, the user position can be finally calculated.

An additional fundamental element in the receiver is the *local oscillators* (LOs), whose function is to set the necessary reference frequency for the down-conversion, the A/D conversion, and code generation. The LOs are derived from a reference oscillator by a frequency synthesizer.

2.2.2 Signal Acquisition

The first signal-processing function in the receiver is detection of the signal coming from a certain satellite, as well as a coarse estimation of its time/frequency (τ, f_D) parameters (in the case detection is positive), or, in a word, acquisition. Extensive discussions of the acquisition problem and related techniques can be found in several textbooks and tutorial publications [8, 14, 27, 36, 39, 48].

The acquisition process consists of a *two-dimensional search* both in time and in the frequency domain. Time uncertainty comes from uncertainty on the user/satellite range (as well as on the satellite time reference). In addition, relatively fast changes in the user/satellite distances over time (caused by satellite and/or receiver motion) and the receiver's oscillator uncertainty imply a frequency uncertainty (Doppler shift) range that needs to be searched as well.

Depending on the a priori information available at the receiver at the beginning of acquisition, we can identify different modes, namely

- cold start
- warm start
- hot start
- assisted.

In the *cold-start* configuration, no a priori information is available about the satellite constellation so that the entire PRN code and the full range of possible Doppler frequencies have to be explored: an exhaustive search over all satellites (i.e., ranging codes), all frequencies (± 10 kHz of maximum Doppler shift on an aircraft) and all time-shifts (± 1 ms of a full C/A code period) is needed. This is typical of the switch-on phase of the receiver.

In the *warm start*, satellite almanac data, coarse user position, and time, among other information, are available. The receiver can thus take advantage of this information, making a restricted search on those satellites that the receivers considers in view and over a small window of time and frequency. The prior information may come for instance from retrieval of a previously saved receiver configuration.

The *hot-start* configuration is defined when satellites have been recently tracked, so that the search is limited, as related to the hold tracking information. This happens when the satellite signal is momentarily lost for a short time (a car entering a tunnel).

Finally, *assisted* acquisition takes advantage of side information such as almanac and approximate time and location given by some external sources. A GNSS receiver embedded in a cellular phone may be assisted by the cellular network in finding its coarse position, and can retrieve the time and almanac via the communication channel of the phone from a centralized server on the Internet.

The notion of a two-dimensional search is represented in Fig. 2.20, where the axes are subdivided in appropriate "bins" that represent the time/frequency resolution of the search.

Usually, the search is carried out in a serial "trial and error" fashion, where each try is one time/frequency combination ("cell"). *Big* cells mean searching through a limited number of possibilities and this translates into a small searching time, but also entails poor sensitivity in the search results. On the contrary, *small* cells increase the sensitivity, but also increase the search time.

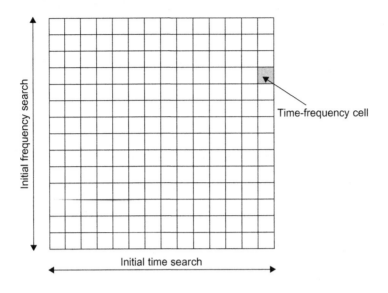

FIGURE 2.20

Time–frequency acquisition search.

A good compromise is a time resolution of half a code chip and a frequency resolution of a few hundred Hz.

The two-dimensional search can be performed following a number of different approaches, depending on the way the decision statistics for each cell is computed. The main methods are the *serial/parallel search* in the time domain and the *FFT-based correlation* in the frequency domain.

In all the acquisition algorithms, the decision statistics are compared with a preset *threshold* to understand whether the signal has/has not been acquired. The value of the threshold affects the probabilities of the *false alarm* and *missed detection* events that may occur as a result of noise and interference [8, 14, 27]. After successful threshold crossing (possibly followed by some *verification* algorithm), acquisition is accomplished and signal tracking starts.

Figure 2.21 presents an example of a pure *serial search* algorithm [14, 48, 51, 52]. The search starts from the a priori most probable Doppler shift value (zero Doppler if no a priori information is available), and with that value, all possible code delays are tested one after the other searching for a threshold crossing. If none is detected, the search is carried over to the next Doppler bin, with the sequence of bins symmetrically alternating on either side of the initial value. In each test, the decision statistic is just the output of the correlation of the received signal with the ranging code of the tested satellite, and after Doppler shift correction of the assumed amount.

The serial search can be substantially sped up by the adoption of a *parallel* acquisition HW, wherein more time bins (up to the full length of a code) are concurrently tested within the same observation period. The performance of the serial and the parallel schemes is the same in terms of false alarm/missed detection probabilities (depending only on the observation time), but the parallel searcher is faster than the pure serial searcher, at the cost of a more complex/costly hardware.

Figure 2.22 presents, on the contrary, the *FFT-based acquisition algorithm*, where a fast Fourier transform (FFT) of the baseband I/Q components of the received signal is first computed [11, 27, 33,

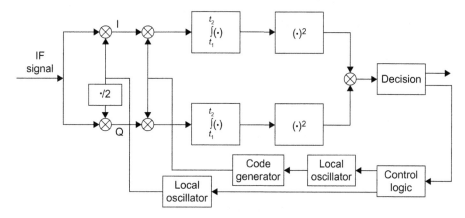

FIGURE 2.21

Serial search acquisition algorithm.

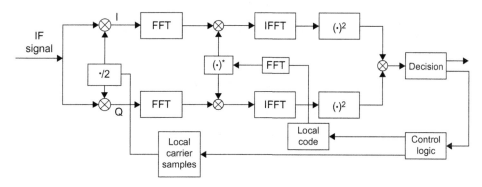

FIGURE 2.22

FFT-based acquisition algorithm.

46]. Then the FFT output is multiplied (frequency component by frequency component) by the complex conjugate corresponding value of the transform of the local code replica. The inverse transform of the product is the cross-correlation function of the two signals, whose peak gives the location in time of the delay of the received signal with respect to the local replica—a complete time search on the code period is automatically carried out for a given Doppler value.

This method is computationally more efficient and faster than the conventional serial search in time-domain technique, but requires the availability of a certain storage capacity in the receiver (input signal buffer before FFT computation and transform of local code replica). A further advantage of the FFT-based algorithm is that the search on the Doppler range does not require any recomputation of the FFT of the incoming signal: a certain frequency shift on the signal can be directly implemented on the FFT output by simply cyclically shifting the FFT samples.

As previously mentioned, once the acquisition process successfully ends (after possible subsequent verification stages), the estimated code offset and carrier Doppler shift are used to initialize the tracking loops to perform finer estimation of the two parameters.

2.2.3 **Signal Tracking**

The fine estimation of the signal delay τ in (2.8) is the key point of a GNSS receiver, as it allows us to estimate the distance of the receiver from the satellite transmitting the signal (*ranging*). This range measure can be accomplished and refined through a fine estimation of the Doppler frequency shift f_D and IF carrier phase ϕ_{IF}. Both the delay and the Doppler frequency and phase are typically extracted from the signal $y_{IF}[n]$ (2.8) using closed-loop synchronization architectures based on the principle of the *null seeker* [14, 27, 36]. Although the null-seeker approach is the most common one in the tracking stage of GNSS receivers, other strategies can be adopted, in particular in the context of software receivers [27].

The most traditional arrangement of the tracking signal processing functions of the receiver has three iterative (closed-loop) estimators for three different signal parameters, namely:

- frequency-locked loop (FLL) for carrier frequency tracking ($f_{IF} + f_D$, where f_{IF} is nominal while f_D is unknown—coarsely estimated in acquisition);
- carrier phase-locked loop (PLL) for carrier phase tracking (ϕ_{IF});
- delay-locked loop (DLL) for code delay tracking (τ).

As is seen in Fig. 2.23, the outputs of the tracking stage are the navigation data and the measurements obtained from the tracking loops (code delay and carrier phase) that are handed over to the navigation processing block. Notice that the architecture shown in Fig. 2.23 is a conceptual scheme, as its functional blocks can be implemented in the actual receiver in several different ways.

The operations of the FLL, PLL, and DLL are governed by the same rationale (the null-seeker architecture), while they differ just for the signal they handle and the corresponding parameter they aim at synchronizing. The so-called null-seeker synchronizer is a general iterative algorithmic architecture, shown in Fig. 2.24, where the input signal $y[n, \xi]$, containing the parameter of interest ξ, is first combined with a locally generated reference signal, $y_{ref}[n, \hat{\xi}[n-1]]$, which has typically the same basic

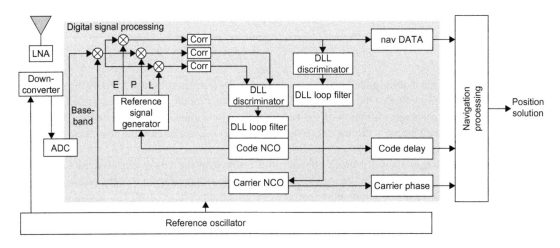

FIGURE 2.23

General architecture of the tracking stage in a GNSS receiver.

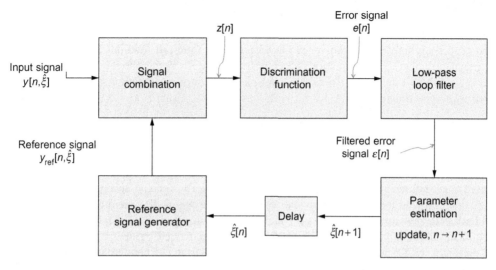

FIGURE 2.24

Basic structure of a null-seeker synchronizer.

structure as the input signal (apart from the presence of noise and other nuisances) and is characterized by the parameter estimated during the previous iteration $\hat{\xi}[n-1]$. Then the combined signal is transformed by a mathematical function (*discrimination function*) into a different metric (*error signal*) $e[n]$, whose value depends on the estimation error. The fundamental property of such a metric is that one of its possible zeros, taken as a function of the parameter to be estimated, corresponds to the searched value. Thus, the key operation of a null seeker is to find a zero of its discrimination function, and this null search is pursued in an iterative way. After the evaluation of the discrimination function, an appropriately defined low-pass filter smooths the error signal, in order to reduce the contribution of the noise $v[n]$ as much as possible, still preserving the reactivity of the loop to the dynamics of the parameter to be estimated. The loop is closed by extracting the estimated parameter $\hat{\xi}[n]$ from the filtered error signal $\epsilon[n]$, so as to generate, in the next iteration, another realization of the reference signal as a function of the new estimate $y_{\text{ref}}[n+1, \hat{\xi}[n]]$.

The device that generates the error signal (*discriminator*) can be nonlinear, but it is convenient to study the overall system in its linearity region, which is approximately correct if the error is small enough (i.e., when the acquisition stage has produced sufficiently accurate estimates of the delay and Doppler frequency values, such that they are within the so-called pull-in ranges of the carrier and code tracking loops). This hypothesis allows the use of standard analysis techniques, valid for linear systems, such as the analysis in the frequency domain and the definition of transfer functions. Furthermore, if the linearity hypothesis is assumed, the error signal becomes linearly proportional to the estimation error.

The null-seeking approach is generally preferred to other synchronization methods, as it is insensitive to the absolute values of the combined signal (the loop is governed by the error signal at the discriminator output, instead of the direct input signal).

The measurements coming from the tracking stage, and in particular the code delay, are evaluated with respect to the (local) receiver references (oscillator), so that they need to be further corrected to find the proper alignment to the universal GNSS reference time (all satellites are kept synchronous by the ground segment). This correction is accomplished by considering the receiver clock offset with respect to the GNSS time as the *fourth* unknown parameter in the navigation solution, in addition to the three user's coordinates.

2.2.3.1 *Carrier-Tracking Loops: the PLL and the FLL*

The principal carrier tracking loop is the PLL: the aim of this carrier-phase tracking loop is to keep the *phase error* between the locally generated carrier and that in the incoming signals as close to zero as possible, in spite of noise, changes in the signal parameters, interference, and so on.

Figure 2.25 outlines the general architecture of a PLL. The (spread-spectrum) incoming signal $y_{IF}[n]$ is first despread through a multiplication with the ranging code generated by the locally controlled DLL code generator (*code wipe-off*), and then it is converted to baseband via I/Q mixing with the two local reference pure sinusoidal signals at $f_{IF} + \hat{f}_D$ generated by the carrier numerically controlled oscillator (NCO) with phase $\hat{\phi}_{IF}$. The carrier NCO acts as the local reference generator as shown in Fig. 2.24. The baseband I/Q signals are integrated by digital *integrate-and-dump* filters (I&D) over a certain observation time. The output signals of the in-phase and quadrature digital I&D,[6] denoted by I_P and Q_P, correspond to the so-called *prompt* correlation values. They can be seen as the signal

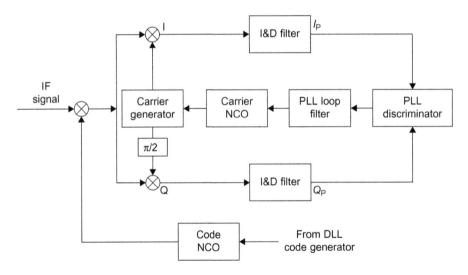

FIGURE 2.25

General architecture of a PLL.

---

[6]We drop here for convenience the dependence on time. The digital IF signal $y_{IF}[n]$ usually runs at a rate equal to 2–4 samples/chip time, while the output of the I&D filters is slower by a factor L, where L is the number of accumulated samples.

$z[n] = I_P + jQ_P$ at the input of the discrimination function in Fig. 2.24, whose aim is to provide a signal that is proportional to the phase error between the local oscillator and the received signal. Moreover, such values are used for data demodulation. The *equilibrium* of the loop is found when the error signal (the output of the discriminator) is on the average zero. The error signal is fed back to the NCO in such a way that any nonzero phase error is reduced (negative feedback).

Depending on the signal structure, different types of discriminators can be chosen. On a pilot signal, the simplest discriminator computes

$$-\arctan\left(\frac{Q_P}{I_P}\right). \tag{2.9}$$

If the signal also carries NAV data, the *dot-product phase discriminator* [37] can be used. Before feeding it back to the NCO, the PLL discriminator output is filtered by a low-pass filter (loop filter) to remove noise components.

The accuracy of the phase estimate provided by the PLL depends on the integration time of the I&D filter, on the discriminator type, on the characteristics of the loop filter, and on the sensitivity of the NCO. With long integration times and a narrowband loop filter bandwidth, the accuracy of the phase estimate is in general increased, but the loop will be very slow to react to possible changes in the received signal because of receiver or satellite motion and/or changing propagation conditions of the radio signal. Such loop parameters therefore have to be set as a trade-off between two conflicting aspects.

The FLL is another carrier-tracking loop capable of tracking the carrier frequency of the incoming signal ignoring its phase term ϕ_{IF}. The FLL is often used to initialize the PLL, by providing a frequency estimate very close to the correct one. Once the FLL has locked its estimated frequency, the PLL can refine the estimation working in a narrower bandwidth around the FLL estimate.

2.2.3.2 *Code Delay Tracking: the DLL*

The code-tracking loop is another null-seeker synchronizer having the structure shown in Fig. 2.24, which estimates the time delay τ experienced by the signal coming from the satellite to the receiver, and therefore estimates also the distance between the two terminals. The fine synchronization of the code is obtained by exploiting the fundamental properties of the pseudorandom noise (PRN) code correlation function. The cross-correlation between two signals (in this case, the incoming satellite code and its locally generated replica) is even as a function of the delay between the two correlated signals. It assumes its maximum when the two signals are perfectly aligned in time. However, recalling that the derivative of a function is null at its minima and maxima, the problem can be transformed again to become a null searching problem.

Figure 2.26 outlines the general DLL structure, which, similar to the PLL, is a specification of the general null-seeker scheme in Fig. 2.24. Here, the IF input signal $y_{IF}[n]$ is first multiplied by the in-phase and quadrature components of the local carrier signal generated by the carrier NCO of the PLL (*carrier wipe-off*), and then the two in-phase and quadrature resulting signals (i.e., the null-seeker input $y[n]$) are correlated not only with the aligned (*prompt*) local replica of the code but also with two additional replicas delayed $\pm\frac{dT_c}{2}$ apart from the prompt code. These two additional instances of the local code are indicated as *early* (E) and *late* (L) codes, respectively. The quantity d represents a fraction of a chip time called the *early–late correlation spacing*. The resulting three components of

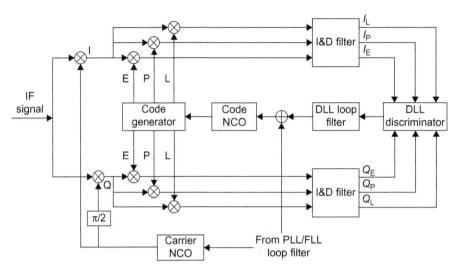

FIGURE 2.26

General architecture of a DLL.

each branch, that is, I_E, I_P, I_L, Q_E, Q_P, and Q_L, enter the DLL discriminator to create the appropriate loop error signal $e[n]$. The following are the fundamental types of discrimination function [14, 36]:

- Early–late coherent: $I_E - I_L$;
- Coherent dot product: $(I_E - I_L)I_P$;
- Transition tracking (aka quasi-coherent dot product):
  $(I_E - I_L)I_P + (Q_E - Q_L)Q_P$;
- Noncoherent: $(I_E{}^2 + Q_E{}^2) - (I_L{}^2 - Q_L{}^2)$.

The coherent and quasi-coherent discriminators do not need an estimate of the carrier phase, so they are more robust in a highly dynamic environment with respect to the coherent dot product, but they are less accurate than the latter in the presence of a high noise level. These discriminators are applicable to a BPSK-modulated data signal. On a pilot signal (i.e., a signal without the navigation data bits) the early–late coherent discriminator is also applicable, with the best performance. It is also the most intuitive architecture and easiest to analyze. It can be shown that the estimation variance of the coherent discriminator is equal to the best that can be obtained by the CRB.

The DLL performance is a function of many parameters: correlation spacing, front-end filter, discriminator, DLL loop filter bandwidth, and code NCO sensitivity. The GPS chip time T_c of the C/A code is roughly equal to 1 μs, which is equivalent to a distance of $c \cdot T_c = 300$ m, where c is the speed of light. Assuming that the loop can estimate the signal delay with an accuracy (standard deviation σ) of 1/100 of chip (a realistic value), the accuracy in the estimation of the user-to-satellite range is of the order of 3 m. The measurement has an intrinsic ambiguity equal to the repetition period 1 ms of the code, or 300 km, which can be easily solved with the aid of the NAV message. As already stated before, the range measurement is affected by errors owing to the unknown time offset between the ultra-stable reference oscillator aboard the satellite, and the local clock signal in the user receiver (this is why it is called the *pseudo*range). This offset is the same for all satellites (since the clocks of all satellites are synchronous), and so it can be considered as a nuisance parameter to be estimated jointly with the user terminal position—and compensated for—by the navigation algorithm.

Finally, once code and carrier lock is achieved so that the receiver can be considered to be in the *tracking* mode, the NAV data bits can be recovered from the prompt correlator output. Message decoding is accomplished after subframe matching to find the start of the frame. From the subframes, the ephemeris data can be extracted, and the pseudoranges between the satellites and the receiver can be determined, and the navigation module is run.

2.2.4 **Navigation Processing**

Navigation processing, also called *data processing*, takes advantage of the carrier and code synchronization obtained once the tracking condition is reached, to compute the user PVT. Figure 2.27 depicts the different processing levels within the receiver.

The value of the pseudorange needed to perform navigation processing is obtained using pieces of information gathered both from the signal processing level (labeled *counters* in Fig. 2.27) and from the decoding of the navigation message. Now the steps leading to the computation of the pseudorange of a satellite with reference to the GPS systems are outlined. Similar concepts also apply to other GNSSs. The readers that are not interested in such details may skip this section—we include this material, which is not so easy to find in the literature on signal processing.

The pseudorange calculation needs a *time-stamp* in the universal GPS time scale on the signals broadcast by the satellites. To understand how this is done, we recall that the GPS navigation message in Fig. 2.3 is organized in pages and subframes, and that each subframe starts with a sequence of two

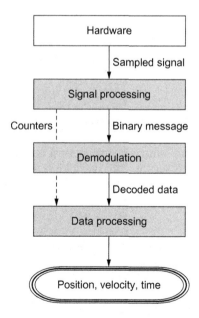

FIGURE 2.27

Logical steps of the receiver signal processing.

specific words. The first two words of each subframe are the telemetry (TLM) and the handover word (HOW). The TLM is made of an 8-bit preamble plus 16 reserved bits and 6 parity bits, and it represents the beginning of the subframe: it is used by the receiver to synchronize the navigation message. The handover word (HOW) is the second word of the subframe and consists of four components:

- Bits 1–17: truncated version of the time of week (TOW) field, a counter representing GPS time in 6 s units;
- Bits 18–19: system state (bit 18, alert bit: index of accuracy; bit 19: anti-spoofing mode);
- Bits 20–22: subframe identification number;
- Bits 23–30: parity bits.

We will see that the content of the handover word (HOW) is essential for pseudorange calculations.

2.2.4.1 *GPS Time Scale*

The GPS time scale is referenced to a UTC (as maintained by the U.S. Naval Observatory) zero time-point, defined as midnight of January 5, 1980/morning of January 6, 1980. The largest unit used in stating GPS time is *one week*, defined as 604,800 s. Let us now examine how the time instants are expressed in GPS time wrt to the zero time point. The *Z-count* is a counter representing the number of 1.5 s epochs (time intervals) elapsed since the zero-time reference, in a special format (see Fig. 2.28). Its 19 LSBs represent the so-called TOW field: the number of 1.5 s epochs within a week, ranging from 1 to $604,800/1.5 = 403,199$; the 10 MSBs of the Z-count are on the contrary the sequential number of the current GPS week. The two fields together uniquely identify the current 1.5 s period from the start of time.

At the end of each GPS week, the TOW is reset to zero, and the "carry" on the counter increases the number of the GPS week.

If we consider only the 17 MSBs of the TOW, we get a time counter in terms of 6-second epochs. Therefore, the HOW represents on the GPS time scale a counter of *subframes*. Each subframe has in fact a duration of

$$\underbrace{10}_{\text{words}} \cdot \underbrace{30}_{\text{bits}} \cdot \underbrace{20}_{\text{code periods in 1 bit}} \cdot \underbrace{1 \text{ ms}}_{\text{code duration}} = 6 \text{ s.} \tag{2.10}$$

Figure 2.29 represents the transmission of the navigation message by the satellites and how they are received by the users. All the GPS satellites are synchronized with the same reference time, so they all send out the navigation messages *at exactly the same time* t_{sub}. The HOW allows determination of the time instant at which the TX of each message subframe starts.

GPS week	Time of week
10 bits	19 bits

FIGURE 2.28

Representation of the Z-count.

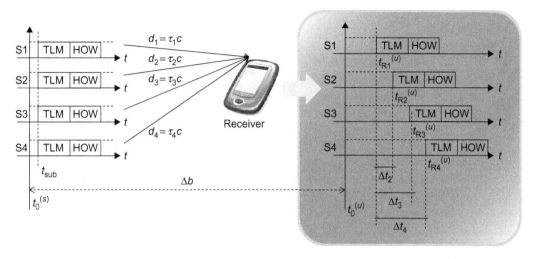

FIGURE 2.29

Data messages: transmission and reception.

Table 2.5 Example of Pseudorange Measurement Based on Absolute Reception Times

CH	Transmit Time	Received Time	Difference	Pseudorange
1	300 s	$t_{R1} = 300.063$ s	63 ms	18,900 km
2	300 s	$t_{R2} = 300.068$ s	68 ms	20,400 km
3	300 s	$t_{R3} = 300.076$ s	76 ms	22,800 km
4	300 s	$t_{R4} = 300.090$ s	95 ms	28,500 km

The satellite messages, generated at the same time, arrive at the RX at different times, because of the different RX/satellite distances. On the basis of the different reception epochs t_{Ri}^u, $i = 1, \ldots, 4$ on the user time scale, the receiver starts estimating the value of the ith pseudorange.

2.2.4.2 *Absolute or Relative Pseudorange Measurements*

Now, the receiver records the reception times, on its own time scale starting at $t_0^{(u)}$, of the satellite TLMs. Every measurement $t_{Ri}^{(u)}$ is affected by the *same* unknown time offset $\Delta b = t_0^{(u)} - t_0^{(s)}$ of the receiver clock versus the GPS time. Table 2.5 shows an example of pseudoranges evaluated in this way.

The procedure can also be based on the measurement of (relative) delta-pseudoranges, taking one of the pseudoranges ($i = 1$) as the reference, and then calculating a time difference, as in Table 2.6. In practice, the measurement of the Δt_i can be performed by means of counters that are reset when the first TLM is received.

None of the pseudoranges appearing in the table corresponds to any actual range, just because of the unknown time offset. The value of the first pseudorange is arbitrarily set by the receiver as the

Table 2.6 Example of Pseudorange Measurement Based on Time Differences

CH	Received Time	Transit Time	Pseudorange
1	0	70 ms	21,000 km
2	$0 + \Delta t_2 = 5\,\text{ms}$	75 ms	22,500 km
3	$0 + \Delta t_3 = 7\,\text{ms}$	77 ms	23,100 km
4	$0 + \Delta t_4 = 12\,\text{ms}$	82 ms	24,600 km

mean satellite-to-earth propagation time, or it is actually measured on the user time scale. Choosing a mean propagation time or setting a chosen received time (setting the value of $t_0^{(u)}$) is equivalent to setting a particular value of Δb.

The computation of position and time is now obtained by solving the following equations in the four unknowns x_u, y_u, z_u (the user position), and the time difference Δb:

$$
\begin{aligned}
\sqrt{(x_1 - x_u)^2 + (y_1 - y_u)^2 + (z_1 - z_u)^2} - c \cdot \Delta b &= \rho_1 \\
\sqrt{(x_2 - x_u)^2 + (y_2 - y_u)^2 + (z_2 - z_u)^2} - c \cdot \Delta b &= \rho_1 + c \cdot \Delta t_2 \\
\sqrt{(x_3 - x_u)^2 + (y_3 - y_u)^2 + (z_3 - z_u)^2} - c \cdot \Delta b &= \rho_1 + c \cdot \Delta t_3 \\
\sqrt{(x_4 - x_u)^2 + (y_4 - y_u)^2 + (z_4 - z_u)^2} - c \cdot \Delta b &= \rho_1 + c \cdot \Delta t_4,
\end{aligned}
\tag{2.11}
$$

where ρ_1 is the first (arbitrary) pseudorange and x_i, y_i, and z_i are the satellite coordinates ($i = 1, \ldots, 4$) derived from the ephemeris.

The computation of position and time can also be repeated at time instants different from the start of a subframe (a typical GNSS receiver performs the computation every second). As depicted in Fig. 2.30, the receiver starts a local counter each time a subframe is received from a certain satellite, and measures the elapsed times δt_i from the ith subframe beginning at the time instant where the position is computed, $t_m^{(u)}$, from which it obtains the differences $\Delta t_i = \delta t_{\max} - \delta t_i$, $i = 2, 3, 4$.

2.2.5 Pseudorange Error Sources

The measurements of the pseudoranges described in the previous subsection are affected by a number of error sources that perturb the stability and accuracy of the DLL. A vast amount of technical literature on this subject is available (e.g., Ref. [14] and references therein). Only the error sources for GPS are discussed now, since the studies related to the other GNSSs are less consolidated or less open to the general public.

The accuracy in the user position estimation depends on a complicated interaction of various factors. Broadly speaking, the GPS accuracy depends on the "quality" of the pseudorange and delta pseudorange measurements as well as of the satellite ephemeris. The uncertainty on the pseudorange values is called user equivalent range error (UERE). The impact of the UERE on the final estimated position is also dependent on some "geometric" factors (the geometric dilution of precision (GDOP)) describing how the satellites are displaced in the sky over the user. The different positions of the satellite in fact affect the accuracy of the numerical procedures that have to be adopted to solve the

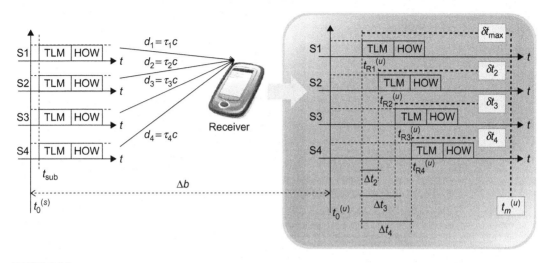

FIGURE 2.30

Data messages: transmission and reception, with respect to a reception time.

positioning equations (2.11). In some cases (i.e., when the four visible satellites are seen from the receiver with a low angular separation) the equations are numerically ill conditioned; that is, small measurement errors on the pseudoranges yield large errors on the final position estimates.

The error sources that affect the pseudoranges can be classified as follows:

1. System control: errors on the ephemeris and on the clocks, measurement errors;
2. Ionospheric delay: the speed of the radio wave depends on the (variable and unknown) density of free electrons along the propagation path;
3. Tropospheric delay and atmospheric attenuation: the propagation delay and attenuation depend on the (variable and unknown) pressure, temperature, and humidity of the air;
4. Multipath: multiple "copies" of the transmitted signal come to the receiver following different propagation paths caused by the presence of reflective surfaces (buildings, hills) and scattering points (small objects like cars, foliage) close to the receiver;
5. Receiver noise and clock instability;
6. Uncompensated relativistic effects;
7. Selective availability (SA): intentional disturbance to the transmitted ranging codes to limit the accuracy of the GPS for civil users (eliminated on May 1, 2000).

Errors caused by the Doppler effect can be neglected in the first instance, since the receiver has anyway to estimate and compensate for the relevant carrier frequency shift.

Assuming that all the previous sources give a random error contribution that is statistically independent of all the others, the total error (standard deviation) on each pseudorange turns out to be

$$\sigma_{\text{UERE}} = \sqrt{\sum_j \sigma_j^2} \quad \text{(m)}, \tag{2.12}$$

Table 2.7 GPS PPS Pseudorange Error Budget

Segment	Error Source	1σ Error (m)
Space	Satellite clock stability	3.0
	Satellite perturbation	1.0
	Other (thermal radiation, etc.)	0.5
Control	Ephemeris prediction error	4.2
	Other (thrusters, etc.)	0.9
User	Ionospheric delay	2.3
	Tropospheric delay	2.0
	Receiver noise and resolution	1.5
	Multipath	1.2
	Other (interchannel bias, etc.)	0.5
UERE	Total (rms)	6.6

Table 2.8 GPS SPS Pseudorange Error Budget

Segment	Error Source	1σ Error (m) with SA	1σ Error (m) without SA
Space	Satellite clock stability	3.0	3.0
	Satellite perturbation	1.0	1.0
	Selective availability (no longer active)	32.3	–
	Other (thermal radiation, etc.)	0.5	0.5
Control	Ephemeris prediction error	4.2	4.2
	Other (thrusters, etc.)	0.9	0.9
User	Ionospheric delay	5.0	5.0
	Tropospheric delay	1.5	1.5
	Receiver noise and resolution	1.5	1.5
	Multipath	2.5	2.5
	Other (interchannel bias, etc.)	0.5	0.5
UERE	Total (rms)	33.3	8.0

where σ_j^2 is the variance of each source. Just to have an idea of the order of magnitude of the UERE in a practical receiver, we present in Tables 2.7 and 2.8 the so-called error budget of the pseudorange measurements for the GPS, PPS, and the SPS, respectively [8, 14].

2.3 AUGMENTATION SYSTEMS AND ASSISTED GNSS

The positioning that can be attained by GNSSs is often accurate enough for most civil services (e.g., car, ship and aircraft navigation in open space), but turns out not to be sufficient for professional, security, or critical services, e.g., ship guidance during harbor entry or airplane landing, as well as indoor navigation where the GNSS precision is poor because of the wall attenuation and multipath propagation. In such situations, where submeter accuracy is required, improvement of GNSS positioning can be achieved basically by the so-called augmentation systems [36]:

1. *Mitigation of measurement errors*: This is enabled by the availability of additional information about the corrections to be applied when solving the positioning equations. Such *differential* information is provided by additional means, for example a reference station located near the terminal (as in differential GPS (DGPS)). Differential corrections are provided by most augmentation systems: satellite-based augmentation systems (SBASs), real-time kinematic (RTK), and assisted GNSS (AGNSS).
2. *Improvement of the satellite geometry*: This is achieved through the use of additional satellites or additional terrestrial transmitters to be considered as pseudo-satellites (the so-called pseudolites). This method is used by SBAS systems and in very few instances by pseudolites.
3. *Use of coarse positioning or timing information provided by independent systems*: This is the case of the AGNSS, wherein the cellular communication network the terminal is connected to provides an assistance message containing a coarse information of the mobile handset position and timing, computed on the basis of measurements obtained by the network infrastructure itself.

Augmentation systems can be satellite-based (SBAS, pseudolites) or terrestrial (e.g., DGPS, AGNSS, network RTK). The former provide information as additional ranging signal(s), which also carries augmentation data (for instance, the ionospheric delay to be used in the positioning equations) in the navigation message. The latter exploit the positioning capability offered by referenced terrestrial stations, as outlined earlier. Augmentation systems are inherently *local* (regional), with the difference that SBASs broadcast their messages over very large areas (thousand of kilometers) via geostationary satellites, while terrestrial services cover local areas (tens of km). For this reason, satellite-based services reach a far wider user population, but ground-based ones offer more accurate differential corrections and faster response [28]. Now, the different techniques and the requirements in terms of system and receiver modifications with respect to what needed in a standard GNSS are sketched (see Table 2.9).

2.3.1 Differential GPS

Differential GPS (DGPS) is an enhancement to GPS that uses a network of fixed, ground-based reference stations to broadcast the difference between the position of a receiver located at each station, as derived instant by instant using the GNSS, and the known fixed positions of the stations themselves. The stations broadcast the difference between the measured satellite pseudoranges and the actual (internally computed) ranges, so that the UT receivers can correct their pseudoranges by the same amount. The underlying hypothesis of DGPS is that any two receivers that are relatively close to each other experience quite the same atmospheric errors. The reference GPS receiver in the ground station, called the *reference station* or the *beacon*, must be set up in a fixed, known and perfectly geo-referenced location. The difference between the GNSS pseudoranges vector and the computed ranges vector represents the differential error which is transmitted to the user terminal. The user terminal, commonly indicated as the *roving receiver*, applies these corrections to its own GPS data.

The differential corrections can be applied either in real time "on the field," or through postprocessing after receiver data storage using special processing software and proper DGPS databases. To apply

Table 2.9 Type of Additional Information Exploited by Different GNSS Augmentation Systems and Achievable Accuracy

Augmentation System	Provided Information	Accuracy	Category	Modifications to Std. GNSS RX
DGPS	Differential corrections on satellite pseudoranges	Meter to decimeter	Terrestrial	Hardware (additional receiver) and firmware
SBAS	Ranging signals, integrity data, ionospheric corrections	Meter	Satellite-based	Firmware
Pseudolites	Ranging signals	Submeter	Terrestrial	Firmware
Network RTK	Precise differential corrections, ambiguity resolution	Centimeter	Terrestrial	Hardware (additional receiver) and firmware
AGNSS	List of visible satellites, doppler frequencies, code phases, navigation messages	Meter	Terrestrial	Hardware (additional receiver) and firmware

The required modifications to a standard GNSS receiver are reported in the last column.

real-time corrections, each reference station broadcasts the calculated corrections for each satellite on a special radio band. The DGPS receiver has to be equipped with the receiver for such differential data. Although real-time and postprocessing methods are based on the same underlying principles, each accesses different data sources and achieves different levels of accuracy.

Several sources of error can be effectively reduced by DGPS: these include the transmission delays in the ionosphere, the satellite position ephemeris data, and clock drift on the satellites. DGPS can provide meter- to decimeter-level accuracy. Its performance inherently decreases with the increasing distance between the user terminal and the reference station, and with the increasing delay between the time the corrections are computed at the reference station and the time they are applied in the terminal.

The radio links between the beacons and the UTs use long-wave radio frequencies between 285 and 325 kHz, commercial AM/FM radio bands, and/or VHF radio links, depending on the installation. The protocol for relaying DGPS correction messages from a reference station to a UT is known as RTCM-104.[7] Each correction message includes data about the station position and health, satellite constellation health, and the corrections to be applied.

For many applications, such as those related to high-precision earth observation, geographical information system (GIS), and mapping applications, extremely precise georeferencing of the roving terminal, responsible for earth measurements, is of utmost importance, but real-time navigation operations are not necessary. The usual procedure is therefore to use a simple on-board GPS receiver that stores GPS measurements and a reference receiver that logs observations at a known location. The corrections to the stored measurements can be computed off-line after the end of the observations

---

[7]The radio technical commission for maritime services (RTCM) is a nonprofit scientific and educational organization that serves all aspects of maritime radio communications, radio navigation, and related technologies.

using a corrections database and a processing software, which computes corrected positions for the rover's file.

2.3.2 Satellite-Based Augmentation Systems

Satellite-based augmentation systems (SBASs) are similar in principle to the DGPS. Instead of a ground station, the correction data are sent via GEO satellites equipped with transponders (but not by signal generators) transmitting in the same band and with the same modulation format as the "core" constellation. Therefore, SBAS signals are decoded by one of the regular channels already present in the GPS receiver so that a different receiver as in DGPS (e.g., FM or VHF) is not required [21, 22, 28, 45].

Because of the direct derivation of these satellite-based systems from DGPS, SBAS is often called wide area differential GPS (WADGPS), particularly when it refers to the American system. The intended coverage of SBASs is worldwide, through the cooperation of different regional systems aiming at augmenting not only GPS, but GNSS systems in general.

SBASs use GEO satellites to broadcast ranging, integrity, and correction information to GNSS users, with the aim of increasing accuracy, reliability, and availability of GNSS positioning. Unlike the standard DGPS, a SBAS satellite does not directly send corrections of the pseudorange data, because the coverage of a GEO satellite is too wide to make this effective. Instead, SBAS estimates the effect of the individual sources of error and sends specific corrections for each one of them (and for each GNSS satellite): clock corrections, ephemeris corrections, and ionospheric corrections.[8] Unlike DGPS, the space segment provides a further augmentation mechanism through the transmission of *additional ranging signals* (optionally used by the UT).

As the RF bands, the modulations, and the family of codes of the SBAS signals are the same as the core constellations, the required modifications of the UT architecture to be compliant with SBAS services are limited. Also, the kind of processing that is required to take into account the SBAS information, although more sophisticated than the basic GNSS one, has an affordable complexity.

A few commercial SBAS systems are currently in operation to provide user services: OmniSTAR (Fugro, The Netherlands), Starfire (NavCom Technology, United States), and Veripos (Subsea 7, United Kingdom). In addition, four noncommercial SBASs are currently under development:

- Wide area augmentation system (WAAS) (United States), which is intended for initial safety-of-life use in 2003;
- European geostationary navigation overlay system (EGNOS) (Europe), whose initial operations started in 2005;
- Multifunctional satellite augmentation system (MSAS) (Japan), for which two multifunctional transport satellite (MTSAT) satellites were launched in 2005–2006;
- GPS-aided GEO augmented navigation (GAGAN) (India), to be completed in 2011.

All SBASs are regional systems, and therefore *interoperability* is a fundamental issue to guarantee full compatibility and reach seamless coverage on much of the Northern Hemisphere. SBAS

---

[8]Tropospheric corrections cannot be broadcast because of the localized nature of this error.

cooperation is currently coordinated through the Interoperability Working Groups EGNOS/MSAS and EGNOS/WAAS and interoperability tests are regularly organized, showing that interoperability is able to improve user performance on SBAS boundaries.

2.3.3 Pseudolites for GNSS

Pseudolites (pseudo-satellites) are ground-based transmitters that generate and transmit GNSS-compatible signals [35]. They can be placed in locations that have poor satellite visibility to improve the availability and accuracy of GNSS-based positioning. In an indoor space where the GNSS signals are heavily attenuated, a set of pseudolites can even be used to replace the whole GNSS constellation. This technology suffers from a number of issues, and is applied in very special applications, so that we will not further consider it.

2.3.4 Network RTK

Real-time kinematic (RTK) is a method to achieve centimeter-level accuracy positioning *in real time*. The basic signal processing function that allows attaining this accuracy is *carrier navigation*, that is, tracking the delay of the GPS carrier (or, equivalently tracking its *phase*) instead of the delay of the ranging code. Since the oscillation period of the GPS carrier $T_{L1} = 1/f_{L1} \simeq 0.635$ ns is much smaller than the ranging code chip time $\simeq 1\,\mu s$, the accuracy in delay estimation and then in ranging is much higher.

The first RTK concepts were developed in the nineties and, similar to DGPS, involved a *reference receiver*, located in a fixed and perfectly georeferenced location, transmitting its raw GPS measurements or observation corrections to a *rover receiver* via some sort of data communication link (e.g., VHF or UHF radio, cellular communication networks) [32]. The rover equipment in turn must be constituted by a GPS receiver, a radio link with the reference station and a software/firmware capable of interpreting the RTK data from the reference station. The data processing at the rover site includes *ambiguity resolution* of the differenced carrier phase data and coordinate estimation of the rover position. The main problem with carrier navigation is in fact that there is no "stamp" on the absolute phase of the L1 carrier, and hence, the measurement of its delay is intrinsically ambiguous and sophisticated signal processing techniques have to be adopted to overcome this issue.

Rapid and reliable resolution of the carrier phase ambiguities requires the deployment of a number of reference stations spaced 20–30 km apart to cover an entire territory. To give a few examples, today the entire Greek territory (mainland and most islands) is covered by an RTK network, maintained by a governmental entity, for geodetic purposes (HEPOS, see Ref. [23]); the 99-station network has been designed and deployed in the years 2005–2006 and has recently become operative.

RTK is a technology essentially implemented with proprietary solutions. The different techniques to transfer the RTK information to the rover receiver, the definition itself of the RTK information to be transferred, and the transfer message format are not standardized, but can vary slightly depending on the system manufacturer [19].

2.3.5 Assisted GNSS

Assisted GNSS is a network-assisted method that integrates GNSS with information provided by a cellular communications network to reduce the so-called time-to-first-fix (TTFF), that is, the time necessary for the GNSS receiver to produce the first estimate of its current position, and, at the same time, increasing the sensitivity of the GNSS receiver [2–4, 24, 34].

During start-up, a GNSS receiver needs to demodulate the whole navigation message to determine the satellite ephemeris, prior to solving the positioning equations (2.11). This can take a few seconds in ideal acquisition and tracking conditions, which easily become some minutes in normal operating conditions (few satellites in visibility, low SNR, etc.). The idea behind AGNSS is to use the cellular network to communicate a copy of the navigation message of each visible satellite to the GNSS receiver, which therefore does not need to wait for the complete signal tracking and data demodulation process before producing its position estimate. This reduced TTFF is clearly essential, for example, in emergency calls.

Network assistance is also fundamental in the case of critical receiving conditions, e.g., in an urban environment with tall buildings and/or indoors, when deep fades in the satellite propagation channel prevent the continuous tracking of the signal, and so the demodulation of the complete navigation message.

In addition, GNSS receiver sensitivity can be enhanced by reducing the acquisition search space in both Doppler shift and code phase domains, thanks to an initial coarse information about the most probable Doppler shift and code phase computed and transmitted by the cellular network to the UT (and based on a coarse estimate of the UT computed by the network itself).

The architecture of an AGNSS service is shown in Fig. 2.31. The main components are

1. the GNSS satellite constellation;
2. the *wide area reference network (WARN),* a network of reference receivers, co-located with the cellular base stations, which collects the navigation messages of all the satellites in view and simultaneously computes differential corrections;

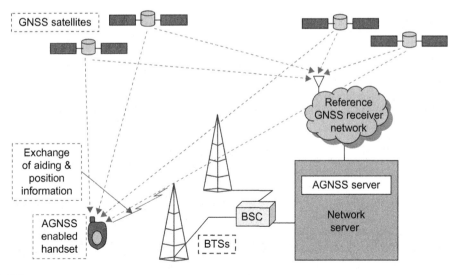

FIGURE 2.31

AGNSS architecture.

3. the *location server*, or *AGNSS server*, which collects and stores data from the WARN, produces the assistance messages for the UTs, and then sends these messages to the UT using the conventional cellular communication link; in the case of UT request, it can also compute and transmit the user's position solution;

4. the *MSs*, that is, the user terminal to be located, equipped with a GNSS receiver and an AGNSS-enabled cellular network receiver.

Assistance data provided to the mobile station (MS) basically include the list of the satellites in view at a particular instant, together with their estimated Doppler shifts: this enables the MS receiver to reduce the number of satellites to test for acquisition, as well as their acquisition search space (Doppler bins) to (considerably) reduce the TTFF. The network can provide the complete list of satellites in view for each MS thanks to the relatively small radius (a few kilometers) of the cell, which ensures that the satellites in view at the base station (the reference location of the WARN) are the same as those visible from the MS.

The MS position can be materially computed either by the MS or by the location server, provided that the MS sends its own computed pseudoranges to the server. This corresponds to two different location approaches:

MS-based mode This is adopted when the location server sends the assistance message to the MS, and then the MS solves the positioning equations on the basis of its own observed pseudoranges and the assistance information. The full GNSS functionality resides in the MS and is just integrated with the procedures and interfaces that enable exploitation of the assistance data. The MS can also work as a conventional, nonassisted GNSS receiver.

MS-assisted mode This approach is adopted when the MS position is determined by the location server and then it is fed back to the MS. In this case, the positioning capability resides in the network server, reducing the computational burden and the total required memory of the MS. The MS device hosts the GNSS antenna, an RF section, the subsequent conversion to baseband or quasi-baseband, and the DSP necessary to perform tracking and pseudorange estimation, but no navigation processing. This heavily limits the autonomy of the MS in terms of nonassisted positioning.

MS-assisted positioning can also be improved by the use of other augmentation approaches implemented at the location server. In particular, SBAS differential corrections can also be obtained via specialized Internet databases such as SISNeT for European EGNOS. This kind of information can also be passed as additional assistance to the MS in the MS-based mode, thus enabling differential corrections to be applied at the MS side. Also typical cellular measurements (time of arrival, time difference of arrival, received signal strength, etc., which are discussed in detail in Chapter 3) may be available at the location server to further improve the accuracy or availability of positioning services.

Three basic types of assistance can be identified (in increasing order of complexity):

- *Acquisition:* Reduces the acquisition search space, and so the TTFF, by providing the MS with a coarse information on the predicted Doppler shift and code phase.
- *Sensitivity assistance:* Improves the sensitivity of the MS GNSS receiver, that is, its ability to acquire and track very weak satellite signals, by providing the list of visible satellites, the coarse

Table 2.10 AGNSS, Possible Assistance Parameters
Assistance Data
Visible satellites
Position of the visible satellites (azimuth, elevation)
Predicted codes phase (and confidence interval)
Code phase confidence intervals ("search window")
Predicted Doppler shifts and Doppler rates
Satellite ephemeris
Satellite almanac
Satellite clock corrections
Navigation time (GNSS time)
Navigation data bits
MS location (either coarse or precise estimation)

acquisition information and the expected navigation message, so as to enable a fast data wipe-off and therefore longer integration times.[9]

- *Network assistance:* Improves the accuracy and integrity of the MS position providing accurate navigation data (e.g., ephemeris, clock corrections) computed via the network resources.

2.3.5.1 *The Assistance Message*

A list of possible assistance data that can be included in the assistance message is shown in Table 2.10. Depending on the operational mode and on the type of assistance, only subsets of this list are provided. For example, in the MS-based mode, the assistance message contains a coarse estimation of the MS location (*reference location*), the list of visible satellites with their ephemeris and clock corrections and the GNSS time; in this case the MS can derive the corresponding Doppler shifts and rates and codes phase for the acquisition. On the contrary, an MS-assisted assistance message contains the list of visible satellites, Doppler information, and expected codes phase.

In the MS-assisted mode, the location server exploits the additional information provided by the cellular network in a number of different ways: knowledge of the reference location altitude and the measurement of the approximate distance between MS and BS can be used to reduce the required minimum number of satellites from 4 to 3. If the reference time is also available, the number of required satellites to solve the positioning problem can be further reduced to 2. The assistance about altitude and reference time do not usually have elevated accuracy, and therefore do not actually increase the

---

[9]The knowledge of the navigation bit signs and boundaries allows one to remove the message from the received signal and to extend the coherent integration time well beyond the one-bit duration, with positive consequences on the acquisition threshold. However, it reduces the size of the Doppler bins in the search space, thus increasing their required number. We recall that extended coherent integration times could be the only means to enable satellite-based positioning in indoor conditions.

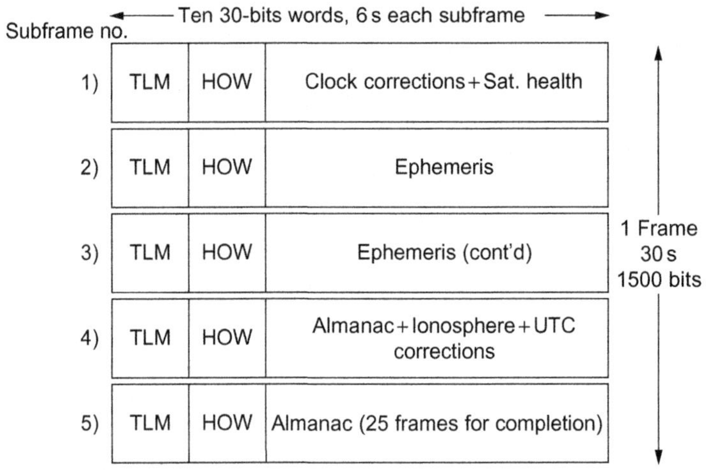

FIGURE 2.32

The GPS navigation message.

final accuracy of the positioning solution; nonetheless, they increase service availability in critical conditions (number of visible satellites).

The assistance message is also "compressed" in the sense that its information content is reduced to the bare minimum (the sections that cannot be locally reconstructed). Consider, for example, the GPS navigation message, whose structure is shown for convenience in Fig. 2.32:

- words 3 to 10 belonging to subframes 1, 2, and 3 are obtainable from the parameters of the navigation model;
- words 3 to 10 belonging to subframes 4 and 5 can be obtained from the almanac;
- parity check control bits are not transmitted, as they can be determined a posteriori;
- subframe preambles are known;
- the 17-bit HOW of every subframe can be determined from the indication of local time.

The assistance message can therefore be limited to contain only the missing bits.

The main telecommunication standardization organizations agreed to include specific protocols and message formats to support AGNSS into their specifications and guarantee *interoperability* across different (cellular) networks. Table 2.11 presents a list of standard specifications where location technologies are addressed. Current regulations in the United States and Europe in terms of emergency management for citizens' safety (E-911, E-112) require wired and wireless telco carriers, including VoIP-based service carriers, to equip their infrastructure with the capability to locate emergency calls, by reporting the telephone number of the originator of the emergency call, the location of the cell site or base station transmitting the call, and the estimated latitude and longitude of the caller. This information is required to meet accuracy standards, established as 100 m for 67% of the calls and 300 m for 95% of the calls for network-based location approaches, and more stringent accuracy requirements for handset-based solutions: namely, 50 m for 67% of calls and 150 m for 95% of calls. For this reason,

Table 2.11 Main Cellular Telecommunication Standard Specifications Addressing Position Location Technologies. *Trilateration and TDOA methods are extensively discussed in Chapter 3

Radio Interface	Specification No.	Location Technology
AMPS	TIA/EIA/IS-817	AGPS
CDMA-2000	IS-801, IS-801-A	AGPS, trilateration*
GSM	3GPP 44.031	AGPS, TDOA*
IS-95	IS-801, IS-801-A	AGPS, trilateration
UMTS (WCDMA)	3GPP 25.331	AGPS, TDOA

AGNSS is not the only caller location approach implemented in cellular networks; rather, positioning is also generally supported by other purely terrestrial, network-based, positioning methods (see Chapter 3 for a detailed discussion).

2.3.5.2 *Assisted GNSS Standards in LTE Networks*

Cellular industry location standards first appeared in the late 1990s, with the Third Generation Partnership (3GPP) Radio Resource Location Services Protocol (RRLP) Technical Specification 44.031 positioning protocol for GSM networks. Today, RRLP is the standardized protocol to carry GNSS assistance data to GNSS-enabled mobile devices [50]. A major update took place in 2007, when RRLP Release 7 added support for assisted Galileo and Release 8 for the rest of the GNSS, including the various SBASs. Both releases provided native assistance data types such as global Klobuchar and NeQuick models for the ionosphere [40, 47]. The same approach was also mapped to 3GPP TS 25.331 Radio Resource Control (RRC) protocol, which defines the positioning procedures and assistance data delivery for UMTS terrestrial radio access (UTRA).

3GPP boosted location services in long-term evolution (LTE) Release 9, frozen in December 2009. According to the LTE location architecture, the evolved serving mobile location center (E-SMLC) is the server component in charge of positioning activities. The mobility management entity (MME) gives the positioning request to the E-SMLC, which then controls the user equipment to be positioned and, possibly, LTE base stations, to perform positioning. The actual positioning and assistance protocol between E-SMLC and the user equipment is called LTE Positioning Protocol (LPP).

RRLP, RRC, and LPP are natively control-plane positioning protocols, that is, they use cellular signaling channels as the transport mechanism for the assistance data and position information. Since signalling channels are not designed to transfer large amounts of information, in 2003 the Open Mobile Alliance (OMA) started to work with Secure User Plane Location (SUPL) 1.0 protocol, which brings the same location capabilities to user plane (and thus the traffic channels) over IP networks as RRLP, RRC, and LPP bring to control plane. SUPL 1.0 is already commercially deployed, and SUPL 2.0 is now being deployed globally. These protocols typically address richer GNSS features for location-based services.

LTE Release 9 introduced extension hooks in LPP messages so that the bodies external to 3GPP could extend the LPP feature set. OMA LPP Extensions (LPPe), supported in SUPL 3.0, build on top

Table 2.12 Assistance Data Content of LPP and LPP Extension

	Common Elements	Generic Elements per GNSS
3GPP LPP	Reference time Reference location Ionosphere model Earth-orientation parameters	Inter-GNSS time model DGNSS Navigation models Real-time integrity Data bit assistance Acquisition assistance Almanac UTC model Auxiliary Information
OMA LPP Extensions	Local ionosphere Local troposphere Altitude assistance Wide area ionosphere common Cross-polar-cap potential common	SV mechanics Differential code biases Navigation models Navigation models degradation Wide area ionosphere Cross-polar-cap potential

of the 3GPP LPP, re-using its procedures and data types as far as possible. This ensures that in the user-plane domain, which dominates in consumer location-based services (LBSs), vendors can utilize exactly the same protocol as in the control plane. This reduces implementation, testing, and deployments costs, and probably will make the LPP/LPPe the de facto standardized positioning protocol in the mobile domain [50]. Table 2.12 summarizes the assistance data content of 3GPP LPP and OMA LPP extensions for AGNSS.

References

[1] United States of America, European Commission, Member States of the European Union, Agreement on the promotion, provision and use of GALILEO and GPS satellite-based navigation systems and related applications, Dromoland Castle, Co. Clare, Ireland, 26 June 2004.

[2] 3GPP Specifications, Requirements for support of assisted global positioning system, 3GPP TS 25.171, v.7.1.0, 2006.

[3] 3GPP Specifications, Radio resource control (RRC); protocol specification (release 7), 3GPP TS 25.331, v.7.8.0, 2008.

[4] 3GPP Specifications, Radio resource LCS protocol (release 7), 3GPP TS 44.031, v.7.8.0, 2008.

[5] NAVSTAR GPS Space Segment/User Segment L5 Interfaces, ARINC Research Corporation, El Segundo, CA, 2002.

[6] NAVSTAR GPS Space Segment/Navigation User Interface, ARINC Research Corporation, El Segundo, CA, 2003.

[7] NAVSTAR GPS Space Segment/User Segment L1C Interfaces, ARINC Research Corporation, El Segundo, CA, 2006.

[8] B.W. Parkinson, J.J. Spilker, Global Positioning System: Theory and Applications, American Institute of Aeronautics, Washington, DC, 1996.

[9] C.A. Williams, Radiodetermination satellite service: Applications in railroad management, IEEE Vehicular Technol. Conf., 36 (May) (1986) pp. 395–397.

[10] Chen Jianyu, Overview of Compass/BeiDou navigation satellite system, in: 2nd Meeting of the International Committee on Global Navigation Satellite Systems (ICG), Bangalore, India, 2007. Available at: http://www.oosa.unvienna.org/oosa/SAP/gnss/icg/icg02/presentations.html (last accessed: June 2011).

[11] A. Coenen, D. Van Nee, Novel fast GPS/GLONASS code-acquisition technique using low update rate FFT, Electron. Lett. 28 (9) (1992) 863–865.

[12] Introduction to the COSPAS-SARSAT System, document c/s G.003, Cospas-Sarsat Secretariat, Montreal, Canada, Issue 6, 1999.

[13] R.C. Dixon, Spread Spectrum Systems, J. Wiley and Sons, New York, 1976.

[14] E.D. Kaplan, Understanding GPS: Principles and Applications, Artech House, Boston, 1996.

[15] The ESA website, http://www.esa.int/esaNA/galileo.html, 2008.

[16] Galileo Open Service, Signal in Space Interface Control Document, European Space Agency / European GNSS Supervisory Authority, Brussels, Belgium, 2008.

[17] G. Gibbons, Russia approves CDMA signals for GLONASS, discussing common signal design, in: Inside GNSS News, Gibbons Media and Research LLC, Eugene, OR, April 28, 2008. Available at: www.insidegnss.com.

[18] G.M. Polischuk, S.G. Revnivykh, Status and development of GLONASS, in: Acta Astronautica, Proceedings of the Plenary Events, vol. 54 (1–120), World Space Congress 2002, 2004, pp. 949–955.

[19] H.-J. Euler, Reference station network information distribution, IAG Working Group 4.5.1: Network RTK, 2005–2008. Available at: http://www.wasoft.de/e/iagwg451/euler/euler.html (last accessed: June 2011).

[20] G.W. Hein, J. Ávila-Rodríguez, S. Wallner, A.R. Pratt, J. Owen, J. Issler, et al., MBOC: The new optimized spreading modulation recommended for GALILEO L1 OS and GPS L1C, in: Proc. IEEE/ION Positioning, Location and Navigation Symposium (PLANS), San Diego, CA, 2006, pp. 883–892.

[21] G.W. Hein, J.A. Avila Rodriguez, S. Wallner, B. Eissfeller, T. Pany, P. Hartl, Envisioning a future GNSS system of systems, GNSS Inside (Jan.–Feb.) (2007) 58–67.

[22] http://www.esa.int/esaNA/GGG63950NDC_egnos_0.html (last accessed: June 2011).

[23] HEPOS website, http://www.hepos.gr (last accessed: June 2011).

[24] S. Sand, GREAT–Galileo Receivers for Mass Market, ICT-Mobile Summary 2008. Conference Proceedings, P. Cunningham, M. Cunningham (Eds.), IIMC, International Information Management Corporation, 2008. ISBN: 978-1-905824-08-3.

[25] http://www.novatel.com/Documents/Papers/GLONASSOverview.pdf (last accessed: June 2011).

[26] ISRO press release, SATNAV industry meet held at ISRO satellite centre. http://www.isro.org/pressrelease/Jul04_2006.htm, 2006 (accessed 04.07.06).

[27] J. Bao-Yen Tsui, Fundamentals of Global Positioning System Receivers: A Software Approach, John Wiley and Sons, Hoboken, New Jersey, 2005.

[28] J. Rife, GNSS solutions: WAAS functions and differential biases, Inside GNSS – GNSS Solutions 3 (4) (May–June) (2008) 18–22.

[29] Japan Aerospace Exploration Agency – QZSS Project Team. Public release of interface specification for qzss – version 1.0 (draft), 2007. http://qzss.jaxa.jp/is-qzss/IS-QZSS_10_E.pdf (last accessed: June 2011).

[30] K.N. Suryanarayana Rao, Presentation of the Indian Regional Navigation Satellite System (IRNSS), in: 2nd Meeting of the International Committee on Global Navigation Satellite Systems (ICG), Bangalore, India, http://www.unoosa.org/pdf/icg/2007/icg2/presentations/03_02.pdf, 2007.

[31] K. Pollpeter, To be more precise: The Beidou satellite navigation and positioning system, China Brief VII (10) (2007).

[32] L. Wanninger, Introduction to network RTK, IAG Working Group 4.5.1: Network RTK, 2004–2008. Available at: http://www.wasoft.de/e/iagwg451/intro/introduction.html (last accessed: June 2011).

[33] H. Lim, S.I. Lee, S.P. Lee, A simple carrier frequency detection algorithm for fine compensation of Doppler shift in direct-sequence code division multiple access mobile satellite communications, in: IEEE International Conference on Communications, New Orleans, LA, vol. 1, 2000, pp. 109–112.

[34] Open Mobile Alliance (OMA), Secure user plane for location (SUPL), OMA-TS-UPL, v.1.0, 2007.

[35] P. Kemmpi, Next generation satellite navigation systems, VTT research notes 2408, ISBN 978-951-38-6961-8, VTT Technical Research Centre of Finland, 2007. Available at: http://www.vtt.fi/inf/pdf/tiedotteet/2007/T2408.pdf (last accessed: June 2011).

[36] P. Misra, P. Enge, Global Positioning System: Signal Measurements and Performance, second ed., Ganga-Jamuna Press, Lincoln, MA, 2006.

[37] P. Ward, Satellite signal acquisition and tracking, in: Understanding GPS: Principles and Applications, Artech House, Boston, 1996.

[38] S. Pace, G. Frost, I. Lachow, D. Frelinger, D. Fossum, D.K. Wassem, et al., The Global Positioning System Assessing National Policies, RAND Corporation, Santa Monica, CA, 1995.

[39] A. Polydoros, S. Glisic, Code synchronization: a review of principles and techniques, in: IEEE Third International Symposium on Spread Spectrum Techniques and Applications ISSSTA, Oulu, Finland, vol. 1, 1994, pp. 115–137.

[40] S.M. Radicella, The NeQuick model genesis, uses and evolution, Ann. Geophys. 52 (3/4) (2009) 417–422.

[41] Minimum Operational Performance Standards for Global Positioning System/Wide Area Augmentation System Airborne Equipment, RTCA Inc., Washington, DC, 2001.

[42] S.G. Revnivykh, GLONASS status and perspectives, in: Civil GPS Service Interface Commettee, Internation Information Subcommettee, Prague, 2005.

[43] S.G. Revnivykh, GLONASS status, development and application, in: 2nd Meeting of the International Committee on Global Navigation Satellite Systems (ICG), Bangalore, India, 2007. http://www.unoosa.org/pdf/icg/2007/icg2/presentations/05.pdf (last accessed: June 2011).

[44] S. Kuzin, S. Revnivykh, S. Tatevyan, GLONASS as a key element of the Russian positioning service, Adv. Space Res. 39 (10) (2007) 1539–1544.

[45] S. Lo, A. Chen, P. Enge, G. Gao, D. Akos, J.-L. Issler, et al., GNSS album. Images and spectral signatures of the new GNSS signals, Inside GNSS 1 (4) (May–June) (2006) 46–56.

[46] D. Van Nee, A. Coenen, New fast GPS code-acquisition technique using FFT, Electron. Lett. 27 (2) (1991) 158–160.

[47] J. Ventura-Traveset, D. Flament (Eds.), EGNOS: The European Geostationary Navigation Overlay System – A Cornerstone to Galileo, European Space Agency, Noordwijk, The Netherlands, 2006.

[48] P. Ward, GPS receiver search techniques, in: IEEE Position Location and Navigation Symposium, Kansas City, MO, 1996, pp. 604–611.

[49] W.D. Wilde, F. Boon, J.-M. Sleewaegen, F. Wilms, More compass points, Inside GNSS 2 (5) (July–August) (2007) 44–48.

[50] L. Wirola, Positioning protocol for next-gen cell phones, GPS World 22 (3) (2011) 8–13.

[51] W. Zhuang, J. Tranquilla, Digital baseband processor for the GPS receiver (parts I and II), IEEE Trans. Aerosp. Electron. Syst. 29 (4) (1993) 1343–1349.

[52] W. Zhuang, J. Tranquilla, Modeling and analysis for the GPS pseudo-range observable, IEEE Trans. Aerosp. Electron. Syst. 31 (2) (1995) 739–751.

CHAPTER

3

Terrestrial Network-Based Positioning and Navigation

Mauricio A. Caceres Duran, Antonio A. D'Amico, Davide Dardari, Mats Rydström, Francesco Sottile, Erik G. Ström, Lorenzo Taponecco

Most terrestrial network-based positioning and navigation systems introduced earlier in Section 1.2.3 were designed with the main objective of building communication networks. After that they have been used as a support to develop positioning applications. On the contrary, satellite-based systems are designed with the primary objective to obtain a positioning network with a wide, possibly global, coverage.

Therefore, for many years, a lot of scientific research has been focused on wireless terrestrial networks to build the basis for positioning and navigation systems. Today these systems are designed to work especially either where the GNSS receiver does not have any satellite visibility (typically in indoor environments) or where satellite visibility is poor (typically in urban areas where the multipath effect may lead the receiver to be synchronized with reflected paths, causing a very high location error).

The principal terrestrial communication systems for which position location capabilities have been deployed are cellular networks, WLANs, WSNs and, recently, wireless networks based on UWB technology. In many cases, these wireless technologies have in common position parameter estimation techniques, such as *received signal strength* (RSS), *time of arrival* (TOA), *time difference of arrival* (TDOA), *angle of arrival* (AOA), and localization algorithms, such as *triangulation, fingerprinting methods,* etc. However, the final location accuracy depends on different aspects, but in particular it is strongly related to the signal format, bandwidth, and propagation properties of the wireless technology used by the location system. This chapter analyzes the positioning and navigation approaches applied on the above-mentioned terrestrial wireless technologies.

3.1 FUNDAMENTALS ON POSITIONING AND NAVIGATION TECHNIQUES IN TERRESTRIAL NETWORKS

This section presents fundamental positioning techniques typically adopted in most terrestrial networks. In general, terrestrial network-based position estimation techniques involve a two-step process [15, 50]. First, a number of position-related signal parameters, such as TOA, RSS, and/or AOA, are estimated (step 1). Then, the position of the MS is determined in a two- or three-dimensional plane based on those signal parameters (step 2). However, recent approaches known as direct position

Satellite and Terrestrial Radio Positioning Techniques. DOI: 10.1016/B978-0-12-382084-6.00003-9

estimation (DPE) claim that there are certain advantages if localization is performed directly in just one rather than two steps [6].

Sections 3.1.1 and 3.1.2 hereafter go through the two process steps, with an overview of the most common methods suitable to perform the position estimation. The discussion of the main sources of error affecting these methods is deferred to Section 3.1.3.

3.1.1 Position-Related Signal Parameter Estimation

As introduced in Section 1.1.2, the starting point of any positioning system is the estimation of a set of position-dependent signal parameters such as TOA, TDOA, and RSS. In some localization systems, a combination of various position-related parameters can be utilized in order to obtain more information about the position. Examples of such *hybrid* schemes include TOA/AOA [24], TOA/RSS [16], TDOA/AOA [24], and TOA/TDOA [140] positioning systems. Also, measurement of multipath power delay profile or channel impulse response related to a received signal can be considered as other types of position-related parameters [122, 175]. In the following, RSS, TOA, TDOA, and AOA estimation methods are presented in more detail.

3.1.1.1 *Received Signal Strength Methods*

RSS localization methods were first introduced in 1969 [45]. Their major advantage compared with other methods is the availability of RSS measurements in practically all systems and the fact that nodes do not have to be time synchronized. To determine the location, RSS measurement must be modeled as a function of MS coordinates. This task can be performed by using a received signal statistical model, which is based either on a propagation model or on measurements [142].

In Ref. [183], Weiss suggested the following model for the received power P_i (in dBm) because of the signal transmitted by the ith BS,

$$P_i = P_\epsilon + P_{\text{ave},i}(\mathbf{x}) + X_\sigma, \quad i = 1, \dots, N_{\text{B}}, \tag{3.1}$$

where

- N_{B} is the number of signals received above a certain threshold out of M base stations.
- P_ϵ is a common error term for signals transmitted from all base stations because of the specific situation of the handset (horizontal or vertical, outdoor or indoor, etc.).
- $P_{\text{ave},i}$ is the average power (in dBm) given by the *propagation model* or measurements or a combination of both.
- $\mathbf{x} = [x, y]^T$ is the vector containing the 2D MS's coordinates.
- X_σ is a zero-mean random error composed of several factors that cause the variation of the measured signal power around the predicted mean (such as multipath fading, shadowing, antenna pattern, thermal noise, and interference). In many scenarios, it is simply modeled as a zero mean Gaussian random variable.

As far as the propagation model is concerned, one can choose any of the several alternatives that exist in the literature such as, for example, the *log-distance* model that has the generic form [133]:

$$P_{\text{ave},i}(\mathbf{x}) = P_0 - 10\alpha \log_{10}(d_i), \tag{3.2}$$

where d_i is the distance between the MS and the ith BS, α is the *path-loss exponent*, and P_0 is the receiver power in dBm at the reference distance of 1 m in free-space conditions. It depends on the transmitter power, antenna gain, and the carrier frequency. RSS measurements can be used to estimate the distance (range) between the MS and BS through (3.1) and (3.2), or to build a *terrain-based model* in fingerprinting positioning techniques as explained in Section 3.2.1.3.

RSS is the most widely used position-parameter estimation method, as it is simple to implement also in cheap devices. However, it does not provide very accurate position/range information, especially in NLOS channel conditions [64].

3.1.1.2 *Time-of-Arrival Methods*

Based on the fact that electromagnetic waves propagate through vacuum, and approximately through any free-from-objects medium, at the constant speed of light ($c \approx 3 \cdot 10^8$ m/s), the distance information between a couple of nodes A and B can be obtained from the measurement of the propagation delay or time of flight (TOF) $\tau_f = d/c$, where d is the actual distance between A and B.

In a first simple scheme shown in Fig. 3.1 (*one-way ranging*), node A emits a packet at time t_1 to a receiving node B. The packet contains the time stamp t_1 at which the transmission started. Node B receives the packet at time t_2. If the nodes were perfectly synchronized to a common reference clock (i.e., sharing the same time reference and time base), it is clear that τ_f would be calculated at node B as $\tau_f = t_2 - t_1$ and the distance estimated. Synchronization error can significantly affect ranging error (this aspect is discussed in Section 3.1.3).

A second scheme which requires less stringent synchronization constraints is *two-way ranging* or *two-way-TOA* (TW-TOA). In this scheme, node A (see Fig. 3.2) emits a packet to node B which, after a fixed response delay τ_d, gives an answer by transmitting back a second acknowledgment (ACK) packet to node A. The round-trip time (RTT) between the node A transmission and response receiving instants is

$$\text{RTT} = 2\tau_f + \tau_d, \tag{3.3}$$

from which the distance can be estimated assuming τ_d is known. Although two-way ranging eliminates the error resulting from imperfect synchronization between nodes, relative clock drift still affects ranging accuracy; even a small clock offset between nodes could correspond to a large error in τ_f estimation because of error accumulation over τ_d (this aspect is discussed further in Section 3.1.3). It is evident that TOA-based techniques require a very accurate time synchronization between all transmitters and receivers, and a time stamp to be included in the signal.

We now investigate the basic scheme to estimate the TOA by considering a simple scenario where a generic signal $g(t)$, with duration T_g, is transmitted and received undistorted, with energy E_g, through an additive white Gaussian noise (AWGN) channel. The received signal can be expressed as

$$r(t) = \sqrt{E_g}\, g(t - \tau) + n(t), \tag{3.4}$$

FIGURE 3.1

One-way ranging.

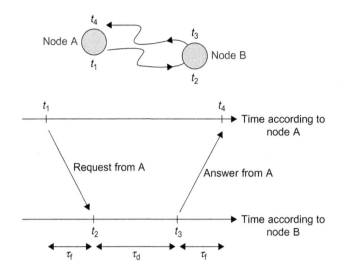

FIGURE 3.2

Two-way ranging.

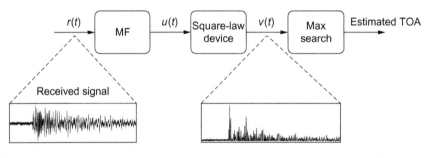

FIGURE 3.3

ML TOA estimator in AWGN.

where $n(t)$ is AWGN with zero mean and two-sided power spectral density $N_0/2$. The parameter τ describes the TOA to be estimated based on the received signal $r(t)$ observed over the interval $[0, T_{ob})$.

In the classical TOA estimation scheme, the received signal is first processed by a filter matched to the signal $g(t)$ (see Fig. 3.3). The TOA estimation is then accomplished simply by observing the instant corresponding to the maximum peak at the output of the matched filter (MF) over the observation interval [174]. This scheme yields an ML estimate, which is asymptotically efficient, that is, for high SNR the estimated TOA, $\hat{\tau}$, becomes unbiased with variance equal to

$$\sigma_{\hat{\tau}}^2 = \frac{1}{8\pi^2 \eta \beta^2}, \tag{3.5}$$

where we have introduced the signal-to-noise ratio $\eta = E_g/N_0$, and the parameter β^2 represents the second moment of the spectrum $G(f)$ of $g(t)$ defined by

$$\beta^2 \triangleq \frac{\int_{-\infty}^{\infty} f^2 |G(f)|^2 \, df}{E_g}. \tag{3.6}$$

As will be shown more in detail in Chapter 4, (3.5) represents the Cramér–Rao lower band (CRB), $\kappa = \sigma_{\hat{\tau}}^2$, that gives the lower bound on the estimation mean square error (MSE) of any unbiased estimation $\hat{\tau}$ of τ, that is,

$$\mathrm{Var}\left\{\hat{\tau}\right\} = \mathbb{E}\left\{(\hat{\tau} - \tau)^2\right\} \geq \kappa. \tag{3.7}$$

Notice that the denominator of (3.5) is proportional to the energy in the signal where the proportionality constant β^2 depends on the shape of the pulse through (3.6). Contrary to RSS-based methods, the ranging capability depends on the signal structure. This reveals that having large values of β^2, that is, a signal with wide bandwidth, is beneficial for ranging. By considering, for example, the typical log-distance path loss model (3.2), the SNR can be expressed as $\eta = \eta_0 d^{-\alpha}$, and hence the distance estimation MSE can be lower bounded by

$$\mathrm{Var}\left\{\hat{d}\right\} \geq \sigma_0^2 \, d^\alpha, \tag{3.8}$$

where η_0 and σ_0^2 are, respectively, the SNR and the CRB on the estimated distance experienced at 1 m.

3.1.1.3 *Time-Difference-of-Arrival Methods*
An efficient way of overcoming the two prerequisites for TOA localization, namely the synchronization between BSs and MSs and the labeling of the signal with time information, is to examine the difference τ between the TOA of the signal at two different BSs, rather than the absolute arrival time.

A straightforward method of estimating that difference in time is to cross-correlate the signals arriving at a pair of BSs. The cross-correlation of two signals $r_1(t)$ and $r_2(t)$, received from BS1 and BS2, respectively, is given by

$$R_{1,2}(\tau) = \frac{1}{T} \int_0^T r_1(t) r_2(t + \tau) dt, \tag{3.9}$$

where T is the observation interval. In the absence of errors (propagation impairments), (3.9) shows a peak for τ equal to the exact TDOA.

3.1.1.4 *Angle-of-Arrival Methods*
In order to estimate the angle of arrival (AOA) of incident signals, a directional antenna, such as an adaptive phased array of two or more antenna elements, is required. The straightforward method is to measure the phase difference between the signals, when impinging on different antenna elements, and convert this to an AOA estimate.

A pictorial representation of this general concept is shown in Fig. 3.4. This method is known as *interferometry.*

Another conceptually simple method to estimate the AOA of a signal is based on *beamforming.* Antenna arrays (also called *smart antennas*) implementing beamforming techniques electronically

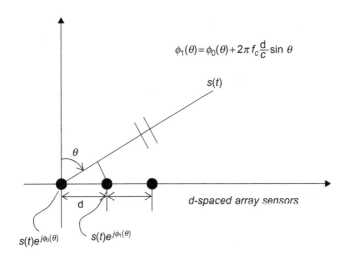

FIGURE 3.4

Signal impinging an antenna array: the phase $\phi_i(\theta)$ seen at each array element i depends on the AOA θ, where f_c is the carrier frequency and c is the speed of light.

steer their main radiation lobe(s) in the direction of arrival of the signal(s) of interest, following appropriate algorithms that somehow measure the input power. A very instructive introduction to AOA estimation techniques can be found in the classic tutorial articles [51, 52, 176], as well as in Ref. [99].

Thanks to their capability of performing such a "spatial filtering," array antenna systems can provide information about the AOA of the signals impinging upon the array. The common array-based, AOA-estimation techniques can be broadly divided into four categories:

- Conventional techniques
- Subspace based techniques
- Maximum likelihood techniques
- Integrated techniques.

Conventional methods are based on classical beamforming techniques and do not exploit any assumption on the model of the received signal nor noise. Conventional AOA-estimation techniques electronically steer beams in all possible directions and look for peaks in the output power [152]. Classic conventional methods are the delay-and-sum method [99] and Capon's minimum variance method [72]. However, in such approaches, the angular resolution is limited by the beamwidth of the array; therefore, a large number of antenna elements is required to achieve high resolution.

Subspace-based methods are high-resolution suboptimal techniques, which exploit the eigenstructure of the input data when organized in a time–space matrix. Historically, the first and most famous subspace-based technique is the multiple signal classification (MUSIC) algorithm [137]. The geometric concepts on which MUSIC is founded form the basis for a broader class of subspace-based algorithms [2]. Another well-known method of this family is the estimation of signal parameters through rotational invariance technique (ESPRIT) [138].

Maximum likelihood techniques are optimal techniques which outperform subspace-based ones even under low signal-to-noise ratio conditions and in correlated signal condition. Their major drawback is that they are often computationally intensive.

Integrated approaches combine property-restore techniques with subspace-based approaches to enhance the final resolution [99, 168].

The main disadvantages of AOA techniques are that they require relatively large and complex hardware and periodic array calibration. The fact that the position estimate based on AOA degrades as the MS moves further from the BSs also makes AOA techniques less attractive than the TOA and the TDOA techniques in many real-life scenarios.

3.1.2 **Position Estimation Techniques**

After obtaining position-related signal parameters, such as RSS, TOA, TDOA, and AOA mentioned earlier, the second step of localization involves the MS position estimation based on those parameters.

3.1.2.1 *Geometric Positioning Techniques*

Geometric techniques estimate the MS position directly from a set of position-related parameters extracted from the received signal. As introduced in Section 1.3, for assessing the MS location, either *deterministic* or *statistical* approaches can be used.

Deterministic Techniques

With the deterministic techniques, the position of the MS is determined by exploiting geometric relationships. These methods are classified as *lateration*, *angulation*, and *hybrid approaches*.

Lateration The lateration approach can use one of the following localization input parameters: TOA, TDOA, and RSS. When either TOA or RSS localization parameter estimation is available, the position of the MS is estimated by measuring its distance (range) from multiple BSs and finding the intersection of circles (spheres in 3D) as shown in Fig. 3.5. Using the formula of the Euclidean distance between two points (x, y) and (x_i, y_i) in a 2D plane and assuming without loss of generality that BS$_1$ is placed in the $(0,0)$ point, we get the following expression in vector form for the unknown MS coordinates (x, y):

$$\begin{bmatrix} x \\ y \end{bmatrix} = \frac{1}{2} \begin{bmatrix} x_2 & y_2 \\ x_3 & y_3 \end{bmatrix}^{-1} \begin{bmatrix} x_2^2 + y_2^2 + d_1^2 - d_2^2 \\ x_3^2 + y_3^2 + d_1^2 - d_3^2 \end{bmatrix}. \tag{3.10}$$

TDOA-based localization systems do not rely on absolute distance estimates between pairs of nodes. Such systems typically use one of two schemes as follows. In the first (Fig. 3.6), multiple signals are broadcast from synchronized fixed nodes (MS or anchors) located at known positions and agents measure the TDOA (similar technique is used by GPS). In the second scheme (Fig. 3.7), a reference signal is broadcast by the MS, and it is received by several BSs. The BSs share their estimated TOA and compute the TDOA. Typically, BSs are synchronized through a wired network connection. In the 2D case, each TDOA measurement can be geometrically interpreted as a hyperbola formed by a set of points with constant range-differences (time-differences) from two BSs rather than a circle [43, 164]. As can be seen from Fig. 3.8, the intersection of two or more hyperboloids, whose generation requires three or more BSs to participate in the measurements, gives the exact location of the MS. In the 3D plane, a minimum of four BSs is required to estimate the MS location.

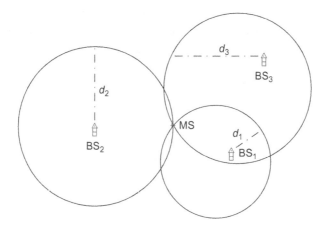

FIGURE 3.5

Lateration technique with three BSs based on range estimation.

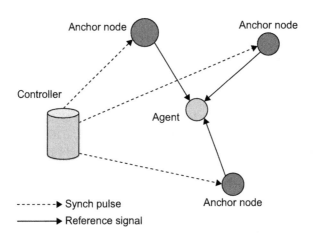

FIGURE 3.6

TDOA scheme 1.

Solving the nonlinear equations produced in the TDOA method can be cumbersome. Many different approaches can be found in the literature. In Ref. [43], an exact solution is presented, but only for the case when the number of TDOA measurements and the number of unknown coordinates are equal. In Ref. [4], the authors present the "divide and conquer" method, whereas Ref. [46] explores the Taylor-series estimation, an iterative technique with good accuracy under the reasonable initial guess assumption. Finally, Ref. [18] presents a noniterative efficient technique that gives an explicit

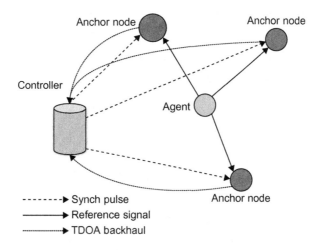

FIGURE 3.7

TDOA scheme 2.

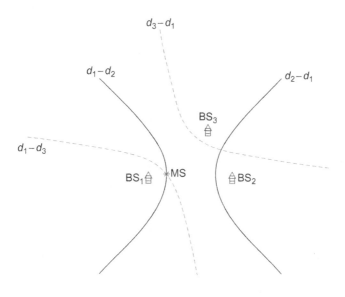

FIGURE 3.8

Lateration technique with three BSs based on TDOA estimation.

solution. In particular, let each difference in distance be $d_{ij} = d_i - d_j = \sqrt{(x - x_i)^2 + (y - y_i)^2} - \sqrt{(x - x_j)^2 + (y - y_j)^2}$. It satisfies the following equation: $d_{ij} = d_{ik} - d_{jk} \ \forall i, j, k \in N_\mathrm{B}$. Using BS_1 as a reference and introducing the terms $K_i = x_i^2 + y_i^2$, $x_{i1} = x_i - x_1$ and $y_{i1} = y_i - y_1$, we get a solution for

x and y that depends on d_1:

$$\begin{bmatrix} x \\ y \end{bmatrix} = - \begin{bmatrix} x_{21} & y_{21} \\ x_{31} & y_{31} \end{bmatrix}^{-1} \left(\begin{bmatrix} d_{21} \\ d_{31} \end{bmatrix} d_1 + \frac{1}{2} \begin{bmatrix} d_{21}^2 - K_2 + K_1 \\ d_{31}^2 - K_3 + K_1 \end{bmatrix} \right). \tag{3.11}$$

Inserting this result into $d_1^2 = (x - x_i)^2 + (y - y_i)^2 = K_1 - 2xx_i - 2yy_i + x^2 + y^2$ a quadratic expression in d_1 is obtained. Substituting the positive root back into (3.11) produces the solution.

Angulation The angulation approach is based on angle measurements. Using the AOA parameter, the location of the MS is found by the intersection of the lines joining the unknown MS and the BS, as shown in Fig. 3.9. To combat inaccuracies, more than two BSs and highly directional antennas might be used for AOA estimation [99]. It is worth noticing that extracting the AOA information requires the use of an antenna array at each MS to be localized, which is often inpractical for reasons of cost and size. The AOA-based methods are therefore less attractive in most cases, when cost and size are the driving constraints.

Hybrid Approach In the hybrid approach, different types of position-related parameters are combined to locate the MS (such as TDOA/AOA, TOA/TDOA, and TOA/AOA). For example, using the hybrid TOA/AOA technique, the object position can be estimated by using a single reference point (see Fig. 3.10).

The main drawback of the geometric techniques is that they provide an estimate of the MS location strongly impaired by the presence of measurement errors in the position-related parameters. This phenomenon is illustrated in Fig. 3.11 for a lateration positioning system. It can be observed that the distances of each BS from the MS are not correctly measured, and the circles do not cross in a single point. Sources of errors and their effects are discussed in Section 3.1.3.

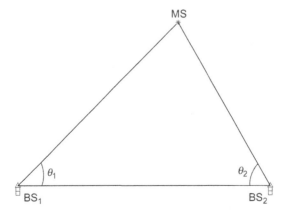

FIGURE 3.9

Angulation technique with two BSs based on AOA estimation.

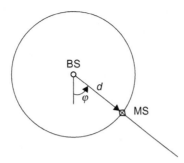

FIGURE 3.10

Hybrid technique with one BS based on TOA/AOA estimation.

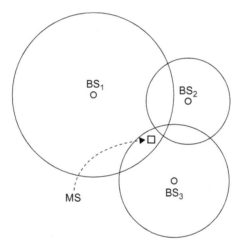

FIGURE 3.11

Lateration technique with noisy parameters.

Statistical Techniques

As already summarized in Section 1.3, statistical techniques provide a theoretical framework that solves the location problem even in the presence of measurement errors. In this way, the probabilistic approach overcomes the drawback of geometric deterministic methods.

Parametric As an example of parametric statistical positioning algorithm, we formulate the ML position estimation problem starting from RSS measurements [30]. Let us consider a scenario populated with N BSs and 1 MS. We assume that $K \leq N$ radio links exist between the MS and the BSs. We denote by $\mathbf{r} = [r_1, \ldots, r_K]$ the vector containing the relative path-loss measurements.[1] The ML estimate is

---

[1] Path-loss measurements are easily obtained by difference once the RSS and the transmitted power in dBm are known.

given by

$$\hat{\mathbf{x}}_{ML} = \arg\max_{\mathbf{x}} p(\mathbf{r}|\mathbf{x}), \tag{3.12}$$

where $\mathbf{x}$ is the MS's position and $p(\mathbf{r}|\mathbf{x})$ is the pdf of the measurements conditioned on the MS's position. By assuming independent fading over each link, we can write the pdf of $\mathbf{r}$ given $\mathbf{x}$ as

$$p(\mathbf{r}|\mathbf{x}) = \prod_{i=1}^{K} p(r_i|\mathbf{x}) = \prod_{i=1}^{K} f(r_i, d_i), \tag{3.13}$$

where d_i is the estimated distance between the MS and the ith BS and $f(r_i, d_i)$ is the pdf of the path-loss measurements. Considering log-normal shadowing, the ML position estimation is given by

$$\hat{\mathbf{x}}_{ML} = \arg\max_{\mathbf{x}} \prod_{i=1}^{K} \frac{1}{\sqrt{2\pi}\sigma_i} \exp\left\{-\frac{(r_i - L_0(d_i))^2}{2\sigma_i^2}\right\}, \tag{3.14}$$

where σ_i represents the standard deviation of the log-normal shadowing of the ith link and $L_0(d)$ is the path loss at the distance of d meters.

In Fig. 3.12, the performance of the ML estimator (3.14) in terms of normalized RMSE (with respect to the scenario dimension) as a function of the standard deviation σ of the channel path-loss fluctuations (assumed equal for all links) is shown. The localization error has been averaged through 1000 different node configurations randomly distributed in a 16×16 m^2 area. The log-distance path-loss model (3.2) has been considered with $P_0 = 40$ dBm and $\alpha = 3.5$. As can be seen, measurement errors strongly affect the final position accuracy which, in general, increases as the number of MSs increases.

Parametric methods for positioning systems are proposed and analyzed in Refs. [9, 19, 25, 30, 36, 66, 106, 130, 135, 181]. For example, in Ref. [130], a *distributed maximum log-likelihood* technique is

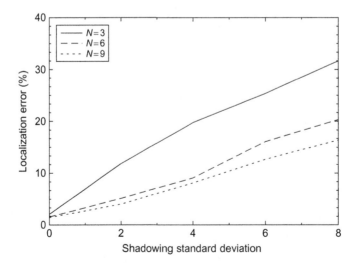

FIGURE 3.12

Localization error percentage as a function of the standard deviation σ for different number N of MSs.

proposed as a positioning solution for UWB-based networks. The network analyzed in the article consists of several BSs, which are set at fixed known locations, and of a large number of MSs. Exploiting TOA estimates, each MS measures its distance from the neighboring devices, which can be either BS or MS terminals. Then, the positioning problem is solved by performing a global maximization of the log-likelihood of all the distances calculated by the devices.

Nonparametric The basic idea behind nonparametric techniques is to estimate a number of position-related parameters greater than the minimum required by a *pure* geometric location algorithm. This redundancy allows one to perform various combinations of location data and, through proper algorithms, to overcome the effects of measurement errors. Generally nonparametric positioning algorithms are based on LS techniques. In the following, we illustrate a possible approach to derive an LS estimator starting from a deterministic geometric approach.

Consider a 2D scenario with $N_B \geq 3$ BSs and 1 MS to be localized. In the presence of ideal range estimates (i.e., $\hat{d}_i = d_i$), the ith BS defines a circle centered in (x_i, y_i) with radius d_i, upon which the MS is located. If the MS has obtained ranges to multiple BSs, then the intersection of the circles corresponds to the position of the target node. In a two-dimensional space, at least three BSs are required. Specifically, the position estimate (x, y) is obtained by solving the system of equations

$$(x_1 - x)^2 + (y_1 - y)^2 = d_1^2$$

$$\vdots$$

$$(x_{N_B} - x)^2 + (y_{N_B} - y)^2 = d_{N_B}^2. \tag{3.15}$$

According to Ref. [15], the system of equations in (3.15) can be linearized by subtracting the last equation from the first $N_B - 1$ equations. The resulting system of linear equations is given by the following matrix form:

$$\mathbf{A} \cdot \mathbf{p} = \mathbf{b}, \tag{3.16}$$

where $\mathbf{p}$ is the MS position,

$$\mathbf{A} \triangleq \begin{pmatrix} 2(x_1 - x_{N_B}) & 2(y_1 - y_{N_B}) \\ \vdots & \vdots \\ 2(x_{N_B-1} - x_{N_B}) & 2(y_{N_B-1} - y_{N_B}) \end{pmatrix}, \tag{3.17}$$

and

$$\mathbf{b} \triangleq \begin{pmatrix} x_1^2 - x_{N_B}^2 + y_1^2 - y_{N_B}^2 + \hat{d}_{N_B}^2 - \hat{d}_1^2 \\ \vdots \\ x_{N_B-1}^2 - x_{N_B}^2 + y_{N_B-1}^2 - y_{N_B}^2 + \hat{d}_{N_B}^2 - \hat{d}_{N_B-1}^2 \end{pmatrix}. \tag{3.18}$$

In a real scenario where estimation errors are present, (3.16) may be inconsistent, that is, the circles do not intersect at one point. In that case, the position can be estimated through a standard linear LS approach as

$$\hat{\mathbf{p}} = (\mathbf{A}^T \mathbf{A})^{-1} \mathbf{A}^T \mathbf{b}, \tag{3.19}$$

with the assumption that $\mathbf{A}^T\mathbf{A}$ is nonsingular and $N_B \geq 3$ [15]. Particular attention must be paid in selecting the BS associated with the last equation in (3.15) and used as reference in (3.16)–(3.18). If the corresponding range measurement is biased, bias will be introduced in all the equations with a consequent performance loss [23].

Some experimental results will now be discussed considering the scenario illustrated in Fig. 3.13, where ranging data were collected using UWB devices [32]. The position estimation RMSE for each location in the grid (node ID) is reported in Fig. 3.14 for the case of $N_B = 3$ (tx1, tx3, tx5) and $N_B = 5$ BSs. BS tx5 is chosen as the reference node. In general, thanks to the good ranging properties of the UWB devices, an accuracy below 1 m can be achieved. However, the presence of several NLOS conditions generates large ranging errors that degrade the positioning accuracy in many locations. In Ref. [32], some mitigation approaches are proposed to improve the performance. It can be noted that the use of a larger number of BSs does not necessarily correspond to better positioning accuracy for all locations. This results from the fact that, in many cases, the added range measurements and/or the chosen reference node are subject to large errors. Moreover, the geometric configuration of the additional BSs may not improve the positioning accuracy in certain locations.

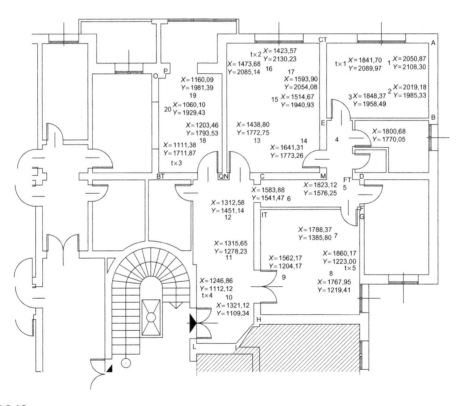

FIGURE 3.13

The measurement environment at the WiLAB, University of Bologna. Coordinates are expressed in centimeters (from Ref. [32]).

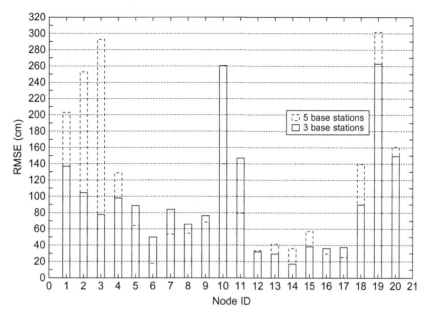

FIGURE 3.14

RMSE as a function of MS ID. $N_B = 3$ (tx1, tx3, tx5) and $N_B = 5$ BSs are considered.

3.1.2.2 *Mapping (Fingerprinting) Techniques*

Mapping techniques operate in two steps: They first collect features (fingerprints) of a scene (environment) and then estimate the location of an object by matching *online* measurements with the closest *a priori* location fingerprint. More precisely, in the first step (offline stage), which is based on a set of training data, the database of location fingerprints is built by performing a survey of the site where the system will be deployed. Essentially, the database can be represented as a collection of pairs of vectors $(\mathbf{f}_n, \mathbf{x}_n)$, where $\mathbf{f}_n$ is the vector of position-related parameters (fingerprints) for the nth training data and $\mathbf{x}_n$ is the corresponding position vector. In the offline phase, most proposed schemes use RSS measurements. However, in practice, the choice of fingerprints largely depends on available parameters and also on their impact on the performance of the localization algorithm. During the second step (online stage), the currently observed signal parameter and the previously collected information are compared to localize the MS device.

The most common positioning algorithms based on fingerprinting are as follows: k-*nearest-neighbor* (k-NN) and *neural networks* (NNs).

1. *k-nearest-neighbor*: The k-NN methods search for the k fingerprints in the database, say $\mathbf{f}_{n_1}, \mathbf{f}_{n_2}, \ldots, \mathbf{f}_{n_k}$, that have the smallest Euclidean distances from the online measured fingerprint $\mathbf{f}$. The estimated position is obtained by the weighted sum of the corresponding location vectors $\mathbf{x}_{n_1}, \mathbf{x}_{n_2}, \ldots, \mathbf{x}_{n_k}$. In general, the weights are determined according to $\mathbf{f}$ and $\mathbf{f}_{n_1}, \mathbf{f}_{n_2}, \ldots, \mathbf{f}_{n_k}$. Different weighting functions can be used, as discussed in Ref. [112]. k-NN techniques provide a sufficient

localization accuracy [124, 182], but they are not appropriate when the size of the database is large. Indeed, the time required to locate the MS is proportional to the dimension of the database.

2. *Neural networks*: With neural networks, the localization problem can be viewed as a function approximation problem consisting of creating a nonlinear mapping from a set of input variables (i.e., the position-related parameters extracted from the received signal) onto the output variables representing the spatial coordinates of the MS. Like the other fingerprinting-based positioning schemes, the NN-based systems operate in two steps: during the offline stage, the neural network is trained by a set of inputs corresponding to known positions of the MS. This phase, where the parameters of the NN are iteratively adjusted to minimize a given performance function (and so creating the mapping from inputs onto outputs), is equivalent to the database construction in other fingerprinting systems. In the real-time phase, the position-related parameters extracted from the received signal are fed to the NN whose output is the estimated value of the MS location. The operation of an NN-based positioning system is depicted in Fig. 3.15. Compared with other finger-printing systems, neural networks provide real-time response and reduce the overhead of database search algorithms that require a considerable computational effort during the online stage [77]. In addition, neural networks are robust against the lack of fingerprint data, thanks to their ability to interpolate the training data (*generalization* property) [65]. NN-based positioning algorithm are discussed in Refs. [77, 122, 166].

3.1.3 Error Sources in Localization

In real environments, several sources of errors may impair the attempt to localize the MS. The impacts of these errors on the accuracy of the localization technique have to be carefully examined, and proper methods to mitigate these effects and measure the obtainable accuracy have to be identified. Theoretical

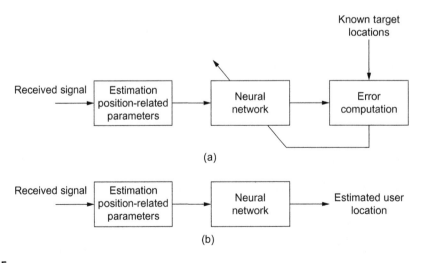

FIGURE 3.15

Neural network technique: (a) offline stage; (b) online stage.

analysis of the accuracy of localization methods will be discussed in Chapter 4, while innovative approaches to overcome the challenges and mitigate the errors will be presented in Chapter 5.

Let us first define a few terms with reference to Fig. 3.16. We refer to a range measurement between two nodes as a *direct path* (DP) measurement if the measured signal traveled in a straight line between the two nodes through a medium with constant and known permittivity (such as in air, as per TX-RX1). A measurement can be *non-DP* if, for example, the DP signal is obstructed (TX-RX2) or if the first arriving path has been reflected by obstacles (TX-RX3). An *LOS measurement* occurs when the first arriving path is along the DP, whereas an *NLOS measurement* comes either from complete DP blockage (in which the direct path is completely blocked by obstacles) or from DP excess delay (in which the signal traverses through different materials in a straight line resulting in additional TOF delays). Time-based ranging in realistic environments is typically corrupted by several joint factors such as multipath, DP excess delay, DP blockage, and clock drift [31, 50].

The two major sources of errors described hereafter are the multipath and the NLOS propagation [15, 84], which mainly affect the geometric techniques. As a matter of fact, fingerprinting techniques have the ability to take advantage of multipath and NLOS propagation rather than the need to alleviate them. Sources of errors for fingerprinting techniques will be presented separately in Section 3.1.3.4. In time-based positioning schemes, internal clock inaccuracy and network synchronization errors

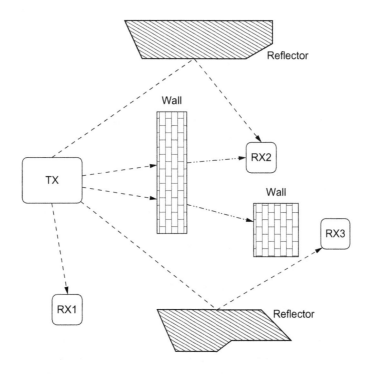

FIGURE 3.16

Possible LOS and NLOS configurations. RX1: LOS condition; RX2: NLOS condition, no DP blockage; RX3: NLOS condition, DP blockage.

may have a fundamental impact on the final localization accuracy. This issue will be investigated in detail in Section 3.1.3.2.

3.1.3.1 *Multipath Propagation*

Multipath fading is caused by the destructive and constructive interference of signals arriving at the receiver through different propagation paths. In RSS techniques, multipath fading increases the variance of the received power because of the reception of multiple copies of the transmitted signal. However, it can be argued that RSS techniques are the least affected by multipath propagation because the fluctuation caused by the multipath fading can be removed by averaging the received power over a time interval.

In TDOA and TOA techniques, the presence of multipath components limits the time resolution of the TDOA/TOA estimate to approximately $1/B_W$, B_W being the signal bandwidth, when the conventional cross-correlation is used. To convert this resolution to accuracy, consider for instance the IS-95 standard for CDMA. The bandwidth of the signal is 1.25 MHz so the signal can propagate $c/B_W = 240$ m in $1/B_W$ seconds. The TDOA/TOA estimation error can be much bigger for signals with narrower bandwidth, like in the GSM standard. Assuming that an LOS path exists between the MS and the BS, multipath propagation is the main cause of inaccuracies [84]. If the receiver is unable to determine which is the first-arriving path in order to identify it as the LOS path, the position error will be very high.

Multipath propagation affects AOA estimates when the receiver cannot resolve closely spaced multipath delays. In flat-fading channels, the delay spread is very small relative to the channel bandwidth B_W, so the multipath components will all be correlated. However, because delay spread and angular spread are interrelated, the receiver must be able to handle small angular spread.

In an attempt to find parameter estimation methods with good accuracy at multipath environments, researchers developed sophisticated methods, such as the ML and the maximum entropy (ME) methods and spatial analysis techniques like the MUSIC [153] and ESPRIT [143] algorithms. In the following, we illustrate how the ML TOA estimator in the presence of multipath can be derived. More advanced schemes can be found in Chapter 5.

ML TOA Estimation in the Presence of Multipath

We consider a scenario in which a unitary energy pulse $g(t)$ is transmitted (with duration T_g) through a channel with multipath and thermal noise. The received signal can be written as

$$r(t) = s(t) + n(t), \tag{3.20}$$

where $s(t)$ is the channel pulse response (CPR) to the transmitted pulse $g(t)$, and $n(t)$ is AWGN with zero mean and two-sided power spectral density $N_0/2$. We consider a frequency-selective fading channel, in which case

$$s(t) = \sqrt{E_g} \sum_{l=1}^{L} \alpha_l g(t - \tau_l) + n(t), \tag{3.21}$$

where L is the number of received multipath components, α_l and τ_l denote the amplitude and delay of the lth path, respectively, with the normalization $\sum_{l=1}^{L} \mathbb{E}\{\alpha_l^2\} = 1$ such that E_g represents the average received energy. The goal is to estimate the TOA $\tau = \tau_1$ of the first path by observing the received signal $r(t)$ in the observation interval T_{ob}. This task can be challenging because of the presence of thermal noise and multipath components.

ML estimators are known to be asymptotically efficient, that is, their performance achieves the CRB in high SNR regions. As discussed in Section 3.1.1.2, a TOA estimate can be obtained by using an MF (or equivalently a correlator) matched to the received signal; the TOA estimate is equal to the time delay that maximizes the MF output. Note that in the presence of multipath, the template in the receiver (also known as the locally generated reference) should be proportional to the CPR $s(t)$ instead of $g(t)$ in the case of an AWGN channel. However, it is difficult to implement this estimator because the received waveform must be estimated. In addition, each received pulse (echo) may have a different shape than the transmitted pulse because of varying antenna characteristics and materials for different propagation paths.

When channel parameters are unknown, TOA estimation in multipath environments is closely related to channel estimation. In this case, path amplitudes and TOAs are jointly estimated using, for example, the ML approach [68, 97, 103, 107, 145, 186]. In the presence of AWGN, the ML criterion is equivalent to the MMSE criterion. Given an observation $r(t)$, the ML estimate of the set $\mathcal{V} = \{\tau_1, \tau_2, \ldots, \tau_L, \alpha_1, \alpha_2, \ldots, \alpha_L\}$ of unknown channel parameters corresponds to the set of values that maximizes the log-likelihood function [145, 199][2]

$$\mathcal{L}(\mathcal{V}) = -\frac{1}{N_0} \int_0^{T_{ob}} \left| r(t) - \sqrt{E_g} \sum_{l=1}^{L} \alpha_l g(t - \tau_l) \right|^2 dt. \tag{3.22}$$

We now outline the derivation for the ML estimate of the path arrival times and their respective amplitudes. Let $\boldsymbol{\tau} = [\tau_1, \tau_2, \ldots, \tau_L]^T$, $\boldsymbol{\alpha} = [\alpha_1, \alpha_2, \ldots, \alpha_L]^T$, and $\boldsymbol{R}(\boldsymbol{\tau})$ be the autocorrelation matrix of $g(t)$ with elements $R_{i,j} = \rho_g(\tau_i - \tau_j)$, $\rho_g(\tau)$ being the autocorrelation function of $g(t)$. It can be shown that the ML estimates of $\boldsymbol{\tau}$ and $\boldsymbol{\alpha}$ are respectively given by [186][3]

$$\hat{\boldsymbol{\tau}} = \arg\max_{\tilde{\boldsymbol{\tau}}} \{\boldsymbol{\chi}^H(\tilde{\boldsymbol{\tau}}) \boldsymbol{R}^{-1}(\tilde{\boldsymbol{\tau}}) \boldsymbol{\chi}(\tilde{\boldsymbol{\tau}})\} \tag{3.23}$$

and

$$\hat{\boldsymbol{\alpha}} = \boldsymbol{R}^{-1}(\hat{\boldsymbol{\tau}}) \boldsymbol{\chi}(\hat{\boldsymbol{\tau}}), \tag{3.24}$$

where

$$\boldsymbol{\chi}(\boldsymbol{\tau}) \triangleq \int_0^{T_{ob}} r(t) \begin{bmatrix} g(t - \tau_1) \\ g(t - \tau_2) \\ \cdot \\ \cdot \\ \cdot \\ g(t - \tau_L) \end{bmatrix} dt \tag{3.25}$$

---

[2]The ML estimation can also be performed by adopting super-resolution techniques which operate in the frequency domain [39, 53, 98, 198].

[3]$\mathbf{X}^H$ denotes the conjugate transpose of the matrix $\mathbf{X}$ (Hermitian operator).

is the correlation between the received signal and differently delayed replicas of the transmitted pulse. From (3.23) and (3.24) we note in general that, when the multipath is not resolvable, the estimate of the first path can depend strongly on other channel parameters.

However, when the channel is resolvable, that is, $|\tau_i - \tau_j| > 2T_g \; \forall i \neq j$, (3.23) and (3.24) simplify to [42, 186]

$$\hat{\tau} = \arg\max_{\tilde{\tau}} \left\{ \chi^T(\tilde{\tau}) \, \chi(\tilde{\tau}) \right\} \tag{3.26}$$

$$= \arg\max_{\tilde{\tau}} \left\{ \sum_{l=1}^{L} \left[\int_0^{T_{\mathrm{ob}}} r(t) \, g(t - \tilde{\tau}_l) \, dt \right]^2 \right\} \tag{3.27}$$

and $\hat{\alpha} = \chi(\hat{\tau})$. In this case, the estimation of the TOA of the direct path is decoupled from the estimation of the other channel parameters and is reduced to a simple process involving a single correlator or an MF matched to the pulse $g(t)$, as in the AWGN case.

As already mentioned, the computational complexity of ML estimators limits their implementation. To alleviate this problem, several practical suboptimal TOA estimators have been proposed in the literature. Several techniques will be addressed in Section 3.4.2.2 and Chapter 5.

3.1.3.2 Clock Drift

Time-based ranging requires precise time interval measurements (e.g., with errors of the order of 1 ns or less when centimeter accuracy is required). To this end, nodes are equipped with an oscillator from which an internal clock reference is derived to measure the true time t. Numerous physical effects cause oscillators to experience independent frequency drifts that result in large timing errors. These errors can be particularly significant in systems with poor oscillators (such as in low-cost WSNs). The local time of a clock in a device can be expressed as a function $C(t)$ of the true time t, where $C(t) = t$ for a perfect clock (see Fig. 3.17). As a consequence, only an estimate $\hat{t} = C(t)$ of the true time t is available. For short time intervals, $C(t)$ can be modeled as [161]

$$C(t) = (1 + \delta)t + \mu \tag{3.28}$$

where δ is the *clock drift* relative to the correct rate, and μ is the *clock offset*.[4] Note that the rate $dC(t)/dt$ of a perfect clock is 1 (i.e., $\delta = 0$).

Clock drift affects the time interval measurement. Specifically, if a single node wants to measure a time interval of true duration $\tau = t_2 - t_1$ seconds, then the corresponding estimated value $\hat{\tau}$ is

$$\hat{\tau} = C(t_2) - C(t_1) = \tau (1 + \delta). \tag{3.29}$$

However, a node effectively generates delay

$$\hat{\tau}_\mathrm{d} = \frac{\tau_\mathrm{d}}{1 + \delta} \tag{3.30}$$

whenever it attempts to generate a delay of τ_d seconds. The following subsections describe the effect of clock drift and offset in one-way and two-way ranging protocols introduced in Section 3.1.1.2 when a couple of nodes, A and B, have clock drifts δ_A, δ_B and offsets μ_A, μ_B, respectively.

---

[4]Another measure of clock performance is part per million (ppm), which is defined as the maximum number of extra (or missed) clock counts over a total of 10^6 counts, that is, $\delta \cdot 10^6$.

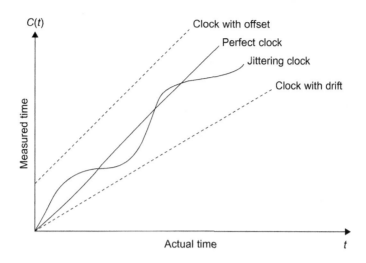

FIGURE 3.17

Relationship between the estimated time and actual time.

Clock Drift in One-Way Ranging Protocols

In one-way ranging, nodes are synchronized to a common time-base by using, for example, a network synchronization protocol [161]. In this case, μ_A and μ_B are the residual offsets caused by imperfect synchronization, with respect to the common time-base, assuming, without loss of generality, that the last network synchronization phase occurred at time $t = 0$.

A packet is transmitted by node A at its local time $C_A(t_1)$, and it is received by node B at its local time $C_B(t_2)$. Node B calculates the estimated propagation delay as[5]

$$\hat{\tau}_f = C_B(t_2) - C_A(t_1) = \tau_f + \delta_B t_2 - \delta_A t_1 + \mu_B - \mu_A. \tag{3.31}$$

Equation (3.31) shows that time estimation is affected both by clock drift and clock offset. However, clock offset could be of the order of several microseconds, and hence ranging accuracy depends highly on network synchronization performance [161]. For this reason, one-way ranging requires stringent network synchronization which may not be feasible in some systems.[6]

Clock Drift in Two-Way Ranging Protocols

In two-way ranging protocols, both accurate delay generation and measurement are important factors in maintaining accurate ranging information. According to (3.30), the effective response delay $\hat{\tau}_d$ generated by node B in the presence of a clock drift δ_B is given by $\hat{\tau}_d = \tau_d/(1 + \delta_B)$, whereas the estimated

---

[5]The time stamp $C_A(t_1)$ is included in the transmitted packet, and hence it is also known by node B.

[6]Network synchronization requirements are reduced when systems use ranging signals, such as ultrasound, with propagation speeds significantly slower than electromagnetic waves. The propagation speed of acoustic waves (≈ 340 m/s) corresponds to propagation delay values that are several orders of magnitude larger than the respective network synchronization errors. This makes one-way ranging techniques attractive for acoustic-based localization systems [67].

RTT denoted by $\hat{\tau}_{RT}$, according to node A's timescale, is

$$\hat{\tau}_{RT} = 2\,\tau_f\,(1+\delta_A) + \frac{\tau_d(1+\delta_A)}{(1+\delta_B)}. \tag{3.32}$$

In the absence of other information, node A can estimate the propagation time $\hat{\tau}_f$ by equating (3.32) with the supposed round-trip time $2\,\hat{\tau}_f + \tau_d$, leading to

$$\hat{\tau}_f = \tau_f\,(1+\delta_A) + \frac{\varepsilon\,\tau_d}{2(1+\delta_A - \varepsilon)}, \tag{3.33}$$

where $\varepsilon \triangleq \delta_A - \delta_B$ is the relative clock offset.

To demonstrate typical ranging errors resulting from the relative clock offset, Fig. 3.18 shows error in TOF estimation $\hat{\tau}_f - \tau_f$ as a function of ε using a typical value $\delta_A = 10^{-5}$ (10 ppm) for different response delays. Figure 3.18 shows that both the relative clock offsets as well as response delay strongly affect ranging accuracy. For example, a target TOF estimation error of 33 ps (about 1 cm) can be satisfied for ε up to 10^{-5} if the response delay τ_d is below 10 μs. For a less pessimistic value of $\varepsilon = 10^{-6}$, the requirement on the maximum response delay relaxes to $\tau_d \approx 100$ μs.

RTT measurement accuracy can be improved by adopting high-precision oscillators (not feasible in low-cost WSNs) or by adopting rapid synchronization techniques at the physical layer [74, 180, 202]. In addition, response packet generation can be implemented at the medium access control (MAC) layer to avoid large delays produced by upper protocol stack layers.

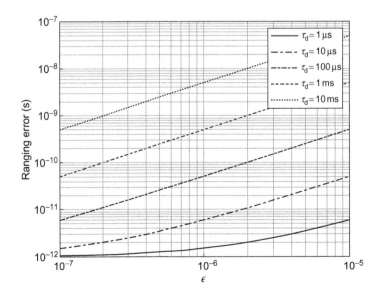

FIGURE 3.18

TOF estimation error in two-way ranging because of relative clock drifts with $\delta_A = 10^{-5}$ and $\tau_f = 100$ ns.

3.1.3.3 *Non-Line-of-Sight (NLOS) Propagation*

All the geometric localization techniques described above are based on the assumption that there exists a DP, commonly referred to as the LOS path, between the transmitter and the receiver. If no such path exists, the signal arrives at the receiver after having been reflected and/or diffracted—possibly more than one time—by one or more objects. The effect of assuming an LOS path in an NLOS propagation environment leads to serious degradation in accuracy.

Even when the DP exists but is partially obstructed by obstacles, an NLOS condition is experienced, and the *DP excess delay* phenomenon incurs by the signal components that must travel through different obstacles like walls in buildings. The propagation time of these components depends not only upon the traveled distance, but also upon the encountered materials. Because the propagation of electromagnetic waves is slower in some materials compared with air, the signal arrives with excess delay (thus with a positive bias in the range estimate).

When the DP to a target node is completely obstructed, then the target receiver will only observe signals from NLOS components, resulting in estimated distances larger than the true distance. An important observation is that the effect of DP blockage and DP excess delay is the same: they both add a positive bias to the true range between ranging devices so that the measured range is larger than the true value.

By neglecting other sources of error, in TOA and TDOA techniques, such a non-negative error term, ϵ_τ, affects the estimated distance as

$$d_i = \sqrt{(x - x_i)^2 + (y - y_i)^2} + c \cdot \epsilon_\tau. \tag{3.34}$$

For AOA positioning, the direction (angle) of the received signal at base station i can be expressed as

$$\tilde{\phi}_i = \phi_i + \epsilon_{\phi_i}, \quad \forall i, \tag{3.35}$$

where ϕ_i is the geometrical angle between the transmitter and the receiver, and ϵ_{ϕ_i} is the NLOS-induced error. $\epsilon_{\phi_i} = 0$ if an LOS path exists. In Fig. 3.19, it can be clearly seen that if the signal arrives at one BS by a reflection, the direction of arrival is misestimated, and the distance error e_d introduced is proportional not only to the angle error ϵ_ϕ but also to the distance d between that BS and the MS. Specifically using the law of sines and having Fig. 3.19 as a reference, we get the following equation for the error of the estimation of BS_1:

$$e_d = \frac{d \sin(\epsilon_{\phi_1})}{\sin(\pi - \tilde{\phi}_1 - \phi_2)}. \tag{3.36}$$

In contrast to other techniques, like TDOA and TOA, little work has been done in improving the accuracy when AOA is used in an NLOS environment (see, e.g., Ref. [192]). This stems from the fact that in the AOA technique, NLOS error can be mitigated only when more than two BSs are used and at least two of them have LOS. Therefore, the great advantage of using the minimum number of just two BSs would be lost.

NLOS Condition Detection

The reliable detection of NLOS propagation is a precursor for any localization algorithm to correct the introduced biases [13, 37, 49, 55, 155, 178, 181, 189].

NLOS condition detection is generally performed by extracting a certain feature from the received waveform that varies with different channel conditions. For example, in Ref. [105], the identification

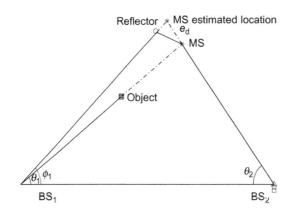

FIGURE 3.19

AOA localization error when there is no LOS path between BS_1 and MS.

is based on the first peak amplitude of the received signal and delay between the first and the strongest path. In Refs. [55, 56], root mean square (RMS) delay-spread, mean excess delay and kurtosis parameters are used for that purpose. The detection can also be conceived without observing the received waveform directly, which is the case of the nonparametric approach proposed in Ref. [49]. This work assumes that multiple and independent TOA measurements between MS and BS are available and that the change in location of the MS during such measurements is negligible. In these conditions, the pdf of distance estimates between MS and BS is obtained from the measurements and is compared with LOS propagation's pdf. If the distance between the pdfs is less than a given threshold, the channel is declared LOS, otherwise it is stated as NLOS. Another recent nonparametric solution based on least-squares support vector machines can be found in Ref. [108].

Most LOS/NLOS identification techniques proposed in the literature can be summarized according to the following classical binary detection scheme, where the detection is performed by extracting a certain number N of features $\gamma = \{\gamma_1, \gamma_2, \ldots, \gamma_N\}$ from the received signal and applying the classical decision theory with a likelihood ratio test (LRT) [35]

$$\frac{p(\gamma|\mathrm{LOS})}{p(\gamma|\mathrm{NLOS})} \underset{\mathcal{H}_1}{\overset{\mathcal{H}_0}{\gtrless}} \frac{p(\mathrm{NLOS})}{p(\mathrm{LOS})}, \tag{3.37}$$

where $p(\gamma|\mathrm{LOS})$ and $p(\gamma|\mathrm{NLOS})$ are, respectively, the joint pdfs of the set of features $\{\gamma_1, \gamma_2, \ldots, \gamma_N\}$ under LOS and NLOS conditions, $p(\mathrm{LOS})$ and $p(\mathrm{NLOS})$ are the prior probabilities of the LOS and NLOS events, respectively, $\mathcal{H}_0$ denotes the hypothesis of an LOS condition, and $\mathcal{H}_1$ denotes the presence of a certain obstruction. Different techniques are then often distinguished by different choice of the set γ of signal features. When more than one parameter is extracted from the signal, for example γ_1 and γ_2, obtaining the joint pdf can be difficult.

A fundamental step in designing (3.37) or (7.8) is the choice of the features γ, extracted by observing the received signal $r(t)$, which are usually more affected by channel conditions. One possibility is the RMS delay spread that captures the temporal energy dispersion of the energy in a signal.

It is defined as

$$\tau_{\text{rms}} = \sqrt{\frac{\int_0^\infty (t - \tau_{\text{m}})^2 |r(t)|^2 dt}{\int_0^\infty |r(t)|^2 dt}}, \tag{3.38}$$

where τ_{m} is the mean excess delay given by

$$\tau_{\text{m}} = \frac{\int_0^\infty t |r(t)|^2 dt}{\int_0^\infty |r(t)|^2 dt}. \tag{3.39}$$

In the case of LOS propagation, the strongest path is typically the first one, whereas in NLOS conditions, it is common to have the strongest path preceded by some other smaller echoes resulting in a larger value of the delay spread. When distributions in (3.37) are unimodal and $p(\text{LOS}) = p(\text{NLOS})$,[7] then (3.37) is equivalent to compare $\gamma = \tau_{\text{rms}}$ with a suitable threshold λ corresponding to the intersection between $p(\gamma | \text{LOS})$ and $p(\gamma | \text{NLOS})$.

Another parameter often considered is the kurtosis, defined by

$$\mathcal{K} = \frac{1}{\sigma_{|r|}^4 T_{\text{ob}}} \int_{T_{\text{ob}}} \left(|r(t)| - \mu_{|r|} \right)^4 dt, \tag{3.40}$$

where $\mu_{|r|} = \frac{1}{T_{\text{ob}}} \int_{T_{\text{ob}}} |r(t)| dt$, $\sigma_{|r|}^2 = \frac{1}{T_{\text{ob}}} \int_{T_{\text{ob}}} (|r(t)| - \mu_{|r|})^2 dt$, and T_{ob} is the observation time. LOS waveforms usually produce a higher value for the kurtosis. For this reason, the decision is taken for NLOS when $\mathcal{K}$ is larger than a certain threshold λ. Parameters τ_{rms} and $\mathcal{K}$ are strongly related to the shape of the waveform.

NLOS Mitigation Methods

Once the LOS/NLOS condition has been properly detected, the most common approach for localizing in NLOS environments is to try to mitigate the NLOS error. Several NLOS mitigation algorithms have been presented in the past; a possible list is reported here below.

- One of the most common approaches consists of correcting the TOA measurements by exploiting the fact that the variance of the TOA measurements is significantly increased in NLOS scenarios [188].
- Works [20, 192] propose methods to discard all BSs which are found to be in NLOS condition.
- In Ref. [76], it is shown that an NLOS measurement carries useful information that might improve the localization accuracy if properly utilized. In Refs. [177, 181], the NLOS information is used to add further constraints in quadratic and linear programming algorithms for NLOS localization.
- Other approaches consist of formulating all possible hypotheses defined by all the possible combinations of the stations under LOS or NLOS condition for a specific instant of time. In Ref. [21], the position is estimated as a linear weighted combination of the partial position estimates associated with each hypothesis, and in Ref. [141], the most likely hypotheses is selected using the ML detection principle. In Ref. [3], the relationship between the timing observations among different snapshots is introduced by applying a trellis search algorithm to track the LOS and NLOS conditions.

---

[7]This assumption is usually considered when no a priori information is available.

- Another method to include NLOS measurements in a localization algorithm is to weight the measurements according to the range estimate accuracy [15, 21, 55, 179].
- Prior knowledge of the environment or cooperation between target nodes are other possible ways to improve the performance in the presence of NLOS conditions [32, 76, 130].
- NLOS can also be dealt with by introducing an appropriate NLOS channel model and using its propagation characteristics to derive new equations that must be satisfied by the MS position's coordinates. In Refs. [5] and [113], single-bounce models (SBM), like the one shown in Fig. 3.20, are efficiently used to improve the accuracy of a location estimate. The single-bounce model describes accurately numerous scenarios, despite the fact that it is very simple. Its wide applicability stems from the fact that in a physical propagation environment, the more bounces, the larger the attenuation will be, not only because the scatterer absorbs some of the signal's energy but also because more bounces usually imply a longer path length. Thus, if a limited number of NLOS signal components with non-negligible energy arrive at the receiver, it is reasonable to assume that they have bounced only once.
- An indirect method to mitigate the effect of NLOS measurements is by introducing memory into the ranging system through the adoption of tracking techniques based, for example, on Bayesian filtering [75, 119].

Some of these techniques will be used in the case studies illustrated in Chapter 5. The reader is referred to Refs. [96, 106, 155] for descriptions of additional NLOS mitigation techniques.

3.1.3.4 *Sources of Error in Fingerprinting Techniques*

The accuracy of the fingerprinting techniques can be greatly affected by either the appropriateness of the chosen statistical channel model, if one is used to create the database, or alternatively by how fast and how much the signal signatures change with time. If the statistical model does not closely match the real propagation environment, the estimated signatures deviate from the real ones, so that during the localization process finding an exact match for the measured signature can be difficult or impossible. If measurements are used instead of modeling, in the creation of the database, the parameters that comprise the signature must be carefully chosen in order to be affected by changes in the environment as little as possible. In addition, the period of the database calibration must be specified and be such that the accuracy does not decrease much from average between two calibrations. In Ref. [87], experiments were performed to evaluate the database's long-term stability. The author defined the long-term time constant τ_{LT} as the time difference Δt for which less than 50% of the pattern's elements are correct in 90% of the cases. For urban environments, he found that τ_{LT} was approximately 42 h, indicating the need for continuous updating of the database.

3.2 POSITIONING IN CELLULAR NETWORKS

Recently, there has been a large interest in exploiting cellular systems to provide positioning services without the necessity to deploy ad hoc and expensive wireless infrastructures. Unfortunately, 2G and 3G cellular standards were designed and optimized having in mind data and voice communication services but not positioning. The simplest method to obtain some row location information is by using the *cell identification (cellID)* as proximity indicator. The localization accuracy will be of the order of the cell size, enough for some applications if picocells are deployed. For larger cells, some more elaborate techniques have to be developed. Potentially, 2G/3G cellular physical layer can provide ranging information through TOA estimation, even though the relatively small bandwidth limits the achievable time resolution (e.g., 1 μs for GSM, about 200 ns for 3G systems).

In the last decade, cellular network standard protocols have allocated resources to carry GNSS assistance data to GNSS-enabled mobile devices, in order to implement assisted GPS (AGPS)/assisted GNSS (AGNSS) services in both GSM and UMTS networks. 3GPP boosted location services in Long-Term Evolution (LTE) release 9, frozen in December 2009 [187]. These aspects have been discussed in Chapter 2, Section 2.3.5.

3.2.1 Positioning and Navigation Approaches

In the following subsections, some geometric and fingerprinting positioning techniques proposed for cellular systems will be explained. The analysis in Sections 3.2.1.1 and 3.2.1.2 is based on the assumption of an ideal error-free environment, unless otherwise stated. The impact of the propagation errors on the techniques and the methods to mitigate them will be described in Section 3.2.1.3.

3.2.1.1 *Geometric Localization Techniques in Cellular Networks*

The main disadvantage of the TDOA and the TOA technique is the so-called hearability problem, that is, the ability of reception of transmitted signals from a sufficient (for localizing) number of BSs. Because cellular wireless systems are designed to minimize the number of BSs receiving a high-strength signal, the lack of signals with adequate SNR can cause a significant variation in the accuracy of these techniques.

In situations where no more than one or two base stations can be used in the localization process, hybrid methods must be applied in order to get the unique location of the mobile station. The work in Ref. [17] introduces a TOA/AOA hybrid scheme that uses only one BS. Assuming the availability of the M-by-L complex valued channel estimation matrix $\mathbf{H}$, where L is the number of taps in the impulse response estimate and M is the number of array elements, the authors suggest premultiplying impulse response matrix $\mathbf{H}$ by an M-by-N_θ matrix of beamformer coefficients $\mathbf{W}$, with N_θ the number of look directions, to obtain the spatial impulse response of the $N_\theta \times L$ antenna array $\mathbf{H}_\theta$:

$$\mathbf{H}_\theta = \mathbf{W}^H \mathbf{H}. \tag{3.41}$$

The squared magnitudes of the entries in $\mathbf{H}_\theta$ represent the received energy as a function of TOA and AOA. Specifically the dominant multipath components are estimated as the N_p highest peaks of the function obtained by taking the squared magnitudes of the entries of $\mathbf{H}_\theta$. Then AOA and TOA estimates are given by the row and column indexes, respectively. This technique, besides having the advantage of using only one BS, achieves relatively good performance. Another interesting approach can be found in Ref. [25], where the TDOA/AOA technique as a hybrid scheme to achieve high location accuracy in a CDMA system is proposed. This method, although requiring a minimum of two BSs, has tremendous advantages over other techniques, mainly because it assumes that AOA is estimated only at the serving base station and there is no need for synchronization as in the TOA/AOA technique.

3.2.1.2 *Fingerprinting Techniques in Cellular Networks*

The signal signature in fingerprinting techniques can be derived from any combination of amplitude, phase, delay, direction, and polarization information of the received signals. Obviously, the more parameters used, the higher the accuracy of the method will be. However, adding parameters also leads to an increase in systems complexity, by requiring more storage space and higher computational power, and in the time needed for the location to be estimated.

A very simplistic approach, presented in Ref. [90], is to use just the received signal strength power which, as already mentioned, is available in almost all wireless systems. Surprisingly, simulations showed that the method has a relatively good performance. The work in Ref. [87] presents the *network measurement report-based* handset localization scheme and uses the framework of the Bayesian network theory to construct the database. The fingerprints considered consist of the power levels received at the MS from different BSs. Thus, the positioning algorithm reduces to a single maximization of a data vector.

A more sophisticated approach is that proposed in Ref. [175], which considers the estimated channel power delay profile (PDP) and also introduces the power spatial delay profile (PSDP) as two alternative fingerprints. Although PDP considers the amplitude and delays of the multipath signals and completely ignores information contained in their phases, the PSDP matrix exploits additional spatial information contained in the phase-shifts. The PSDP matrix is defined as

$$PSDP(\tau) = \mathbb{E}_A \left\{ \mathbf{h}_{x,y}(t,\tau) \mathbf{h}_{x,y}^H(t,\tau) \right\} \tag{3.42}$$

where the received impulse response vector $\mathbf{h}_{x,y}(t,\tau)$ is given by

$$\mathbf{h}_{x,y}(t,\tau) = \left(\sum_{l=1}^{L} A_l(t) \mathbf{a}_l(\theta_l) s(t - t_l) \right) e^{j2\pi f_c \tau}, \quad \in \mathbb{C}^{M \times 1} \tag{3.43}$$

L is the number of multipath components, $s(t)$ is the convolution of the transmit and receive filters, $A_l(t)$ is the complex fading envelope, $\mathbf{a}_l(\theta_l)$ is the vector array response to the lth path on direction θ_l and f_c is the carrier frequency. The expectation is taken over the fading coefficients A_l. Assuming independent fading between paths (3.42) becomes

$$PSDP(\tau) = \sum_{l=1}^{L} \sigma_{A_l}^2 |s(\tau - \tau_l)|^2 \mathbf{a}_l(\theta_l) \mathbf{a}_l^H(\theta_l), \tag{3.44}$$

where $\sigma_{A_l}^2$ is variance of the fading coefficient A_l.

Once the database containing the fingerprints $PDP_{x,y}(\tau)$ or $PSDP_{x,y}(\tau)$ for each location (x,y) of interest is available, during the operative phase, the estimation of the actual MS's position (x,y) is obtained by minimizing a suitable cost function. The cost function, sometimes referred to as *matching score function*, has to be carefully chosen in order to fully take advantage of the fingerprinting technique. For example, in Ref. [175], the cost function for the PDP is defined as

$$C(x,y) = \int_\tau \left\| PDP(\tau) - PDP_{x,y}(\tau) \right\|^2 d\tau, \tag{3.45}$$

which is the Euclidean distance between the estimated PDP and that stored in the database for each location (x,y) of interest.

The solution of ambiguities that may appear when the cost function has more than one minima is an important aspect which might be present in fingerprinting techniques. This problem can be mitigated, for example, by adding more parameters to the signature, measuring the MS signature in more than one BS, or by using tracking motion systems.

3.2.1.3 *Positioning in Cellular Networks in the Presence of Propagation Errors*

As discussed in Section 3.1.3, multipath and NLOS propagation are the main causes of impairment for positioning. In the presence of multipath, high-resolution estimation techniques are required for better accuracy. In Ref. [39], the Root-MUSIC algorithm is used to resolve multipath components that arrive within one chip interval of one another in spread spectrum systems. In Ref. [146], the total least square ESPRIT (TLS-ESPRIT) algorithm is applied to produce unbiased accurate time delay estimations. The author converted the TDOA estimation problem into an estimation of frequencies of complex sinusoids in a white nonstationary noise and then applied the TLS-ESPRIT to estimate the unknown frequencies. Minimum variance (MV) and normalized minimum variance (NMV) approaches were successfully developed and applied to location systems in Refs. [73] and [121], achieving an accurate estimation of the first delay, even in the presence of dense multipath and when the LOS signal is highly attenuated. In the MV and NMV solutions, the a priori knowledge of the number of propagation paths is not required. Efficient implementations of the MV and NMV high-resolution TOA estimators based on FFT computation have been presented in Ref. [89], allowing a significant reduction of the computational cost. Furthermore, new approaches, which transform the traditional grid search of the MV and NMV estimation in a polynomial rooting procedure, are proposed. The polynomial root finding estimators provide less bias and variance than the maximum search ones. In Ref. [83], a concatenation of matched filtering, set-theoretic deconvolution, and autoregressive modeling is used to estimate the parameters of the multipath channel, while in Refs. [47, 48], the background theory of cyclostationary signals is presented and applied to the same problem. Recently in Ref. [134], Qi et al. considered a totally different approach for positioning in multipath environments and derived results for the TOA method. Instead of estimating TOA, based solely on the first arriving signal—which propagates through the LOS path, when one exists—they investigated the enhancement in performance when multipath components are also being processed. They showed that the signal strength of those components and the variance of their delays play an important role in the enhancement.

In the case of NLOS propagation, Ref. [113] proposes a framework to jointly estimate the MS and the L scatterers location, by assuming knowledge of the channel-dependent parameters contained in the vector $\theta = [\phi^T, \psi^T, \mathbf{d}^T]^T$, where the subvectors contain the angle of departure (AOD) ψ, the AOA ϕ, and the path lengths $\mathbf{d}$, for all the N_B different BSs. In their technique, the authors first describe the MS position with a straight-line equation and then consider the LS and the ML algorithms to estimate it. The LS algorithm yields the following solution:

$$(\hat{x}, \hat{y})_{LS} = \arg\min_{(x,y)} \sum_{i=1}^{N_B} (k_i x + b_i - y)^2, \qquad (3.46)$$

where N_B is the number of BSs, and k_i and b_i are given by

$$k_i = \frac{\cos\psi_i + \cos\phi_i}{\sin\psi_i + \sin\phi_i} \qquad (3.47)$$

$$b_i = -k_i(x_i - d\sin\psi_i) + y_i - d\cos\psi_i. \qquad (3.48)$$

For the ML algorithm, all three kinds of estimated parameters (AOD, AOA, distances) are assumed to be Gaussian distributed independent random variables. This yields the ML solution

$$\widehat{\mathbf{x}} = \arg\min_{\mathbf{x}} \{\mathcal{L}\}, \qquad (3.49)$$

where **x** is the vector of the MS and L scatterers locations, $\mathbf{x} \triangleq [(x_{MS}, y_{MS}), (x_{s_1}, y_{s_1}), \ldots, (x_{s_L}, y_{s_L})]^T$, and

$$\mathcal{L} \triangleq \mathcal{L}(\boldsymbol{\theta}(\mathbf{x})) = \frac{1}{2}\left(\widehat{\boldsymbol{\theta}} - \boldsymbol{\theta}(\mathbf{x})\right)^T \mathbf{C}^{-1}\left(\widehat{\boldsymbol{\theta}} - \boldsymbol{\theta}(\mathbf{x})\right), \tag{3.50}$$

where **C** is the covariance matrix of the estimated vector $\widehat{\boldsymbol{\theta}}$. Through simulations of the above two algorithms, the authors showed that the ML algorithm not only outperforms the LS algorithm, but also can achieve optimal RMSE performance, that is, the CRB.

In Ref. [125], the authors investigated how the information about the movement of a mobile terminal can be integrated in traditional geometrical localization techniques to improve their accuracy. Such information is available through the Doppler frequency shifts. Instead of considering a static channel snapshot, they considered a set of adjacent snapshots, separated by a very small amount of time. By assuming linear movement of the MS, the aforementioned amount of information can be utilized in the estimation procedure and only two extra parameters, namely the projections of the speed vector along the two axes, need to be jointly estimated. As shown below, an ML solution was formulated for the case when estimated values for the AOAs, AODs, delays and frequency shifts f_d are available at no more than one BS. The extension to the case when the above quantities have been estimated in more BSs is trivial. With respect to Fig. 3.20, using subscripts i, j, respectively, for the parameters at

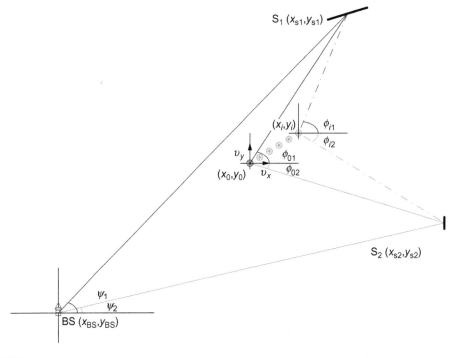

FIGURE 3.20

Single-bounce model.

time instant t_i, $0 \le i < N_t$, and corresponding to path (or scatterer) j, $1 \le j \le L$ and letting $x_0 = x_{MS}(t_0)$, $y_0 = y_{MS}(t_0)$, $dt_{i0} = t_i - t_0$, the channel parameters are given by

$$\phi_{ij} = \begin{cases} \tan^{-1} \dfrac{y_{sj} - (y_0 + \upsilon_y dt_{i0})}{x_{sj} - (x_0 + \upsilon_x dt_{i0})}, & \frac{y_{sj} - (y_0 + \upsilon_y dt_{i0})}{x_{sj} - (x_0 + \upsilon_x dt_{i0})} > 0 \\ \pi + \tan^{-1} \frac{y_{sj} - (y_0 + \upsilon_y dt_{i0})}{x_{sj} - (x_0 + \upsilon_x dt_{i0})}, & \frac{y_{sj} - (y_0 + \upsilon_y dt_{i0})}{x_{sj} - (x_0 + \upsilon_x dt_{i0})} < 0 \end{cases} \tag{3.51}$$

$$\psi_{ij} = \psi_j = \begin{cases} \tan^{-1} \dfrac{y_{sj} - y_{BS}}{x_{sj} - x_{BS}}, & \frac{y_{sj} - y_{BS}}{x_{sj} - x_{BS}} > 0 \\ \pi + \tan^{-1} \frac{y_{sj} - y_{BS}}{x_{sj} - x_{BS}}, & \frac{y_{sj} - y_{BS}}{x_{sj} - x_{BS}} < 0 \end{cases} \tag{3.52}$$

$$d_{ij} = \sqrt{(y_{sj} - (y_0 + \upsilon_y dt_{i0}))^2 + (x_{sj} - (x_0 + \upsilon_x dt_{i0}))^2}$$
$$+ \sqrt{(y_{sj} - y_{BS})^2 + (x_{sj} - x_{BS})^2} \tag{3.53}$$

$$f_{d,ij} = \frac{f_c}{c} \frac{\upsilon_x(x_{sj} - (x_0 + \upsilon_x dt_{i0})) + \upsilon_y(y_{sj} - (y_0 + \upsilon_y dt_{i0}))}{\sqrt{(y_{sj} - (y_0 + \upsilon_y dt_{i0}))^2 + (x_{sj} - (x_0 + \upsilon_x dt_{i0}))^2}} \tag{3.54}$$

The solution is again given by (3.49), (3.50), but the vectors containing the channel-dependent parameters and the parameters to be estimated are in this case $\boldsymbol{\theta} = [\boldsymbol{\phi}^T, \boldsymbol{\psi}^T, \boldsymbol{d}^T, \mathbf{f_d}^T]^T$ and $\mathbf{x} = [x_0, y_0, \upsilon_x, \upsilon_y, x_{s1}, y_{s1}, \ldots, x_{sN_s}, y_{sN_s}]^T$, respectively.

3.3 POSITIONING IN WIRELESS LANs

Nowadays, wireless local area network (WLAN) technology is widely used both in private and in public environments, such as companies, universities, corporations, airports, museums, shopping malls, and so on, providing a ubiquitous network coverage. With the growing adoption of the WLAN technology, lots of interest has focused on the possibility of developing positioning systems based on WLAN RF signals, especially working in in-building environments. At the same time, the research community has also focused on developing services architectures for these location-aware systems, such as navigation in museums, tracking of assets fleet management and other applications. WLAN-based location systems have also been proposed to locate vehicles both in regions without GPS visibility (such as indoor parking areas and tunnels) and where GPS signals are corrupted, typically in urban areas where the effects of urban canyons limit the direct visibility of satellites.

Most WLAN-based location systems are based on RSS measurements because it is easy to implement and no additional hardware is required. Because of the fact that RSS sharply decreases in a nonlinear and unpredicted fashion with distance, many WLAN location algorithms are based on fingerprinting techniques (presented earlier in Section 3.1.2) where a mapping between RSS values and predefined positions has to be created first. An example system using extensive calibration phase is the RADAR system [7], presented in 2000, which was one of the first approaches proposing an indoor positioning system based on WLAN. Another approach is to use the propagation time of radio signals.

TOA increases linearly with distance in free air, but it is difficult to implement in current hardware because it requires high-resolution timing.

The remainder of this section is organized as follows. Sections 3.3.1 and 3.3.2 present the WLAN architecture and the IEEE 802.11a/b/g wireless communications standards, respectively. Finally, Section 3.3.3 goes through positioning and navigation approaches based on WLAN systems.

3.3.1 Architecture of a WLAN

A typical WLAN architecture includes a number of access points (APs), or BSs, and a group of MSs served by those APs. The AP unit serves a small group of users within a typical range of up to a few hundred meters, and additionally stands for the interface between the WLAN and the wired network infrastructure. What is illustrated above (depicted in Fig. 3.21) is the so-called infrastructure network, in which the MSs can communicate with each other only under the direct control of a single AP, similarly to what happens in a cell of a cellular communications network.

The other possible scenario in WLAN applications depicted in Fig. 3.22 is based on the *ad hoc network*, wherein any MS equipped with a wireless networking interface card can establish a direct communications session with any other unit without the requirement of a centralized AP managing all data traffic.

The range within which a wireless connection can be maintained between the AP and MS is limited and can vary depending on indoor or outdoor environmental conditions and the performance that is expected. Typical indoor ranges are 50–100 m, but they get shorter if the building construction interferes with radio transmissions, whereas outdoor ranges can be up to a few kilometers depending on the

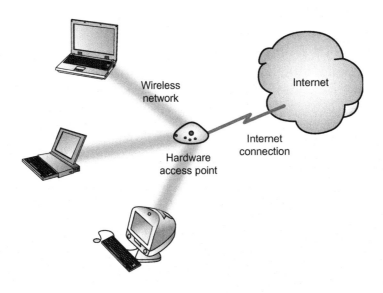

FIGURE 3.21

WLAN infrastructure.

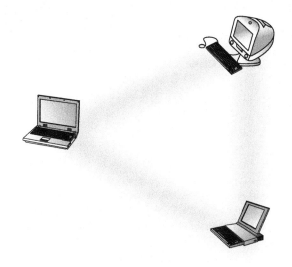

FIGURE 3.22

Ad hoc WLAN.

antennas used, the environment, and the local regulations on emissions. Therefore, when a given area is too large to be covered by a single AP, multiple APs can be used.

3.3.2 IEEE 802.11a/b/g Standards

The first WLAN standard, indicated as 802.11, was approved and released worldwide by IEEE in July 1997. It defines WLAN operations at data rates up to 2 Mb/s in the 2.4–2.4835 GHz, unlicensed, ISM band specifying a common MAC sublayer and three types of physical layer (PHY): frequency hopping spread spectrum (FHSS), direct sequence spread spectrum (DSSS), and infrared [69].

The maximum data rate of 2 Mb/s rapidly turned out to be too low to satisfy the user demand of wireless connectivity, and so in September 1999, the IEEE approved two high-rate standard extensions called IEEE 802.11a and IEEE 802.11b. The 802.11a specifies a WLAN operating in the U-NII 5-GHz band with a maximum data rate of 54 Mb/s using orthogonal frequency division multiplexing (OFDM) modulation. The 802.11b standard, popularly known as Wi-Fi, is defined for operation in the 2.4-GHz band and can achieve up to 11 Mb/s using complementary code keying (CCK) and also DSSS for backward compatibility with 802.11. Also of note is the fact that 802.11b products (resulting from an easier implementation) appeared on the market first, starting in late 1999, and since then have been widely deployed, as shown by the number of Wi-Fi hotspots currently present in residential and public locations, while the first 802.11a-based products were shipped only in early 2002. By moving to the 5-GHz band, the 802.11a standard enables increase of both the maximum data rate per channel (from 11 to 54 Mb/s) and the number of nonoverlapping channels. Indeed, the UNII band is actually made up of three subbands, namely UNII1 (5.15–5.25 GHz), UNII2 (5.25–5.35 GHz), and

Table 3.1 IEEE 802.11 a/b/g: Main Parameters of the PHY Layer

Standard	PHY Layer	Carrier Frequency (GHz)	Data Rate (Mbit/s)
IEEE 802.11	FHSS-DSSS	2.4	1–2
IEEE 802.11a	OFDM	5	Up to 54
IEEE 802.11b	CCK-DSSS	2.4	5.5–11
IEEE 802.11g	OFDM-CCK	2.4	Up to 54

UNII3 (5.725–5.825 GHz), and so up to eight nonoverlapping channels are available when UNII1 and UNII2 are both used, versus three in the 83.5-MHz frequency range of the 2.4-GHz ISM band. A larger total bandwidth of 300 MHz enables a larger number of simultaneous users being supported without problems of interference, but the price to be paid is that the 802.11b and 802.11a standards are not compatible, and the higher operating frequency of 802.11a means a relatively shorter range, or equivalently, a larger number of APs to cover the same area.

In June 2003, the IEEE 802.11 working group released the third high-speed WLAN standard, 802.11g, for operations in the 2.4-GHz band with a data rate up to 54 Mb/s. The same OFDM-based PHY of 802.11a allows higher speeds than 802.11b, and at the same time backward compatibility up to 11 Mb/s is guaranteed using DSSS and CCK modulations. As drawbacks, the 802.11g is limited to three nonoverlapping bands like its predecessor 802.11b, unlike the eight available in the 5-GHz band, and does not alleviate the congestion problem that the 2.4-GHz band currently undergoes. Over the years, the portfolio of IEEE 802.11 standard has been increased by including new versions such as the IEEE 802.11n with the purpose of achieving higher bit rates. The base standard, encompassing all the amendments from 1999 on that defined the versions a, b, d, e, g, h, i, j, is completely defined in the current document IEEE 802.11-2007 [69]. The basic features of the three high-speed WLANs of 802.11 standard described so far, namely 802.11 b/a/g, are summarized in Table 3.1.

3.3.3 Positioning and Navigation Approaches

Section 3.1.2 already analyzed the typical positioning estimation schemes: geometric and mapping techniques. In this section, some applicative approaches will be discussed. Following the notation used so far, the term BSs will be adopted to indicate the infrastructure access points, while the MSs term will refer to the agent node to be located through its RF interface.

3.3.3.1 Position-Related Signal Parameters

In a WLAN, two different measurement architectures can be considered according to system requirements and constraints: *active* and *passive* MSs.

In the active MS architecture (Fig. 3.23(a)), the MS to be localized is the one that broadcasts the stimulus signal, and the BSs are in charge of sensing them. This approach assures the simultaneity of the measurements without synchronization, a key factor that improves the accuracy when the mobile station is in motion. The major drawback is the scalability of a system like this because the channel

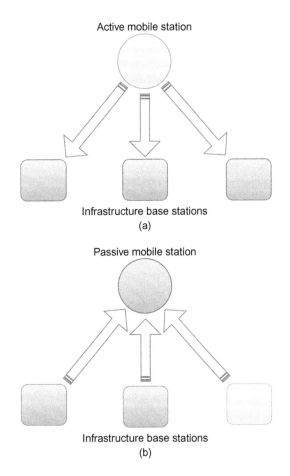

Active mobile station

Infrastructure base stations

(a)

Passive mobile station

Infrastructure base stations

(b)

FIGURE 3.23

Possible measurement architectures. (a) Active MS. (b) Passive MS.

use is proportional to the users on the system. Here a key aspect is the privacy of the MSs because they decide when to send the signal to get localized or not [163].

The passive MS architecture (Fig. 3.23(b)), uses the BSs to send the beacons and each MS receives them. The advantage of this method is its scalability because the channel use is proportional to the number of BSs. However, some accuracy can be lost because the BSs do not transmit the beacons synchronized, and in most cases, the MSs usually have to perform a scan for BSs sequentially on different channels, thus resulting in slow positioning rate [163].

Within WLAN-based positioning systems, the most used signal parameters are RSS, TOA, and TDOA, introduced in Section 3.1.1. Hereafter, how these parameters can be measured using the WLAN technology is described in more detail.

Received Signal Strength (RSS)

RSS is the main measurement quantity used in WLAN positioning systems. Because this signal has to be collected from different BSs which usually are allocated on different channels to avoid interference, three methods can be used in order to measure the RSS.

- *Passive scan*: The MS is configured into *promiscuous mode* and sniffs all the traffic it senses. The drawback of this technique is a possible overload in processing *all* the packets present on the channel. The advantage is that each MS could use the information obtained by receiving the packets coming from other MSs in order to improve the localization accuracy by means, for example, of cooperative localization algorithms.
- *Active scan*: The MS broadcasts a scan request packet, and waits for either the scan response packet or a periodic beacon from the available BSs. This scan is usually performed by the wireless interface's driver, so the actual implementation is subject to the operative system and the hardware manufacturer.
- *Ad hoc packet*: An ad hoc packet that contains specific positioning messages is used. However, this approach requires middleware running on both the BSs and MSs in order to generate/process those messages.

Time of Arrival (TOA)

TOA estimation techniques can be applied in WLAN-based positioning systems if a very accurate timer is used to measure the propagation time between the transmitted and the received RF signal. In fact, because the speed of the electromagnetic wave is about $c \approx 3 \cdot 10^8$ m/s, in order to obtain a distance estimator based on TOA with an accuracy of 1 m, a clock with 3.33 ns of accuracy is needed, which is a very strong requirement. Current WLAN products use hardware time stamp with a resolution of 1 μs, corresponding to 300 m of distance error. Therefore, in terms of the achievable accuracy, this discrete time resolution is not precise enough for ranging. However, time resolution can be increased when averaging numerous observations [71].

In order to avoid the need to synchronize the MS with the BSs, TOA can be estimated from a two-way ranging scheme described earlier by measuring the RTT. As can be seen in Fig. 3.24, RTT is estimated by measuring the time elapsed between two consecutive frames under the IEEE 802.11 standard: a link layer data frame sent by the transmitter (MS) and the reception of the corresponding link layer acknowledgment ACK from the receiver (i.e., the BS). The ready-to-send (RTS) and clear-to-send (CTS) frames of the 802.11 standard can also be used for this purpose. As described in Fig. 3.25, the receiver replies with the ACK message after waiting for a short interframe spacing (SIFS) duration which ensures the highest priority over other data packets.

In Ref. [71], the authors propose to measure the RTT by using the internal clock module (clock frequency $f_c = 44$ MHz) of a WLAN card. The module starts counting cycles when it detects the end of transmission of a data frame, and it stops when the corresponding ACK frame arrives. The RTT is time-variant because of the variability of the radio channel multipath, the clock quantization error, delays due to electronics of the hardware module, and the relative clock drift. In order to mitigate this uncertainty, the RTT is estimated as the average value of different RTT samples.

In the first phase, RTT is estimated by deploying the MS and BS at zero distance. The corresponding measure RTT_0 returns the end-to-end device processing time, which can be removed from the measured

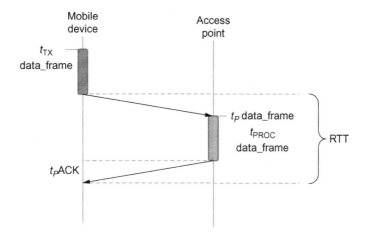

FIGURE 3.24

RTT measurements using IEEE 802.11 data/ACK frames.

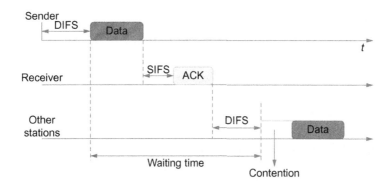

FIGURE 3.25

The IEEE 802.11 protocol.

RTT when the nodes are located at a different distance. Specifically, the correct RTT is

$$RTT = RTT_a - RTT_0, \tag{3.55}$$

where RTT_a is the RTT measured at a certain distance d between the transmitter and receiver which also includes the processing time.

In a real implementation, the RTT can be measured as the average number of clock cycles η divided by the clock frequency f_c. Therefore, the estimated distance d expressed in meters can be obtained as follows:

$$d = c\,(\eta_a - \eta_0)\,/\,(2f_c). \tag{3.56}$$

During the development process carried out in Ref. [71], it was observed that all the distance estimates were longer than the actual distance. Therefore, the corrected distance estimator is the previous one multiplied by an empirical coefficient $k < 1$:

$$d = kc(\eta_a - \eta_0)/(2f_c). \tag{3.57}$$

The empirical coefficient k is justified by the different sources of error commented on before, which can increase the theoretical expected RTT. To estimate k, the linear regression method applied to a set of exact distances and the corresponding set of measured RTT_a can be used. The authors in Ref. [71] obtained $k = 0.694$. After the calibration phase at 0 m, they estimated that the distance error distribution at a fix distance of 10 m fits the Gaussian distribution with mean and standard deviation values of $\mu = 1.12$ m and $\sigma = 0.84$ m, respectively.

Time Difference of Arrival (TDOA)

In Ref. [191], the authors presented a method to estimate the TDOA by using the IEEE 802.11 link layer frames without the need to have the BSs synchronized.

Suppose we have three BSs (BS_0, BS_1, and BS_2) at known locations. Firstly, the BS_0 sends a data frame to the MS at time t_0, then the MS replies with an ACK message after it receives the data. Meanwhile BS_1 and BS_2 monitor the communication between BS_0 and MS and measure the time delays between the arriving time of the data frame and the ACK message, that is, τ_{11} and τ_{21} as shown in Fig. 3.26 and described hereafter.

BS_1 and BS_2 receive the data frame at t_{10} and t_{20}, and ACK message at t_{11} and t_{21}, respectively. The delays τ_{10} and τ_{20} are TOAs from BS_0 to BS_1 and to BS_2, respectively, while the delays τ_1 and

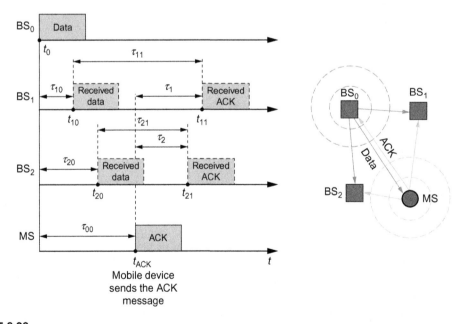

FIGURE 3.26

TDOA method for IEEE 802.11 WLAN.

τ_2 are (unknown) TOAs from MS to BS_1 and to BS_2, respectively. Because the distances from BS_0 to BS_1 and to BS_2 are known, the TOAs τ_{10} and τ_{20} can be accurately estimated. Therefore, the TDOA from MS to BS_1 and BS_2 can be obtained as follows:

$$
\begin{aligned}
\text{TDOA}_{21} &= \tau_2 - \tau_1 \\
&= [(\tau_{20} + \tau_{21}) - \tau_{00}] - [(\tau_{10} + \tau_{11}) - \tau_{00}] \\
&= (\tau_{20} + \tau_{21}) - (\tau_{10} + \tau_{11}) \,.
\end{aligned} \tag{3.58}
$$

Using this method, BS_1 and BS_2 have only to measure time delays τ_{11} and τ_{21}, respectively, that is, the delay between arriving times of data frame and ACK message. The same principle of TDOA measuring method can be used for systems using the optional RTS/CTS mechanism. Because this measurement is a time delay, it is not necessary to have BS_1 and BS_2 synchronized to a common reference time. However, it should be noted that to measure the time delay accurately, a high-precision timer is needed at each BS. In fact, because the chipping rate is 11 MHz for 802.11 WLAN systems, an alignment of the received PN code with the local PN code by using conventional correlation techniques is not sufficient for accurate ranging.

3.3.3.2 *Positioning and Navigation Approaches*

Depending on the availability of RSS or TOA/TDOA measurements, different positioning approaches can be implemented. As remarked in the previous section, obtaining accurate TOA measurement using currently available hardware is not an easy task. For this reason, most localization systems proposed for WLANs rely on RSS measurements. Apart from proximity techniques which can be used when the precision of the system is not very constrained, like in area or room locating systems [41], generally geometric and fingerprinting approaches are preferred.

Geometric Techniques

Low-cost solutions use *lateration*, estimating distances from the measured RSS between the MS and the BSs, based on some propagation channel model, usually the log-normal shadow path loss (3.2) [129]. If additional hardware is available, *angulation* or *hybrid* techniques can be used (see Section 3.1.2.1). Geometric techniques require the knowledge of the BSs' positions, as well as the best fit propagation channel model parameters, found after a previous site analysis or dynamically estimated [67, 101, 151].

Mapping Techniques

Certainly the most popular technique used in WLAN positioning systems is fingerprinting, in which the online locating system usually measures the RSSs and uses them to locate the MS, searching a database to find the k-nearest-neighbor [7, 8] or using an approximated function by a trained neural network [10, 11]. More sophisticated approaches are *probabilistic fingerprinting*, where the location of the MS is estimated as a pdf using statistical search techniques [82, 85].

Navigation systems can also be designed based on WLAN technologies. Basic algorithms used for tracking will be explained in Chapter 6. Dedicated schemes for WLANs can be found in Refs. [82, 85, 86, 159].

3.4 POSITIONING IN WIRELESS SENSOR NETWORKS

The principle of WSNs has been introduced in Section 1.2.3.4 by emphasizing that the potential range of applications, requirements, and network topologies is very wide. This led to the development of a large variety of localization algorithms, each one proposing a different approach and obtaining a different level of performance in order to meet the requirements imposed.

In the following, we first introduce the IEEE 802.15.4 standard, which is the most common physical layer technology adopted in WSNs today, especially in ZigBee networks [180]. We also briefly discuss the IEEE 802.15.4a UWB physical layer because of the relevance this technology has assumed in the last few years, thus gaining the space for an autonomous treatment. Finally, we list some of the more important information sources for positioning that can be exploited at the WSN physical layer.

3.4.1 Physical Layers for WSNs

3.4.1.1 *The IEEE 802.15.4 ZigBee Physical Layer*

ZigBee is a communication technology intended for WSN applications that have relatively low requirements on throughput and latency [1, 180]. The key features of ZigBee are low complexity, low power consumption, and low data rate transmissions, which are to be supported by low-cost stationary or moving devices.

The ZigBee physical layer operates in three different unlicensed bands (and with different modalities) according to the geographical area where the system is deployed. The three frequency band modes, described in detail in Ref. [1], are as follows:

- The 868-MHz band mode, ranging from 868.0 to 868.6 MHz, is used in the European area. This mode adopts a raised-cosine-shaped BPSK modulation format, with DSSS at chip rate of 300 kchip/s. The ideal transmission range, not considering wave reflection, diffraction, and scattering, is approximately 1 km.
- The 915-MHz band mode, ranging between 902 and 928 MHz, is used in the North American and Pacific area. This mode also adopts a raised-cosine-shaped BPSK modulation format, with DSSS, but at chip rate of 600 kchip/s. The ideal transmission range is again approximately 1 km.
- The 2.4-GHz industrial scientific medical (ISM) band mode, which extends from 2400 to 2483.5 MHz, is used worldwide. This mode adopts a half-sine-shaped offset quadrature phase-shift keying (OQPSK) modulation format, with DSSS at 2 Mchip/s. The ideal transmission range is approximately 220 m, that is, significantly less than the other two modes above.

Because the 2.4-GHz band is shared with many other services, the other two available bands can be used as an alternative. The standard allows some devices to operate with both the transmitter and the receiver inactive for more than 99% of time. This low duty cycle allows for a very low energy consumption, and hence an extended battery lifetime.

To overcome the limited transmission range, multihop self-organizing network topologies are required. These can be implemented considering that IEEE 802.15.4 defines two types of devices: the full function device (FFD) and the reduced function device (RFD). The FFD contains the complete set of MAC services and can operate as either a network coordinator or as a simple network device (or "node"). The RFD contains a reduced set of MAC services and can operate only as a network node.

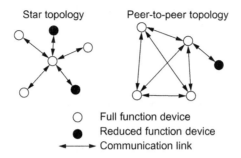

Star topology   Peer-to-peer topology

○ Full function device
● Reduced function device
◄─────► Communication link

FIGURE 3.27

IEEE 802.15.4-compliant star and peer-to-peer network topologies.

Two basic topologies are allowed, but not completely described by the standard because definition of higher layer functionalities is beyond the scope of IEEE 802.15.4. The two topologies are a *star topology*, formed around an FFD acting as a personal area network (PAN) coordinator, which is the only node allowed to form links with more than one device, and a *peer-to-peer topology*, where each device is able to form multiple direct links to other devices so that redundant paths are available. An example of both IEEE 802.15.4-compliant network topologies is shown in Fig. 3.27.

The *star* topology is preferable in the case of small coverage area and if low latencies are required by the application. In this topology, communication is controlled by the PAN coordinator that acts as network master, sending packets (*beacons*) for synchronization and managing device association. Network devices are allowed to communicate only with the PAN coordinator and any FFD may establish its own network by becoming a PAN coordinator according to a predefined policy. A network device wishing to join a star network listens for a beacon message and, after receiving it, can send an *association request* back to the PAN coordinator, which allows the association or not. Star networks also support a nonbeacon-enabled mode. In this case, beacons are used for association purposes only, whereas synchronization is achieved by polling the PAN coordinator for data on a periodic basis. Star networks operate independently of any neighboring networks.

A *peer-to-peer* topology is preferable if a large area needs to be covered and latency is not a critical issue. This topology allows the formation of more complex networks and permits any FFD to communicate with any other FFD beyond its transmission range through multihop communication. Each device in a peer-to-peer structure needs to proactively search for other network devices. Once a device is found, the two devices can exchange parameters to recognize the type of services and features each supports. However, the introduction of multihop requires additional device memory for routing tables.

IEEE 802.15.4 can also support other network topologies, such as cluster, mesh, and tree topologies [78]. These topology options are not part of the IEEE 802.15.4 standard, although the tree topology is described in the ZigBee Alliance specifications [1]. All devices belonging to a particular network, and regardless of the type of topology, use their unique IEEE 64-bit addresses and a short 16-bit address is allocated by the PAN coordinator to uniquely identify the network. The availability of a unique ID for each node simplifies the positioning problem somewhat. The choice of topology, that is, star, mesh, tree, and so on, may have a significant impact on the positioning capabilities of the system.

The IEEE 802.15.4 standard does not directly support ranging between nodes for localization purposes. However, ranging can always be performed using RSS measurements if made available by the physical layer hardware [129, 159], as further discussed below.

3.4.1.2 *Position-Related Parameters*

As in WLAN, the most common position-related parameters used as input measurements to localize the agent node is the RSS. Because of the potential of WSNs to deploy a large number of low-cost nodes, anchor-free localization schemes based on proximity become an interesting option with the potential of achieving discrete localization accuracy as will be discussed in Section 3.4.3.

Interferometry

One innovative source of position information that has shown good results in testbed implementations, at least in outdoor applications, is the so-called interferometric ranging technique. The idea behind interferometric ranging is to utilize two spatially separated transmitters at known positions. They transmit sinusoids at slightly different frequencies $f_1 = f_c + \epsilon$ and $f_2 = f_c - \epsilon$. Because of the slight difference in the transmitted frequencies, the envelope of the received composite signal after band-pass filtering varies slowly over time. The phase offset of this envelope can be measured using cheap low-precision RF chips through the observation of RSS zero crossing events, and it contains information about the difference in distance of the two links. Because there is often no time synchronization between the transmitters and the receiver, and the phase offset also depends on the initial phase of the two transmitted sinusoids, one measured phase offset is of little use. However, if a similar measurement is made by another node in a different location, the difference in phase offset measured at the two receivers depends only on the distance between the first node and transmitters, and the second node and transmitters, that is, four distances. Hence, given a number of "phase-offset-difference" measurements, the unknown locations of the two receivers may be inferred. The approach is illustrated in Fig. 3.28, where $r_C(t)$ and $r_D(t)$ are the received signals at nodes C and D, respectively, and λ_c is the wavelength at the center frequency f_c.

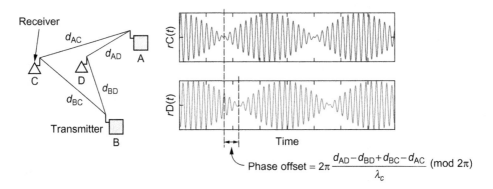

FIGURE 3.28

The interferometric technique.

The main strength of interferometry as a source of position information is that it does not require very sophisticated hardware. In fact, a test-bed implementation of an interferometric outdoor positioning system, described in Ref. [88], has shown that submeter positioning accuracy is achievable using very simple RF circuitry based on RSS measurements. This success is partly because of the fact that the measured envelope, in general, varies slowly over time, which means that (small) synchronization errors and a slow clock drift can be tolerated.

On the down-side, interferometry requires fixed infrastructure and has been shown to perform poorly in severe multipath, for example in indoor environments. Also, positioning based on interferometric data is not straightforward, and positioning algorithms based on interferometry tends to be quite complex. For instance, in Ref. [110], a large-scale genetic algorithm was used.

Sensor Data Measurements

Finally, the data reported by sensors connected to the WSN nodes can be used for positioning purposes. Suppose for instance that sensors are monitoring the temperature throughout an area, and that the temperature variations are spatially smooth. Then, if two sensor nodes measure widely different temperatures, it can be assumed that these sensors are not located close to each other. Similar to connectivity information, this source of information does not require any additional hardware other than the sensors already mounted on the sensor node boards, and only gives coarse-grained position information [127].

3.4.1.3 *The IEEE 802.15.4a UWB Physical Layer*

UWB systems employ pulses of very short durations (subnanosecond) with very-low-power spectral densities. UWB signals are robust to channel fading and have very good time-domain resolution allowing for many location and tracking applications [147]. In addition, they can facilitate design of low-complexity and low-cost transceivers.

UWB signals are characterized by their very large bandwidths relative to those of conventional narrowband/wideband signals [50]. According to the U.S. Federal Communications Commission (FCC), a UWB signal is defined as a signal that has an absolute bandwidth of at least 500 MHz or a fractional (relative) bandwidth larger than 20% [44].

Because UWB signals occupy a very large portion in the spectrum, they must coexist with the incumbent systems without causing considerable interference. Therefore, a set of regulations is imposed on UWB transmissions. The FCC in the United States limits UWB systems to transmit below certain power levels in order not to cause significant interference to the legacy systems in the same frequency spectrum. Specifically, the average power spectral density (PSD) cannot exceed -41.3 dBm/MHz over the frequency band from 3.1 to 10.6 GHz, and it must be even lower outside this band, depending on the specific application [44]. The FCC limits for indoor communications systems are shown in Fig. 3.29 as an example. After the FCC allowed the limited use of UWB signals in the United States, considerable amount of effort has been put into development and standardization of UWB systems [40, 70]. In addition, both Japan and Europe recently allowed the use of UWB systems under certain regulations [81, 169].

Regulation and standardization are major issues for the industrial and commercial future of these technologies. The convoluted story of regulation in Europe has resulted in decisions which allow UWB "generic" emissions in certain bands, within a rather complicated spectrum mask [169]. Two bands appear relevant for real-time locating system (RTLS), which are 3.1–4.8 GHz on one hand and

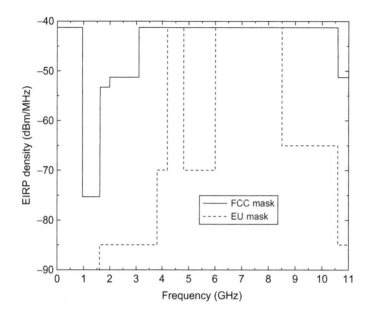

FIGURE 3.29

FCC and EU spectral masks, respectively, for indoor commercial UWB systems in the absence of appropriate mitigation techniques.

6–8.5 GHz on the other. Although the latter allows −41.3 dBm/MHz for the PSD, the former is limited to −70 dBm/MHz at most, unless specific interference limitation techniques are used. These techniques are *low duty cycle* (LDC) for devices emitting on rare occasions, or detect and avoid (DAA) in the general case. DAA requires the UWB device to sense the use of the band by primary services, such as Broadband Wireless Access or Radars in S-band, and apply transmit power control if necessary to protect incumbent services.

A common technique to implement UWB systems is called *impulse radio* (IR) UWB [185]. In an IR UWB system, a number of UWB pulses are transmitted per information symbol and information is usually conveyed by the timings or the polarities of the pulses. In addition to communications, such pulse-based UWB systems can also provide accurate positioning as a result of high-time resolution of UWB pulses. In particular, short-range WSNs, which combine low-to-medium data rate communications with positioning capability, are an emerging application of UWB signals [50]. Positioning applications of UWB WSNs include the following areas [147, 162]: medical (e.g., wireless body area networking), military (personnel tracking), smart homes (e.g., home security and control of home appliances), and inventory control.

Indeed the potentials of UWB systems triggered the formation of the IEEE 802.15 low rate alternative PHY task group (TG4a) in 2004 with the objective of providing an amendment to IEEE 802.15.4 [1] for an alternative physical layer. Compared with the existing IEEE 802.15.4 standard for WSNs, the main interest was in high precision ranging (lower than meter accuracy) and ultra low power consumption. In January 2005, 26 companies turned in proposals to be considered in the standard and,

SYNCH Preamble (16, 64, 1024, 4096) symbols	Start of frame delimiter, SFD (8, 64) symbols	PHY header (PHR)	Data field (PSDU)

FIGURE 3.30

The IEEE 802.15.4a packet structure.

in March 2005, all proposals were merged and a baseline specification was selected. The baseline is constituted by two optional physical layers consisting of a UWB impulse radio (operating in unlicensed UWB spectrum, that is, 250–750 MHz, 3244–4742 MHz, and 5944–10234 MHz) and a chirp spread spectrum (CSS) system (operating in unlicensed 2.4-GHz spectrum) [70]. Due to the large bandwidth, the former will be able to deliver high-precision propagation-delay-based ranging in addition to communications. In particular, the UWB PHY supports an over-the-air mandatory data rate of 851 kbit/s, with optional data rates of 110 kbit/s, 6.81 Mbit/s, and 27.24 Mbit/s. The modulation combines both BPSK and pulse position modulation (PPM) signaling so that both coherent and low-complexity noncoherent receivers can be used to demodulate the signal.

According to the IEEE 802.15.4a specification, the packet structure shown in Fig. 3.30 is exchanged between devices. Each packet consists of a synchronization header (SHR) preamble, a physical layer header (PHR), and a data field. The SHR preamble consists of a ranging preamble and a start of frame delimiter (SFD). The ranging preamble is used for acquisition, channel sounding, and leading edge detection (i.e., TOA), whereas the SFD notifies the end of the preamble [147]. The PHR contains data rate and frame length information, and the data field is the part that carries the communication data. The ranging preamble is the part that is used for range estimation in IEEE 802.15.4a systems. The main ranging protocol adopted by the standard is two-way ranging, but it also enables the use of *TDOA* and *symmetric double sided* (SDS) ranging protocols [63, 147].

3.4.2 Position-Related Signal Parameters Using UWB

Because time-based approaches are quite popular for UWB positioning systems as a result of their relatively low implementation complexity and good accuracy, the remaining part of the section focuses on UWB ranging.

3.4.2.1 *TOA Estimation Algorithms for UWB Signals*

IR-UWB systems can provide centimeter accuracy in ranging and can be exploited in a variety of applications aimed at localizing valuable assets in hospitals, industrial areas, airports, and so on [31, 50].

In an AWGN channel, IR-UWB TOA can be measured within the accuracy of the Cramér–Rao lower bound (CRB) given in (3.5) by correlating the received signal with a template waveform shaped as the transmitted pulse and looking for the time shift of the template that yields the largest correlation, as explained in Section 3.1.1.2. Unfortunately, an AWGN channel is a poor model for a typical indoor environment where propagation takes place with hundreds of distinct trajectories [186] and the received pulses are distorted because of frequency-selective effects resulting from scattering [136] and frequency variations of the antenna pattern [154]. Distortions may affect even the DP if building materials are present between transmitter and receiver [115].

When using UWB signals, the effect of DP excess delay on TOA estimation occurring when the DP is partially obstructed could be particularly relevant. Specifically, when an obstacle is present, the degree of ranging error resulting from DP excess delay can be characterized as follows: An electromagnetic wave traveling in a homogeneous material is slowed down by a factor of $\sqrt{\epsilon_r}$ with respect to the speed of light c, where ϵ_r is the relative electrical permittivity of the material. The theoretical extra delay $\Delta\tau$ introduced by the material with thickness d_{mat} is given by [120, 193]:

$$\Delta\tau = \left(\sqrt{\epsilon_r} - 1\right)\frac{d_{mat}}{c}. \tag{3.59}$$

Experiments have shown that this delay significantly affects ranging accuracy of commercial UWB devices in common office environments. For instance, Ref. [32] shows that the distance estimate bias resulting from excess delay may be of the same order as the thickness of the obstacles, such as walls, between two transmitting nodes. Furthermore, in certain scenarios, excess delay can dominate ranging error even in the presence of multipath propagation.

In summary, the transmission of a single pulse (monocycle) generates a train of echoes at the receiver, with different shapes and random delays over tens of nanoseconds, with the possibility to be partially overlapped that makes TOA estimation in UWB channels a challenging task. Intensive research activity has been carried out in this area.

The estimation of TOA is a particular case of channel estimation, for which the ML solution is known given by (3.22) [103]. However, the ML estimator has strong practical limitations because of the requirements of very high sampling rates and associated computational load. Although different ML approaches have appeared in the literature that manage to reduce complexity considerably [102], there still exist practical limitations for their use in positioning applications.

Reference [93] approaches TOA estimation adopting a simplified version of the generalized ML criterion. The received pulses are considered undistorted, and their common shape is used for correlation with the incoming signal. As the direct path does not need to be the strongest, the estimation process is split into two parts. The TOA of the strongest pulse is found first, looking for the highest peak in the correlator output. If such a pulse is not the earliest that arrives at the receiver, then TOAs and amplitudes of the prior pulses are computed in succession, proceeding backward in time until the last pulse (the one from the direct path) is reached.

However, the use of frequency domain estimation methods has attracted considerable interest for ranging and positioning applications in UWB, motivated by the reduced sampling constraints when applying a channelization approach to the receiver architecture, allowing operation at sub-Nyquist sampling rate. There exist practical implementations, both analog [117] and digital [100], for such receivers front-ends.

The approach in Refs. [50, 195, 196] is based on the idea of a "noisy template." No assumption is made on the pulse shapes, and distortions are implicitly taken into account. The template is derived from delay-and-average operations on the received signal $r(t)$. Correlations with $r(t)$ are taken on suitable windows at symbol distance from each other. Next, the correlations are squared, and their arithmetic mean is computed. TOA is estimated as the windows' position that corresponds to the maximum mean.

Assessing pros and cons of the above methods is a complex task that involves performance comparisons and implementation considerations. In most applications, the goal is to achieve ranging accuracies of a meter with low-power and low-complexity algorithms. In the following, some TOA estimation schemes are illustrated in more detail and some performance comparisons using experimental data are provided. More advanced schemes will be shown in Chapter 5.

Peak Detection-Based Estimators

With reference to Figs. 3.31 and 3.32, we consider three algorithms based on peak detection of the signal after the MF, called *(i) Single Search, (ii) Search and Subtract*, and *(iii) Search, Subtract, and Readjust*, proposed in Ref. [42]. As will be clearer in the following, they are characterized by an increasing level of complexity. These algorithms essentially involve the detection of the N largest positive and negative values of the MF output, where N is the number of paths considered in the search, and the estimation of the corresponding time locations $\hat{\tau}_1, \hat{\tau}_2, \ldots, \hat{\tau}_N$. While these algorithms are equivalent, that is, give the same delay and amplitude estimates when the multipaths are separable, the last two algorithms consider the effects of a nonseparable channel.

Single Search   The delay $\{\hat{\tau}_i\}_{i=1}^N$ and amplitude vectors $\{\alpha_i\}_{i=1}^N$ are estimated with a single look. The delay estimate $\hat{\tau}$ of the TOA of the direct path is set as the minimum of the set $\{\hat{\tau}_i\}_{i=1}^N$.

Search and Subtract   This algorithm provides a way to detect multipath components in a nonseparable channel. First, the sample $\alpha_{\max}$ at time instant $\hat{\tau}_1$ corresponding to the largest peak of the absolute value of the MF output is determined. Then the estimated contribution of the strongest path is subtracted out from the received signal as shown in the example of Fig. 3.33, and a new observation signal is calculated from which the next largest peak is detected. The same process is repeated until the N strongest paths are found. At the end, the minimum of $\{\hat{\tau}_i\}_{i=1}^N$ is selected as the estimate $\hat{\tau}$ of the TOA of the direct path.

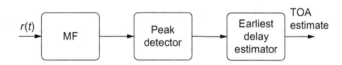

FIGURE 3.31

Single Search TOA estimator.

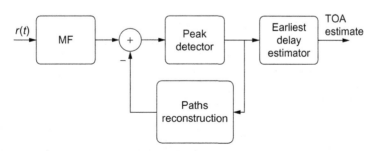

FIGURE 3.32

Search Subtract (and Readjust) TOA estimator.

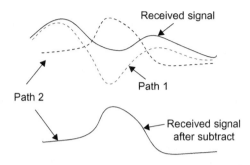

FIGURE 3.33

Path subtracting in the Search and Subtract estimator.

Search Subtract and Readjust　In the previous scheme, the delay and amplitude of each path are estimated separately at each step; here all the amplitudes of the $K \leq N$ paths selected as the strongest ones are jointly estimated at each step using an ML approach, as described in more detail in Ref. [42]. It is worth noting that, when $K = N$, this scheme corresponds to the ML multipath estimator described in Section 3.1.3.1.

In the above three algorithms, the parameter N has to be determined with an optimization process. Moreover, the choice of N also affects the computational complexity of the strategy adopted.

Thresholding-Based Estimator

Probably the simplest TOA estimator is that based on the threshold detection algorithm. It consists of comparing the absolute signal at the output of the MF to a fixed threshold λ. The first threshold crossing event is taken as the estimation $\hat{\tau}$ of the TOA [29]. The choice of threshold is important. With a small threshold, the probability of detecting peaks due to noise, that is, false alarm, and thus estimating the position of an erroneous path that comes earlier than the actual direct path is high, whereas with a large threshold, the probability of missing the direct path and thus estimating the position of an erroneous path that comes later, that is, missed detection, is high. Some criteria for properly set λ are suggested in Ref. [29].

We show now some performance results obtained using the collected data from the UWB experiment illustrated in Ref. [186]. Multipath profiles data were collected in 14 rooms and along the hallways on one floor of the building.[8] In each room, the measurements are made at 49 different points that are located at a fixed height on a 7×7 square grid, covering 90×90 centimeters (cm) with 15 cm spacing between measurement points. With reference to the laboratory environment map reported in Ref. [186], we focus our attention on the measured signals coming from the following four locations:

- Room F1 represents a typical "direct line-of-sight (LOS)" UWB signal transmission environment. The transmitter and the receiver are located in the same room, without any blockage in between.
- Room P represents a typical "high SNR" UWB signal transmission environment. The approximate distance between the transmitter and the receiver is 6 m.

---

[8]A detailed floor plan of the building where the measurement experiment was performed can be viewed in Ref. [186].

- Room H represents a typical "low SNR" UWB signal transmission environment. The approximate distance between the transmitter and the receiver is 10 m.
- Room B represents a typical "extreme-low SNR" UWB signal transmission environment. The approximate distance between the transmitter and the receiver is 17 m.

Figure 3.34 shows some representative examples of the received waveforms measured in the different locations. Table 3.2 shows the numerical results for mean and standard deviation of TOA and distance estimation error obtained from rooms F1, P, H, and B. It summarizes the performance of the four algorithms illustrated above in the optimum operating point of the parameters λ and N.

In general, only slight differences can be observed in the performance of the algorithms. However, it is interesting to note that in the "direct LOS" and "high SNR" cases, the *Threshold* and *Single Search* algorithms give better results than the other algorithms; while, in the "extreme-low SNR" and "low SNR" cases, the *Search and Subtract* and *Search Subtract and Readjust* are superior, probably due to the larger presence of overlapped paths. Thus, when the operating environment is good in terms of SNR, it is sufficient to use the *Threshold and Search* or the *Single Search* algorithms, which have very low complexity. However, in an environment with worse SNR conditions, the *Search and Subtract* and *Search Subtract and Readjust* can be used to reach a reasonable ranging accuracy, in spite of the higher complexity.

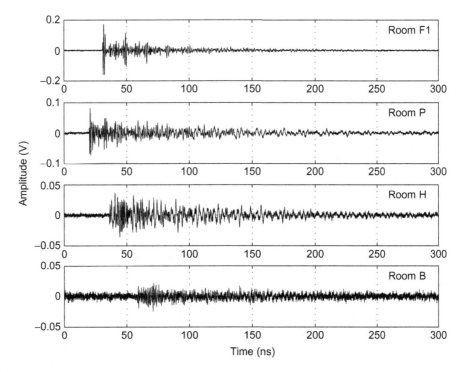

FIGURE 3.34

Multipath profile measured at the center point (4,4) of the grid in room F1, room P, room H, and room B (from Ref. [42]).

Table 3.2 TOA and Distance Estimation Error in Each Room for the Four Algorithms

Room	Algorithm	μ_ϵ [ns]	σ_ϵ [ns]	μ_ρ [cm]	σ_ρ [cm]
F1 $d = 9.49$m	Threshold	−0.10	0.15	−3.0	5.7
	Single Search	−0.045	0.19	−1.3	5.7
	Search and Subtract	0.17	0.24	5.1	7.2
	Search Subtract and Readjust	0.42	0.20	12.6	5.6
P $d = 5.77$m	Threshold	−0.082	0.20	−2.46	5.6
	Single Search	−0.10	0.25	−3.0	7.5
	Search and Subtract	0.22	0.25	6.6	7.5
	Search Subtract and Readjust	0.24	0.24	7.2	7.2
H $d = 10.13$m	Threshold	−0.20	0.43	−5.6	12.9
	Single Search	−0.38	0.44	−11.4	13.2
	Search and Subtract	0.11	0.30	3.3	9.0
	Search Subtract and Readjust	0.086	0.34	2.6	10.2
B $d = 16.91$m	Threshold	−0.17	0.56	−5.1	16.8
	Single Search	−0.19	0.66	−5.7	19.8
	Search and Subtract	−0.11	0.40	−3.3	12.0
	Search Subtract and Readjust	0.013	0.45	0.39	13.5

3.4.2.2 *Energy Detection-Based TOA Estimation*

A primary barrier to the implementation of ML- or MF-based estimators is that they usually require implementation at the Nyquist sampling rate or higher, and these sampling rates can be difficult to achieve because of the large bandwidth of UWB signals. When dealing with complexity, a common warning is that a digital approach involves ADC converters operating in the multi-GHz range. Their usage in low-power operations is problematic, especially if high-bit resolutions are required. However, an analog approach may be challenging.

As an alternative, subsampling TOA estimation schemes based on *energy detector* (ED) have recently received significant attention [22, 27, 28, 31, 33, 54, 57–60, 62, 139, 148]. These TOA estimation techniques rely on the energy collected at sub-Nyquist rate over several time slots. Indeed, they do not require expensive pulse-shape estimation algorithms and represent a good solution for low-power and low-complexity systems.

With the aim of illustrating the main practical TOA ED estimators that have been proposed in the literature, we outline a typical signal structure allowing the coexistence of several users. The following constraints are placed into the problem statement: without loss of generality, we consider the transmission of a preamble from the generic user u at the beginning of each packet. The preamble, for acquisition, synchronization, and ranging, consists of N_{sym} symbols each with duration T_s. Each symbol is an unmodulated time-hopping (TH) signal and divided in time intervals T_f called *frames*, which are further decomposed into smaller time slots T_c called *chips*. A single unitary energy pulse $g(t)$, with duration $T_g < T_c$, is transmitted in each frame in a position specified by a user-specific pseudorandom

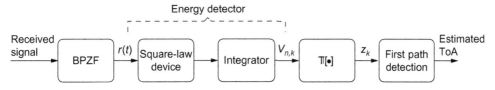

FIGURE 3.35

Threshold-based TOA estimator with ED scheme.

TH sequence $\left\{c_k^{(u)}\right\}$ having period N_s, that is, the number of frames per symbol [184]. Hence the preamble is composed of $N_t = N_{\text{sym}} \cdot N_s$ pulses. Without loss of generality, $u = 1$ denotes the desired user.

A typical ED-based TOA estimator scheme is shown in Fig. 3.35. The received signal is first passed through a band-pass zonal filter (BPZF) with bandwidth B_W and center frequency f_c around the signal band to eliminate the out-of-band noise. The output of BPZF can be written as

$$r(t) = s(t) + d(t) + n(t), \tag{3.60}$$

where

$$s(t) = \sum_{m=0}^{N_t-1} w^{(1)}\left(t - c_m^{(1)} T_c - m T_f\right) \tag{3.61}$$

and

$$w^{(u)}(t) = \sqrt{\frac{E_s^{(u)}}{N_s}} \sum_{l=1}^{L} \alpha_l^{(u)} g\left(t - \tau_l^{(u)}\right), \tag{3.62}$$

where L is the number of multipath components, and where $\left\{\tau_1^{(u)}, \tau_2^{(u)}, \ldots, \tau_L^{(u)}, \alpha_1^{(u)}, \alpha_2^{(u)}, \ldots, \alpha_L^{(u)}\right\}$ is a set of parameters composed of delays $\tau_l^{(u)}$ and path gains $\alpha_l^{(u)}$ related to the uth user. The TOA to be estimated is $\tau \triangleq \tau_1^{(1)}$. If the normalization $\sum_{l=1}^{L} \mathbb{E}\left\{\left[\alpha_l^{(u)}\right]^2\right\} = 1$ for all users is considered, then $E_s^{(u)}$ represents the average received symbol energy per user. The component $d(t)$ in (3.60) represents the interfering term eventually present whose expression depends on the nature of the interference, and $n(t)$ is an AWGN component.

We assume that the receiver has acquired the sequence of the desired user to estimate the TOA τ of the direct path associated with the first user based on the observation of the received signal $r(t)$.[9] The estimator utilizes a section of $r(t)$ consisting of N_t subintervals $I_m \triangleq [c_m^{(1)} T_c + m T_f, c_m^{(1)} T_c + m T_f + T_{\text{ob}}]$, each of duration $T_{\text{ob}} < T_f$, with $m = 0, 1, \ldots, N_t - 1$.

The observed signal forms the input to the ED, whose output is sampled at every T_{int} seconds (integration time), thus $K = \lfloor T_{\text{ob}}/T_{\text{int}} \rfloor$ samples (with indexes $[0, 1, \ldots, K - 1]$ corresponding to K time

---

[9]The IEEE 802.15.4a standard proposes, for robust sequence acquisition, a preamble of only amplitude modulated pulses (i.e., $c_m = 0$) with a length-31 ternary sequence [95], which is characterized by an ideal periodic autocorrelation function, to support both coherent and ED TOA estimators.

slots) are collected in each subinterval. The true TOA τ is contained in the time slot $n_{TOA} = \lfloor \tau/T_{int} \rfloor$. In the absence of other information, the system can assume that τ is uniformly distributed in the interval $[0, T_a)$, with $T_a < T_{ob}$; as a consequence, the discrete random variable (RV) n_{TOA} is uniformly distributed on the integers $0, 1, \ldots, N_{TOA} - 1$, where $N_{TOA} = \lfloor T_a/T_{int} \rfloor$. Note that the interval corresponding to the first n_{TOA} energy samples contains noise and possibly interference (called the *noise region*), whereas the interval corresponding to the remaining $K - n_{TOA} + 1$ samples may also contain echoes of the useful signal (the *multipath region*), in addition to the noise and interference. Because of the presence of the ED, the integration time T_{int} determines the resolution in estimating the TOA, and thus, the minimum achievable RMSE on TOA estimation is given by $T_{int}/\sqrt{12}$.

The $N_t \cdot K$ collected samples (N_t intervals with K samples each) at the output of the ED can be arranged, for further processing convenience as suggested in Ref. [148], to form an $N_t \times K$ matrix $\mathbf{V}$ with elements

$$v_{m,k} = [\mathbf{V}]_{m,k} = \int_{kT_{int}}^{(k+1)T_{int}} |r_m(t)|^2 dt \quad m = 0, \ldots, N_t - 1; \quad k = 0, \ldots, K - 1, \tag{3.63}$$

where $r_m(t)$ is the portion of the received signal after the dehopping process in the subinterval I_m.

Prior to the TOA estimator, the collected samples can be preprocessed or filtered with the aim of reducing the interference effects and to improve the detection of the first path [33, 148]. This can be carried out by introducing a generic transformation $\mathbb{T}[\cdot]$ (see Fig. 3.35), whose output $\mathbf{z} = \{z_k\}$ is a vector used by the TOA estimator for determining the TOA of the first path. In general,

$$z_k = \mathbb{T}[\{v_{m,k}\}], \quad k = 0, \ldots, K - 1, \tag{3.64}$$

where $\mathbb{T}[\cdot]$ can be either a 2D linear or a nonlinear transformation. A variety of filtering schemes can be adopted. The conventional way to obtain the decision vector $\mathbf{z}$ is through simple column averaging (*averaging filter*), that is [33],

$$z_k = \sum_{m=0}^{N_t-1} v_{m,k} \quad k = 0, \ldots, K - 1. \tag{3.65}$$

In the following, we describe some techniques recently proposed in the literature to detect the first path from $\mathbf{z}$. For completeness, further techniques can be found in Refs. [26, 91, 95, 109, 111, 135, 167, 170–173, 194, 200, 201].

First Path Detection
1. **Max**

    The Max criteria is based on the selection of the strongest sample in $\mathbf{z}$ [167]. If the time index corresponding to this sample is k_{max} (see Fig. 3.36), the TOA of the received signal is estimated as

    $$\hat{\tau} = T_{int} \cdot k_{max} + \frac{T_{int}}{2}. \tag{3.66}$$

    This criterion has the advantage of not requiring extra parameters to be tuned accordingly to the received signal (channel model and/or noise level). However, as will be clear later, it suffers from performance degradation when NLOS propagation reduces the energy of the first path so that the strongest path is not necessarily the first.

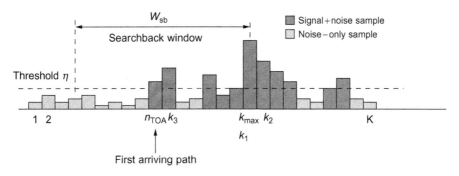

FIGURE 3.36

Illustration of the Max, *P*-Max, simple thresholding, JBSF, SBS, and SBSMC algorithms (from Ref. [31]).

2. Peak-Max

   The Peak-Max (*P*-Max) criterion is based on the selection of the earliest sample among the *P* strongest in **z**. It represents the discrete time version of the Max Search algorithm presented earlier. If the time indices corresponding to these samples are given by $k_1, k_2, \ldots k_P$ (see Fig. 3.36), the TOA of the received signal is estimated as [42]

$$\hat{\tau} = T_{\text{int}} \cdot \min_i \{k_i\} + \frac{T_{\text{int}}}{2}. \tag{3.67}$$

   The TOA estimation performance depends on the parameter *P*, which has to be optimized according to channel characteristics.

3. Simple thresholding

   The simple thresholding (ST) takes an estimate of n_{TOA}, and hence τ, by comparing each element of **z** within the observation interval with a fixed threshold *h* [28]. For further convenience, we define the threshold-to-noise ratio $\text{TNR} \triangleq \lambda/N_0$, and so the TOA estimate is taken as the first threshold crossing event (see Fig. 3.36). Note that this threshold value can be optimized according to operating conditions and channel statistics. The main advantage of simple thresholding is that it can be implemented completely in analog hardware; such implementation is particularly attractive for low-cost battery devices such as WSNs [28, 29, 34, 94, 104, 180].

   The performance of this detection process and of this TOA estimator depends on the choice of the threshold λ through the threshold-to-noise ratio (TNR). In Ref. [29], a simple criterion to determine the suitable threshold based on the evaluation of the probability of early detection is proposed. The probability of early detection P_{ed} is given by

$$P_{\text{ed}} = 1 + \frac{(1 - q_0)^{N_{\text{TOA}}} - 1}{N_{\text{TOA}} q_0}, \tag{3.68}$$

where

$$q_0 \triangleq \exp(-\text{TNR}) \sum_{i=0}^{M/2-1} \frac{\text{TNR}^i}{i!} \tag{3.69}$$

and $M = 2N_s T_{int} B_W$. This expression can be used to evaluate the TNR corresponding to a target P_{ed}. In Ref. [29], it is shown that the evaluation of TNR through P_{ed} does not lead to significant performance degradation with the main advantage that it does not require any prior channel knowledge.

4. Jump back and search forward (JBSF)

The jump back and search forward (JBSF) TOA estimation scheme is based on the detection of the strongest sample and a forward search procedure [57]. In particular, it assumes that the receiver is synchronized to the strongest path and the leading edge of the signal is searched element-by-element in a window of length W_{sb} samples preceding the strongest one. The search begins from the sample in **z**, with index $k_{max} - W_{sb}$, and the search proceeds forward until the sample-under-test is above a threshold h within the window (see Fig. 3.36). Formally, the TOA of the received signal is estimated as

$$\hat{\tau} = T_{int} \cdot \min\left\{k \in \{k_{max}, k_{max} - 1, \ldots, k_{max} - W_{sb}\} | z_k > h\right\} + \frac{T_{int}}{2}. \tag{3.70}$$

Note that the optimal selection of W_{sb} and h depends on channel characteristics as well as SNR, both of which are usually unknown a priori.

5. Serial backward search (SBS)

In Ref. [57], other TOA estimation schemes based on the detection of the strongest sample and a search-back procedure are proposed. One scheme is the serial backward search (SBS) where the leading edge of the signal is backward searched element-by-element. The search begins from the strongest sample in **z**, with index k_{max}, and the search proceeds sequentially backward until the sample-under-test goes below a threshold h within a search-back window of length W_{sb} (see Fig. 3.36). Formally, the TOA of the received signal is estimated as

$$\hat{\tau} = T_{int} \cdot \max\left\{k \in \{k_{max}, k_{max} - 1, \ldots, k_{max} - W_{sb}\} | z_k > h \text{ and } z_{k-1} < h\right\} + \frac{T_{int}}{2}. \tag{3.71}$$

Some criteria to calculate suitable values of h and W_{sb} are presented in Ref. [57].

The received multipath components in typical UWB channels usually arrive at the receiver in multiple clusters that are separated by noise-only samples. In this case, the SBS algorithm may lock to a sample that arrives later than the leading edge. The clustering problem may be handled by allowing a number of D consecutive occurrences of noise samples while continuing the backward search [61]. This is the key idea implemented in the serial backward search for multiple clusters (SBSMC) algorithm where the leading edge estimation is then modified as follows:

$$\hat{\tau} = T_{int} \cdot \max\left\{k \in \{k_{max}, \ldots, k_{max} - W_{sb}\} | z_k > h \right. \tag{3.72}$$

$$\left. \cap \max\{z_{k-1}, \ldots, z_{max\{k-D, k_{max} - W_{sb}\}}\} < h\right\} + \frac{T_{int}}{2}. \tag{3.73}$$

In Ref. [61], some criteria to calculate suitable values of h, W_{sb}, and D are presented.

To compare the performance of the discussed schemes, we consider dense multipath channels based on the IEEE 802.15.4a channel models [116] and root raised cosine/radio resource control (RRC) bandpass pulse with $f_c = 4\,GHz$ and bandwidth $W = 1.6\,GHz$. The other system parameters are $T_f = 128\,ns$, $T_{ob} = 120\,ns$, $T_a = 100\,ns$, $T_{int} = 2\,ns$, and the RMSE is evaluated through Monte Carlo simulation.

The corresponding RMSE in distance estimation is obtained by multiplying that of TOA by the speed of light c.

We compare the performance of the ED-based TOA estimators described previously. Based on the pulse's frequency spectrum, we adopt an ideal BPZF with bandwidth $W = 1.6$ GHz. From Ref. [31], Fig. 3.37 shows the RMSE as a function of the SNR for Max, P-Max, ST, JBSF, SBS, and SBSMC. As already mentioned, all these techniques require the proper tuning of parameters (e.g., h, W_{sb}, D, or P) accordingly to channel conditions. In the simple thresholding scheme, this problem can be avoided by adopting the suboptimal criteria given in (3.68) by setting $P_{ed} = 10^{-4}$ [29]. A blind approach, where the noise power knowledge is not required and there are no parameters to set, has recently been proposed based on model order selection criteria [203]. Note that the Max and P-Max schemes exhibit an asymptotic floor higher than the other techniques. In addition, the maximum number of peaks P to be considered is a critical parameter to be tuned. The SBS and SBSMC estimation algorithms offer better performance in the intermediate SNR region, but even with optimal parameters, they exhibit an error floor for high SNR values. The best performance is obtained with either the ST or JBSF scheme. For large SNRs, they achieve the theoretical performance limit for ED-based TOA estimators, that is, $T_{int}/\sqrt{12} = 0.57$ ns.

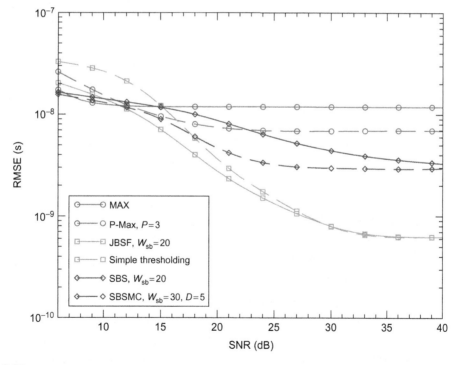

FIGURE 3.37

RMSE as a function of SNR for different ED-based TOA estimation schemes with $N_{sym} = 400$ and $N_s = 4$ (from Ref. [31]).

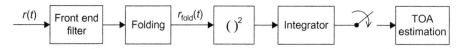

FIGURE 3.38

Energy-based TOA estimation folding scheme proposed in Ref. [27].

Folding Preprocessing

A different approach is taken in Ref. [27], in which an ED TOA estimation algorithm is obtained as the solution of an LS problem. The resulting scheme is depicted in Fig. 3.38, where $r_{fold}(t)$ is the sum of delayed versions of $r(t)$. It is seen that the algorithm proposed in Ref. [27] involves a preprocessing (the folding operation) of the received signal followed by energy measurements. This is in contrast to the scheme in Fig. 3.36 where the measurements are directly made on $r(t)$. The advantage of the preprocessing phase is that the noise component is reduced while the signal component keeps the same timing information, so that the estimation accuracy is enhanced. The implementation of the delays must be very accurate, say within a fraction of the pulse width. As this accuracy cannot be achieved with analog delay lines, a digital implementation of the estimator is proposed.

3.4.2.3 *AOA Estimation Algorithms*

In addition to TOA, UWB systems can also provide accurate AOA information. The estimation of the AOA in UWB systems has been addressed in Ref. [80] considering subspace techniques, which provide accurate estimates. The subspace approach considered involves an eigenvalue decomposition as well as requiring focalization (focusing) techniques to address array processing when dealing with wideband signals. The method is of high complexity and incurs large estimation errors in rich multipath scenarios, as is the case for UWB signals. Frequency domain approaches for direction finding in UWB were considered in Ref. [100]. In this case, the output of the array is split into multiple frequency bands and then down-converted into a much lower frequency to alleviate sampling requirements. Then, the frequency samples of received signal are used to compute a projection matrix and AOA is estimated by solving a minimization problem in the projected noise subspace. Besides the complexity involved in subspace methods, the AOA estimator reported difficulties in providing accurate estimates in a multipath environment. Other approaches that consider the estimation of the AOA based on temporal delays can be found [131], where the authors propose to estimate the AOA by evaluating the propagation delays impinging from each element in the array. This approach suffers from the constraints associated with high sampling requirements when the TOA estimates are obtained based on a time domain correlation approach. In Ref. [114], the use of a low-complexity frequency domain approach for joint high-resolution estimation of TOA and AOA has been proposed. AOA estimates are obtained from TOA estimates whose high temporal resolution yields accurate estimates.

3.4.3 Positioning Approaches for WSNs

The positioning problem in WSNs can be coarsely divided into absolute and relative coordinate (anchor-free) estimation problems, the difference being whether a global coordinate system can be established, for instance, using fixed reference nodes (absolute positioning), or not (relative

Table 3.3 Main Positioning Approaches for WSNs

Method	Acronyms	Basic Approach	References
Absolute Coordinate Estimation			
Least-squares, weighted least-squares, maximum likelihood	LS, WLS, ML	Coordinates are estimated by (often numerical) optimization of an objective function derived from a model of available data	[79, 128, 144]
Projections onto convex sets	POCS, Min–Max	Coordinates are estimated by finding the intersection of a number of convex sets	[12, 144, 150]
Multihop position estimation	e.g., N-hop	Nodes cooperatively estimate coordinates using multihop communication. Incremental, inference-based and semi-definite programming algorithms	[92, 123, 149, 150, 156, 158, 190]
Range-free localization	e.g., DV-hop	Position information is inferred from connectivity or proximity information	[14, 38, 123]
Relative Coordinate Estimation			
Anchor-free localization	AFL	Two-phase approach consisting of an initialization followed by a "force-directed" relaxation	[132]
Robust quadrilaterals	ORQ, ERQ, RLE	A network map is stitched together by pieces (local coordinate estimates) called "robust quadrilaterals"	[118, 165, 197]

positioning). The high degree of potential topology configurations led to a variegate range of positioning algorithms for WSNs over the last few years. Below, we describe the main approaches developed for positioning in WSNs. The discussed algorithms are briefly summarized in Table 3.3.

The four principal approaches that derive the agent's coordinates in the presence of fixed reference nodes are as follows (see Table 3.3):

- LS family
- Projections onto convex sets (POCS) methods
- Multihop methods
- Range-free methods.

They are presented in the following paragraphs.

Weighted Least-Squares

The most commonly occurring estimators of absolute coordinates in the WSN positioning literature are based on weighted least-squares (WLS). If a Gaussian measurement error model is assumed, the WLS estimator coincides with the ML estimator [79, 128]. Assume we have access to a noisy vector $\hat{\mathbf{d}} = \mathbf{d}(\mathbf{x}) + \mathbf{n} \in \mathbb{R}^M$ of distance estimates between nodes in a WSN, where M is the number of measurements. The corresponding WLS estimate of coordinates **x** of the agent is

$$\hat{\mathbf{x}} = \arg\min_{\mathbf{x}} \left\{ [\hat{\mathbf{d}} - \mathbf{d}(\mathbf{x})]^T \mathbf{W}^{-1}[\hat{\mathbf{d}} - \mathbf{d}(\mathbf{x})] \right\}, \tag{3.74}$$

where **W** is a weighting matrix. In the case of Gaussian noise, it is often taken as an estimate of the noise covariance matrix **V**.

On the upside, this approach allows us to do positioning in networks with an unlimited number of agents, as long as we have access to a sufficiently large number of measurements. Different types of measurements, for example TOF, RSS, and so on, are easily incorporated in this solution strategy. Further, the estimation of additional unknown parameters, for example nuisance parameters such as clock offsets, is straightforward.

On the downside, Eq. (3.74) represents a nonlinear optimization problem that cannot, in general, be solved analytically. In most cases, numerical optimization techniques that often only approximate the WLS estimate are necessary. Further complicating the optimization procedure is the common occurrence of local minima, to which numerical optimization algorithms may erroneously converge. This is exemplified in Fig. 3.39, where an example objective function taken from Ref. [144] has been plotted against the coordinates of one agent. Another issue related to the practical implementation of WLS-type coordinate estimators is the complexity associated with the numerical minimization of (3.74), which can grow quickly with an increasing number of agents, especially if **W** is not a diagonal matrix. Hence, scalability is sometimes an issue with WLS-type coordinate estimators.

Projections onto Convex Sets

Unlike the WLS estimator in (3.74) that requires a global optimization of a least-squares type objective function, POCS sequentially projects a point in the space onto convex sets, formed from, for example, distance data, until a point in the intersection of sets has been found. This point is then taken as an estimate of the unknown node coordinates. If distance estimates are available, then convex discs can be used in POCS, or spheres if the problem is three-dimensional (3D). Hyperbolas, ellipsoids, and cuboids

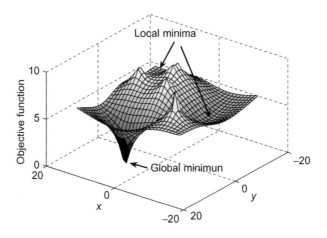

FIGURE 3.39

A WLS objective function with multiple local minima, taken from Ref. [144].

are some examples of other convex sets that have been used together with POCS for positioning. Algorithms based on POCS are often of very low complexity and can easily be distributed over the network. Note that a complete description of the POCS methodology for the positioning problem is available in Chapter 5.

One of the main advantages with POCS, when used in a network with fixed reference nodes, is a resilience to the problem with local minima in the WLS objective function [12]. Another interesting property of POCS-based algorithms that uses projections onto discs is a robustness to overestimated distances, for example due to NLOS or multipath propagation in TOF-based wireless ranging. This robustness to overestimated distances is because of the fact that an overestimated distance gives an oversized disc, and because the projection of a point already on the disc onto the disc is the point itself, such oversized discs will not affect POCS iterations [144]. The main drawback with POCS when used only with discs is that the agent must be located inside the reference node perimeter, or convex hull. If this is not the case, then the intersection of discs will be large, and consequently, the estimation error will also be large.

An example of POCS iterations on discs with one overestimated distance is shown in Fig. 3.40. Note that in this scenario, regardless of where in $\mathbb{R}^2$ we initialize POCS iterations, the algorithm will

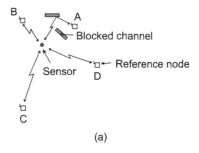

(a)

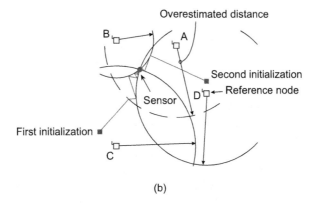

(b)

FIGURE 3.40

2D example of circular POCS where one link distance has been overestimated (dashed line). (a) Network layout. (b) POCS iterations.

always converge to the coordinates of the sensor node, cf. the two initialization points marked in Fig. 3.40. The benefits and drawbacks of the POCS algorithm are discussed in more detail in Chapter 5 and in Refs. [12, 144].

3.4.3.1 *Multihop Position Estimation*

In many situations, some nodes in a network may not be able to interact with a sufficient number of reference nodes (at least three are needed for a two-dimensional fix). As a consequence, cooperation among nodes may be required to estimate node positions through *multihop cooperative* localization algorithms.

An overview of fundamental cooperative localization algorithms based on the principles of estimation theory and statistical inference can be found in Ref. [160], where a framework for the systematic design of inference algorithms, capitalizing on the theory of factor graphs and the sum–product algorithm, is developed. Advanced cooperative localization techniques will be presented in Chapter 5.

A quantitative comparison between several multihop cooperative algorithms is reported in Ref. [92], where a common three-phase structure is identified. During the first phase, agents determine in a multihop fashion their distance to anchor nodes. In the next phase, each node derives an estimate of its position from determined distances using some multilateration scheme (e.g., WLS) [78]. In phase three, each estimated node position is refined using information about the distance to, and positions of, neighboring nodes.

The *N-hop multilateration* algorithm was introduced by Savvides et al. [150]. In this algorithm, the distances to the anchor nodes are approximated by adding the ranges encountered at each hop during a network flooding. In particular, the anchor nodes send a beacon message including their identity, position, and path length accumulator set to 0. Each receiving node adds the measured range from the previous node to the path length field and broadcasts the new message to the other nodes. If multiple messages about the same anchor node are received, the node keeps and forwards only the one containing the minimum value of path length. One of the main disadvantages of this approach is that ranging errors tend to accumulate over multiple hops. This cumulative error can become significant in large networks with few reference nodes or poor ranging hardware (e.g., based on RSS measurements).

Once distances to a sufficiently large number of anchor nodes have been estimated, any multilateration scheme, such as the WLS method mentioned above, can be used. However, we again emphasize that solving LS equations (3.74) is quite expensive because complex matrix floating point operations are required that often are not available in typical WSN devices.

A much simpler method, presented as a part of the *N*-hop multilateration algorithm in Ref. [150], is the *Min–Max* approach. The idea behind this approach is to construct a bounding box starting from each anchor located at known position (x_i, y_i) and distance measurement d_i (see Fig. 3.41). In particular, the bounding box corners of node i are

$$(x_i - d_i, y_i - d_i) \times (x_i + d_i, y_i + d_i). \tag{3.75}$$

The intersection of bounding boxes can be computed by simply taking the maximum of all coordinate minima and the minimum of all maxima, that is,

$$[\max_i(x_i - d_i), \max_i(y_i - d_i)] \times [\min_i(x_i + d_i), \min_i(y_i + d_i)]. \tag{3.76}$$

An estimate of position is then obtained as the center of the intersection of the bounding boxes, that is, the average of corner coordinates. The advantage of the Min–Max method is that it requires only

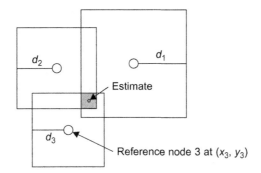

FIGURE 3.41

Example of Min–Max multilateration.

low-complexity sum and compare operations. It is also interesting to note that the Min–Max approach to positioning is very similar to the POCS method discussed earlier in this section. In fact, the only thing that separates Min–Max from a "true" POCS algorithm is the last step where the center of the intersection is taken to be the position estimate.

In Ref. [92], an extensive simulation campaign is presented where several combinations as regards phases one and two are compared in terms of coverage and localization accuracy. The scenario considered consists of a network of 225 nodes placed in a square area with side 100 distance units. The anchor nodes represent a fraction of 5% and are placed in a regular grid. The link budget has been set to have radio connectivity in the range of 14 units (no shadowing effects are considered). On average, less than one anchor node is seen by each node and the connectivity degree is about 12–13 nodes. The standard deviation of the (Gaussian) ranging error is set to 10% of the radio range. As an example, Fig. 3.42 gives the average position error of the six combinations, proposed in Ref. [92], for phases one and two for varying range error variance (top), radio range (middle), and anchors fraction (bottom). The solid lines denote the multilateration variants, whereas the dashed lines denote the Min–Max variants. The first observation is that the sensitivity to the anchors fraction is quite similar for all combinations and that all combinations are quite sensitive to the radio range (connectivity). The Euclidean/Lateration combination clearly outperforms the others in the absence of range errors. However, the Min–Max is more insensitive to distance errors than multilateration, but it requires good anchor node placement. It can be concluded that *no particular algorithm gives the best performance but different behaviors against range errors, anchors fraction, placement, and coverage have been found.*

Incremental Algorithms

These algorithms usually start with a very limited core of three or four nodes with assigned coordinates. In incremental algorithms, nodes surrounding anchor nodes cooperatively establish position estimates that are successively propagated to more distant nodes, allowing them to estimate their position without direct reference node visibility [149]. At each iteration step, once a node with unknown position hears N nodes with known (anchors) or estimated positions, it would be able to estimate its position starting from the measured distances and known positions if $N \geq 3$ (see Fig. 3.43). These coordinate calculations are based on either simple trigonometric equations or some local optimization scheme.

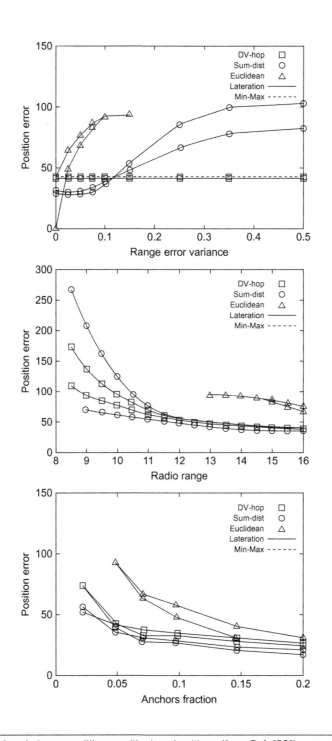

FIGURE 3.42

Performance comparison between multihop positioning algorithms (from Ref. [92]).

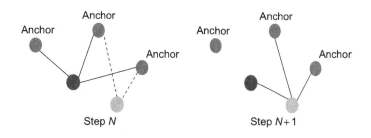

FIGURE 3.43

Example of incremental localization algorithm.

A drawback of incremental algorithms is that they propagate measurement errors, resulting in poor overall coordinate assignments. Some incremental approaches apply a later global optimization phase to balance such errors, but it remains difficult to jump out of local minima introduced by the local optimization in the incremental phase.

Semidefinite Programming Algorithms

Another interesting approach is proposed in Refs. [157, 158], named constrained semidefinite programming localization algorithm (CSDPLA). It is based on a different approach to the problem; instead of trying to minimize the error in distance estimation between a pair of nodes, it exploits a semidefinite programming (SDP) tool, combined with information about connectivity, with the purpose to make positioning errors less sensitive to single ranging errors (especially if positively biased). A comparison between this approach and other multihop localization algorithms can be found in Ref. [156]. Results show that the CSDPLA algorithm is well suited to scenarios characterized by a large number of hops.

3.4.3.2 *Range-Free Positioning*

As stated in Section 1.1.2, position information can also be inferred from simple connectivity or proximity information. Positioning algorithms based on proximity are often called *range-free* localization algorithms. Consider a scenario where m anchor nodes are present with coordinates $\mathbf{b} = (x_1, y_1, x_2, y_2, \ldots, x_m, y_m)^T$ and the positions $\mathbf{x} = (x_{m+1}, y_{m+1}, \ldots, x_n, y_n)^T$ of the remaining $n - m$ nodes are unknown (agents). The problem is to find $\mathbf{x}$ such that all proximity constraints (or as many as possible) are satisfied. One very simple model for the proximity constraints can be obtained from the so-called circular radio coverage model, where transmission range is modeled by a circle with fixed radius r_0. As the number of constraints increases, the feasible region of solutions for $\mathbf{x}$, given by the intersection of individual constraints, becomes smaller. As an example, in Fig. 3.44, the feasible set (shaded region) of solutions of an agent (white node) is shown for an increasing number of anchor nodes (black nodes), $m = 1, 2, 3$. From scenarios (a)–(c), the intersection region decreases for each added constraint.

In Ref. [14], anchor nodes with overlapping regions of coverage are placed on a regular grid with distance d, and they act as beacons. A generic agent i infers proximity to a collection $\mathcal{C}$ of anchor nodes for which the connectivity metric exceeds a certain threshold (no cooperation among agents nodes is considered here). The agent localizes itself to the region given by the intersection of the connectivity regions of radius r_0 and centered on anchor nodes position. The centroid of this region is taken as position estimate $(\hat{x}_i, \hat{y}_i)$ for the agent i. Obviously, by increasing the ratio r_0/d (i.e., the range overlap of the anchor nodes), the accuracy of the location estimate improves.

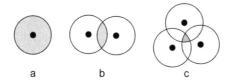

FIGURE 3.44

The feasible region of solutions for the unknown node position (gray area) as the number of anchor nodes (black nodes) increases.

The more general case where agents also cooperate in the positioning process is more challenging. The work in Ref. [38] presents a centralized methodology to solve this problem as a linear or semidefinite program. It is shown that the position estimation error can be dramatically reduced as the network connectivity increases.

The *DV-Hop* scheme proposed in Ref. [123] is actually a range-free positioning algorithm because distance estimation between agents and anchor nodes is performed by hop counting. The DV-Hop algorithm is similar to the *N*-hop multilateration procedure. In the DV-Hop algorithm, anchor nodes first flood the network with so-called beacon packets. Each receiving node maintains the minimum counter value per anchor node of all beacons it receives and ignores those beacons with higher hop-count values as done in the classical distance vector routing scheme. In this way, each node in the network has a rough distance information, in terms of hops, to every anchor node. To enable the conversion from number of hops and physical distance, anchor nodes evaluate the average single hop distance, d_{hop}, starting from the hop-count information and known position of all other anchor nodes inside the network. Once calculated, anchor nodes broadcast the estimated average hop size information. Agents can evaluate the estimated distance to anchor node i by multiplying the counted hops by the average hop size d_{hop}. Finally, those agents that obtain the distance estimation to at least three references can estimate their location by using multilateration (e.g., the simple Min–Max algorithm).

3.4.3.3 *Anchor-Free Positioning*

In some WSN application scenarios, none of the nodes in the network has a priori position information, that is, no reference (anchor) nodes are available. In this case, only relative coordinates among nodes can be determined.

The relative coordinate estimation problem can be formulated as follows: Given a set of agents with unknown position and range measurements between some nodes, determine the (relative) coordinates of all (or a subset of) nodes in the network. Unfortunately, this problem is *NP*-hard. In addition, distributed algorithms are often necessary. Some heuristic algorithms have been proposed to attack this difficult problem, mainly based on the model shown in Fig. 3.45: Cooperative localization is analogous to finding the resting point of masses (representing the nodes) connected by springs (with length proportional to distance measurements) [126]. Springs apply "forces" to the nodes which cause the nodes to shift, and the system will eventually reach an equilibrium state. The equilibrium point of masses represents the minimum-energy localization estimate. Force-directed relaxation methods can be used to converge toward a minimum-energy configuration. However, such methods are susceptible to local minima, and a reasonably accurate initialization of the system is therefore required.

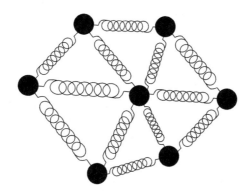

FIGURE 3.45

Anchor-free positioning as spring embedder system.

Recalling Table 3.3, the principal methodologies for the relative coordinate estimation are described in the following paragraphs.

The Anchor-Free Localization (AFL) Algorithm

In Ref. [132], a fully decentralized algorithm called anchor-free localization (AFL) is proposed. It is based on the observation that many false minima are caused because nodes operating on local information converge falsely to configurations where groups of nodes are topologically folded with respect to the true configuration.

It is composed of two phases. First, an initial layout is computed to alleviate problems with local minima. Specifically, AFL seeks to first configure nodes into a "fold-free" configuration that is a scaled-up, unfolded, and locally distorted version of the true configuration. The second phase consists of a force-directed relaxation method based on the masses and springs analogy described above, taking care to not seriously violate fold-freedom. The result is a correct solution for a large class of input networks in practice. The algorithm has been tested through simulations for different node connectivity and density levels, as well as distance estimation error conditions. It has been shown to give better results in terms of precision and robustness than incremental algorithms such as the one presented in Ref. [149].

The Robust Quad Algorithm

This section introduces the original robust quad (ORQ) anchor-free localization algorithm, which is based on robust quadrilateral theory presented later.

One of the most popular relative localization algorithms is the one proposed by Moore et al. [118], called the *robust quad algorithm*, here denoted ORQ algorithm. The ORQ is a distributed, linear-time algorithm for localizing nodes in a WSN without reference nodes and in the presence of range measurement noise. The algorithm is based on the concept of robust quadrilaterals (RQs), which are sets of four nodes that are fully connected and "well-spaced" such that even in the presence of measurement noise, their relative positions are unambiguous, as shown in Fig. 3.46. The RQs are the main building blocks of the ORQ algorithm, where they are used as a way to avoid *flip ambiguities* that otherwise corrupt localization computations [157]. By observing Fig. 3.47, flip ambiguity occurs when Vertex A

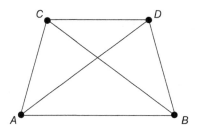

FIGURE 3.46

A robust four-vertex quadrilateral.

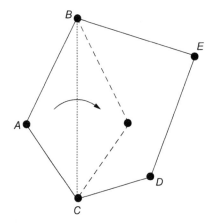

FIGURE 3.47

An example of flip ambiguity.

can be reflected across the line connecting B and C with no change in the distance constraints. The main drawback of ORQ is that in conditions of low node connectivity or high measurement error, the algorithm often only locates a small percentage of nodes in a WSN.

The ORQ algorithm has the following four-step structure [118]:

- *Step 1: Cluster localization.* Each node is elected as the head of a cluster composed of the node itself and all its one-hop neighbors. The cluster head first searches for all RQs in its cluster (step 1a), and then finds the largest subgraph composed only of overlapping RQs [118] and estimates the coordinates of its neighbors with respect to a local coordinate system defined within the cluster (step 1b).
- *Step 2: Cluster optimization (optional).* Node positions estimated within each cluster are refined, for example using numerical optimization methods.
- *Step 3: Cluster transformation.* Nodes shared between pairs of adjacent clusters are used to shift, rotate, and reflect local coordinate systems, in order to refer all estimated node coordinates to a global reference system common to the whole WSN.

- *Step 4: Final refinement (optional).* Location estimates of all nodes in the WSN can be further refined, again using numerical optimization methods.

In the following, only step 1 is explained because it represents the most important step where first step 1a finds the quads robustness set and then step 1b uses this set to localize nodes, hence minimizing both the probability of flip and localization error.

ORQ-Step 1a: Quad robustness test   The robustness of each quad in each cluster head's neighborhood is computed in step 1a, prior to performing any location calculation. To assess robustness, the ORQ algorithm uses a test called the *original quad robustness test* (OQRT).

First of all, the concept of *flip error* is introduced for a generic quad that may occur when a node of the quad is estimated by trilateration from the other three. Referring to Fig. 3.48, assume that nodes A, B, C, and D define a quadrilateral of which the coordinates of nodes A, B, and C are known exactly, and the coordinates of D must be estimated by means of a trilateration algorithm. Due to the definition of an RQ, all internode distance measurements are available.

In a first phase, the trilateration algorithm ignores node C and pinpoints two ambiguous locations of D, D_{AB} and D'_{AB}, at the intersections between circles $A\left(A, \hat{d}_{AD}\right)$ and $B\left(B, \hat{d}_{BD}\right)$, selecting between D_{AB} and D'_{AB}, the one whose distance to node C is closer to the measured distance $\hat{d}_{DC}$. Because of the presence of measurement errors, this selection criterion is subject to a flip error that occurs when the measured distance $\hat{d}_{DC}$ is closer to the distance $d_2 = \mathrm{dist}\{C, D'_{AB}\}$, between C and the incorrect solution, than to the distance $d_1 = \mathrm{dist}\{C, D_{AB}\}$, between C and the correct solution. From a mathematical point of view, the triangulation algorithm performs the following selection:

$$\begin{cases} D_{AB} & \text{if } \left(\left| d_1 - \hat{d}_{CD} \right| < \left| d_2 - \hat{d}_{CD} \right| \right), \\ D'_{AB} & \text{otherwise.} \end{cases} \tag{3.77}$$

It can be proven [118] that for a zero-mean random additive measurement error e_{DC} in the estimation of $\hat{d}_{CD}$, a flip error occurs if e_{DC} is at least equal to $\delta = \frac{1}{2}(d_2 - d_1)$, hence with probability

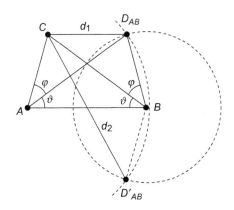

FIGURE 3.48

A diagram of a quadrilateral for deriving the worst-case probability of flip error.

$P\{\text{"flip"}\} = P\{e_{DC} \geq \delta\}$. This flip probability decreases as δ increases; thus, it can in principle be made smaller than an arbitrarily small value P_f, provided that the distribution of e_{DC} is known and δ is larger than a *robustness threshold* $d_{RT,O}$ such that $P\{e_{DC} \geq d_{RT,O}\} = P_f$.[10] By considering the worst-case flip probability, the authors in Ref. [118] found that the minimum value of δ is $\delta_{min} = \hat{d}_{AB}\sin^2(\theta)$, where θ is the smallest angle of the triangle. In conclusion, considering all the four nodes to be located, a quad $ABCD$ is assessed *robust* if and only if the inequality

$$b\sin^2(\theta) > d_{RT,O} \qquad (3.78)$$

holds for all triangles $\triangle ABC$, $\triangle ABD$, $\triangle ACD$, $\triangle BCD$, where b is the shortest side.

ORQ-Step 1b: Node locations estimation   The location estimation in step 1b is initiated within the first RQ. Let node A be the cluster head, nodes B, C, and D be the remaining nodes of the first RQ and $\hat{d}_{AB}$, $\hat{d}_{AC}$, $\hat{d}_{AD}$, $\hat{d}_{BC}$, $\hat{d}_{BD}$, $\hat{d}_{CD}$ the corresponding six noisy internode estimates.

First, the cluster head coordinates are defined to be at the origin of a local coordinate system, as shown in Fig. 3.49, and then node B is placed along the x-axis, and node C is estimated inside the first two quadrants, that is,

$$\hat{A} = (0,0), \quad \hat{B} = \left(\hat{d}_{AB}, 0\right), \quad \hat{C} = \left(\hat{d}_{AC}\cos(\beta), \hat{d}_{AC}\sin(\beta)\right), \qquad (3.79)$$

where

$$\beta = \arccos\frac{\hat{d}_{AC}^2 + \hat{d}_{AB}^2 - \hat{d}_{BC}^2}{2\hat{d}_{AC}\hat{d}_{AB}}$$

is obtained by applying the rule of cosines to the triangle $\triangle ABC$. Finally, the fourth node D is localized using a simple trilateration algorithm.

After completing the localization of nodes in the first RQ, the location of the remaining neighbors of the cluster head are estimated incrementally by following the chain of RQs sharing three nodes

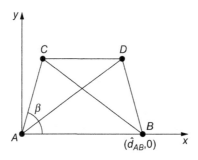

FIGURE 3.49

Example of the first RQ localization in a local coordinate system.

---

[10]In practice, however, since the coordinates of nodes A, B, and C are not known exactly, but estimated using noisy ranges, it is hard to derive the threshold $d_{RT,O}$ given a distribution for e_{DC} and a target flip probability P_f.

Table 3.4 The ORQ (a), ERQ (b) and RLE (c) Positioning Algorithms

(a) Original Robust Quad (ORQ)

The ORQ algorithm is based on the concept of RQs in order to avoid flip ambiguities that corrupt localization computations.

Algorithm structure:

- *Cluster localization.*
    - Assess the robustness of all quads in each cluster of nodes.
    - Estimate by triangulation the coordinates of nodes following the largest chain of RQs.
- *Cluster transformation.* Refer all node coordinates estimated in different clusters to a global reference system common to the whole WSN.

(b) Enhanced Robust Quad (ERQ)

The ERQ algorithm has the same structure as the ORQ algorithm. However, it enhances the performance by using a new test to assess the robustness of quadrilaterals and a new trilateration algorithm.

Algorithm structure:

- *Cluster localization.*
    - Assess the robustness of all quads by using a *new test of robustness*.
    - Estimate the coordinates of nodes following the largest chain of RQs by using a *new trilateration algorithm*.
- *Cluster transformation.* Exactly the same as the ORQ algorithm.

(c) Robust Location Estimation (RLE)

The RLE algorithm estimates only a single cluster called CC by using a trilateration algorithm called robust trilateration (RT) that is based on the concepts of RQ applied directly into the trilateration procedure.

Algorithm structure:

- *Central cluster localization.* The cluster head node tries to estimate the location of as many adjacent nodes as possible of the CC using the RT algorithm.
- *Complete localization.* Coordinates of nodes that have not been localized yet are estimated by using the RT algorithm recursively.

with an RQ already located (for additional details, see Ref. [118]) and estimating the fourth node's coordinates, again using a simple trilateration algorithm.

Enhanced versions of the ORQ algorithm are the ERQ and RLE introduced in Ref. [165]. Table 3.4 summarizes the main characteristics that distinguish these three algorithms.

References

[1] IEEE 802.15.4 Standard, Part 15.4: Wireless Medium Access Control (MAC) and Physical Layer (PHY) Specifications for Low-Rate Wireless Personal Area Networks (LR-WPANS), IEEE Press, Piscataway, NJ, 2003.

[2] A. Paulraj, et al., Handbook of Statistics, vol. 10, Chapter Subspace Methods for Direction of Arrival Estimation, Elsevier, 1993, pp. 693–640.

[3] J.R.A. Urruela, H. Morata, NLOS mitigation based on trellis search for wireless location, in: IEEE Workshop on Signal Processing Advances in Wireless Communications, 2005, pp. 665–669.

[4] J.S. Abel, J.O. Smith, Source range and depth estimation from multipath range difference measurements, IEEE Trans. Acoust. Speech Signal Process. 37 (8) (1989) 1157–1165.

[5] S. Al-Jazzar, J.J. Caffery, ML and Bayesian TOA location estimators for NLOS environments, in: Proceedings of the 56th IEEE Vehicular Technology Conference, vol. 2, 2002, pp. 1178–1181.

[6] A. Amar, A.J. Weiss, Advances in direct position determination, in: Proceedings of the Sensor Array and Multichannel Signal Processing Workshop, 2004, p. 584.

[7] P. Bahl, V.N. Padmanabhan, Radar: An in-building RF-based user location and tracking system, in: INFOCOM, vol. 2, 2000, pp. 775–784.

[8] P. Bahl, V.N. Padmanabhan, A. Balachandran, Enhancements to the radar user location and tracking system, Technical report, Microsoft Research, 2000.

[9] R. Barton, D. Rao, Performance capabilities of long-range UWB-IR TDOA localization systems, EURASIP J. Adv. Signal Process. (2008) 1–17.

[10] R. Battiti, N.T. Le, A. Villani, Location-aware computing: A neural network model for determining location in wireless LANs. Technical report, Feb. 01 2002.

[11] R. Battiti, A. Villani, T.L. Nhat, Neural network models for intelligent networks: Deriving the location from signal patterns, in: First Annual Symposium on Autonomous Intelligent Networks and Systems (AINS), 2002.

[12] D. Blatt, A.O. Hero, Energy-based sensor network source localization via projection onto convex sets, IEEE Trans. Signal Process. 54 (9) (2006) 3614–3619.

[13] J. Borras, P. Hatrack, N.B. Mandayam, Decision theoretic framework for NLOS identification, in: Proceedings of the IEEE Veh. Technol. Conf. (VTC), vol. 2, Ontario, Canada, 1998, pp. 1583–1587.

[14] N. Bulusu, J. Heidemann, D. Estrin, GPS-less low-cost outdoor localization for very small devices, IEEE [see also IEEE Wireless Communications] Pers. Commun. 7 (5) (2000) 28–34.

[15] J.J. Caffery, G.L. Stuber, Overview of radiolocation in CDMA cellular systems, IEEE Commun. Mag. 36 (4) (1998) 38–45.

[16] A. Catovic, Z. Sahinoglu, The Cramer–Rao bounds of hybrid TOA/RSS and TDOA/RSS location estimation schemes, IEEE Commun. Lett. 8 (2004) 626–628.

[17] F. Cesbron, R. Arnott, Locating GSM mobiles using antenna array. Electron. Lett. 34 (16) (1998) 1539–1540.

[18] Y. Chan, K. Ho, A simple and efficient estimator for hyperbolic location, IEEE Trans. Signal Process. 42 (8) (1994) 1905–1915.

[19] Y.T. Chan, H.Y.C. Hang, P.C. Ching, Exact and approximate maximum likelihood localization algorithms, IEEE Trans. Veh. Technol. 55 (1) (2006) 10–16.

[20] Y.-T. Chan, W.-Y. Tsui, H. So, P.-C. Ching, Time-of-arrival based localization under NLOS conditions, IEEE Trans. Veh. Technol. 55 (1) (2006) 17–24.

[21] P.C. Chen, A non-line-of-sight error mitigation algorithm in location estimation, in: IEEE Wireless Communications and Networking Conference, pp. 316–320.

[22] P. Cheong, A. Rabbachin, J. Montillet, K. Yu, I. Oppermann, Synchronization, TOA and position estimation for low-complexity LDR UWB devices, in: IEEE International Conference on Ultra-Wideband, Zurich, Switzerland, 2005, pp. 480–484.

[23] C.C. Chong, F. Watanabe, I. Guvenc, H. Inamura, NLOS identification and weighted least-squares localization for UWB systems using multipath channel statistics, EURASIP J. Adv. Signal Process. (2008) 1–14.

[24] L. Cong, W. Zhuang, Hybrid TOA/AOA mobile user location for wideband CDMA cellular systems, IEEE Trans. Wireless Commun. 1 (3) (2002) 439–447.

[25] L. Cong, W. Zhuang, Non-line-of-sight error mitigation in mobile location, IEEE Trans. Wireless Commun. 4 (2005) 560–573.

[26] A. D'Amico, U. Mengali, L. Taponecco, TOA estimation with pulses of unknown shape, in: Proc. IEEE Int. Conf. Commun. (ICC), Glasgow, UK, 2007, pp. 4287–4292.

[27] A. D'Amico, U. Mengali, L. Taponecco, Energy-based TOA estimation, IEEE Trans. Wireless Commun. 7 (3) (2008) 838–847.

[28] D. Dardari, C.C. Chong, M.Z. Win, Analysis of threshold-based TOA estimator in UWB channels, in: Proc. European Sig. Processing Conf. (EUSIPCO), Florence, Italy, European Association for Signal Processing (EURASIP), 2006.

[29] D. Dardari, C.-C. Chong, M.Z. Win, Threshold-based time-of-arrival estimators in UWB dense multipath channels, IEEE Trans. Commun. 56 (8) (2008) 1366–1378.

[30] D. Dardari, A. Conti, A sub-optimal hierarchical maximum likelihood algorithm for collaborative localization in ad-hoc networks, in: First IEEE International Conference on Sensor and Ad Hoc Communications and Networks, SECON 2004, Santa Clara, CA, 2004, pp. 425–429.

[31] D. Dardari, A. Conti, U. Ferner, A. Giorgetti, M.Z. Win, Ranging with ultrawide bandwidth signals in multipath environments, Proc. IEEE, Special Issue on UWB Technology & Emerging Applications, 97 (2) (2009) 404–426.

[32] D. Dardari, A. Conti, J. Lien, M. Win, The effect of cooperation on localization systems using UWB experimental data, EURASIP J. Adv. Signal Process. 513873 (2008) 1–11.

[33] D. Dardari, A. Giorgetti, M.Z. Win, Time-of-arrival estimation of UWB signals in the presence of narrowband and wideband interference, in: Proc. IEEE Int. Conf. on Ultra Wideband (ICUWB), Singapore, 2007.

[34] D. Dardari, M.Z. Win, Threshold-based time-of-arrival estimators in UWB dense multipath channels, in: Proc. IEEE Int. Conf. Commun. (ICC), vol. 10, Istanbul, Turkey, 2006, pp. 4723–4728.

[35] N. Decarli, D. Dardari, S. Gezici, A. D'Amico, LOS/NLOS detection for UWB signals: A comparative study using experimental data, in: IEEE International Symposium on Wireless Pervasive Computing (ISWPC 2010), Modena, Italy, 2010, pp. 1–5.

[36] B. Denis, N. Daniele, NLOS ranging error mitigation in a distributed positioning algorithm for indoor UWB ad-hoc networks, in: Proc. International Workshop on Wireless Ad-Hoc Networks, 2004, pp. 356–360.

[37] B. Denis, J. Keignart, N. Daniele, Impact of NLOS propagation upon ranging precision in UWB systems, in: Proc. IEEE Conf. Ultrawideband Syst. Technol. (UWBST), Reston, VA, 2003, pp. 379–383.

[38] L. Doherty, K.S. Pister, L.E. Ghaoui, Convex position estimation in wireless sensor networks, in: INFOCOM 2001. Twentieth Annual Joint Conference of the IEEE Computer and Communications Societies, Proceedings IEEE, vol. 3, 2001, pp. 1655–1663.

[39] L. Dumont, M. Fattouche, G. Morrison, Super-resolution of multipath channels in a spread spectrum location system, Electron. Lett. 30 (19) (1994) 1583–1584.

[40] ECMA-368. High rate ultra wideband PHY and MAC standard, first ed., Available from: http://www.ecma-international.org/publications/files/ECMA-ST/ECMA-368.pdf.

[41] E. Elnahrawy, X. Li, R. Martin, Using area-based presentations and metrics for localization systems in wireless LANS, in: LCN '04: Proceedings of the 29th Annual IEEE International Conference on Local Computer Networks, IEEE Computer Society, Washington, DC, 2004, pp. 650–657.

[42] C. Falsi, D. Dardari, L. Mucchi, M.Z. Win, Time of arrival estimation for UWB localizers in realistic environments, EURASIP J. Appl. Signal Process. (2006) 1–13.

[43] B.T. Fang, Simple solutions for hyperbolic and related position fixes, IEEE Trans. Aerosp. Electron. Syst. 26 (3) (1990) 748–753.

[44] Federal Communications Commission. First Report and Order 02-48, 2002.

[45] W.G. Figel, N.H. Shepherd, W.F. Trammel, Vehicle location by a signal attenuation method, IEEE Trans. Aerosp. Electron. Syst. 18 (3) (1969) 245–251.

[46] W.H. Foy, Position-location solutions by Taylor-series estimation, IEEE Trans. Aerosp. Electron. Syst. 12 (2) (1976) 187–194.

[47] W. Gardner, C.-K. Chen, Signal-selective time-difference-of-arrival estimation for passive location of man-made signal sources in highly corruptive environments, part I: Theory and method, IEEE Trans. Signal Process. 40 (5) (1992) 1168–1184.

[48] W. Gardner, C.-K. Chen, Signal-selective time-difference-of-arrival estimation for passive location of man-made signal sources in highly corruptive environments, part II: Algorithms and performance, IEEE Trans. Signal Process. 40 (5) (1992) 1185–1197.

[49] S. Gezici, H. Kobayashi, H. Poor, Nonparametric non-line-of-sight identification, in: Vehicular Technology Conference, 2003. VTC 2003-Fall. 2003 IEEE 58th, vol. 4, 2003, pp. 2544–2548.

[50] S. Gezici, H.V. Poor, Position estimation via ultra-wideband signals, Proc. IEEE (Special Issue on UWB Technology and Emerging Applications) 97 (2) (2009) 386–403.

[51] L.C. Godara, Application of antenna arrays to mobile communications, Part I: Performance improvement, feasibility and system considerations, Proc. IEEE 85 (7) (1997) 1029–1060.

[52] L.C. Godara, Application of antenna arrays to mobile communications, Part II: Beam forming and direction-of-arrival considerations, Proc. IEEE 85 (8) (1997) 1193–1295.

[53] Y. Guan, Y. Zhou, C.L. Law, High-resolution UWB ranging based on phase-only correlator, in: IEEE International Conference on Ultra-Wideband (ICUWB), Singapore, 2007, pp. 100–104.

[54] I. Guvenc, H. Arslan, Comparison of two searchback schemes for non-coherent TOA estimation in IR-UWB systems, in: Proc. IEEE Sarnoff Symp., Princeton, NJ, 2006.

[55] I. Guvenc, C.-C. Chong, F. Watanabe, NLOS identification and mitigation for UWB localization systems, in: Wireless Communications and Networking Conference, 2007. WCNC 2007. IEEE, 2007, pp. 1571–1576.

[56] I. Guvenc, C.-C. Chong, F. Watanabe, H. Inamura, NLOS identification and weighted least-squares localization for UWB systems using multipath channel statistics, EURASIP J. Adv. Signal Process. (2008).

[57] I. Guvenc, Z. Sahinoglu, Threshold-based TOA estimation for impulse radio UWB systems, in: IEEE International Conference on Ultra-Wideband, pp. 420–425.

[58] I. Guvenc, Z. Sahinoglu, Multiscale energy products for TOA estimation in IR-UWB systems, in: Proc. IEEE Global Telecommun. Conf. (GLOBECOM), vol. 1, St. Louis, MO, 2005, pp. 209–213.

[59] I. Guvenc, Z. Sahinoglu, Threshold selection for UWB TOA estimation based on kurtosis analysis, IEEE Commun. Lett. 9 (12) (2005) 1025–1027.

[60] I. Guvenc, Z. Sahinoglu, TOA estimation with different IR-UWB transceiver types, in: Proc. IEEE Int. Conf. UWB (ICU), Zurich, Switzerland, 2005, pp. 426–431.

[61] I. Guvenc, Z. Sahinoglu, A.F. Molisch, P. Orlik, Non-coherent TOA estimation in IR-UWB systems with different signal waveforms, in: Proc. IEEE Int. Workshop on Ultrawideband Networks (UWBNETS), Boston, MA, 2005, pp. 245–251.

[62] I. Guvenc, Z. Sahinoglu, P. Orlik, TOA estimation for IR-UWB systems with different transceiver types, IEEE Trans. Microwave Theory Tech. (Special Issue on Ultrawideband) 54 (4) (2006) 1876–1886.

[63] R. Hach, Symmetric double sided two-way ranging, 2005. doc: IEEE 15-05-0334-00-004a.

[64] A. Hatami, K. Pahlavan, M. Heidari, F. Akgul, On RSS and TOA based indoor geolocation — A comparative performance evaluation, in: Wireless Communications and Networking Conference (WCNC), vol. 4, 2006, pp. 2267 –2272.

[65] S. Haykin, Neural Networks: A Comprehensive Foundation, second ed., Prentice-Hall Inc., Upper Saddle River, NJ, 1999.

[66] L. He, D. Denis, L. Ouvry, A flexible distributed maximum log-likelihood scheme for UWB indoor positioning, in: Proc. WPNC 2007, 2007.

[67] J. Hightower, G. Borriello, Location systems for ubiquitous computing, IEEE Comput. 34 (8) (2001) 57–66.

[68] J.P. Iannello, Large and small error performance limits for multipath time delay estimation, IEEE Trans. Acoust. Speech Signal Process. ASSP-34 (2) (1986) 245–251.

[69] IEEE Std 802.11-2007 (2007 revision). IEEE-SA. IEEE 802.11: Wireless LAN Medium Access Control (MAC) and Physical Layer (PHY) Specifications, 2007.

[70] IEEE Std 802.15.4a-2007 (Amendment to IEEE Std 802.15.4-2006). IEEE standard for information technology – telecommunications and information exchange between systems – local and metropolitan area networks – specific requirement part 15.4: Wireless medium access control (MAC) and physical layer (PHY) specifications for low-rate wireless personal area networks (WPANs), 2007.

[71] F. Izquierdo, M. Ciurana, F. Barceló, J. Paradells, E. Zola, Performance evaluation of a TOA-based trilateration method to locate terminals in WLAN, in: Proc. IEEE ISWPC, 2006, pp. 217–222.

[72] J. Capon, High-resolution frequency-wavenumber spectrum analysis, Proc. IEEE 57 (8) (1969) 1408–1418.

[73] R.J.J. Vidal, M. Nájar, High resolution time-of-arrival detection for wireless positioning systems, in: IEEE Vehicular Technology Conference, Vancouver, Canada, 2002.

[74] Y. Jiang, V. Leung, An asymmetric double sided two-way ranging for crystal offset, in: Proc. International Symposium on Signals, Systems and Electronics (ISSSE), 2007, pp. 525–528.

[75] D. Jourdan, J.J. Deyst, M. Win, N. Roy, Monte Carlo localization in dense multipath environments using UWB ranging, in: IEEE International Conference on Ultra-Wideband, 2005, pp. 314–319.

[76] D.B. Jourdan, D. Dardari, M.Z. Win, Position error bound for UWB localization in dense cluttered environments, IEEE Trans. Aerosp. Electron. Syst. 44 (2) (2008) 613–628.

[77] N. Kandil, S. Affes, A. Taok, S. Georges, Fingerprinting localization using ultra-wideband and neural networks, in: Proc. International Symposium on Signals, Systems and Electronics, 2007, pp. 529–532.

[78] H. Karl, A. Willig, Protocols and Architectures for Wireless Sensor Networks, John Wiley and Sons Ltd., Chichester, 2006.

[79] S.M. Kay, Fundamentals of Statistical Signal Processing: Estimation Theory, Prentice Hall PTR, Upper Saddle River, NJ, 1993.

[80] H. Keshavarz, Weighted signal-subspace direction-finding of ultra-wideband sources, in: IEEE International Conference on Wireless and Mobile Computing, Networking and Communications, 2005. (WiMob'2005), Montreal, Canada, 2005, pp. 23–29.

[81] R. Kohno, Interpretation and future modification of Japanese regulation for UWB, IEEE P802.15-06/261r0.

[82] P. Kontkanen, P. Myllymki, T. Roos, H. Tirri, K. Valtonen, H. Wettig, Probabilistic methods for location estimation in wireless networks, Technical report, Helsinki Institute for Information Technology (HIIT), 2004.

[83] Z. Kostic, M. Sezan, E. Titlebaum, Estimation of the parameters of a multipath channel using set-theoretic deconvolution, IEEE Trans. Commun. 40 (6) (1992) 1449–1452.

[84] K.J. Krizman, T.E. Biedka, T.S. Rappaport, Wireless position location: Fundamentals, implementation strategies, and sources of error, in: Proc. 47th IEEE Vehicular Technology Conference, vol. 2, 1997, pp. 919–923.

[85] J. Krumm, Probabilistic inferencing for location, in: Workshop on Location-Aware Computing (Part of UbiComp 2003), Microsoft Research, Seattle, WA, 2003.

[86] J. Krumm, L. Williams, G. Smith, SmartMoveX on a graph — an inexpensive active badge tracker, Lect. Notes Comput. Sci. 2498 (2002) 343–350.

[87] H. Kunczier, Mobile Handset Localization by Received Signal Level Pattern Matching, PhD thesis, Technical University of Vienna, Vienna, Austria, 2006.

[88] B. Kusy, J. Salla, G. Balogh, A. Ledeczi, V. Protopopescu, J. Tolliver, et al., Radio interferometric tracking of mobile wireless nodes, in: Proc. ACM International Conference on Mobile Systems, Applications, and Services, San Juan, Puerto Rico, ACM, New York, 2007, pp. 139–151.

[89] M.N.L. Blanco, J. Serra, Low complexity TOA estimation for wireless location, in: IEEE Global Telecommunication Conference, San Francisco, 2006.

[90] H. Laitinen, J. Lahteenmaki, T. Nordstrom, Database correlation method for GSM location, in: Proc. 53rd IEEE Vehicular Technology Conference, vol. 4, 2001, pp. 2504–2508.

[91] Y. Lang, G.B. Giannakis, Timing ultra-wideband signals with dirty templates, IEEE Trans. Commun. 53 (11) (2005) 1951–1963.

[92] K. Langendoen, N. Reijers, Distributed localization in wireless sensor networks: A quantitative comparison, Comput. Networks Elsevier 43 (2003) 499–518.

[93] J.-Y. Lee, R.A. Scholtz, Ranging in a dense multipath environment using an UWB radio link, IEEE J. Sel. Areas Commun. 20 (9) (2002) 1677–1683.

[94] J.Y. Lee, S. Yoo, Large error performance of UWB ranging in multipath and multiuser environments, IEEE Trans. Microw. Theory Tech. 54 (4) (2006) 1887–1895.

[95] Z. Lei, F. Chin, Y.-S. Kwok, UWB ranging with energy detectors using ternary preamble sequences, in: Proc. IEEE Wireless Commun. Networking Conf. (WCNC), Las Vegas, NV, 2006, pp. 872–877.

[96] B. Li, A. Dempster, C. Rizos, H.K. Lee, A database method to mitigate NLOS error in mobilephone positioning, in: Proc. IEEE Position, Location, and Navigation Symp. (PLANS), San Diego, CA, 2006, pp. 173–178.

[97] J. Li, R. Wu, An efficient algorithm for time delay estimation, IEEE Trans. Signal Process. 46 (8) (1998) 2231–2235.

[98] X. Li, K. Pahlavan, Super-resolution TOA estimation with diversity for indoor geolocation, IEEE Trans. Wireless Commun. 3 (1) (2004) 224–234.

[99] J.C. Liberti, T.S. Rappaport, Smart Antennas for Wireless Communications: IS-95 and Third Generation CDMA Applications, Prentice Hall Ptr, Upper Saddle River, NJ, 1999.

[100] J. Lie, C.M. See, B.P. Ng, Ultra wideband direction finding using digital channelization receiver architecture, IEEE Commun. Lett. 10 (2) (2006) 85–87.

[101] H. Liu, H. Darabi, P. Banerjee, J. Liu, Survey of wireless indoor positioning techniques and systems, IEEE Trans. Syst. Man Cybern. C 37 (6) (2007) 1067–1080.

[102] J. López-Salcedo, G. Vázquez, NDA maximum-likelihood timing acquisition of UWB signals, in: IEEE Workshop on Signal Processing Advances in Wireless Communications, New York, 2005.

[103] V. Lottici, A. D'Andrea, U. Mengali, Channel estimation for ultra-wideband communications, IEEE J. Select. Areas Commun. 20 (9) (2002) 1638–1645.

[104] Z.N. Low, J.H. Cheong, C.L. Law, W.T. Ng, Y.J. Lee, Pulse detection algorithm for Line-of-Sight (LOS) UWB ranging applications, IEEE Antennas Wireless Propag. Lett. 4 (2005) 63–67.

[105] A. Maali, H. Mimoun, G. Baudoin, A. Ouldali, A new low complexity NLOS identification approach based on UWB energy detection, in: IEEE Radio and Wireless Symposium, 2009. RWS '09, 2009, pp. 675–678.

[106] M. Maman, B. Denis, L. Ouvry, Overhead and sensitivity to UWB ranging models within a distributed Bayesian positioning solution, IEEE Trans. Microwave Theory Tech. 54 (2006) 1896–1911.

[107] T.G. Manickam, R.J. Vaccaro, D.W. Tufts, A least-squares algorithm for multipath time-delay estimation, IEEE Trans. Signal Process. 42 (11) (1994) 3229–3233.

[108] S. Maranò, W.M. Gifford, H. Wymeersch, M.Z. Win, Nonparametric obstruction detection for UWB localization, in: IEEE Global Telecommunications Conference, 2009. GLOBECOM '09, 2009.

[109] I. Maravic, M. Vetterli, Low-complexity subspace methods for channel estimation and synchronization in ultra-wideband systems, in: Proc. International Workshop on Ultra Wideband Systems (IWUWBS), 2003.

[110] M. Maroti, P. Völgyesi, S. Dora, B. Kusy, A. Nadas, A. Ledeczi, et al., Radio interferometric geolocation, in: Proc. ACM Conference on Embedded Networked Sensor Systems, San Diego, ACM, New York, 2005, pp. 1–12.

[111] C. Mazzucco, U. Spagnolini, G. Mulas, A ranging technique for UWB indoor channel based on power delay profile analysis, in: Proc. IEEE Veh. Technol. Conf. (VTC), Los Angeles, CA, 2004, pp. 2595–2599.

[112] M. McGuire, K.N. Plataniotis, A.N. Venetsanopoulos, Location of mobile terminals using time measurements and survey points, IEEE Trans. Veh. Technol. 52 (4) (2003) 999–1011.

[113] H. Miao, K. Yu, M.J. Juntti, Positioning for NLOS propagation: Algorithm derivations and Cramer–Rao bounds, in: Proc. IEEE 2006 International Conference on Acoustics, Speech and Signal Processing, vol. 4, 2006, pp. 1045–1048.

[114] M. Navarro, M. Nájar, Joint estimation of TOA and DOA in IR-UWB, in: IEEE Workshop on Signal Processing Advances in Wireless Communications, Helsinki, Finland, 2007.

[115] A.F. Molisch, Ultrawideband propagation channels – theory, measurement, and modeling, IEEE Trans. Veh. Technol. 54 (5) (2005) 1528–1545.

[116] A.F. Molisch, D. Cassioli, C.-C. Chong, S. Emami, A. Fort, B. Kannan, et al., A comprehensive standardized model for ultrawideband propagation channels, IEEE Trans. Antennas Propag. 54 (11) (2006) 3151–3166.

[117] A. Mollfulleda, M. Najar, P. Miskovsky, J.A. Leyva, L. Berenguer, C. Ibars, et al., QUETZAL: Qualified ultra-wideband testbed for reduced data-rates and location, in: 2nd International Conference on Testbeds and Research Infrastructures for the Development of Networks and Communities, Tridentcom, Barcelona, Spain, IEEE (Xplore), 2006.

[118] D. Moore, J. Leonard, D. Rus, S. Teller, Robust distributed network localization with noisy range measurements, in: SenSys '04: Proceedings of the 2nd International Conference on Embedded Networked Sensor Systems, Baltimore, MD, ACM, New York, 2004, pp. 50–61.

[119] C. Morelli, M. Nicoli, V. Rampa, U. Spagnolini, Hidden Markov models for radio localization in mixed LOS/NLOS conditions, IEEE Trans. Signal Process. 55 (4) (2007) 1525–1542.

[120] A. Muqaibel, A. Safaai-Jazi, A. Bayeam, A.M. Attiya, S.M. Riad, Ultrawideband through-the-wall propagation, IEEE Proc. Microw. Antennas Propag. 152 (6) (2005) 581–588.

[121] M. Nájar, J.M. Huerta, J. Vidal, J.A. Castro, Mobile location with bias tracking in non-line-of-sight, in: IEEE International Conference on Acoustics, Speech and Signal Processing, Montreal, Canada, 2004, pp. 956–959.

[122] C. Nerguizian, C. Despins, S. Affes, Geolocation in mines with an impulse response fingerprinting technique and neural networks, IEEE Trans. Wireless Commun. 5 (3) (2006) 603–611.

[123] D. Niculescu, B. Nath, Ad hoc positioning system (APS), in: Global Telecommunications Conference, 2001. GLOBECOM '01. IEEE, vol. 5, 2001, pp. 2926–2931.

[124] J.J. Pan, S.J. Pan, V.W. Zheng, Q. Yang, Digital wall: A power-efficient solution for location-based data sharing, in: Proc. IEEE International Conference on Pervasive Computing and Communications, 2008, pp. 645–650.

[125] K. Papakonstantinou, D. Slock, NLOS mobile terminal position and speed estimation, in: Proc. 3rd International Symposium on Communications, Control and Signal Processing, 2008.

[126] N. Patwari, J. Ash, S. Kyperountas, A. Hero, R. Moses, N. Correal, Locating the nodes: Cooperative localization in wireless sensor networks, IEEE Signal Process. Mag. 22 (4) (2005) 54–69.

[127] N. Patwari, A.O. Hero, Manifold learning algorithms for localization in wireless sensor networks, in: Proc. IEEE International Conference on Acoustics, Speech and Signal Processing, vol. 3, 2004, pp. 857–860.

[128] N. Patwari, A.O. Hero III, M. Perkins, N.S. Correal, R.J. O'Dea, Relative location estimation in wireless sensor networks, IEEE Trans. Signal Process. 8 (51) (2003) 2137–2148.

[129] T. Pavani, G. Costa, M. Mazzotti, D. Dardari, A. Conti, Experimental results on indoor localization technique through wireless sensors network, in: Proc. IEEE Vehicular Tech. Conf. (VTC 2006-Spring), Melbourne, Australia, 2006.

[130] J.B. Pierrot, B. Denis, C. Abou-Rjeily, Joint distributed time synchronization and positioning in UWB ad-hoc networks using TOA, IEEE Trans. MTT, Special Issue on Ultra Wideband, 54 (2006) 1896–1911.

[131] L. Pierucci, P.J. Roig, UWB localization on indoor MIMO channels, in: IEEE International Conference on Wireless and Mobile Computing, Networking and Communications, 2005 (WiMob'2005), Montreal, Canada, 2005 pp. 890–894.

[132] N.B. Priyantha, H. Balakrishnan, E. Demaine, S. Teller, Anchor-free distributed localization in sensor networks, Tech Report 892, MIT Laboratory for Computer Science, April 2003. http://nms.lcs.mit.edu/cricket.

[133] J.G. Proakis, Digital Communications, fourth ed., McGraw-Hill, New York, 2001.

[134] Y. Qi, H. Kobayashi, H. Suda, Analysis of wireless geolocation in a non-line-of-sight environment, IEEE Trans. Wireless Commun. 5 (3) (2006) 672–681.

[135] Y. Qi, H. Kobayashi, H. Suda, On time-of-arrival positioning in a multipath environment, IEEE Trans. Veh. Technol. 55 (5) (2006) 1516–1526.

[136] R.C. Qiu, C. Zhou, Q. Liu, Physics-based pulse distortion for ultra-wideband signals, IEEE Trans. Veh. Technol. 54 (5) (2005) 1546–1555.

[137] R.O. Schimdt, Multiple emitter location and signal parameter estimation, IEEE Trans. Antennas Propag. AP-34 (3) (1986) 276–280.

[138] R. Roy, T. Kailath, ESPRIT – estimation of signal parameters via rotational invariant techniques, IEEE Trans. Acoust. Speech Signal Process. 37 (1989) 984–995.

[139] A. Rabbachin, I. Oppermann, B. Denis, ML time-of-arrival estimation based on low complexity UWB energy detection, in: Proc. IEEE Int. Conf. Ultra-Wideband (ICUWB), Waltham, MA, 2006, pp. 598–604.

[140] R.I. Reza, Data Fusion for Improved TOA/TDOA Position Determination in Wireless Systems, Ph.D. dissertation, Virginia Polytechnic Institute and State University, 2000.

[141] J. Riba, A. Urruela, A non-line-of-sight mitigation technique based on ML detection, in: IEEE International Conference on Acoustics, Speech and Signal Processing, Montreal, Canada, 2004, pp. 153–156.

[142] T. Roos, P. Myllymaki, H. Tirri, A statistical modeling approach to location estimation, IEEE Trans. Mobile Comput. 1 (1) (2002) 59–69.

[143] R. Roy, T. Kailath, ESPRIT-estimation of signal parameters via rotational invariance techniques, IEEE Trans. Acoust. Speech Signal Process. 37 (7) (1989) 984–995.

[144] M. Rydström, E.G. Ström, A. Svensson, Robust sensor network positioning based on projections onto circular and hyperbolic convex sets (POCS), in: Proc. IEEE Workshop on Signal Processing Advances in Wireless Communications, 2006, pp. 1–5.

[145] H. Saarnisaari, ML time delay estimation in a multipath channel, in: IEEE 4th International Symposium on Spread Spectrum Techniques and Applications, vol. 3, 1996, pp. 1007–1011.

[146] H. Saarnisaari, TLS-ESPRIT in a time delay estimation, in: Proc. 47th IEEE Vehicular Technology Conference, 1997.

[147] Z. Sahinoglu, S. Gezici, I. Guvenc, Ultra-Wideband Positioning Systems: Theoretical Limits, Ranging Algorithms, and Protocols, Cambridge University Press, New York, 2008.

[148] Z. Sahinoglu, I. Guvenc, Multiuser interference mitigation in noncoherent UWB ranging via nonlinear filtering, in: EURASIP J. on Wireless Commun. and Networking, Pages Article ID 56849, 2006, p. 10.

[149] C. Savarese, J.M. Rabaey, J. Beutel, Locationing in distributed ad-hoc wireless sensor networks, in: Proc. IEEE International Conference on Acoustics, Speech and Signal Processing, vol. 4, 2001, pp. 2037–2040.

[150] A. Savvides, H. Park, M. Srivastava, The bits and flops of the N-hop multilateration primitive for node localization problems, in: First ACM Workshop on Wireless Sensor Networks and Application (WSNA), Atlanta, GA, ACM, New York, 2002, pp. 112–121.

[151] A. Sayed, A. Tarighat, N. Khajehnouri, Network-based wireless location: Challenges faced in developing techniques for accurate wireless location information, Signal Process. Mag. IEEE 22 (4) (2005) 24–40.

[152] S.V. Schell, W.A. Gardner, High Resolution Direction Finding, vol. 10, chapter Handbook of Statistics, Elsevier, 1993, pp. 755–817.

[153] R. Schmidt, Multiple emitter location and signal parameter estimation, IEEE Trans. Antennas Propag. 34 (3) (1986) 276–280.

[154] R.A. Scholtz, D.M. Pozar, W. Namgoong, Ultra-wideband radio, EURASIP J. Appl. Signal Process. 2005 (3) (2005) 252–272.

[155] J. Schroeder, S. Galler, K. Kyamakya, T. Kaiser, Three-dimensional indoor localization in non line of sight UWB channels, in: Proc. IEEE Int. Conf. on Ultra Wideband (ICUWB), Singapore, 2007.

[156] S. Severi, G. Abreu, D. Dardari, A quantitative comparison on multihop algorithms, in: Proc. Workshop on Positioning, Navigation and Communication (WPNC 10), Dresden, Germany, 2010, pp. 1–6.

[157] S. Severi, G. Abreu, G. Destino, D. Dardari, Understanding and solving flip-ambiguity in network localization via semidefinite programming, in: IEEE Global Communications Conference (GLOBECOM 2009), Honolulu, Hawaii, 2009, pp. 1–6.

[158] S. Severi, D. Dardari, G. Destino, G. Abreu, Efficient and accurate localization in multihop networks, in: Proc. of the Asilomar Conference on Signals, Systems, and Computers, Pacific Grove, CA, 2009, pp. 1–6.

[159] S. Severi, G. Liva, M. Chiani, D. Dardari, A new low-complexity user tracking algorithm for WLAN-based positioning systems, in: 16th IST Mobile and Wireless Communications Summit, Budapest, Hungary, IEEE (Xplore), 2007.

[160] Y. Shen, H. Wymeersch, M. Win, Fundamental limits of wideband localization – part II: Cooperative networks, Inf. Theory IEEE Trans. 56 (10) (2010) 4981–5000.

[161] F. Sivrikaya, B. Yener, Time synchronization in sensor networks: A survey, IEEE Network 18 (4) (2004) 45–50.

[162] K. Siwiak, J. Gabig, IEEE 802.15.4IGa informal call for application response, contribution#11. Doc.: IEEE 802.15-04/266r0, 2003.

[163] A. Smith, H. Balakrishnan, M. Goraczko, N.B. Priyantha, Tracking moving devices with the cricket location system, in: MobiSys, 2004.

[164] J. Smith, J. Abel, Closed-form least-squares source location estimation from range-difference measurements, IEEE Trans. Acoust. Speech Signal Process. 35 (12) (1987) 1661–1669.

[165] F. Sottile, M.A. Spirito, Robust localization for wireless sensor networks, in: Proceedings of the IEEE Conference on Sensor and Ad Hoc Communications and Networks (SECON), San Francisco, 2008, pp. 50–61.

[166] M.R.M. Stella, D. Begusic, Location determination in indoor environment based on RSS fingerprinting and artificial neural network, in: Proc. International Conference on Telecommunications, 2007, pp. 301–306.

[167] L. Stoica, A. Rabbachin, I. Oppermann, A low-complexity noncoherent IR-UWB transceiver architecture with TOA estimation, IEEE Trans. Microw. Theory Tech. 54 (4) (2006) 1637–1646.

[168] T.S. Rappaport, Smart Antennas: Adaptive Arrays, Algorithms, and Wireless Position Location – Selected Readings, IEEE Press, Piscataway, NJ, 1998.

[169] The Commission of the European Communities. Commission Decision of 21 February 2007 on allowing the use of the radio spectrum for equipment using ultra-wideband technology in a harmonised manner in the Community, Official Journal of the European Union, 2007/131/EC, Feb. 23, 2007.

[170] Z. Tian, G.B. Giannakis, A GLRT approach to data-aided timing acquisition in UWB radios – Part I: Algorithms, IEEE Trans. Wireless Commun. 4 (6) (2005) 2956–2967.

[171] Z. Tian, G.B. Giannakis, A GLRT approach to data-aided timing acquisition in UWB radios – Part II: Training sequence design, IEEE Trans. Wireless Commun. 4 (6) (2005) 2994–3004.

[172] Z. Tian, V. Lottici, Efficient timing acquisition in dense multipath for UWB communications, in: Proc. IEEE Veh. Technol. Conf. (VTC), vol. 2, Orlando, FL, 2003, pp. 1318–1322.

[173] Z. Tian, L. Wu, Timing acquisition with noisy template for ultra-wideband communications in dense multipath, EURASIP J. Appl. Sig. Process. (3) (2005) 439–454.

[174] H.L.V. Trees, Detection, Estimation, and Modulation Theory: Part I, second ed., John Wiley & Sons, Inc., New York, 2001.

[175] M. Triki, D.T.M. Slock, V. Rigal, P. Francois, Mobile terminal positioning via power delay profile fingerprinting: Reproducible validation simulations, in: Proc. 64th IEEE Vehicular Technology Conference, 2006.

[176] B.D. Van Veen, K. Buckley, Beamforming: A versatile approach to spatial filtering, in: IEEE Acoustic, Speech and Signal Processing Magazine, 1988, pp. 4–24.

[177] S. Venkatesh, R.M. Buehrer, A linear programming approach to NLOS error mitigation in sensor networks, in: Proc. IEEE Int. Symp. Information Processing in Sensor Networks (IPSN), Nashville, Tennessee, 2006.

[178] S. Venkatraman, J. Caffery, A statistical approach to non-line-of-sight BS identification, in: Proc. Int. Symp. on Wireless Personal Multimedia Commun., IEEE Conference, Honolulu, HI, 2002, pp. 296–300.

[179] S. Venkatraman, J.J. Caffery, H.-R. You, A novel ToA location algorithm using LoS range estimation for NLoS environments, IEEE Trans. Veh. Technol. 53 (5) (2004) 1515–1524.

[180] R. Verdone, D. Dardari, G. Mazzini, A. Conti, in: Wireless Sensor and Actuator Networks: Technologies, Analysis and Design, Elsevier, 2008.

[181] X. Wang, Z. Wang, B.O. Dea, A TOA based location algorithm reducing the errors due to non-line-of-sight (NLOS) propagation, IEEE Trans. Veh. Technol. 52 (1) (2003) 112–116.

[182] G.I. Wassi, C. Despins, D. Grenier, C. Nerguzian, Indoor location using received signal strength of IEEE 802.11b access point, in: Proc. Canadian Conference on Electrical and Computer Engineering, 2005, pp. 1367–1370.

[183] A. Weiss, On the accuracy of a cellular location system based on RSS measurements, IEEE Trans. Veh. Technol. 52 (6) (2003) 1508–1518.

[184] M. Win, Ultra-wideband bandwidth time-hopping spread-spectrum impulse radio for wireless multiple-access communications, IEEE Trans. Commun. 48 (4) (2000) 679–691.

[185] M.Z. Win, R.A. Scholtz, Impulse radio: How it works, IEEE Commun. Lett. 2 (2) (1998) 36–38.

[186] M.Z. Win, R.A. Scholtz, Characterization of ultra-wide bandwidth wireless indoor communications channel: A communication theoretic view, IEEE J. Select. Areas Commun. 20 (9) (2002) 1613–1627.

[187] L. Wirola, Positioning protocol for next-gen cell phones, GPS World 22 (3) (2011) 1–4.

[188] M. Wylie, J. Holtzman, The non-line of sight problem in mobile location estimation, in: IEEE International Conference on Universal Personal Communications, 1996, pp. 827–831.

[189] M.P. Wylie-Green, S.S.P. Wang, Robust range estimation in the presence of the non-line-of-sight error, in: Proc. IEEE Semiannual Veh. Technol. Conf., vol. 1, 2001, pp. 101–105.

[190] H. Wymeersch, J. Lien, M.Z. Win, Cooperative localization in wireless networks, in: Proc. of IEEE, Special Issue on UWB Technology & Emerging Applications, 2009.

[191] M.Y.X. Li, K. Pahlavan, M. Latva-aho, Comparison of indoor geolocation methods in DSSS and OFDM wireless LAN systems, in: IEEE VTS-Fall VTC 2000. 52nd, vol. 6, 2000, pp. 3015–3020.

[192] L. Xiong, A selective model to suppress NLOS signals in Angle-of-Arrival (AOA) location estimation, in: Proc. 9th IEEE International Symposium on Personal, Indoor and Mobile Radio Communications.

[193] C.-F. Yang, C.-J. Ko, B.-C. Wu, A free space approach for extracting the equivalent dielectric constants of the walls in buildings, in: Proc. Antennas and Propagation Society International Symposium, 1996, pp. 1036–1039.

[194] L. Yang, G.B. Giannakis, Optimal pilot waveform assisted modulation for ultrawideband communications, IEEE Trans. Wireless Commun. 3 (4) (2004) 1236–1249.

[195] L. Yang, G.B. Giannakis, Ultra-wideband communications: An idea whose time has come, IEEE Signal Process. Mag. 21 (6) (2004) 26–54.

[196] L. Yang, G.B. Giannakis, Timing ultra-wideband signals with dirty templates, IEEE Trans. Commun. 53 (11) (2005) 1952–1963.

[197] A. Youssef, M. Younis, M. Youssef, A. Agrawala, On the accuracy of multi-hop relative location estimation in wireless sensor networks, in: IWCMC '07: Proceedings of the 2007 International Conference on Wireless Communications and Mobile Computing, Honolulu, Hawaii, ACM, New York, 2007, pp. 481–486.

[198] H. Zhan, J. Ayadi, J. Farserotu, J.-Y. Le Boudec, High-resolution impulse radio ultra wideband ranging, in: IEEE International Conference on Ultra-Wideband (ICUWB), 2007, pp. 568–573.

[199] J. Zhang, R.A. Kennedy, T.D. Abhayapala, Cramer–Rao lower bounds for the time delay estimation of UWB signals, in: Proc. IEEE Int. Conf. Commun. (ICC), Paris, France, 2004, pp. 3424–3428.

[200] Z. Zhang, C.L. Law, Y.L. Guan, BA-POC-based ranging method with multipath mitigation, IEEE Antennas Wireless Propag. Lett. 4 (2005) 492–495.

[201] Z. Zhang, C.L. Law, Y.L. Guan, BA-POC-based ranging method with multipath mitigation in the NLOS environment, Microwave Opt. Tech. Lett. 47 (4) (2005) 318–320.

[202] B. Zhen, H.-B. Li, R. Kohno, Clock management in ultra-wideband ranging, in: Proc. Mobile and Wireless Communications Summit, 2007, pp. 1–5.

[203] A. Giorgetti, M. Chiani, A new approach to time-of-arrival estimation based on information theoretic criteria, in: Proc. 2011 Conference on Ultra-Wideband (ICUWB 2011), Bologna, Italy, Sept. 2011.

Fundamental Limits in the Accuracy of Wireless Positioning

Giacomo Bacci, Pau Closas, Antonio D'Amico, Davide Dardari, Carles Fernández-Prades, Diana Fontanella, Sinan Gezici, Marco Luise, Achraf Mallat, Umberto Mengali, Montse Nájar, Monica Nicoli, Claude Oestges, Luc Vandendorpe

In the coming years, the field of radio positioning will have the seamless integration of both satellite and terrestrial positioning techniques to attain ubiquitous coverage, including hostile radio propagation environments such as "urban canyons" (roads sided with tall buildings) and interior of buildings (possibly with sub-meter accuracy). This specific requirement of high-definition situation-aware (HDSA) applications calls for the definition of new clever signal structures, the introduction of enhanced transmission technologies and paradigms such as UWB and cognitive radio, and the application of smart and powerful signal processing techniques.

Bearing in mind this perspective, this chapter deals with the study of *fundamental positioning techniques* and the derivation of the relevant *theoretical performance limits* to understand whether and in what situations the most stringent requirements of the mentioned HDSA applications can be met.

4.1 ACCURACY BOUNDS IN PARAMETER ESTIMATION AND POSITIONING

Estimation error bounds play a fundamental role in parameter estimation in general, and in positioning in particular, since they serve as useful benchmarks to assess the performance of practical algorithms, techniques, and estimators. The workhorse in this respect is the well-known Cramér–Rao lower bound (CRB) that gives the performance limit of any unbiased estimator in terms of MSE. Due to its popularity, the CRBs have been derived for a number of parameter estimation problems in the last decades. However, the introduction of new technologies such as UWB, multiple-input multiple-output (MIMO), and cognitive radio (CR) systems, as well as new network and propagation scenarios (e.g., ad hoc networks, hybrid positioning systems, etc.), has introduced new problems for which the relevant bounds were still to be derived.

In particular, UWB technology achieves high ranging accuracy even in harsh environments [22, 25] due to its ability to resolve multipath and penetrate obstacles [13, 42, 75], but it poses new challenges in estimator design and its performance evaluation. This chapter presents new CRB expressions applied to TOA estimation for UWB signals (Section 4.2.1), along with an analysis of the theoretical effect of dense multipath environments with path overlap (Section 4.2.2). Unfortunately, it is well known that the CRB is not accurate at low and moderate SNRs or for short observation intervals, and cannot take into account possible a priori side information about the parameter to be estimated, in particular a finite range of variation. Other bounds, more complicated to compute but tighter than the CRB, have been proposed in the literature. Among them, the Ziv–Zakai lower bound (ZZB), with its improved versions such as the Bellini–Tartara bound [10] and the Chazan–Zakai–Ziv bound [14], can be applied to a wider range of SNRs. Such bounds are not analytically tractable in many cases of interest or require more complicated numerical computations compared with the CRB, especially in the presence of nuisance parameters such as those emerging when operating in the presence of multipath. This particular issue (bounds for UWB on a multipath channel) will also be discussed in the following (Section 4.3).

We have focused until now on TOA estimation for a good reason: We already mentioned in Chapter 3 that position estimation is typically achieved using a pragmatic two-step approach, in which some physical parameters, such as the TOA, are estimated initially, and then a particular localization algorithm (e.g., triangulation) is applied to the estimated parameter values (e.g., ranges) to obtain the final position estimate. From a theoretical point of view, this two-step approach does not lead to optimum *position* estimators: a one-step *direct position estimation* (DPE) gives in general an improved performance. The analysis presented in Section 4.4.1 derives and compares the CRB for DPE with GNSS signals with that of a conventional approach.

In conventional approaches for terrestrial localization, once physical parameters such as TOA/AOA or RSS are estimated, the position of the *agent* is determined by combining measurements coming from different nodes depending on network topology. When multiple targets are present in the same scenario, an interesting possibility is to let nodes *cooperate* with each other to (mutually) improve the position estimation. The derivation of fundamental bounds for cooperative localization is a quite new research topic that is addressed in this chapter in Section 4.4.2, whereas practical schemes will be illustrated in Chapter 5.

Another new and exciting research field is positioning in cognitive radio (CR) systems, wherein the received signal may be scattered over a number of carriers possibly in noncontiguous bands. The relevant bounds for TOA estimation in CR-based positioning problems is also presented in this chapter (Section 4.4.3). Based on the bound, we will investigate criteria to optimize the distribution of the signal power across different carriers for best positioning in Chapter 5.

Once theoretical limits are established, practical positioning algorithms have to face a number of constraints in terms of power consumption, complexity, delay, flexibility, etc. One of the most popular approaches to devise parameter estimators is the ML criterion, which is known to provide asymptotically efficient estimators. Unfortunately, the asymptotical regime would not be feasible in practice (e.g., because of power emission constraints), and the design of optimal estimators for low-to-medium SNRs is still *terra incognita*. Another aspect is that the ML estimator is often too complex so that suboptimal schemes must be considered. A lot of research is being done on the design of practical techniques that can trade performance and complexity according to the specific application requirements. This will be the main topic addressed in Chapter 5.

4.1.1 Fundamental Limits in TOA Ranging with UWB Signals

In positioning systems based on ranging measurements, the key aspect is estimation of the TOA of the radio signal(s). As a consequence, analysis of the theoretical performance limits or bounds of TOA estimators is important to design practical TOA estimators. We will focus our discussion on the transmission of a single baseband UWB impulse-radio pulse $g(t)$ with unitary energy. Assuming a simple AWGN channel, the received signal is given by

$$r(t) = \sqrt{E_g}\, g(t - \tau) + v(t), \tag{4.1}$$

where $v(t)$ is AWGN with zero mean and two-sided power spectral density $N_0/2$, and E_g represents the received signal energy. Our aim is to estimate the TOA τ of $g(t)$ by observing the received signal $r(t)$ within the observation interval $[0, T_{\rm ob})$. This is a classic nonlinear parameter estimation problem that can be solved using a receiver based on an MF that yields an ML estimate as explained in Chapter 3 [67].

The Cramér–Rao Lower Bound (CRB)

Assuming that the observation time is larger than the duration of the pulse $g(t)$, the estimation MSE of any unbiased estimation $\hat{\tau}$ of τ can be lower bounded by the CRB as follows:

$$\mathrm{Var}\left\{\hat{\tau}\right\} = \mathbb{E}\left\{\xi^2\right\} \geq \kappa(\tau), \tag{4.2}$$

where $\xi = \hat{\tau} - \tau$ is the estimation error, and the CRB $\kappa(\tau)$ is given by [67]

$$\kappa(\tau) = \frac{N_0/2}{(2\pi)^2 E_g\, \beta^2} = \frac{1}{8\pi^2 \beta^2 \eta_g}. \tag{4.3}$$

Here $\eta_g \triangleq E_g/N_0$ is the SNR, and the parameter β^2 represents the second moment of the spectrum $G(f)$ of $g(t)$, that is,

$$\beta^2 \triangleq \frac{\int_{-\infty}^{\infty} f^2 |G(f)|^2\, df}{\int_{-\infty}^{\infty} |G(f)|^2\, df}, \tag{4.4}$$

where β is also called the *effective bandwidth*, the *(normalized) RMS bandwidth*, or the *Gabor bandwidth*. The best achievable accuracy of a *range* estimate $\hat{d}$ as derived from a TOA estimate satisfies

$$\mathrm{Var}\left\{\hat{d}\right\} \geq \frac{c^2}{8\pi^2 \beta^2 \eta_g}, \tag{4.5}$$

where c is the speed of light. Notice that the denominator of (4.5) is proportional to the energy in the signal, whereas the constant β^2 depends on the shape of the pulse through (4.4). This shows that, in general, having large values of the RMS bandwidth β^2 is beneficial for ranging.

In UWB positioning systems, the usual elementary pulse that is adopted is the nth derivative of the basic Gaussian pulse $g_0(t) = \exp\left(-2\pi t^2/\tau_g^2\right)$ [60]. In this case, we have

$$g(t) = g_0^{(n)}(t) \sqrt{\frac{(n-1)!}{(2n-1)!\,\pi^n \tau_g^{(1-2n)}}}\; \cos(2\pi f_c t), \quad \text{for } n > 0, \tag{4.6}$$

where τ_g is the *time duration parameter* (that also affects the pulse bandwidth), $g_0^{(n)}(t)$ denotes the nth-order time derivative of $g_0(t)$, the factor under the square root is a normalization constant, and where we also introduced for generality an (optional) carrier frequency f_c. It is easy to show that the effective (baseband, i.e., not considering the carrier) bandwidth $\beta^{(n)}$ of this Gaussian pulse (as defined by (4.4)) is given by

$$\beta^{(n)} = \sqrt{\frac{2n+1}{2\pi\tau_g^2}}. \tag{4.7}$$

From (4.7), we can observe that a lower CRB can be achieved, as is clear, by decreasing τ_g but also by increasing n.

For data communications, the IEEE 802.15.4 standard [1] also suggests the following band-pass pulse with center frequency f_c and root-raised-cosine (RRC) envelope:

$$g(t) = \frac{4\varrho\sqrt{2}}{\pi\sqrt{\tau_g}} \frac{\cos\left((1+\varrho)\pi t/\tau_g\right) + \frac{\sin((1-\varrho)\pi t/\tau_g)}{4\varrho t/\tau_g}}{1-(4\varrho t/\tau_g)^2} \cdot \cos(2\pi f_c t), \tag{4.8}$$

where the parameter τ_g and the roll-off factor ϱ determine the bandwidth $B_W = (1+\varrho)/\tau_g$ of the pulse. Two different values of τ_g are recommended: $\tau_g = 1$ ns and $\tau_g = 3.2$ ns with $\varrho = 0.6$, corresponding to two different pulse bandwidths, respectively, $B_W = 1.6$ GHz and $B_W = 500$ MHz. The same pulses can also be used for positioning applications within the data-communication standard.

The Ziv–Zakai Lower Bound (ZZB)

As already mentioned, the CRB is not accurate at low and moderate SNR and/or when the observation time is short. The MSE performance of a TOA estimator can be characterized by its behavior in different SNR regions: low, medium, and high SNRs. In particular, two *threshold* boundaries between the different regimes can be identified and have been studied in a variety of contexts, for example Refs. [31, 38, 39, 69, 70, 72]. In the low-SNR region (also known as the *a priori region*), signal observation provides very little additional information a priori, and the MSE is close to that obtained solely from the a priori information about the TOA (i.e., a finite possible range of variation). In the high-SNR region (also known as the *asymptotic region*), the MSE is quite accurately described by the CRB. Between these two extremes, there may be an additional region (also known as the *transition region* or *ambiguity region*) where observations are subject to ambiguities that are not considered by the CRB [67]. Therefore other bounds, which are more complicated but tighter than the CRB, have been introduced, such as the *Barankin lower bound* [78, 79], the *Ziv–Zakai lower bound* (ZZB) [4, 9, 14, 20, 30, 54], and the *Weiss–Weinstein bound* (WWB) [70]. All of these bounds are more accurate than the CRB over a wider SNR interval and take into account the presence of ambiguities. However, their analytical evaluation is often more complicated than the evaluation of the CRB.

The ZZB can be derived starting from the following general identity for MSE estimation [14]:[1]

$$\text{Var}\{\hat{\tau}\} = \mathbb{E}\{\xi^2\} = \frac{1}{2}\int_0^\infty z\cdot\mathcal{P}\left\{|\xi|\geq\frac{z}{2}\right\}dz, \tag{4.9}$$

where $\xi = \hat{\tau} - \tau$ represents the estimation error. A lower bound on $\mathcal{P}\{|\xi|\geq z/2\}$ is now found considering that such probability is related to the *error probability* of a classical *binary detection scheme* with

---

[1] Here the expectation is to be taken with respect to τ and the noise component in $r(t)$.

equally probable hypotheses[2]

$$\mathcal{H}_1 : r(t) \sim p\{r(t)|\tau\} \tag{4.10}$$

$$\mathcal{H}_2 : r(t) \sim p\{r(t)|\tau + z\}.$$

As a consequence, (4.9) can be lower bounded using the error probability of the optimum likelihood ratio test (LRT) decision rule [14]

$$\Lambda\left(r(t)\right) = \frac{p\{r(t)|\tau\}}{p\{r(t)|\tau + z\}}. \tag{4.11}$$

When τ is uniformly distributed in $[0, T_a)$, the ZZB is given by [14]

$$ZZB(\tau) = \frac{1}{T_a} \int_0^{T_a} z\,(T_a - z)\,\mathcal{P}_{min}(z)\,dz, \tag{4.12}$$

where $\mathcal{P}_{min}(z)$ is the error probability corresponding to the optimum decision rule.

On the AWGN channel, the minimum attainable probability of error is [50]:

$$\mathcal{P}_{min}(z) = Q\left[\sqrt{\eta_g\left(1 - \rho_g(z)\right)}\right], \tag{4.13}$$

where $Q[\cdot]$ is the complementary normalized Gaussian distribution function, $\rho_g(z)$ is the autocorrelation function of $g(t)$, and as before, $\eta_g = E_g/N_0$ [50]. In more complex propagation scenarios, the main challenge in (4.12) is finding the (optimum) binary detection scheme based on (4.11) and deriving a (manageable) expression for its probability of error $\mathcal{P}_{min}(z)$ [26].

Figure 4.1 shows the CRB and ZZB for the TOA RMSE using the band-pass RRC pulse $g(t)$ (4.8) with center frequency $f_c = 4$ GHz, roll-off $\varrho = 0.6$, $\tau_g = 1$ ns, and $\tau_g = 3.2$ ns [22]. The three regions of the bound (and in particular the thresholds) in the ZZB curve are apparent. For low SNR values, the received signal is too noisy so that the estimator only relies on the prior information, that is, τ being uniformly distributed in $[0, T_a)$: the ZZB approaches the variance of the uniformly distributed random variable, $T_a^2/12$ (no such effect is present in the CRB). For high SNR, the estimator "locks" onto the correct peak at the output of the MF with no ambiguity and very small probability of error $\mathcal{P}_{min}(z)$s so that the estimation error approaches that predicted by the CRB. In between, there is ambiguity, and the estimator may give an indication that still has some resemblance to the true value, but is affected by occasional large errors (ambiguity) that lead to a large estimation variance.

The band-pass RRC pulse with $\tau_g = 3.2$ ns bears an enlarged intermediate region between 12 dB and 30 dB. This depends on the presence of the carrier that produces additional ambiguity on the autocorrelation function of the received signal with respect to the baseband pulse. In fact, the RMSE gap depends on the ratio between the central frequency f_c and the signal bandwidth B_W as outlined in Refs. [54, 69]. Decreasing τ_g to 1 ns the intermediate region disappears because the carrier period is comparable to the pulse duration (same kind of ambiguity).

One of the interesting properties of the ZZB is that it turns the performance evaluation of an estimation problem into a binary detection problem. This property is exploited in Refs. [20, 23, 26] to

---

[2]The notation $\sim p\{\cdot\}$ stands for "following the statistical distribution $p\{\cdot\}$."

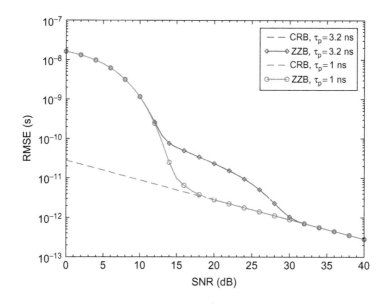

FIGURE 4.1

CRB and ZZB as a function of SNR in AWGN channel with $T_a = 100$ ns. Root raised cosine/radio resource control (RRC) band-pass pulses with $f_c = 4$ GHz, $\varrho = 0.6$, $\tau_g = 1$ ns, and $\tau_g = 3.2$ ns are considered.

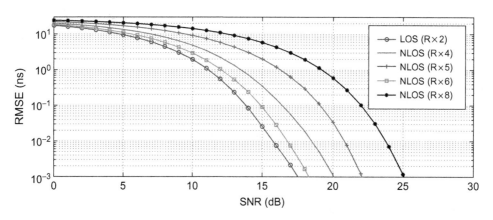

FIGURE 4.2

ZZB as a function of SNR using experimental data in Ref. [20], where $T_a = 100$ ns.

simplify the bound evaluation in a rich multipath environment. In Section 4.3, some improved ZZBs applied to more realistic channel models will be presented.

Figure 4.2 illustrates the ZZB of TOA as a function of the SNR obtained using real data [20] collected on a multipath radio channel under different propagation conditions. As can be seen, the curves shift to the right while moving from LOS to NLOS conditions, which implies that the TOA estimation error degrades under the NLOS condition. In general, given a certain target RMSE, for example 1 ns (corresponding to a ranging accuracy of 30 cm), it can be observed from the numerical results that the required SNR strongly depends on the multipath conditions.

4.2 VARIATIONS ON THE CRAMÉR–RAO BOUNDS

Let us assume that we intend to estimate more than just one unknown parameter characterizing the shape of the received signal. The multiple-parameter CRB states that for any unbiased estimate of a generic, real-valued parameter vector $\boldsymbol{\xi}$, the covariance matrix of the estimates, $\mathbf{C}(\hat{\boldsymbol{\xi}})$, is bounded as

$$\mathbf{C}(\hat{\boldsymbol{\xi}}) \geq \mathbf{J}^{-1}(\boldsymbol{\xi}) = \kappa(\boldsymbol{\xi}), \tag{4.14}$$

where $\mathbf{J}(\boldsymbol{\xi})$ is commonly referred to as the Fisher information matrix (FIM), whose inverse is the CRB matrix $\kappa(\boldsymbol{\xi})$ [36]. The matrix inequality in (4.14) means that $\kappa(\boldsymbol{\xi}) = \mathbf{C}(\hat{\boldsymbol{\xi}}) - \mathbf{J}^{-1}(\boldsymbol{\xi})$ is a non-negative definite matrix. Indicating by $\Lambda_{\mathbf{x}}(\boldsymbol{\xi})$ the log-likelihood function of the parameters under estimation, the (u,v)th FIM element is defined as

$$[\mathbf{J}(\boldsymbol{\xi})]_{u,v} = -\mathbb{E}\left\{\frac{\partial^2 \Lambda_{\mathbf{x}}(\boldsymbol{\xi})}{\partial \xi_u \partial \xi_v}\right\}. \tag{4.15}$$

As a "side-effect" this general result provides the desired lower bound on the variance of any unbiased estimator of the parameter ξ_u:

$$\mathbb{E}\left\{\left(\hat{\xi}_u - \xi_u\right)^2\right\} \geq \left[\mathbf{J}^{-1}(\boldsymbol{\xi})\right]_{u,u}. \tag{4.16}$$

Let us consider the following generic "single snapshot" AWGN observation model:

$$\mathbf{x}(t_k) = \boldsymbol{\mu}_x(t_k, \boldsymbol{\xi}) + \mathbf{n}(t_k) \in \mathbb{C}^N, \tag{4.17}$$

where $\mathbf{n}(t_k)$ is an additive complex Gaussian noise with covariance matrix $\mathbf{Q}$ [76], and t_k is the observation time. Its log-likelihood function is

$$\Lambda_x(\boldsymbol{\xi}) = -\left(\mathbf{x}(t_k) - \boldsymbol{\mu}_x(t_k, \boldsymbol{\xi})\right)^H \mathbf{Q}^{-1}\left(\mathbf{x}(t_k) - \boldsymbol{\mu}_x(t_k, \boldsymbol{\xi})\right) - \ln\det\left(\pi^N \mathbf{Q}\right). \tag{4.18}$$

Under these assumptions, using Slepian–Bang's formula [35, 62], we know that the (u,v)th element of the FIM for the generic K-snapshots case is[3]

$$[\mathbf{J}(\boldsymbol{\xi})]_{u,v} = K\,\mathrm{Tr}\left\{\mathbf{Q}^{-1}\frac{\partial \mathbf{Q}}{\partial \xi_v}\mathbf{Q}^{-1}\frac{\partial \mathbf{Q}}{\partial \xi_u}\right\} +$$

$$+ 2\Re\mathfrak{e}\left\{\sum_{k=0}^{K-1}\frac{\partial\left(\boldsymbol{\mu}_{\mathbf{x}}(t_k, \boldsymbol{\xi})\right)^H}{\partial \xi_u}\mathbf{Q}^{-1}\frac{\partial \boldsymbol{\mu}_{\mathbf{x}}(t_k, \boldsymbol{\xi})}{\partial \xi_v}\right\}. \tag{4.19}$$

---

[3]Tr $\{\cdot\}$ denotes the *trace* operator.

This is our starting point for further derivations of the bound in particular cases. An extensive analysis of CRBs applied to positioning can be found in Refs. [33, 57, 59, 73].

4.2.1 Cramér–Rao Bounds on TOA Estimation in the UWB Multipath Channel

In this section, we derive the CRB for TOA estimation under realistic channel propagation conditions, in particular extending previous results with nonoverlapping pulses [25], that turn out to be optimistic. The derivation is carried out in the *frequency domain*.

The general expression of the multipath fading propagation channel is as follows:

$$h(t) = \sum_{l=0}^{L-1} h_l \delta(t - \tau_l), \tag{4.20}$$

where h_l and τ_l are the amplitude and the delay of the lth path. With no loss of generality, we assume $\tau_0 < \tau_1 < \cdots < \tau_{L-1}$, where τ_0 is the TOA that is to be estimated.

Generalizing, we consider K_s observations of the received signal. Specifically, the kth observation is the sum of multiple delayed and attenuated replicas of the transmitted pulse $s(t)$:

$$r(t,k) = \sum_{l=0}^{L-1} h_l(k) s(t - \tau_l) + v(t,k). \tag{4.21}$$

The received pulse from each lth path has the same waveform but experiences a different channel coefficient (for each observation), $h_l(k)$, and time delay, τ_l. The additive noise[4] $v(t,k) \sim \mathcal{N}(0, N_0)$ is modeled as Gaussian circularly symmetric.

The frequency domain expression of the received signal is obtained after Fourier transforming (4.21):

$$R(\omega,k) = \sum_{l=0}^{L-1} h_l(k) S(\omega) e^{-j\omega\tau_l} + V(\omega,k), \tag{4.22}$$

where $R(\omega,k)$ is the noisy frequency domain observed signal, $S(\omega)$ is the transmitted signal in frequency domain, and $V(\omega,k)$ is the additive noise. Sampling (4.22) at $\omega_n = \omega_0 n$ for $n = 0, 1, \ldots, N-1$ and $\omega_0 = 2\pi/N$ and rearranging the frequency domain samples $R[n,k]$ into the vector $\mathbf{R}_k \in \mathbb{C}^{N \times 1}$ yields

$$\mathbf{R}_k = \mathbf{SE}_\tau \mathbf{h}(k) + \mathbf{V}_k, \tag{4.23}$$

where the matrix $\mathbf{R} \in \mathbb{C}^{N \times N}$ is diagonal, with components equal to the frequency samples of $S(\omega)$, and the matrix $\mathbf{E}_\tau \in \mathbb{C}^{N \times L}$ contains the delay-signature vectors (harmonic components) associated with each arriving delayed signal (paths),

$$\mathbf{E}_\tau = [\mathbf{e}_{\tau_0} \quad \cdots \quad \mathbf{e}_{\tau_j} \quad \cdots \quad \mathbf{e}_{\tau_{L-1}}], \tag{4.24}$$

---

[4]Note that, by writing the expression $\mathcal{N}(0, N_0)$, we are implicitly assuming that the two-sided receiver bandwidth is normalized to 1. This is in line with the following assumption on the sampling pulsation $\omega_0 = 2\pi/N$, which corresponds to a unitary sampling frequency.

with $\mathbf{e}_{\tau_j} = \left[1, e^{-j\omega_0 \tau_j} \dots e^{-j\omega_0 (N-1)\tau_j} \right]^T$. The channel fading coefficients are arranged in the vector $\mathbf{h}(k) = [h_0(k) \cdots h_{L-1}(k)]^T \in \mathbb{C}^{L \times 1}$, and the noise samples in vector $\mathbf{V}_k \in \mathbb{C}^{N \times 1}$.

The derivation of the CRB for TOA estimation, considering the frequency domain signal model given by (4.23), is a direct application of the CRB for spectral estimation presented in [63]. Let us first define the vector of the real parameters to be estimated as

$$\tau = [\tau_0 \dots \tau_{L-1}]^T. \tag{4.25}$$

The CRB is derived after taking into account the following assumptions:

1. Vectors $\mathbf{e}_{\tau_j}$ are linearly independent for different values of the parameter τ_i, and the number of frequency samples is greater than the number of paths, $N > L$.
2. The additive noise is Gaussian with $\mathbb{E}\{\mathbf{V}_k\} = 0$, $\mathbb{E}\{\mathbf{V}_k \mathbf{V}_k^H\} = \sigma_n^2 \mathbf{I}$, σ_n^2 being the noise samples variance.
3. The noise is uncorrelated: $\mathbb{E}\{\mathbf{V}_k \mathbf{V}_l^H\} = 0$, $\forall k \neq l$.

Considering K_s observations, the CRB is then given by

$$\kappa(\tau) = \sigma_n^2 \cdot \left(\sum_{k=0}^{K_s-1} \Re e \left\{ \mathbf{H}_k^H \mathbf{D}^H \mathbf{S}^H \left(\mathbf{I} - \mathbf{S}\mathbf{E}_\tau \left(\mathbf{E}_\tau^H \mathbf{S}^H \mathbf{S}\mathbf{E}_\tau \right)^{-1} \mathbf{E}_\tau^H \mathbf{S}^H \right) \mathbf{S}\mathbf{D}\mathbf{H}_k \right\} \right)^{-1}, \tag{4.26}$$

where $\mathbf{D} = \left[\dfrac{\partial \mathbf{e}_{\tau_0}}{\partial \tau_0} \cdots \dfrac{\partial \mathbf{e}_{\tau_{L-1}}}{\partial \tau_{L-1}} \right]$ and $\mathbf{H}$ is a diagonal matrix whose elements are the unknown channel fading coefficients,

$$\mathbf{H}_k = \begin{bmatrix} h_0(k) & & 0 \\ & \ddots & \\ 0 & & h_{L-1}(k) \end{bmatrix}. \tag{4.27}$$

For sufficiently large number of observations, the expression is bounded by

$$\kappa(\tau) = \frac{\sigma_n^2}{K_s} \left(\Re e \left\{ \left[\mathbf{D}^H \mathbf{S}^H \left(\mathbf{I} - \mathbf{S}\mathbf{E}_\tau \left(\mathbf{E}_\tau^H \|\mathbf{S}\|^2 \mathbf{E}_\tau \right)^{-1} \mathbf{E}_\tau^H \mathbf{S}^H \right) \mathbf{S}\mathbf{D} \right] \odot \bar{\mathbf{H}}^H \right\} \right)^{-1}, \tag{4.28}$$

where $\bar{\mathbf{H}}$ is the diagonal matrix obtained from

$$\bar{\mathbf{H}} = \mathbb{E}\left\{ \mathbf{H}_k \mathbf{H}_k^H \right\} \tag{4.29}$$

and where $\odot$ denotes the matrix Hadamard product (component-by-component product). If the delays between channel paths are greater than the pulse duration (nonoverlapping pulses), the delay-signature vectors $\mathbf{e}_{\tau_i}$ are orthogonal, and the following matrix products are diagonal [25]:

$$\mathbf{E}_\tau^H \|\mathbf{S}\|^2 \mathbf{E}_\tau = \sum_{n=0}^{N-1} |S[n]|^2 \mathbf{I} \tag{4.30}$$

$$\mathbf{D}^H \|\mathbf{S}\|^2 \mathbf{D} = \sum_{n=0}^{N-1} (n\omega_0)^2 |S[n]|^2 \mathbf{I} \tag{4.31}$$

$$\mathbf{D}^H \|\mathbf{S}\|^2 \mathbf{E}_\tau = j \sum_{n=0}^{N-1} n\omega_0 |S[n]|^2 \mathbf{I} \tag{4.32}$$

so that the CRB reduces to

$$\kappa(\tau) = \frac{\sigma_n^2}{K_s \left(\sum_{n=0}^{N-1} (n\omega_0)^2 |S[n]|^2 - \left| j \sum_{n=0}^{N-1} n\omega_0 |S[n]|^2 \right|^2 \left| \sum_{n=0}^{N-1} |S[n]|^2 \right|^{-1} \right)} \cdot \begin{bmatrix} \frac{1}{\mathbb{E}\{|h_0|^2\}} & & 0 \\ & \ddots & \\ 0 & & \frac{1}{\mathbb{E}\{|h_{L-1}|^2\}} \end{bmatrix} \tag{4.33}$$

or, simplifying notation,

$$\kappa(\tau_i) = \frac{\sigma_n^2}{K_s \left(E_p'' - E_p'^2/E_p \right) \mathbb{E}\{|h_i|^2\}}, \tag{4.34}$$

where the pulse-energy-related parameters are defined as follows:

$$E_p \triangleq \sum_{n=0}^{N-1} |S[n]|^2, \quad E_p'^2 \triangleq \left| j \sum_{n=0}^{N-1} n\omega_0 |S[n]|^2 \right|^2 \tag{4.35}$$

$$E_p'' \triangleq \sum_{n=0}^{N-1} (n\omega_0)^2 |S[n]|^2. \tag{4.36}$$

In general, the matrix defined in (4.28) is *not* diagonal, and the nonoverlapping CRB defined in (4.34) turns out to be optimistic. For positioning applications, we are interested only in the estimation of the first delay, defined as $\kappa(\tau_0) = \kappa_{1,1}(\tau)$. The condition of nonoverlapping pulses would be preferable (clearly leading to better time-of-arrival estimation), but unfortunately it is not realistic. Let us now try to evaluate numerically the impact of the overlap.

4.2.2 CRBs for UWB Multipath Channel Estimation: Impact of the Overlapping Pulses

In this section, we show the impact of the overlap of pulses on the performance of the multipath parameter estimation when IR-UWB signals are used. We show that for a pulse width sufficiently smaller than the average multipath component rate of arrival (ROA), the probability to have more than three overlapping multipath components is relatively small. We also show that instead of considering all multipath components together, we can split the received signal into *consecutive nonoverlapping blocks* and study each block separately.

4.2.2.1 *System Model*

In this paragraph, we describe the system model assumed for the addressed analysis. We consider an LOS multipath+AWGN radio propagation channel. The received signal can be written as usual as

$$r(t) = \sum_{l=1}^{L} h_l s(t - \tau_l) + v(t),$$ (4.37)

where $s(t)$ is the signal sent by the transmitter, h_l and τ_l are the gain and the time delay of the lth multipath component, respectively, L is the total number of multipath components, and $v(t)$ is the AWGN. Let $S_n = N_0/2$ be the two-sided PSD of $v(t)$. $s(t)$ is an impulse-radio (IR) UWB signal. Both pulse amplitude modulation (PAM) and PPM modulations can be used, and both delay spread (DS) and TH spreading and multiple access codes as well. Then $s(t)$ can be written as

$$s(t) = \sum_{i=-\infty}^{+\infty} a_i \sum_{j=0}^{N_c-1} c_j g_{T_w}(t - iT_s - jT_c - d_j T_h - b_i \epsilon),$$ (4.38)

where $g_{T_w}(t)$ is the transmitted pulse, T_w represents its width, a_i and b_i are the PAM and PPM symbols (if any), respectively, c_j and d_j are the DS and TH codes, respectively, N_c is the code length, T_s is the symbol period, $T_c = T_s/N_c$ is the chip period, T_h is the time-hop period, and ϵ the PPM time shift.

We will evaluate the numerical results considering a doublet (second derivative of a Gaussian pulse) UWB pulse:

$$g_{T_w}(t) \propto \left[1 - 4\pi \left(\frac{t}{T_w} \right)^2 \right] e^{-2\pi \left(\frac{t}{T_w} \right)^2}.$$ (4.39)

We denote by $S_s(f)$ and $P_s = \int_0^{+\infty} S_s(f) \, df$ the unilateral PSD and the power of $s(t)$, respectively, with T the integration time, $\gamma_p^{(l)}$ the loss due to the propagation through the lth path ($\gamma_p^{(1)} = 4\pi d^2$ where d is the transmitter-to-receiver distance), $A = c^2/(4\pi f_c^2)$ (f_c the central frequency) the receiver aperture gain, γ_{il} the implementation loss, η_t the transmitted signal to received noise ratio, and $\rho_r^{(l)}$ the SNR of multipath component l. Based on this, η_t and $\rho_r^{(l)}$ can be written as

$$\eta_t = \frac{P_s T}{S_n}, \quad \rho_r^{(l)} = \frac{P_s T A}{\gamma_p^{(l)} \gamma_{il} S_n}.$$ (4.40)

In Fig. 4.3, we show $\rho_r^{(l)}$ with respect to d (η_t is independent of d) when a doublet is used. We take $S_s^{peak} = -41.3$ dBm/MHz, $T_w = 0.13$ and 1.3 ns, $T = 50$ and 200 ns, $N_0/2 = -107.4$ dBm/MHz, and $\gamma_{il} = 2.5$ dB. For $T_w = 0.13$ ns, $T = 50$ ns, and $d = 9$ m, we obtain $\rho_r^{(1)} \approx 20$ dB and $\eta_t \approx 90$ dB.

Statistical Channel Model

Based on IEEE 802.15.3a and IEEE 802.15.4a channels models [11, 41], we assume that the multipath components arrive at the receiver in *consecutive clusters*. The TOA of clusters is a Poisson point

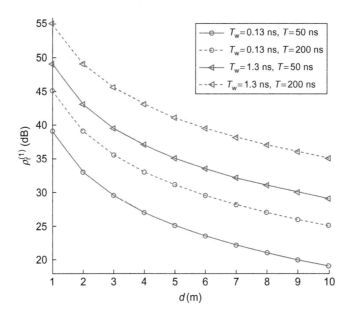

FIGURE 4.3

Transmitted signal to received noise ratio $\rho_r^{(1)}$ with respect to d (transmitter-to-receiver distance) (from Ref. [80]).

process of ROA $1/\Lambda$, so that the cluster interarrival times are exponentially distributed as follows:

$$p\left(\tau_{i,1}|\tau_{i-1,1}\right) = \Lambda e^{-\Lambda(\tau_{i,1}-\tau_{i-1,1})}. \tag{4.41}$$

The TOA of a multipath component within a cluster follows a mixture of two Poisson processes of ROAs $1/\lambda_1$ and $1/\lambda_2$. The (conditional) pdf of the TOA of the jth multipath component within the ith cluster is then

$$p\left(\tau_{i,j}|\tau_{i,j-1}\right) = k\lambda_1 e^{-\lambda_1(\tau_{i,j}-\tau_{i,j-1})} + (1-k)\lambda_2 e^{-\lambda_2(\tau_{i,j}-\tau_{i,j-1})}, \tag{4.42}$$

where $\tau_{i,j}$ denotes the TOA (taking as reference the time of transmission) of the jth multipath component within the ith cluster, and $k \in [0,1]$. Table 4.1 reports the cluster ROA, multipath component ROAs, and average multipath component ROA $1/\lambda$ (expectation of $\tau_{i,j} - \tau_{i,j-1}$ in (4.42)) in a few typical environments [11, 41].

Now we compute the probability of overlapping between neighboring multipath components. Assuming the chip period T_c (4.38) larger than the maximum time delay, we can consider without loss of generality the transmission of only one pulse. Given the channel statistical model just described, the received signal can be split into *consecutive nonoverlapping blocks of overlapping pulses* (i.e., the blocks are not overlapping, while neighboring multipath components within a block are overlapping) if the pulse width is sufficiently smaller than the average multipath component ROA.

Table 4.1 Standard Parameters of UWB Channels in Some Types of Environments

	Residential LOS	Office LOS	Outdoor LOS	Short-Range LOS
Range [m]	7–20	3–28	5–17	0–4
$1/\Lambda$ [ns]	21.28	62.5	208.3	42.92
$1/\lambda_1$ [ns]	0.65	5.26	3.7	0.4
$1/\lambda_2$ [ns]	6.67	0.34	0.41	–
$1/\lambda$ [ns]	6.1	0.43	0.44	0.4

Let $\mathcal{P}^{t_0}_{\Delta\tau}$ be the probability that $\Delta\tau^{(l,l-1)} \leq t_0$ (where $\Delta\tau^{(l',l)} \triangleq \tau_{l'} - \tau_l$). Assuming that the $(l-1)$th and lth multipath components fall in the same cluster, we can write

$$\mathcal{P}^{t_0}_{\Delta\tau} = \mathcal{P}\left\{\Delta\tau^{(l,l-1)} \leq t_0\right\} = \int_0^{t_0} \left\{k\lambda_1 e^{-\lambda_1 t} + (1-k)\lambda_2 e^{-\lambda_2 t}\right\} dt$$

$$= k(1 - e^{-\lambda_1 t_0}) + (1-k)(1 - e^{-\lambda_2 t_0}). \qquad (4.43)$$

Let us now indicate by L_b the random variable indicating the number of multipath components falling within an *estimation block*, that is, within one observation time of the TOA estimator, and let $\mathcal{P}^{l_b}_{L_b}$ be the probability that the *first* block contains just l_b components (i.e., the probability that $L_b = l_b$). Assuming that the first block BK_1 (BK_i denotes the ith estimation block) falls entirely within the first cluster, we can write

$$\mathcal{P}^{l_b}_{L_b} = \mathcal{P}\{L_b = l_b\} = \mathcal{P}\left\{\left(\Delta\tau^{(l,l-1)} \leq T_w, l = 2, l_b\right), \left(\Delta\tau^{(l_b+1,l_b)} > T_w\right)\right\}$$

$$= \left(\mathcal{P}^{T_w}_{\Delta\tau}\right)^{l_b-1}\left(1 - \mathcal{P}^{T_w}_{\Delta\tau}\right). \qquad (4.44)$$

Note that even if BK_1 does not entirely fall in the first cluster, (4.44) can be considered a good approximation if T_w and $1/\lambda$ are sufficiently smaller than $1/\Lambda$. The probability that an arbitrary block contains l_b multipath components is slightly different from that given in (4.44). For brevity, we considered only BK_1.

In Fig. 4.4, we show $\mathcal{P}^{l_b}_{L_b}$ with respect to the *pulse width to average multipath component rate of arrival ratio* λT_w for $l_b = 1, 2, 3$ and $l_b \leq 3$ as a function of λT_w and for residential LOS and office LOS environments. We can see that $\mathcal{P}^1_{L_b} > \mathcal{P}^2_{L_b} > \mathcal{P}^3_{L_b}$, $\forall \lambda T_w$. We can also see that for $\lambda T_w = 0.5$, the probability of having up to three multipath components is greater than 0.9 for both residential and office environments.

Taking into consideration the last result, we can simplify our study by considering blocks of one, two, and three multipath components instead of considering the whole channel impulse response. When the pulse width becomes larger than the average multipath component ROA, blocks of more than three multipath components must be considered. As an instructive example, we limit our study to blocks of up to three multipath components.

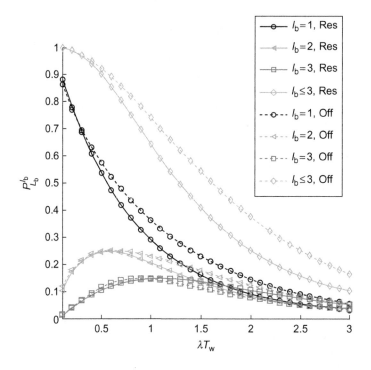

FIGURE 4.4

$\mathcal{P}^{l_b}_{L_b}$ with respect to λT_w for $l_b = 1, 2, 3$ and $l_b \leq 3$ in residential LOS (Res) and office LOS (Off) environments (from Ref. [80]).

4.2.2.2 Computation of the Bounds

We now derive the FIM for the *joint* estimation of the channel parameters (delays and amplitudes) $\boldsymbol{\xi} \triangleq [h_1, \tau_1, \ldots h_L, \tau_L]$. The FIM is given by

$$J(\boldsymbol{\xi}) = -\mathbb{E}\left\{\left(\frac{\partial^2 \boldsymbol{\Lambda}(\boldsymbol{\xi})}{\partial \xi_i \partial \xi_j}\right)_{i,j}\right\}, \tag{4.45}$$

where $\boldsymbol{\Lambda}(\boldsymbol{\xi})$ is the log-likelihood function, which is given by

$$\boldsymbol{\Lambda}(\boldsymbol{\xi}) = \mathcal{C} - \frac{1}{2S_n}\int_T \left\{r(t) - \sum_{l=1}^{L} h_l s(t - \tau_l)\right\}^2 dt, \tag{4.46}$$

where $\mathcal{C}$ is a constant. With reference to a generic signal $x(t)$, we denote now by $\dot{x}(t)$ its time derivative, with $X(f)$ its Fourier transform, E_x its energy in the time interval T, $\beta_x^2 \triangleq \int 4\pi^2 f^2 |X(f)|^2 df / E_x$ its mean quadratic (Gabor) bandwidth, and $R_x(\tau)$ its autocorrelation function. We also indicate by $X_{x,y}(\tau) = $ the cross-correlation function between $x(t)$ and $y(t)$, with $\check{R}_x(\tau) = R_x(\tau)/E_x$ and $\check{X}_{x,y}(\tau) = X_{x,y}(\tau)/\sqrt{E_x E_y}$ the normalized auto- and cross-correlation functions, respectively. Taking the expectation of the second partial derivatives of $\boldsymbol{\Lambda}(\boldsymbol{\xi})$ leads to the following expressions for the different elements of $J(\xi)$:

$$f^{h_l,h_{l'}} = \frac{R_s^{(l',l)}}{S_n} = \eta_t \hat{R}_s^{(l',l)} \tag{4.47}$$

$$f^{\tau_l,\tau_{l'}} = \frac{h_l h_{l'} R_s^{(l',l)}}{S_n} = \sqrt{\rho_r^{(l)} \rho_r^{(l')}} \beta_s^2 \hat{R}_s^{(l',l)} \tag{4.48}$$

$$f^{h_l,\tau_{l'}} = \frac{-h_{l'} X_{s\dot{s}}^{(l',l)}}{S_n} = -\sqrt{\eta_t \rho_r^{(l')}} \beta_s \hat{X}_{s\dot{s}}^{(l',l)}, \tag{4.49}$$

where for the generic auto- or cross-correlation function (normalized or not) Z, the short-hand notation $Z^{(l',l)}$ indicates $Z(\Delta\tau^{(l',l)})$, and where $f^{h_l,\tau_{l'}} = f^{\tau_{l'},h_l}$. Taking $l = l'$ and assuming $s^2(t) = 0$ if $t \notin [0,T]$, we obtain the elements of $J(\xi)$ corresponding to the lth multipath component as

$$f^{h_l,h_l} = \eta_t, \quad f^{\tau_l,\tau_l} = \rho_r^{(l)} \beta_s^2, \quad f^{h_l,\tau_l} = 0. \tag{4.50}$$

Now, we can write $J(\xi)$ as

$$J(\xi) = \begin{pmatrix} J^{(1,1)} & J^{(1,2)} & \cdots \\ J^{(2,1)} & J^{(2,2)} & \cdots \\ \vdots & \vdots & \ddots \end{pmatrix}, \tag{4.51}$$

where

$$J^{(l,l')} = \begin{pmatrix} f^{h_l,h_{l'}} & f^{h_l,\tau_{l'}} \\ f^{\tau_l,h_{l'}} & f^{\tau_l,\tau_{l'}} \end{pmatrix}. \tag{4.52}$$

If the lth multipath component does not overlap with any other multipath component, $J^{(l,l')}$ becomes null for all $l' \neq l$. Given that $J^{(l,l)}$ is diagonal, $\kappa(\tau_l)$ and $\kappa(h_l)$ can be derived by inverting f^{h_l,h_l} and f^{τ_l,τ_l}, respectively. The CRBs obtained in this case are the *smallest* which can be obtained. They correspond to the CRBs obtained in an AWGN channel.

We have stated that the received signal can be split into nonoverlapping blocks of overlapping multipath components when the pulse width is sufficiently smaller than the average multipath component ROA. Given that $X_{x,y}(\tau) = 0$ (respectively $R_x(\tau) = 0$) if $x(t+\tau)$ and $y(t)$ (respectively $x(t)$) are not overlapping, $f^{h_l,h_{l'}}$, $f^{\tau_l,\tau_{l'}}$, and $f^{h_l,\tau_{l'}}$ become null if the lth and l'th multipath components belong in two different blocks. Let $J(\xi)_{b_i}$ be the FIM of BK_i. $J(\xi)$ can be written as

$$J(\xi) = \begin{pmatrix} J(\xi)_{b_1} & 0 & \cdots \\ 0 & J(\xi)_{b_2} & \cdots \\ \vdots & \vdots & \ddots \end{pmatrix}. \tag{4.53}$$

We understand from (4.53) that the CRBs of the parameters of a given block can be obtained by inverting only the FIM of the concerned block. This means that different blocks can be considered separately, and by studying all possible cases of an arbitrary block we can characterize the whole channel. We have already mentioned that only blocks of up to three multipath components will be considered.

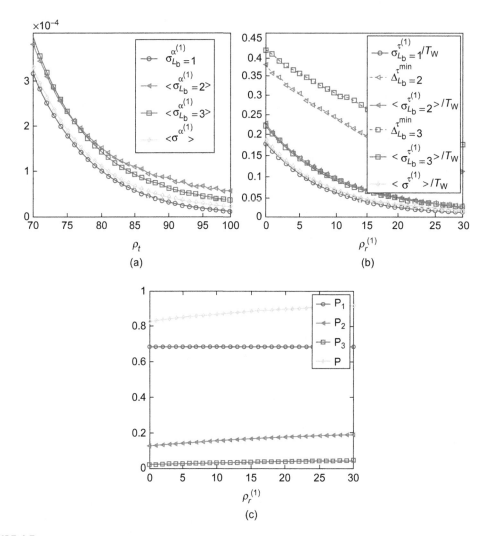

FIGURE 4.5

(a) Bounds $\sigma_{L_b=1}^{h^{(1)}}$, $\left\langle \sigma_{L_b=l_b}^{h^{(1)}} \right\rangle$, $l_b = 2, 3$, and $\left\langle \sigma^{h^{(1)}} \right\rangle$ with respect to η_t. (b) Bounds $\sigma_{L_b=1}^{\tau^{(1)}}/T_w$, $\Delta\tau_{L_b=l_b}^{min}/T_w$ and $\left\langle \sigma_{L_b=l_b}^{\tau^{(1)}} \right\rangle/T_w$, $l_b = 2, 3$, and $\left\langle c^{\tau^{(1)}} \right\rangle/T_w$ with respect to $\rho_r^{(1)}$. (c) Probabilities $\mathcal{P}_1$, $\mathcal{P}_2$, $\mathcal{P}_3$, and $\mathcal{P}$ with respect to $\rho_r^{(1)}$ (from Ref. [80]).

Skipping the details of the computation of the bounds for blocks of one, two, and three multipath components, we jump directly to the computation of the *average* of the square roots of the CRBs of the gain and the TOA of the first multipath component when the square root of the CRB of the TOA is smaller than the intercomponent delay. Since this average does not consider all possible cases, we also compute the probability of the averaged cases. In Fig. 4.5(a), we show the rms bounds $\sigma_{L_b=1}^{h^{(1)}}$,

$\left\langle \sigma_{L_b=l_b}^{h^{(1)}} \right\rangle$, $l_b = 2,3$, and $\left\langle \sigma^{h^{(1)}} \right\rangle$ with respect to η_t when a doublet of $T_w = 0.13$ ns is used in office LOS environments ($\lambda T_w = 0.3824$). In Fig. 4.5(b), we show $\sigma_{L_b=1}^{\tau^{(1)}}/T_w$, $\Delta \tau_{L_b=l_b}^{\min}/T_w$, and $\left\langle \sigma_{L_b=l_b}^{\tau^{(1)}} \right\rangle/T_w$, $l_b = 2,3$, and $\left\langle c^{\tau^{(1)}} \right\rangle/T_w$ with respect to $\rho_r^{(1)}$. Finally, in Fig. 4.5(c) we show $\mathcal{P}_i$, $i = 1,2,3$, and $\mathcal{P}$ with respect to $\rho_r^{(1)}$.

We can see that the average bounds for the estimation of the gain and the TOA are very close to the bounds obtained in the nonoverlapping case (which are the smallest bounds that can be obtained). We can also see that the probability of the averaged cases is greater than 0.8 (respectively 0.9) for $\rho_r^{(1)} = 0$ dB (respectively $\rho_r^{(1)} \geq 23$ dB). This means that the approximate bounds can be considered a good approximation of the bounds obtained for joint multipath component parameter estimation. A different analysis on the effect of multipath overlapping can be found in Ref. [57] and references therein.

4.3 VARIATIONS ON THE ZIV–ZAKAI BOUND

In dense multipath channels, it may happen that the first path is not the strongest, making estimation of the TOA particularly challenging [21]. Derivation of the relevant estimation bounds [10, 14] in such conditions is particularly relevant, since they serve as useful performance benchmarks for the design of TOA estimators.

As an example, Ref. [77] evaluates the ZZB for Gaussian signals assuming perfect channel knowledge at the receiver. A few results for the case in which the receiver has a partial or no knowledge about the channel are reported in Refs. [4, 5, 20, 26]. In particular, in Ref. [20] the ZZB using measured data as well as Monte Carlo-generated CPRs is investigated, whereas Ref. [4] derives the ZZB using a second-order nonparametric approach by modeling the received signal as a nonstationary Gaussian random process.

In many cases, assuming complete a priori information about the channel (perfect or even partial channel status information) is not realistic. Much more realistic may be assuming knowledge of some *statistical* information about the channel. In this section, we derive analytical expressions for the ZZB on TOA estimation in wideband and UWB ranging systems operating in realistic multipath environments, assuming the receiver has a priori statistical knowledge about the multipath characteristics [23, 26].

4.3.1 Signal and Channel Models for UWB Scenarios

We assume that a single pulse $g(t)$ with duration T_g, energy E_g, and normalized autocorrelation function $\rho_g(\tau)$ is transmitted.[5] Generally, $g(t)$ is a bandpass signal with bandwidth[6] B_W and central frequency f_c as defined in Section 4.1, (4.6), and (4.8):

$$g(t) = \sqrt{2E_g}\, b(t) \cos(2\pi f_c t), \tag{4.54}$$

where $b(t)$ is a unit-energy baseband pulse.

---

[5]In general, $g(t)$ can represent multiple-access signals, such as direct sequence or TH [74].
[6]Since a band-limited signal cannot also be time limited, we assume it can be truncated in time to duration T_g with some small approximation error.

When $g(t)$ is sent through an AWGN channel with multipath, the received signal is

$$r(t) = r_0(t - \tau) + v(t), \tag{4.55}$$

where $r_0(t)$ is the channel pulse response (CPR), τ is the TOA of the received signal to be estimated, and $v(t)$ is AWGN with zero mean and two-sided spectral density $N_0/2$. We consider a typical L-path multipath channel model

$$r_0(t) = \sqrt{2E_g} \sum_{l=1}^{L} \Re \left\{ \alpha_l b(t - \tau_l) e^{j 2\pi f_c (t - \tau_l)} \right\}, \tag{4.56}$$

where L is the number of multipath components, α_ls are the complex path gains and τ_ls are the *differential delays* with respect to the TOA of the first path (i.e., $\tau_1 = 0$). Note that with respect to (4.20), we have here

$$h_l = \alpha_l e^{-j 2\pi f_c \tau_l}. \tag{4.57}$$

For a band-limited transmitted pulse, the received signal can also be cast into a different form, following the *tapped delay line* channel model:

$$r_0(t) = \sum_{l=1}^{L} \left[a_l g_I(t - \tau_l) + a_{L+l} g_Q(t - \tau_l) \right], \tag{4.58}$$

where a_l and a_{L+l} for $l = 1, 2, \ldots, L$ are, respectively, the real and imaginary parts of α_l, $g_I(t) \triangleq g(t) = \sqrt{2E_g} b(t) \cos(2\pi f_c t)$ and $g_Q(t) \triangleq -\sqrt{2E_g} b(t) \sin(2\pi f_c t)$, and where $\tau_l = \Delta \cdot (l - 1)$, with Δ denoting the width of the resolvable time bin, and $T_d = \Delta \cdot (L - 1)$, the dispersion (maximum delay) of the channel [13, 42]. We further consider *resolvable* multipath, that is, $|\tau_l - \tau_i| \geq T_g$ for $i \neq l$.[7] The average power gain of the lth path is $\Lambda_l = \mathbb{E}\left\{ |a_l|^2 + |a_{L+l}|^2 \right\}$. Without loss of generality, we assume $\sum_{l=1}^{L} \Lambda_l = 1$, so that E_g represents the average received energy per transmitted pulse. We denote by $p_A(\mathbf{a})$ the (joint) pdf of vector $\mathbf{a}$. For each set of channel parameters $\mathbf{a} = \{a_1, a_2, \ldots, a_{2L}\}$, we obtain a sample $r_0(t|\mathbf{a})$ of the random process $r_0(t)$.

In the Rice/Rayleigh multipath channel model (the simplest to treat), the a_ls are independent Gaussian random variables with mean

$$\mu_l = \mathbb{E}\{a_l\} = \begin{cases} s_l & 1 \leq l \leq L, \\ 0 & L < l \leq 2L \end{cases} \tag{4.59}$$

and variance $\mathrm{Var}\{a_l\} = \mathrm{Var}\{a_{L+l}\} = \sigma_l^2$ for $1 \leq l \leq L$, and the pdf $p_A(\mathbf{a})$ is multivariate Gaussian. The parameters s_l and σ_l^2 represent the amplitude of the specular (reflected) component (when present), and the gain of the scattered component related to the lth path. The average power gain of the lth path is $\Lambda_l = \mu_l^2 + 2\sigma_l^2$. Consideration of more complicated distributions, like Nakagami-m fading, does not add much insight to the problem. Evidence of this is the fact that Nakagami channels can be

---

[7]When the multipath is not resolvable, the following analysis gives weaker lower bounds on the achievable performance. A more general analysis can be found in Ref. [26].

well approximated by Rice statistics with a suitable mapping of parameters [42]. Given the Nakagami parameters m_N and Λ, the Rice parameters $K_{\text{Rice}} = s^2/(2\sigma^2)$ (the power ratio between the specular and the scattered signals) and σ can be evaluated through the following relationships [1]:

$$K_{\text{Rice}} = \frac{\sqrt{m_N - 1}}{1 - \sqrt{m_N - 1}} \tag{4.60}$$

$$\sigma^2 = \frac{\Lambda}{2(K_{\text{Rice}} + 1)}. \tag{4.61}$$

4.3.2 Derivation of the Ziv–Zakai Lower Bound

As mentioned in Section 4.1, the key to compute the ZZB for TOA is the derivation of the error probability $\mathcal{P}_{\min}(z)$ for optimum binary decision in the presence of multipath. According to our multipath channel model, the conditional pdf of $r(t)$ (4.55) is [75]

$$p\{r(t)|\tau,\mathbf{a}\} = \mathcal{C} \exp\left\{ -\int_0^{T_{\text{ob}}} [r(t) - r_o(t - \tau|\mathbf{a})]^2 \, dt \right\}, \tag{4.62}$$

with $\mathcal{C}$ denoting a normalization constant, whose exact value does not affect the final result. Using (4.58) into (4.62), and assuming resolvable multipath components, we get

$$p\{r(t)|\tau,\mathbf{a}\} = \exp\left\{ -\eta_g \sum_{k=1}^{2L} \left(a_k^2 - 2a_k q_k(\tau) \right) \right\}, \tag{4.63}$$

where

$$q_k(\tau) \triangleq \begin{cases} \frac{1}{E_g} \int_0^{T_{\text{ob}}} r(t) \, g_I(t - \tau - \tau_k) \, dt & 1 \le k \le L \\ \frac{1}{E_g} \int_0^{T_{\text{ob}}} r(t) \, g_Q(t - \tau - \tau_{k-L}) \, dt & L < k \le 2L \end{cases} \tag{4.64}$$

and $\eta_g = \frac{E_g}{N_0}$ is the SNR.

To derive the bound (4.12), we further need the marginal pdf $p\{r(t)|\tau\}$ that, for Rayleigh/Rice fading, is found to be

$$p\{r(t)|\tau\} = \mathbb{E}\{p\{r(t)|\tau,\mathbf{a}\}\} \propto \prod_{k=1}^{2L} \frac{1}{\sqrt{1 + \eta_g 2\sigma_k^2}} \cdot \exp\left\{ \frac{\left(\frac{\mu_k}{2\sigma_k^2} + \eta_g q_k(\tau) \right)^2}{\left(\frac{1}{2\sigma_k^2} + \eta_g \right)} \right\}. \tag{4.65}$$

The corresponding log-likelihood ratio (LLR) is [23]

$$l(r(t)) = \ln \frac{p\{r(t)|\tau\}}{p\{r(t)|\tau + z\}} = \sum_{k=1}^{2L} A_k \cdot \left[\left(\frac{\mu_k}{2\sigma_k^2} + \eta_g q_k(\tau) \right)^2 - \left(\frac{\mu_k}{2\sigma_k^2} + \eta_g q_k(\tau + z) \right)^2 \right], \tag{4.66}$$

where $A_k \triangleq 2\sigma_k^2/(1+2\sigma_k^2 n_g)$. The LLR (4.66) leads to the optimal binary PPM (partially coherent) Rake receiver depicted in Fig. 4.6.

The probability of error $\mathcal{P}_{\min}(z)$, required to evaluate the ZZB in (4.12), is given by

$$\mathcal{P}_{\min}(z) = \frac{1}{2}\mathcal{P}_{e|\mathcal{H}_1}(z) + \frac{1}{2}\mathcal{P}_{e|\mathcal{H}_2}(z), \qquad (4.67)$$

where $\mathcal{P}_{e|\mathcal{H}_1}(z)$ and $\mathcal{P}_{e|\mathcal{H}_2}(z)$ are the error probabilities conditioned to hypotheses $\mathcal{H}_1$ and $\mathcal{H}_2$, respectively. From symmetry, $\mathcal{P}_{e|\mathcal{H}_1}(z) = \mathcal{P}_{e|\mathcal{H}_2}(z)$ so that it is sufficient to evaluate $\mathcal{P}_{e|\mathcal{H}_1}(z)$. From (4.66), under hypothesis $\mathcal{H}_1$, the decision variable y is

$$y = l(r(t)|\mathcal{H}_1) = \sum_{k=1}^{2L} A_k\left[\left(X_k^{(1)}\right)^2 - \left(X_k^{(2)}\right)^2\right], \qquad (4.68)$$

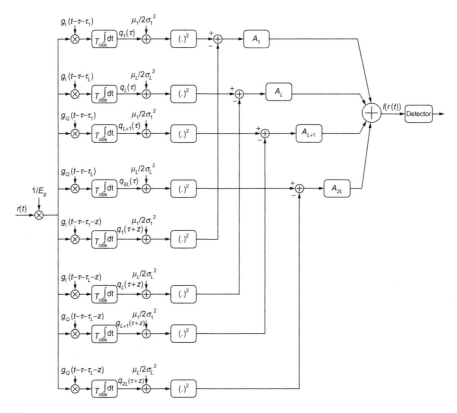

FIGURE 4.6

Optimum binary detector in the presence of Rayleigh/Rice fading (from Ref. [23]).

where

$$X_k^{(1)} \triangleq \frac{\mu_k}{2\sigma_k^2} + \eta_g q_k^{(1)} \tag{4.69}$$

$$X_k^{(2)} \triangleq \frac{\mu_k}{2\sigma_k^2} + \eta_g q_k^{(2)} \tag{4.70}$$

and $q_k^{(1)} \triangleq q_k(\tau)|_{\mathcal{H}_1}$ and $q_k^{(2)} \triangleq q_k(\tau + z)|_{\mathcal{H}_1}$ for $k = 1, 2, \ldots, 2L$. The probability of error required to compute the ZZB in (4.12) is therefore

$$\mathcal{P}_{\min}(z) = \mathcal{P}_e|_{\mathcal{H}_1}(z) = \mathcal{P}\{y < 0|\mathcal{H}_1\}. \tag{4.71}$$

It is easy to see that y in (4.68) is a noncentral quadratic form in real nonidentically distributed Gaussian-correlated RVs. The required probability cannot be found in closed form, but can be numerically evaluated through the moment-generating function method with saddlepoint integration [23, 26].

4.3.3 Numerical Results in the Presence of Multipath

We consider for our numerical examples a band-pass pulse with an RRC envelope and with center frequency $f_c = 4$ GHz, roll-off factor $\varrho = 0.6$, and $\tau_g = 1$ ns. In Figs 4.7 and 4.8, the RMSE for the ZZB and the CRB versus the SNR are plotted for uniform PDP ($\Lambda_l = 1$ for each l), with $\Delta = 2$ ns, in different multipath conditions, and assuming that the receiver has a priori statistical channel knowledge. In particular, Fig. 4.7 assumes that all paths are affected by Rayleigh fading. It is seen that the ZZB decreases on increasing the number of paths L, especially at large SNR, due to the diversity gain achieved by the estimator. The presence of multipath has a strong detrimental effect on the lower bound compared with the AWGN case. This effect is not reflected by the CRB that appears to be *extremely looser*, even for high SNRs.

Figure 4.8 considers Ricean fading for the first path with $K_{\text{Rice}} = 9.3$ dB. Again, there is a distinct diversity gain that leads to rapid convergence of the ZZB toward the CRB for large SNR when $L > 2$. Still, the CRB fails to accurately predict the MSE of TOA for a wide range of practical SNR values.

We now consider a multipath channel with exponential PDP with average power gains given by

$$\Lambda_l = \frac{(e^{\Delta/\varepsilon_\Lambda} - 1)e^{-\Delta(l-1)/\varepsilon_\Lambda}}{e^{\Delta/\varepsilon_\Lambda}(1 - e^{L\Delta/\varepsilon_\Lambda})} \tag{4.72}$$

and Nakagami-m fading with Nakagami severity parameter $m_l = m_1 e^{-(l-1)/\varepsilon_m}$ for $l = 1, 2, \ldots, L$ [13, 42]. The parameter ε_Λ describes the multipath spread of the channel. We consider $L = 8$, $\varepsilon_\Lambda = 6$ ns, $\varepsilon_m = 4$, and $m_1 = 3$. The comparison between different channel models is reported in Fig. 4.9 where the ZZB in AWGN and in multipath channels with exponential PDP (4.72), and Rayleigh uniform PDP is shown. The ZZB accounts for the impact of the fading statistic as well as the multipath PDP. Rayleigh fading provides the worst performance at high SNR.

The conclusion of this section is that, for practical SNR values, the classical CRB turns out to be too optimistic. In particular, the CRB fails to properly account for the effect of multipath, whereas the ZZB provides much improved bounds, especially in the presence of Ricean and/or rich multipaths.

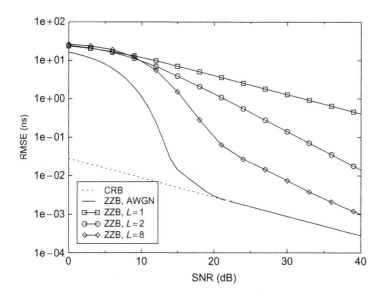

FIGURE 4.7

ZZB and CRB on the RMSE as a function of SNR for uniform PDP with Rayleigh fading. Band-pass pulse with RRC envelope, $f_c = 4$ GHz, $\varrho = 0.6$, $\tau_g = 1$ ns, and $T_a = 100$ ns considered.

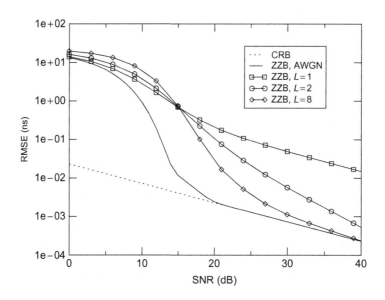

FIGURE 4.8

ZZB and CRB on the RMSE as a function of SNR for uniform PDP with Ricean fading for the first path with $K = 9.3$ dB. Band-pass pulse with RRC envelope, $f_c = 4$ GHz, $\varrho = 0.6$, $\tau_g = 1$ ns, and $T_a = 100$ ns considered.

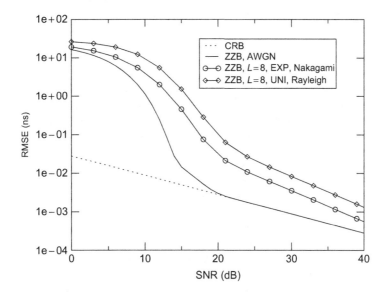

FIGURE 4.9

ZZB and CRB on the RMSE as a function of SNR. Comparison between different channel models. Band-pass pulse with RRC envelope, $f_c = 4$ GHz, $\varrho = 0.6$, $\tau_g = 1$ ns and $T_a = 100$ ns considered.

4.4 INNOVATIVE POSITIONING ALGORITHMS AND THE RELEVANT BOUNDS

Up to now, the context in which positioning algorithms were applied was pretty standard: the end-user terminal performs a number of ranging measurements, and from those it derives its own position via some kind of trilateration algorithm. This conventional setup is being revised and improved following a radically different approach. On one side, many researchers are investigating *direct positioning* algorithms that strive to directly derive the position of the user terminal with the application of a single-step algorithm that does not bear explicitly the two former disjoint functions of TOA estimation and trilateration. On the other side, a number of techniques are being developed to perform *cooperative positioning* of a whole set of user terminals that can exchange data during the function of positioning and, by doing so, mutually improve the accuracy of their own position computation. Finally, the emergence of the concept of cognitive radio inspired the study of *cognitive positioning* algorithms for cognitive terminals that can sense the interference and noise level of their environment and adapt their emitted signals to such conditions. This section presents the latest results about the accuracy bounds computation for such innovative approaches.

4.4.1 Theoretical Bounds for Direct Position Estimation in GNSS

As known, the conventional approach to GNSS positioning is based on a two-step procedure. First, the receiver estimates the distance between the receiver and the satellites. Then, these distances are used to obtain user position by a trilateration procedure. Recent results produced in the context of GNSS receivers following the software defined radio (SDR) paradigm envisage a different approach to the positioning problem. In Ref. [17], a *direct position estimation* (DPE) approach was presented: whereas conventional receivers require estimates of time difference of arrival (TDOA) of satellite signals to geometrically solve user's coordinates, DPE focuses on the estimation (directly from the received and sampled IF signal) of the position coordinates. Position coordinates are indeed the parameters of interest to the end user. The avoidance of intermediate estimation steps helps to partially overcome some limitations of current approaches, such as the degradation in position accuracy due to multipath and severe channel fading conditions [19], especially in multi-antenna receivers. This section focuses on the computation of bounds and algorithms following this general approach.

Recently, Ref. [71] introduced the notion of *direct positioning*: merge the two-step *ranging-plus-trilateration*, conventional procedure into a single estimation step to derive the position of narrowband *emitters*. Although specular to the GNSS problem, which is the positioning of a passive receiver instead of an emitter, the signal processing methodology of both applications is similar. In Ref. [3] this principle was also applied to the radiolocation of emitters using Doppler measurements.

The ML direct position estimator was first derived in Ref. [17]. The resulting criterion calls for the maximization of a multivariate nonconvex cost function, so that a number of different algorithms can be implemented in order to obtain efficient implementation architectures for DPE. The results obtained so far show a performance improvement of DPE with respect to the conventional approach: Ref. [18] provides an interesting proof showing that the conventional two-step approach cannot outperform a direct estimation. In the following we will try to perform a comparison of the two methods.

A single-antenna GNSS receiver observes a superposition of plane waves corrupted by noise and, possibly, interference and multipath. The resulting signal is the sum of N_{sat} scaled, time-delayed and Doppler-shifted signals with known structure coming from the N_{sat} different satellites in view. The received complex baseband signal (complex envelope) is

$$r(t) = \sum_{i=1}^{N_{sat}} a_i s_i(t - \tau_i) \exp\{j2\pi f_{d_i} t\} + v(t) = \sum_{i=1}^{N_{sat}} a_i d_i(t) + v(t), \tag{4.73}$$

where $s_i(t)$ is the ith transmitted navigation signal, a_i is its complex amplitude (amplitude/phase), τ_i is the time delay, f_{d_i} the Doppler shift, and $v(t)$ is zero-mean complex additive Gaussian noise with PSD $2N_0$ [24, 55, 56].

In a multiple-antenna receiver, an M-element antenna array receives the same N_{sat} scaled, time-delayed and Doppler-shifted signals. Indeed, each antenna element receives a replica of the complex baseband signal modeled by Eq. (4.73), with a different amplitude/phase depending on the array geometry and the direction of arrival (DOA) [32, 43, 65]. Then, the column-vector output of the antenna array can be expressed as

$$\mathbf{r}(t) = \mathbf{G}(\boldsymbol{\theta}, \boldsymbol{\phi})\mathbf{A}\mathbf{d}(t, \boldsymbol{\tau}, \mathbf{f}_d) + \mathbf{v}(t), \tag{4.74}$$

where each row corresponds to one antenna and

- $\mathbf{r}(t) \in \mathbb{C}^{M \times 1}$ is the observed signal vector,
- $\mathbf{G}(\boldsymbol{\theta}, \boldsymbol{\phi}) \in \mathbb{C}^{M \times N_{sat}}$ is the *spatial signature matrix*, related to the array geometry and the DOA of the impinging signals; $\boldsymbol{\theta}, \boldsymbol{\phi} \in \mathbb{R}^{N_{sat} \times 1}$ stand for the azimuth and elevation vectors of the N_{sat} sources, respectively,

- $\mathbf{A} \in \mathbb{C}^{N_{\text{sat}} \times N_{\text{sat}}}$ is a diagonal matrix with the elements of complex amplitude vector $\mathbf{a} = [a_1, \ldots, a_{N_{\text{sat}}}]^T \in \mathbb{C}^{N_{\text{sat}} \times 1}$ along its diagonal,
- $\boldsymbol{\tau}, \mathbf{f}_d \in \mathbb{R}^{N_{\text{sat}} \times 1}$ are column vectors that contain time delays and Doppler shifts of each satellite,
- $\mathbf{d}(t, \boldsymbol{\tau}, \mathbf{f}_d) = [d_1, \ldots, d_{N_{\text{sat}}}]^T \in \mathbb{C}^{N_{\text{sat}} \times 1}$, where each component is defined by $d_i = s_i(t - \tau_i) \exp\{j 2\pi f_{d_i} t\}$ as in (4.74), and
- $\boldsymbol{\nu}(t) \in \mathbb{C}^{M \times 1}$ represents all noise sources (AWGN, interference, multipath). This term is considered complex, zero-mean, Gaussian with (possibly nondiagonal) covariance matrix $\mathbf{Q}$.

This model is built on the *narrowband array* assumption, in which the time required for the signal to propagate along the array aperture is much smaller than the inverse of the signal bandwidth. Thus, a nonfrequency-selective phase-shift can be used to describe the difference in signal propagation from one antenna to another. In the same way, we have assumed that the Doppler effect can be modeled by a pure frequency shift with no further distortion of the signal, which is commonly referred to as the narrowband signal assumption.

Assume now that the vector signal is sampled K times at a suitable sampling rate $f_s = 1/T_s$, so that K "snapshots" of the impinging signal are taken. Then, the observed data are

$$\mathbf{Y} = \mathbf{G}(\boldsymbol{\theta}, \boldsymbol{\phi}) \mathbf{A} \mathbf{D}(\boldsymbol{\tau}, \mathbf{f}_d) + \mathbf{V}, \tag{4.75}$$

using the following definitions:

- $\mathbf{Y} = [\mathbf{r}(t_0), \ldots, \mathbf{r}(t_{K-1})] \in \mathbb{C}^{M \times K}$ is the matrix of sampled signals, also called the *spatiotemporal data* matrix.
- $\mathbf{D}(\boldsymbol{\tau}, \mathbf{f}_d) = [\mathbf{d}(t_0, \boldsymbol{\tau}, \mathbf{f}_d), \ldots, \mathbf{d}(t_{K-1}, \boldsymbol{\tau}, \mathbf{f}_d)] \in \mathbb{C}^{N_{\text{sat}} \times K}$ is the signal matrix, known as the *basis-function* matrix.
- $\mathbf{V} = [\boldsymbol{\nu}(t_0), \ldots, \boldsymbol{\nu}(t_{K-1})] \in \mathbb{C}^{M \times K}$ is a matrix containing all noise contributions to $\mathbf{X}$. The matrix is zero-mean Gaussian, with a covariance matrix $\mathbf{Q}$.

We assume that the parameters of interest of the model and the covariance matrix of the error are piecewise constant during the observation interval, that is, they do not change over K/f_s seconds. This assumption is reasonable in GNSS as long as $K/f_s < 1$ ms [46].

In a conventional receiver, the estimates of $\boldsymbol{\tau}$ obtained after observation of (4.75) are used to compute the pseudoranges and from those to derive the user position [27, 66]. For the convenience of the reader, we recall here the procedure to estimate the receiver position from pseudorange measurements, which was already discussed in Chapter 2, and we re-cast it in a slightly different notation that is more suited to a DPE approach and to the relevant CRB derivation.

The signal propagation time between the ith satellite to the user is estimated by the signal processing section of the receiver via code correlation (the location of the peak of the cross-correlation between the received signal and the local code replica). This time-delay estimate (denoted by $\hat{\tau}_i$) provides an estimation of the distance between the ith satellite and the user, that is, the pseudorange $\rho_i = c\hat{\tau}_i$. Each pseudorange represents a (nonlinear) relation between the user's position ($\mathbf{x} = [x, y, z]^T$) and the estimated time delay of each satellite:

$$\rho_i = \Delta_i + c\,(\delta t - \delta t_i) + \epsilon_i, \tag{4.76}$$

where c is the speed of light, $\Delta_i = \| \mathbf{x}_i - \mathbf{x} \|$ is the geometric distance between the receiver and the ith satellite, δt is the *bias* of the receiver clock with respect to the GNSS time, which is unknown, δt_i is the clock bias of the ith satellite with respect to GNSS time, which is known from the navigation message and/or negligible in a first approximation, $\mathbf{x}_i = [x_i, y_i, z_i]^T$ are the coordinates of the ith satellite in the ECEF coordinate system, which can be computed from the ephemeris, after decoding the low-rate NAV message, and finally the term ϵ_i cumulates all the unknown errors from various sources such as atmospheric delays, multipath biases, ephemeris mis-modeling, and relativistic effects among others.

The carrier Doppler shifts f_{d_i} are possibly used for the estimation of the user *velocity*. These frequency shifts are in fact caused by user-satellite relative motion, as well as by frequency errors and drifts in user and satellite clocks. The Doppler shift due to the relative motion of the user and the ith satellite is [46]

$$f_{d_i} = - (\mathbf{v}_i - \mathbf{v})^T \mathbf{u}_i \frac{f_c}{c}, \tag{4.77}$$

where $\mathbf{v} = [v_x, v_y, v_z]^T$ and $\mathbf{v}_i = [v_{x_i}, v_{y_i}, v_{z_i}]^T$ are the velocity vectors of the user and the ith satellite, respectively, $\mathbf{u}_i$ is the unitary direction vector of the ith satellite relative to the user,

$$\mathbf{u}_i = \frac{\mathbf{x}_i - \mathbf{x}}{\| \mathbf{x}_i - \mathbf{x} \|}, \tag{4.78}$$

and f_c is the nominal carrier frequency. In the following, we assume that the Doppler shifts are perfectly estimated and compensated for by the tracking section of the receiver before computing the user position.

The positioning solution used in conventional GNSS receivers is based on the linearization of a geometrical problem [12, 27, 34, 46, 64]. The problem is to compute the user's position and clock offset from a set of N_{sat} estimated pseudoranges (where $N_{\text{sat}} \geq 4$). Thus, from Eq. (4.76), we obtain the following set of equations:

$$\| \mathbf{x}_i - \mathbf{x} \| + c\delta t = \rho_i + c\delta t_i - \epsilon_i$$
$$i = \{1, \ldots, N_{\text{sat}}\} \text{ s.t. } N_{\text{sat}} \geq 4, \tag{4.79}$$

in the four unknowns $x, y, z, \delta t$. The equations are nonlinear, and the set is overdetermined when $N_{\text{sat}} \geq 4$. The equations are usually solved through an iterative method wherein the nonlinear equations are linearized in the neighborhood of the solution at the previous iteration or in the neighborhood of an initial guess $\mathbf{x}^o = [x^o, y^o, z^o]^T$ of the position altogether:

$$\Delta_i \approx \Delta_i^o + \frac{x_i - x^o}{\Delta_i^o} \delta_x + \frac{y_i - y^o}{\Delta_i^o} \delta_y + \frac{z_i - z^o}{\Delta_i^o} \delta_z, \tag{4.80}$$

where $\delta_x = x^o - x$, $\delta_y = y^o - y$, $\delta_z = z^o - z$, and $\Delta_i^o = \| \mathbf{x}_i - \mathbf{x}^o \|$.

When the system is overdetermined, we can solve it considering the unknown error term ϵ_i as zero-mean equi-variance random variable and searching for the minimum mean square error (MMSE) solution:

$$\hat{\boldsymbol{\delta}} = \arg\min_{\boldsymbol{\delta}} \left\{ \mathbb{E} \| \mathbf{y} - \mathbf{H}\boldsymbol{\delta} \|^2 \right\}, \tag{4.81}$$

where the unknown is now the "offset" vector $\boldsymbol{\delta} = \begin{bmatrix} \delta_x, \delta_y, \delta_z, \delta t \end{bmatrix}^T$, and

$$
\mathbf{y} = \begin{bmatrix} \rho_1 + c\delta t_1 - \epsilon_1 - \Delta_1^o \\ \vdots \\ \rho_{N_{\mathrm{sat}}} + c\delta t_{N_{\mathrm{sat}}} - \epsilon_{N_{\mathrm{sat}}} - \Delta_{N_{\mathrm{sat}}}^o \end{bmatrix}
$$

is the observables array, while the linearized matrix is

$$
\mathbf{H} = \begin{bmatrix} \frac{x_1 - x^o}{\Delta_1^o} & \frac{y_1 - y^o}{\Delta_1^o} & \frac{z_1 - z^o}{\Delta_1^o} & 1 \\ \vdots & \vdots & \vdots & \vdots \\ \frac{x_{N_{\mathrm{sat}}} - x^o}{\Delta_{N_{\mathrm{sat}}}^o} & \frac{y_{N_{\mathrm{sat}}} - y^o}{\Delta_{N_{\mathrm{sat}}}^o} & \frac{z_{N_{\mathrm{sat}}} - z^o}{\Delta_{N_{\mathrm{sat}}}^o} & 1 \end{bmatrix}
$$

so that the MMSE solution is found via the Moore–Penrose pseudoinverse $(\mathbf{H}^\dagger)$:

$$
\hat{\boldsymbol{\delta}} = \mathbf{H}^\dagger \mathbf{y} \triangleq \left(\mathbf{H}^H \mathbf{H} \right)^{-1} \mathbf{H}^H \mathbf{y}. \tag{4.82}
$$

The final estimate that is provided by the navigation receiver is then $\hat{\mathbf{x}} = \mathbf{x}^0 + \hat{\boldsymbol{\delta}}$.

The MMSE solution presented in Eq. (4.82) can be improved by using side information, for instance the a priori known accuracy of the different pseudoranges, as derived via the measured received signal strength or via the geometry of the constellation. This is tantamount to assuming different variances and/or nonzero cross-correlation for the different errors ϵ_i on the different pseudoranges and is reflected in the MMSE solution through the introduction of a *weighting matrix* $\mathbf{W}$:

$$
\hat{\boldsymbol{\delta}} = \left(\mathbf{H}^H \mathbf{W} \mathbf{H} \right)^{-1} \mathbf{H}^H \mathbf{W} \mathbf{y}, \tag{4.83}
$$

where $\mathbf{W} = \mathbf{Q}^{-1}$, $\mathbf{Q}$ being the correlation matrix of the measurement errors ϵ_i. In most cases, such errors are (considered to be) uncorrelated, so that the weighting matrix turns out to be diagonal, $\mathbf{W} = \mathrm{diag}\left\{ 1/\sigma_{\epsilon_1}^2, \ldots, 1/\sigma_{\epsilon_{N_{\mathrm{sat}}}}^2 \right\}$.

4.4.1.1 *Direct Position Estimation*

The DPE approach is based on a simple fact: The synchronization parameters of each satellite can be expressed as functions of the *same* common parameters, above all the user's position, so that the observables are *not* independent from one satellite to another. In particular,

$$
\boldsymbol{\tau} \triangleq \boldsymbol{\tau}(\boldsymbol{\gamma})
$$

$$
\mathbf{f}_d \triangleq \mathbf{f}_d(\boldsymbol{\gamma}), \tag{4.84}
$$

where $\boldsymbol{\gamma} \in \mathbb{R}^{n_\gamma}$ is a vector gathering all considered motion parameters, and whose size is given by $n_\gamma = \dim\{\boldsymbol{\gamma}\}$. Including also the user velocity, the simplest configuration is $\boldsymbol{\gamma} = \begin{bmatrix} \mathbf{x}^T, \mathbf{v}^T, \delta t \end{bmatrix}^T$ and $n_\gamma = 7$, but more parameters could be included as well [8, 16, 28].

The main difference with respect to Eq. (4.73) is that the latter is a time-frequency parameterization of the incoming signals that does not take into account the mutual dependence of the different signals, while the model considered in the DPE approach is directly parameterized on γ. Taking this remark into consideration, we can then write

$$r(t) = \sum_{i=1}^{N_{\text{sat}}} a_i s_i(t - \tau_i(\gamma)) \exp\{j2\pi f_{d_i}(\gamma)t\} + v(t), \tag{4.85}$$

and the DPE counterpart of the array model in (4.75) is

$$\mathbf{Y} = \mathbf{G}(\boldsymbol{\theta}, \boldsymbol{\phi})\mathbf{A}\mathbf{D}(\gamma) + \mathbf{V}, \tag{4.86}$$

with the same previous definitions of parameters.

The optimal, ML solution to the DPE problem was introduced in Ref. [17] for the single antenna case and in Ref. [19] for the antenna array receiver. In both cases, it was seen that the solution requires the optimization of a multivariate nonconvex function. We will report here the solution for the antenna array, which is the one that provides the largest performance improvement with respect to the conventional two-step approach, especially in the presence of multipath propagation.

For the general case of an antenna array receiver, the maximum likelihood estimator (MLE) of γ reads as

$$\hat{\gamma}_{\text{ML}} = \arg\min_{\gamma} \left\{ \ln\left(\det\left(\mathbf{Y}^H\left(\mathbf{I} - \boldsymbol{\Pi}(\gamma)\right)\mathbf{Y}\right)\right)\right\},$$

where $\boldsymbol{\Pi}(\gamma) \triangleq \mathbf{D}(\mathbf{D}^H\mathbf{D})^{-1}\mathbf{D}^H$ is the *projection matrix* of the observed signal onto the subspace spanned by the columns of $\mathbf{D}(\gamma)$. A Bayesian solution that takes into account possible prior information was also derived in Ref. [16] and further refined in Ref. [15]. As already stated, synchronization parameters have a strong constraint: $\boldsymbol{\tau}$ and $\mathbf{f}_d$ are functions of the *same* user position. However, the conventional approach ignores this restriction and estimates synchronization parameters of each satellite independently; this is the ultimate cause of the suboptimality of the two-step approach. Another assumption underlying DPE is that the transmitters are perfectly synchronous. This assumption is reasonable in GNSS applications, since satellites are continuously monitored for that purpose. In addition, an accurate estimation of the ith satellite clock bias (δt_i)—performed by the control segment—is broadcasted in the navigation message to further correct and refine this possible source of error.

Nevertheless, the potential benefits of DPE do not come at no cost. The computational complexity of DPE is higher than that of the two-step approach. Whereas in the former a single multivariate nonconvex optimization problem has to be faced, the latter splits the solution into several lower dimensional problems (which in addition can be efficiently handled by a simple correlation processing). At present (2011), there is still an effort to be made to find efficient algorithms to implement DPE in real-time GNSS receivers.

4.4.1.2 CRBs for the Conventional Approach

In this section, we derive the CRB for the conventional two-step approach with a multi-antenna receiver (4.83), and in its most general formulation. Let us consider the generic single-snapshot signal model, expressed as

$$\mathbf{r}(t_k) = \boldsymbol{\mu}_r(t_k, \boldsymbol{\xi}) + \mathbf{v}(t_k) \in \mathbb{C}^{M \times 1} \tag{4.87}$$

with $\text{Cov}(\boldsymbol{v}(t_k)) = \mathbf{Q}$. The noiseless received signal is $\boldsymbol{\mu}_r(t_k, \boldsymbol{\xi}_A) = \mathbf{G}(\boldsymbol{\theta}, \boldsymbol{\phi}) \cdot \mathbf{Ad}(t_k, \boldsymbol{\tau}, \mathbf{f}_d)$, where the vector of unknown parameters is

$$\boldsymbol{\xi}_A = \left[\boldsymbol{\alpha}^T, \boldsymbol{\psi}^T, \boldsymbol{v}^T \right]^T, \tag{4.88}$$

which can be split into three subvectors containing the received complex amplitudes $\boldsymbol{\alpha}$, the direction-of-arrival angles (DOAs) $\boldsymbol{\psi}$, and the usual synchronization parameters $\boldsymbol{v}$:

$$\boldsymbol{\alpha} = \left[\mathfrak{Re}\{\mathbf{a}\}^T, \mathfrak{Im}\{\mathbf{a}\}^T \right]^T \in \mathbb{R}^{2N_{\text{sat}} \times 1}$$
$$\boldsymbol{\psi} = \left[\boldsymbol{\theta}^T, \boldsymbol{\phi}^T \right]^T \in \mathbb{R}^{2N_{\text{sat}} \times 1}$$
$$\boldsymbol{v} = \left[\boldsymbol{\tau}^T, \mathbf{f}_d^T \right]^T \in \mathbb{R}^{2N_{\text{sat}} \times 1}. \tag{4.89}$$

From this model, the FIM can be decomposed into $2N_{\text{sat}} \times 2N_{\text{sat}}$ submatrices as follows:

$$\mathbf{J}(\boldsymbol{\xi}_A) = \begin{pmatrix} \mathbf{J}_\alpha & \mathbf{J}_{\psi\alpha}^T & \mathbf{J}_{v\alpha}^T \\ \mathbf{J}_{\psi\alpha} & \mathbf{J}_{\psi\psi} & \mathbf{J}_{v\psi}^T \\ \mathbf{J}_{v\alpha} & \mathbf{J}_{v\psi} & \mathbf{J}_{vv} \end{pmatrix}. \tag{4.90}$$

The elements of the different submatrices can be computed considering that $\mathbf{Q}$ is independent of $\boldsymbol{\xi}_A$ and that we assume a K-snapshot observation interval. For any $u, v \in \{1, \ldots, 2N_{\text{sat}}\}$ we get

$$\mathbf{J}_{\alpha_u \alpha_v} = 2\mathfrak{Re} \left\{ \sum_{k=0}^{K-1} \mathbf{d}(t_k)^H \frac{\partial \mathbf{A}^H}{\partial \alpha_u} \mathbf{G}^H \mathbf{Q}^{-1} \mathbf{G} \frac{\partial \mathbf{A}}{\partial \alpha_v} \mathbf{d}(t_k) \right\} \tag{4.91}$$

$$\mathbf{J}_{\psi_u \alpha_v} = 2\mathfrak{Re} \left\{ \sum_{k=0}^{K-1} \mathbf{d}(t_k)^H \mathbf{A}^H \frac{\partial \mathbf{G}^H}{\partial \psi_u} \mathbf{Q}^{-1} \mathbf{G} \frac{\partial \mathbf{A}}{\partial \alpha_v} \mathbf{d}(t_k) \right\} \tag{4.92}$$

$$\mathbf{J}_{v_u \alpha_v} = 2\mathfrak{Re} \left\{ \sum_{k=0}^{K-1} \frac{\partial \mathbf{d}(t_k)^H}{\partial v_u} \mathbf{A}^H \mathbf{G}^H \mathbf{Q}^{-1} \mathbf{G} \frac{\partial \mathbf{A}}{\partial \alpha_v} \mathbf{d}(t_k) \right\} \tag{4.93}$$

$$\mathbf{J}_{\psi_u \psi_v} = 2\mathfrak{Re} \left\{ \sum_{k=0}^{K-1} \mathbf{d}(t_k)^H \mathbf{A}^H \frac{\partial \mathbf{G}^H}{\partial \psi_u} \mathbf{Q}^{-1} \frac{\partial \mathbf{G}}{\partial \psi_v} \mathbf{Ad}(t_k) \right\} \tag{4.94}$$

$$\mathbf{J}_{v_u \psi_v} = 2\mathfrak{Re} \left\{ \sum_{k=0}^{K-1} \frac{\partial \mathbf{d}(t_k)^H}{\partial v_u} \mathbf{A}^H \mathbf{G}^H \mathbf{Q}^{-1} \frac{\partial \mathbf{G}}{\partial \psi_v} \mathbf{Ad}(t_k) \right\} \tag{4.95}$$

$$\mathbf{J}_{v_u v_v} = 2\mathfrak{Re} \left\{ \sum_{k=0}^{K-1} \frac{\partial \mathbf{d}(t_k)^H}{\partial v_u} \mathbf{A}^H \mathbf{G}^H \mathbf{Q}^{-1} \mathbf{GA} \frac{\partial \mathbf{d}(t_k)}{\partial v_v} \right\}, \tag{4.96}$$

where $\partial \mathbf{A} / \partial \alpha_u$ is an all-zero $2N_{\text{sat}} \times 2N_{\text{sat}}$ matrix except for a 1 in the u, u position in the case of $1 \le u \le N_{\text{sat}}$, and an all-zero $2N_{\text{sat}} \times 2N_{\text{sat}}$ matrix except for a $j = \sqrt{-1}$ in the u, u position for the case of $N_{\text{sat}} < u \le 2N_{\text{sat}}$. The case $1 \le u \le N_{\text{sat}}$ addresses the derivative with respect to the elements of $\mathfrak{Re}\{\mathbf{a}\}$, and $N_{\text{sat}} < u \le 2N_{\text{sat}}$ indicates the derivative with respect to the elements of $\mathfrak{Im}\{\mathbf{a}\}$.

Now, we define $\mathbf{G}^T = \exp\{j\pi\,\mathbf{KP}\}$, where $\mathbf{K} \in \mathbb{R}^{N_{sat} \times 3}$ is the *wavenumber matrix*

$$\mathbf{K} = \begin{pmatrix} \cos(\theta_1)\cos(\phi_1) & \sin(\theta_1)\cos(\theta_1) & \sin(\phi_1) \\ \vdots & \vdots & \vdots \\ \cos(\theta_{N_{sat}})\cos(\phi_{N_{sat}}) & \sin(\theta_{N_{sat}})\cos(\theta_{N_{sat}}) & \sin(\phi_{N_{sat}}) \end{pmatrix} \qquad (4.97)$$

and θ_i is the (azimuth) angle of the ith satellite defined anticlockwise from the x axis on the xy plane, and ϕ_i the (elevation) angle with respect to the xy plane. Also,

$$\mathbf{P} = \begin{pmatrix} p_{x_1} & \cdots & p_{x_M} \\ p_{y_1} & \cdots & p_{y_M} \\ p_{z_1} & \cdots & p_{z_M} \end{pmatrix} \qquad (4.98)$$

is the matrix of the positions (coordinates) of the antenna elements, all normalized to a half-wavelength unit. Then, the derivative of $\mathbf{G}$ is

$$\frac{\partial \mathbf{G}}{\partial \psi_u} = j\pi\,\mathbf{P}^T \frac{\partial \mathbf{K}^T}{\partial \psi_u} \odot \exp\{j\pi\,\mathbf{P}^T\mathbf{K}^T\}, \qquad (4.99)$$

where the derivatives of $\mathbf{K}$ are

$$\frac{\partial \mathbf{K}}{\partial \psi_u} = \left(-\sin(\theta_u)\cos(\phi_u),\, \cos^2(\theta_u) - \sin^2(\theta_u),\, 0\right) \quad 1 \le u \le N_{sat} \qquad (4.100)$$

for the uth row, and zeros otherwise; and

$$\frac{\partial \mathbf{K}}{\partial \psi_u} = \left(-\cos(\theta_{(u-N_{sat})})\sin(\phi_{(u-N_{sat})}),\, 0,\, \cos(\phi_{(u-N_{sat})})\right)$$

$$N_{sat} < u \le 2N_{sat} \qquad (4.101)$$

for the $(u - N_{sat})$-th row, and zeros otherwise. Finally,

$$\frac{\partial \mathbf{d}(t_k)}{\partial v_u} = \left[\mathbf{0}^T,\, -\dot{s}_u(t_k - \tau_u)e^{j2\pi f_{d_u}t_k},\, \mathbf{0}^T\right]^T \quad 1 \le u \le N_{sat} \qquad (4.102)$$

and

$$\frac{\partial \mathbf{d}(t_k)}{\partial v_u} = \left[\mathbf{0}^T,\, j2\pi t_k s_{(u-N_{sat})}(t_k - \tau_{(u-N_{sat})})e^{j2\pi f_{d_{(u-N_{sat})}}t_k},\, \mathbf{0}^T\right]^T$$

$$N_{sat} < u \le 2N_{sat} \qquad (4.103)$$

stand for the derivatives with respect to the elements of $\boldsymbol{\tau}$ and $\mathbf{f}_d$, respectively ($\dot{s}_i(t)$ is the time derivative of $s_i(t)$).

Substituting Eqs (4.99)–(4.103) into (4.91)–(4.96), the FIM is completely defined, and therefore the CRB for all the parameters can be directly computed by inverting (4.90).

The conventional two-step approach just establishes a deterministic mapping between the time-delay estimates and the positioning solution, as provided by Eq. (4.83). Therefore, its CRB can be directly obtained from the CRB of the time estimates. According to Ref. [35], if the desired estimate can be expressed as $[\hat{\mathbf{x}}^T, \hat{\delta t}]^T = g(\hat{\boldsymbol{\tau}})$, then its covariance matrix is bounded by

$$\mathbf{C}(\hat{\mathbf{x}}, \hat{\delta t}) \succeq \frac{\partial g(\boldsymbol{\tau})}{\partial \boldsymbol{\tau}} \mathbf{J}_{\tau\tau}^{-1} \frac{\partial g(\boldsymbol{\tau})^T}{\partial \boldsymbol{\tau}}, \tag{4.104}$$

where $\mathbf{J}_{\tau\tau}$ is the FIM of the time-delay parameter that can be obtained from (4.96) with $1 \le \{u, v\} \le N_{\text{sat}}$. In this case, we know from (4.83) that

$$g(\boldsymbol{\tau}) = \begin{pmatrix} \mathbf{x}^o \\ 0 \end{pmatrix} + \left(\mathbf{H}^H \mathbf{W} \mathbf{H}\right)^{-1} \mathbf{H}^H \mathbf{W} \mathbf{y}, \tag{4.105}$$

where its Jacobian matrix is

$$\frac{\partial g(\boldsymbol{\tau})}{\partial \boldsymbol{\tau}} = c \left(\mathbf{H}^H \mathbf{W} \mathbf{H}\right)^{-1} \mathbf{H}^H \mathbf{W}, \tag{4.106}$$

where we assumed that the linearization point is the true position, that is, $\mathbf{x}^o = \mathbf{x}$.

From the above it follows that the covariance matrix of position and clock offset estimates in the conventional approach is lower bounded by the CRB matrix as

$$\mathbf{C}(\hat{\mathbf{x}}, \hat{\delta t}) \succeq \underbrace{c^2 \left(\left(\mathbf{H}^H \mathbf{W} \mathbf{H}\right)^{-1} \mathbf{H}^H \mathbf{W}\right) \mathbf{J}_{\tau\tau}^{-1} \left(\left(\mathbf{H}^H \mathbf{W} \mathbf{H}\right)^{-1} \mathbf{H}^H \mathbf{W}\right)^T}_{\overset{\triangle}{=} \kappa_{\text{conv}}(\mathbf{x}, \delta t)}, \tag{4.107}$$

where $\mathbf{J}_{\tau\tau}^{-1}$ is expressed in units of time, and $\kappa_{\text{conv}}(\mathbf{x}, \delta t)$ is the CRB matrix on the position solution for the conventional GNSS approach.

4.4.1.3 *CRB for the DPE Approach*

In this case, $\boldsymbol{\mu}_r(t_k, \boldsymbol{\xi}_B) = \mathbf{G}(\boldsymbol{\theta}, \boldsymbol{\phi}) \, \mathbf{Ad}(t_k, \boldsymbol{\gamma})$, and the vector of unknown parameters is

$$\boldsymbol{\xi}_B = \left[\boldsymbol{\alpha}^T, \boldsymbol{\psi}^T, \boldsymbol{\gamma}^T\right]^T, \tag{4.108}$$

where $\boldsymbol{\gamma}$ can include any parameter related to the position and motion of the receiver and the receiver clock drift. The FIM can be cast into the following form:

$$\mathbf{J}(\boldsymbol{\xi}_B) = \begin{pmatrix} \mathbf{J}_\alpha & \mathbf{J}_{\psi\alpha}^T & \mathbf{J}_{\gamma\alpha}^T \\ \mathbf{J}_{\psi\alpha} & \mathbf{J}_{\psi\psi} & \mathbf{J}_{\gamma\psi}^T \\ \mathbf{J}_{\gamma\alpha} & \mathbf{J}_{\gamma\psi} & \mathbf{J}_{\gamma\gamma} \end{pmatrix}, \tag{4.109}$$

where $\mathbf{J}_\alpha$, $\mathbf{J}_{\psi\alpha}$, and $\mathbf{J}_{\psi\psi}$ are $2N_{\text{sat}} \times 2N_{\text{sat}}$ as defined in (4.91), (4.92), and (4.94), respectively. What is still to be computed is the value of $\mathbf{J}_{\gamma\alpha}, \mathbf{J}_{\gamma\psi} \in \mathbb{R}^{\dim\{\gamma\}\times 2N_{\text{sat}}}$ and $\mathbf{J}_{\gamma\gamma} \in \mathbb{R}^{\dim\{\gamma\}\times\dim\{\gamma\}}$:

$$\mathbf{J}_{\gamma_u\alpha_v} = 2\Re\left\{ \sum_{k=0}^{K-1} \frac{\partial \mathbf{d}(t_k)^H}{\partial \gamma_u} \mathbf{A}^H \mathbf{G}^H \mathbf{Q}^{-1} \mathbf{G} \frac{\partial \mathbf{A}}{\partial \alpha_v} \mathbf{d}(t_k) \right\} \tag{4.110}$$

$$\mathbf{J}_{\gamma_u\psi_v} = 2\Re\left\{ \sum_{k=0}^{K-1} \frac{\partial \mathbf{d}(t_k)^H}{\partial \gamma_u} \mathbf{A}^H \mathbf{G}^H \mathbf{Q}^{-1} \frac{\partial \mathbf{G}}{\partial \psi_v} \mathbf{A}\mathbf{d}(t_k) \right\} \tag{4.111}$$

$$\mathbf{J}_{\gamma_u\gamma_v} = 2\Re\left\{ \sum_{k=0}^{K-1} \frac{\partial \mathbf{d}(t_k)^H}{\partial \gamma_u} \mathbf{A}^H \mathbf{G}^H \mathbf{Q}^{-1} \mathbf{G}\mathbf{A} \frac{\partial \mathbf{d}(t_k)}{\partial \gamma_v} \right\}, \tag{4.112}$$

where in particular the value of $\partial \mathbf{d}(t_k)/\partial \gamma_u$ is still not known.

Recalling the model in (4.74), we can express the basis-function vector as

$$\mathbf{d}(t_k, \gamma) = \mathbf{s}(t_k - \boldsymbol{\tau}(\gamma)) \odot \exp\{j2\pi \mathbf{f}_d(\gamma)t_k\}, \tag{4.113}$$

with $\mathbf{s}(t_k - \boldsymbol{\tau}(\gamma)) \in \mathbb{C}^{N_{\text{sat}}\times 1}$ being the vector containing $s_i(t_k - \tau_i(\gamma))$ in its ith row. Then, after differentiating with respect to γ_u and rearranging terms, we get

$$\frac{\partial \mathbf{d}(t_k)}{\partial \gamma_u} = \left(\sum_{i=1}^{N_{\text{sat}}} \frac{\partial \mathbf{d}(t_k)}{\partial \tau_i}\right) \odot \frac{\partial \boldsymbol{\tau}(\gamma)}{\partial \gamma_u} + \left(\sum_{i=1}^{N_{\text{sat}}} \frac{\partial \mathbf{d}(t_k)}{\partial f_{d_i}}\right) \odot \frac{\partial \mathbf{f}_d(\gamma)}{\partial \gamma_u},$$

where $\partial \mathbf{d}(t_k)/\partial \tau_i$ and $\partial \mathbf{d}(t_k)/\partial f_{d_i}$ are calculated using Eqs (4.102) and (4.103), respectively.

In the sequel, we assume that $\gamma \triangleq [\mathbf{x}^T, \mathbf{v}^T, \delta t]^T = [x, y, z, v_x, v_y, v_z, \delta t]^T$ and then $\dim\{\gamma\} = 7$. In that case, the ith row of the matrices $\partial \boldsymbol{\tau}(\gamma)/\partial \gamma$ and $\partial \mathbf{f}_d(\gamma)/\partial \gamma$ are

$$\frac{\partial \tau_i(\gamma)}{\partial \gamma} = \left(\begin{array}{c} \frac{\partial \tau_i(\gamma)}{\partial \mathbf{x}} \\ \frac{\partial \tau_i(\gamma)}{\partial \mathbf{v}} \\ \frac{\partial \tau_i(\gamma)}{\partial \delta t} \end{array}\right)^T, \quad \frac{\partial f_{d_i}(\gamma)}{\partial \gamma} = \left(\begin{array}{c} \frac{\partial f_{d_i}(\gamma)}{\partial \mathbf{x}} \\ \frac{\partial f_{d_i}(\gamma)}{\partial \mathbf{v}} \\ \frac{\partial f_{d_i}(\gamma)}{\partial \delta t} \end{array}\right)^T, \tag{4.114}$$

where the derivatives can be computed as follows. To deal with the derivative of $\tau_i(\gamma)$ with respect to vector $\gamma = [x, y, z, v_x, v_y, v_z, \delta t]^T$, we first recall from Eq. (4.76) that

$$\tau_i(\gamma) = \frac{1}{c} \| \mathbf{x}_i - \mathbf{x} \| + (\delta t - \delta t_i) + \epsilon_i. \tag{4.115}$$

Then, applying basic linear algebra we obtain

$$\frac{\partial \tau_i(\gamma)}{\partial \mathbf{x}} = -\frac{\mathbf{u}_i^T}{c}$$

$$\frac{\partial \tau_i(\gamma)}{\partial \mathbf{v}} = \begin{cases} \mathbf{0}^T & \text{if } \mathbf{a} = \mathbf{0} \\ -\frac{\mathbf{u}_i^T}{c}\left(\mathbf{v} \otimes \tilde{\mathbf{a}}^T\right) & \text{otherwise} \end{cases}$$

$$\frac{\partial \tau_i(\gamma)}{\partial \delta t} = 1, \tag{4.116}$$

where we took into account that

$$\frac{\partial \| \mathbf{x}_i - \mathbf{x} \|}{\partial \mathbf{x}} = \frac{1}{2 \| \mathbf{x}_i - \mathbf{x} \|} \frac{\partial \left((\mathbf{x}_i - \mathbf{x})^T (\mathbf{x}_i - \mathbf{x}) \right)}{\partial \mathbf{x}}$$

$$= -\frac{(\mathbf{x}_i - \mathbf{x})^T}{\| \mathbf{x}_i - \mathbf{x} \|} \triangleq -\mathbf{u}_i^T \tag{4.117}$$

$$\frac{\partial \| \mathbf{x}_i - \mathbf{x} \|}{\partial \mathbf{v}} = -\frac{(\mathbf{x}_i - \mathbf{x})^T}{\| \mathbf{x}_i - \mathbf{x} \|} \frac{\partial \mathbf{x}}{\partial \mathbf{v}} = -\mathbf{u}_i^T \frac{\partial \mathbf{x}}{\partial \mathbf{v}} \tag{4.118}$$

and

$$\frac{\partial \mathbf{x}}{\partial \mathbf{v}} = \begin{pmatrix} \frac{\partial x}{\partial v_x} & \frac{\partial x}{\partial v_y} & \frac{\partial x}{\partial v_z} \\ \frac{\partial y}{\partial v_x} & \frac{\partial y}{\partial v_y} & \frac{\partial y}{\partial v_z} \\ \frac{\partial z}{\partial v_x} & \frac{\partial z}{\partial v_y} & \frac{\partial z}{\partial v_z} \end{pmatrix} = \mathbf{v} \otimes \tilde{\mathbf{a}}^T$$

$$\tilde{\mathbf{a}} \triangleq \left[\frac{\partial t}{\partial v_x}, \frac{\partial t}{\partial v_y}, \frac{\partial t}{\partial v_z} \right]^T = \left[\frac{1}{a_x}, \frac{1}{a_y}, \frac{1}{a_z} \right]^T. \tag{4.119}$$

Notice that (4.118) is only valid when the components of the acceleration vector ($\mathbf{a} = \left[a_x, a_y, a_z \right]^T$) are not null, otherwise the derivative is equal to $\mathbf{0}^T$.

The derivative of Eq. (4.77) with respect to $\mathbf{x}$ is

$$\frac{\partial f_{d_i}(\boldsymbol{\gamma})}{\partial \mathbf{x}} = \begin{cases} (\mathbf{v}_i - \mathbf{v})^T \frac{\mathbf{I} - \mathbf{u}_i \mathbf{u}_i^T}{\|\mathbf{x}_i - \mathbf{x}\|} \frac{f_c}{c} & \text{if } \mathbf{v} = \mathbf{0} \\ -\frac{\mathbf{u}_i^T}{c} \left(\mathbf{a} \otimes \tilde{\mathbf{v}}^T \right) \frac{f_c}{c} + (\mathbf{v}_i - \mathbf{v})^T \frac{\mathbf{I} - \mathbf{u}_i \mathbf{u}_i^T}{\|\mathbf{x}_i - \mathbf{x}\|} \frac{f_c}{c} & \text{otherwise,} \end{cases} \tag{4.120}$$

where we defined

$$\frac{\partial \mathbf{v}}{\partial \mathbf{x}} = \begin{pmatrix} \frac{\partial v_x}{\partial x} & \frac{\partial v_x}{\partial y} & \frac{\partial v_x}{\partial z} \\ \frac{\partial v_y}{\partial x} & \frac{\partial v_y}{\partial y} & \frac{\partial v_y}{\partial z} \\ \frac{\partial v_z}{\partial x} & \frac{\partial v_z}{\partial y} & \frac{\partial v_z}{\partial z} \end{pmatrix} = \mathbf{a} \otimes \tilde{\mathbf{v}}^T$$

$$\tilde{\mathbf{v}} \triangleq \left[\frac{\partial t}{\partial x}, \frac{\partial t}{\partial y}, \frac{\partial t}{\partial z} \right]^T = \left[\frac{1}{v_x}, \frac{1}{v_y}, \frac{1}{v_z} \right]^T. \tag{4.121}$$

The derivatives with respect to the rest of the considered parameters in $\boldsymbol{\gamma}$ are

$$\frac{\partial f_{d_i}(\boldsymbol{\gamma})}{\partial \mathbf{v}} = \begin{cases} \mathbf{u}_i^T \frac{f_c}{c} & \text{if } \mathbf{a} = \mathbf{0} \\ \mathbf{u}_i^T \frac{f_c}{c} + \frac{\partial f_{d_i}(\boldsymbol{\gamma})}{\partial \mathbf{x}} \left(\mathbf{v} \otimes \tilde{\mathbf{a}}^T \right) & \text{otherwise} \end{cases}$$

$$\frac{\partial f_{d_i}(\boldsymbol{\gamma})}{\partial \delta t} = 0, \tag{4.122}$$

where we considered that the derivative of the direction vector with respect to the position of the user is

$$
\frac{\partial \mathbf{u}_i}{\partial \mathbf{x}} = \frac{\| \mathbf{x}_i - \mathbf{x} \| \frac{\partial (\mathbf{x}_i - \mathbf{x})^T}{\partial \mathbf{x}} - (\mathbf{x}_i - \mathbf{x}) \frac{\partial \| \mathbf{x}_i - \mathbf{x} \|}{\partial \mathbf{x}}}{\| \mathbf{x}_i - \mathbf{x} \|^2}
$$

$$
= \frac{- \| \mathbf{x}_i - \mathbf{x} \| + (\mathbf{x}_i - \mathbf{x}) \mathbf{u}_i^T}{\| \mathbf{x}_i - \mathbf{x} \|^2}
$$

$$
= \frac{\mathbf{u}_i \mathbf{u}_i^T - \mathbf{I}}{\| \mathbf{x}_i - \mathbf{x} \|} \tag{4.123}
$$

$$
\frac{\partial \mathbf{u}_i}{\partial \mathbf{v}} = \frac{\partial \mathbf{u}_i}{\partial \mathbf{x}} \frac{\partial \mathbf{x}}{\partial \mathbf{v}}. \tag{4.124}
$$

We have two expressions for $\partial f_{d_i}(\gamma)/\partial \mathbf{x}$ depending on whether the receiver is static or it has no null velocity. Similarly, $\partial f_{d_i}(\gamma)/\partial \mathbf{v}$ is computed differently if the receiver has constant velocity or if it has a certain acceleration.

Now, all the elements in Eq. (4.109) can be evaluated, so that the CRB of the user position following the DPE approach can be computed after inversion of the FIM matrix in (4.109):

$$
\boldsymbol{\kappa}_{\text{DPE}}(\boldsymbol{\alpha}, \boldsymbol{\psi}, \boldsymbol{\gamma}) = \mathbf{J}^{-1}(\boldsymbol{\xi}_B). \tag{4.125}
$$

4.4.1.4 *Numerical Results*

We now apply the expressions for the CRB of both approaches to a case study based on GPS (although our results can be applied to any GNSS signal set). We assume that the GPS C/A signal is filtered with a 2.2 MHz bandwidth filter, down-converted to baseband and then digitized at a sampling frequency of $f_s = 5.714$ MHz. The observation time is 1 ms, which corresponds to $K = 5714$ samples. Two receiver architectures are considered: single antenna ($M = 1$) and multi-antenna ($M = 8$), in particular with an 8-element circular antenna array. The receiver is static in both cases, that is, $\mathbf{v} = \mathbf{0}^T$ for simplicity.

The recreated scenario corresponds to a realistic constellation geometry. In particular, two versions are simulated. The first one consists of $N_{\text{sat}} = 7$ satellites, whose azimuth and elevation angles are, respectively (in degrees)

$$
\boldsymbol{\theta} = [288.9, 215.2, 87.9, 295.4, 123.5, 46.1, 130.6]^T
$$

$$
\boldsymbol{\phi} = [46.9, 24.5, 29.1, 32.1, 71.5, 24.4, 60.7]^T. \tag{4.126}
$$

The corresponding pseudorandom noise (PRN) code numbers are $\{9, 12, 17, 18, 26, 28, 29\}$. The second considered scenario is one with $N_{\text{sat}} = 4$ satellites, with PRN numbers $\{9, 12, 17, 18\}$, and with the same geometry as in (4.126). In order to avoid plotting the bounds for the three position coordinates, the following figures depict the CRB of the three-dimensional position vector, defined as

$$
\kappa_{\mathbf{x}}^2 \triangleq \mathbb{E}\left\{ (\mathbf{x} - \hat{\mathbf{x}})^T (\mathbf{x} - \hat{\mathbf{x}}) \right\} = \kappa_x^2 + \kappa_y^2 + \kappa_z^2, \tag{4.127}
$$

where κ_x^2, κ_y^2, and κ_z^2 are the CRBs of each coordinate, as computed by the corresponding CRB. Recall that the conventional approach is the positioning solution given by first estimating synchronization parameters independently and then using those estimates to compute the user's location.

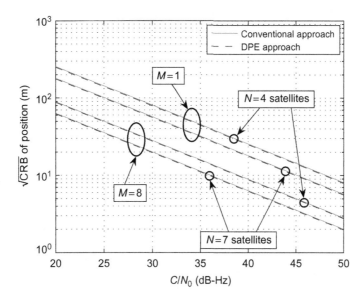

FIGURE 4.10

CRB versus C/N_0 of the satellites. $M = 1$ corresponds to a single-antenna receiver and $M = 8$ corresponds to an eight-element circular antenna array (from Ref. [81]).

In some cases, we also compare the performance of the conventional approach without or with a priori information (Eqs (4.82) and (4.83), respectively). In particular, the diagonal entries in the "weighting" matrix $\mathbf{W}$ are the C/N_0 of the corresponding satellites, normalized to the highest C/N_0 value.[8] Thus, if $(C/N_0)_i$ denotes the C/N_0 of the ith satellite, the weighting matrix of the WLS problem in (4.83) is constructed as

$$\mathbf{W} = \frac{1}{\Gamma} \, \mathrm{diag}\left\{(C/N_0)_1, \dots, (C/N_0)_{N_{\mathrm{sat}}}\right\}, \tag{4.128}$$

with

$$\Gamma = \max\left\{(C/N_0)_1, \dots, (C/N_0)_{N_{\mathrm{sat}}}\right\}. \tag{4.129}$$

With this setup, Fig. 4.10 shows a comparison between the CRBs of position as a function of the C/N_0 of the visible satellites, for the two approaches. For simplicity, we also assume that all satellites have the same C/N_0. The advantage of DPE in this case is *not* apparent at all—both positioning alternatives obtain the same accuracy in such an *ideal* scenario.

Where DPE really shines is in the case of a larger diversity between different satellites. This advantage in terms of *satellite diversity* was already noted in Ref. [19] and in Refs. [2, 3] for the (reverse) radiolocation problem. Figure 4.11(a) refers to a scenario wherein all but one satellites have the same

---

[8]The Gauss–Markov theorem [49] shows that when errors are uncorrelated with each other and with the independent variables and have equal variance, LS is the best linear unbiased estimator (BLUE). If measurements are uncorrelated but have different variances, WLS is BLUE if each weight is equal to the reciprocal of the variance of the measurement.

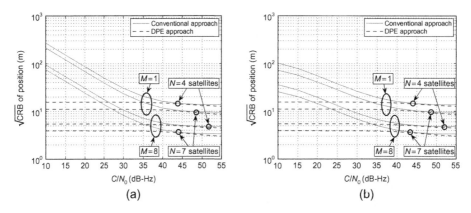

FIGURE 4.11

CRB as a function of the C/N_0 of one of the satellites and $C/N_0 = 45$ dB-Hz for the rest. (a) DPE versus conventional approach with LS. (b) DPE versus conventional approach with WLS (from Ref. [81]).

$C/N_0 = 45$ dB-Hz, and the different one has the (variable) C/N_0 shown on the x-axis. The DPE appears robust against shadowing of one of the satellites in the constellation (i.e., when its C/N_0 gets very low). This is because the remaining visible satellites are used to estimate the user's position, automatically overcoming a weaker power level of the faded satellite. On the contrary, if that satellite is used to compute the user's position in the conventional approach, it contributes to the higher error variances as seen in Fig. 4.11.

Of course the conventional approach can be improved by the use of a WLS (algorithm 4.11(b)). However, the conventional approach is still unable to attain the performance of a direct estimation approach. Surprisingly, below a certain C/N_0 threshold, the use of a single-antenna receiver implementing DPE overcomes the results of an antenna array receiver with the conventional positioning approach (in the specific setup in Fig. 4.11, this threshold is of the order of 25 dB-Hz).

Multipath propagation is known to be one of the most detrimental effects in GNSS receivers, and probably the dominant source of error in high-precision applications. The most annoying situation is when the extra path covered by the reflected wave(s) is shorter than one chip period (about 300 m for a chip rate of 1.023 Mcps), so that the non-LOS components cannot be completely *resolved*. As a test situation, we assumed a two-ray (LOS + reflected) multipath channel for satellite ♯9 with a signal-to-multipath ratio (SMR) 3 dB (LOS component 3 dB larger than the reflected one). The relative azimuth of the replica with respect to the LOS signal was 180°, with the same elevation angle. All satellites have the same power levels, $C/N_0 = 45$ dB-Hz. Figure 4.12 shows the squared root CRBs as a function of the relative delay between the multipath replica and the LOS signal of satellite ♯9, normalized to the chip period T_c. The results shown in Fig. 4.12 are in agreement with those in Ref. [17], in which DPE provided the receiver with improved multipath mitigation capabilities with respect to the conventional approach. Even in the single-antenna case, the receiver virtually eliminates the multipath effect when implementing DPE.

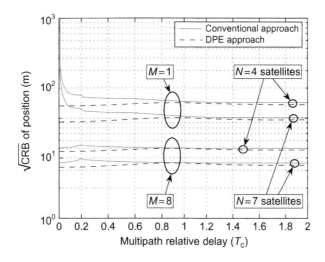

FIGURE 4.12

CRB as a function of the relative multipath delay (from Ref. [81]).

Finally, we compare the performance of real-world estimators with our results to understand how a receiver can actually get close to the bounds. In particular, we consider the case of $M = 1$ and $N_{\text{sat}} = 7$. Both the conventional and the DPE ML estimators were implemented using the accelerated random search (ARS) algorithm, proposed in Ref. [6]. ARS is an iterative algorithm used to optimize non-convex/multivariate functions that improve the convergence rate of the pure random search algorithm. The results of each position estimator are shown in Fig. 4.13(a) for the same scenario as in Fig. 4.10. The estimator attains the bound when the C/N_0 is medium to high. This may be due to the known inaccuracy of the CRB for low C/N_0.

The same setup considered in Fig. 4.11 was used to obtain Fig. 4.13(b), that is, variable C/N_0 for satellite ♯9, with the rest at 45 dB-Hz. As expected, the two-step WLS solution outperforms the LS. An interesting effect is seen after satellite ♯9 reaches $C/N_0 = 45$ dB-Hz, that is, the power level of the rest of the satellites. At that point, the overall position RMSE degrades due to an increase of the multiple-access interference (MAI) that satellite ♯9 induces to the rest. Since the estimation of synchronization parameters is performed independently, a conventional receiver is not immune to MAI. In contrast, DPE provides a way to jointly process all signals, similarly to what is done in multiuser receivers for code-division multiple-access signals [68]. Again, the CRB is not tight enough for low C/N_0 values. Finally, Fig. 4.14 presents the results for the two-ray multipath channel as in Fig. 4.12. As already known (see Ref. [61] for instance), time-delay ML estimates are biased for close multipath. That bias is propagated through the LS-based positioning solution. The same effect also applies to the DPE solution. However, the degradation is mitigated in the latter, as depicted in Fig. 4.14(a). The results correspond to the setup where $C/N_0 = 45$ dB-Hz for all satellites, $C/N_0 = 25$ dB-Hz for satellite ♯9 and the SMR between the LOS signal of ♯9 and its multipath replica is −20 dB, as can be seen in Fig. 4.14(b).

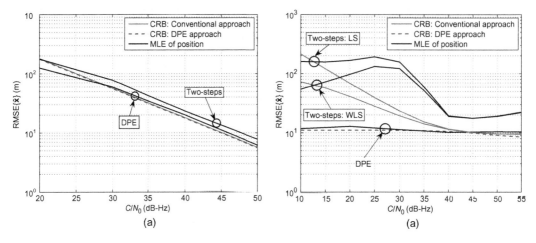

FIGURE 4.13

RMSE{$\hat{\mathbf{x}}$} of MLE of position against the corresponding CRB. (a) Versus C/N_0 of the satellites. (b) Versus C/N_0 of satellite ♯9 (from Ref. [81]).

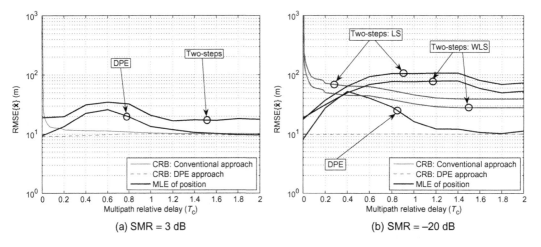

FIGURE 4.14

RMSE{$\hat{\mathbf{x}}$} of MLE of position against CRBs as a function of the relative multipath delay (from Ref. [81]).

4.4.2 Theoretical Performance Limits in Cooperative Localization

The problem of cooperatively localizing an object that emits some types of characteristic (signature) signal has been widely studied over the past few decades, both for military and industrial applications. Emitted signals of interest include, among others, acoustic and radio-frequency waveforms [8, 44, 48, 53]. In this section, we will consider and evaluate the theoretical accuracy limits inherent to this problem, focusing in particular on the recent field of cooperative localization for wireless sensor networks (WSNs).

In a typical WSN application, only a few nodes have a known position (*reference nodes* (RNs) or anchors), while the others (*unknown nodes* (UNs) or agents) have to be localized by the network itself, in particular those that are out of the range of any reference. In such a situation, localization can be obtained only through the *cooperation* of all nodes, using not only measurements from RNs but also measurements between pairs of UNs [47]. If nodes are equipped with UWB transmission capabilities, accurate internode range measurements are achievable through time-of-arrival (TOA) estimation [25].

CRB results for cooperative localization based on different types of internode measurements can be found in Refs. [7, 25, 33, 37, 47, 51, 58, 59, 73]. The localization error depends on several elements, in particular on the measurement reliability (which may be impaired by multipath, non-line-of-sight, synchronization errors, etc.) and the *topology* of the network. In this section, we will perform a CRB analysis for TOA-based cooperative localization, also including the path-loss effect [45]. The fundamental bound is then analyzed for simple one-dimensional (1D) regular network topologies, referring the reader to Ref. [45] for the more complicated analysis for 2D regular networks. The aim is to obtain some understanding of the achievable positioning performance for different system settings. We show in particular that the positioning RMSE is related to the number of known/unknown nodes and to the degree of network connectivity. Based on this analysis, novel approaches for cooperatively solving the localization problem in WSNs can be envisaged, for example those employing principles of belief propagation (BP). This aspect will be thoroughly presented and analyzed in Chapter 5.

4.4.2.1 *Signal Model*

Consider a network composed of N nodes that are deployed along a straight line. This may be a convenient model for vehicles on a road, and/or other transportation systems. The kth node location in the 1D space is indicated by $x_k \in \mathbb{R}$, $k = 1, \ldots, N$. For convenience of notation, the first N_u positions $\mathbf{x} = [x_1 \cdots x_{N_u}]^T$ are assumed to be unknown, while the remaining $N_B = N - N_u$ are the RN locations $\mathbf{x}_r = [x_{N_u+1} \cdots x_N]^T$. Cooperative estimation of the N_u parameters $\mathbf{x}$ is obtained from pairwise range measurements $\{\hat{d}_{k,\ell}\}$ made between any pair of (either known or unknown) nodes k and ℓ. Let Ω_k be the subset of nodes with which node k makes measurements; we indicate by $I_{\Omega_k}(\ell)$ the *visibility function* such that $I_{\Omega_k}(\ell) = 1$ if $\ell \in \Omega_k$, or $I_{\Omega_k}(\ell) = 0$ if not. For $I_{\Omega_k}(\ell) = 1$, the range measured from node k to node ℓ is assumed to be obtained through the estimation of the TOA between the two nodes, which follows Gaussian statistics. Specifically, $\hat{d}_{k,\ell} \sim \mathcal{N}\left(d_{k,\ell}, \sigma_{k,\ell}^2\right)$, whose mean $d_{k,\ell}$ is equal to the actual distance between nodes k and ℓ, $d_{k,\ell} = |x_k - x_\ell|$, and variance $\sigma_{k,\ell}^2$. Range measurements associated with different links are considered uncorrelated; we also assume that both $\hat{d}_{k,\ell}$ and $\hat{d}_{\ell,k}$ are available for each pair of nodes k, ℓ, so that clock biases can be canceled as for round-trip measurements. According to the CRB on TOA estimates [25] and the path-loss law, the variance of the distance estimate can be evaluated as

$$\sigma_{k,\ell}^2 = \sigma_0^2 \cdot \left(\frac{d_{k,\ell}}{d_0}\right)^\alpha = \sigma_0^2 \cdot \left(\frac{|x_k - x_\ell|}{d_0}\right)^\alpha, \tag{4.130}$$

where α is the path-loss exponent, $\sigma_0 = c/(2\sqrt{2}\pi \sqrt{\eta_0}\beta)$ is the distance estimate accuracy at a reference distance d_0 (as defined in (4.5)), η_0 is the corresponding signal-to-noise ratio, and β is the effective (Gabor) bandwidth of the radio signal used for TOA estimation, as defined in (4.4).

To proceed with our analysis, we arrange all measurements into the vector $\hat{\mathbf{d}} = [z_1 \cdots z_M]^T$. Based on the assumptions above, $\hat{\mathbf{d}} \sim \mathcal{N}(\mathbf{d}(\mathbf{x}), \mathbf{Q}(\mathbf{x}))$, where $\mathbf{d}(\mathbf{x})$ is the vector collecting the terms $\{d_{k,\ell}\}$, while $\mathbf{Q}(\mathbf{x})$ is the diagonal covariance matrix having as diagonal entries the variances $\{\sigma_{k,\ell}^2\}$ that depend on the unknown parameters $\mathbf{x}$ according to (4.130). The total number of measurements is $K \leq N(N-1) - N_B(N_B - 1)$, and it is maximum in the case of *full coverage*, that is, when each node makes measurements with all of the others (excluding measurements between two RNs). More realistic scenarios assume that each node has a limited coverage and makes measurements only to nodes located within a radius r from itself, that is, $\Omega_k = \{x_\ell : |\Delta x_{k,\ell}| \leq r\}$.

4.4.2.2 Analysis of the CRB

We can derive the generic element of the $N_u \times N_u$ FIM $\mathbf{J}(\mathbf{x})$ for the estimation of $\mathbf{x}$ as follows:

$$[\mathbf{J}(\mathbf{x})]_{k,\ell} = [\mathcal{D}(\mathbf{x})^T \mathbf{Q}(\mathbf{x})^{-1} \mathcal{D}(\mathbf{x})]_{k,\ell} + \frac{1}{2} \text{tr}\left(\mathbf{Q}(\mathbf{x})^{-1} \frac{\partial \mathbf{Q}(\mathbf{x})}{\partial x_k} \mathbf{Q}(\mathbf{x})^{-1} \frac{\partial \mathbf{Q}(\mathbf{x})}{\partial x_\ell}\right), \qquad (4.131)$$

where $\mathcal{D}(\mathbf{x}) = \frac{\partial \hat{\mathbf{d}}(\mathbf{x})}{\partial \mathbf{x}}$ is the $M \times N_u$ gradient matrix. Starting from the FIM the CRB $\kappa(\mathbf{x})$ is obtained through (4.14).

For TOA-based cooperative localization, the CRB has been derived in Refs. [47], [25], and [37] under the assumption that there is no a priori information about the measurement accuracies $\mathbf{Q}(\mathbf{x})$ on the locations $\mathbf{x}$, that is, by neglecting the second term in (4.131). If we assume such knowledge, the result is (see Ref. [45] for the derivation):

$$\mathbf{J}(\mathbf{x}) = \begin{bmatrix} \sum_{\ell=1}^{N_u} \gamma_{1,\ell} & -\gamma_{1,2} & \cdots & -\gamma_{1,N_u} \\ -\gamma_{2,1} & \sum_{\ell=1}^{N_u} \gamma_{2,\ell} & & -\gamma_{2,N_u} \\ \vdots & & \ddots & \\ -\gamma_{N_u,1} & -\gamma_{N_u,2} & \cdots & \sum_{\ell=1}^{N_u} \gamma_{N_u,\ell} \end{bmatrix}, \qquad (4.132)$$

where $\gamma_{k,\ell} = I_{\Omega_k}(\ell)\left(\frac{2}{\sigma_{k,\ell}^2} + \zeta \frac{\alpha^2}{d_{k,\ell}^2}\right)$ denotes the information for the link (k, ℓ). The parameter ζ is the indicator of the a priori knowledge of (4.130): $\zeta = 1$ in the case of a priori knowledge, and $\zeta = 0$ otherwise. In the former case, the term added to each element of the FIM leads to a lower estimate error. For increasing SNR ($\sigma_0^2 \to 0$), this contribution becomes negligible and the FIM elements tend to those for $\zeta = 0$. This result is confirmed by Fig. 4.15(a) that compares the two CRBs ($\zeta = 0$ and $\zeta = 1$) for different values of the SNR η_0 in a certain network configuration that we will describe later on.

The performance metrics that we will take into consideration are the lower bound on the kth location estimate accuracy

$$\mathsf{MSE}_k = \text{Cov}\left(\hat{x}_k\right) \geq [\kappa(\mathbf{x})]_{k,k}, \qquad (4.133)$$

and the *average* localization accuracy for the whole network:

$$\mathsf{MSE} = \frac{1}{N_u} \sum_{k=1}^{N_u} \mathsf{MSE}_k \geq \frac{1}{N_u} \text{tr}(\kappa(\mathbf{x})). \qquad (4.134)$$

As a simple case study, we investigate the bounds for a uniformly spaced network of N nodes on a straight line at positions $x_k = (k-1)\Delta$, $k = 1, \ldots, N$, with inter-distance Δ. Of these nodes, N_B are

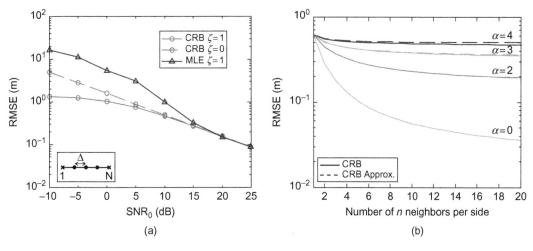

FIGURE 4.15

CRB of the location accuracy: (a) w.r.t. η_0 and (b) w.r.t. the number of neighbors per side N_n. The CRB is computed with $\zeta = 1$ (continuous line) and $\zeta = 0$ (dashed line) for the network deployment exemplified in the box with $N_B = 2$ and $\Delta = 5$ m ("•" denotes unknown node, "x" is an anchor). (a) $N_u = 10$, $\alpha = 4$, $d_0 = 1$ m, $r = 10$ m. (b) $N = 100$, $N_B = 2$, $\eta_0 = -10$ dB at $d_0 = 5$ m.

RNs placed uniformly among the UNs (see the deployment on the bottom of Fig. 4.15(a) where "x" denotes the references, "•" the unknowns). The coverage radius ranges from $r = \Delta$ (each node makes measurements to two neighbors only) to $r \to \infty$ (full coverage). The level of cooperation depends on the coverage radius r and on the path-loss exponent α. Indeed, the number of cooperating neighbors on each side of the generic node is $N_n = \lfloor r/\Delta \rfloor$ (neglecting edge effects), while the reliability of the information provided by the neighbor at distance $m \cdot \Delta$ is proportional to $1/m^\alpha$, $m = 1, \ldots, N_n$.

The effect of parameters r and α on the localization performance is investigated in Fig. 4.15(b), which shows the average accuracy versus the number of cooperating nodes N_n for a uniform linear network with $N = 100$, $\Delta = 5$ m, $\eta_0 = -10$ dB at $d_0 = 5$ m and $\beta = 500$ MHz. As expected, the localization accuracy increases for large N_n and small α. For $\alpha \neq 0$, the positioning error is shown to reach a floor for increasing number of cooperating neighbors, just because the information provided by nodes that are too far away is not accurate enough to further improve performance. The figure also shows a comparison between the true CRB ($\zeta = 1$, continuous line) and the approximated one for $\zeta = 0$ (dashed line). For high SNR, the two bounds tend to be the same, especially when α is small. In the next analyses, we will consider the approximation with $\zeta = 0$.

The localization error along the network is shown in Fig. 4.16(a) for a few different configurations: (1) an array of $N = 34$ nodes with $N_B = 4$; (2) three separate (noncooperating) arrays each having $N = 12$ nodes, and (3) $N_B = 2$ reference nodes at the ends (dashed line). For $r \leq 3\Delta$, the performance of the 12-node array (2) is almost the same as that of the 34-node array (1), insofar as the RN density is constant. Thereby, the localization problem can be simplified to a single subarray deployment for limited node coverage. Figure 4.16(b) shows the localization error as a function of the "density" of RNs.

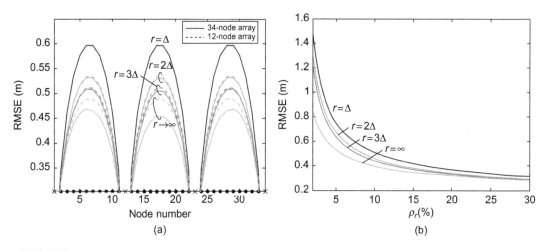

FIGURE 4.16

(a) CRB for 12-node (dashed line) and 34-node (solid line) sensor arrays, for different values of coverage radius r. (b) CRB w.r.t. the fraction of RNs for $N = 100$. For both figures, $\eta_0 = 10$ dB at $d_0 = 1$, $\alpha = 4$, $\Delta = 5$ m.

There is no convenience in having too many RNs, since the information coming from additional RNs is comparable with that coming from UNs through cooperation.

We now focus on case (3), that is, a single regular array with $N_B = 2$ RNs at the two ends. Due to the uniform deployment $d_{k,\ell} = |k - \ell|\Delta$, and so, for large N, the FIM tends to be a *Toeplitz* matrix. The matrix is also *banded* with bandwidth $2N_n + 1$ due to the limited transmission coverage. A closed form of the average network CRB (averaged over all positions) can be derived through the eigenanalysis of the approximated banded-Toeplitz FIM [45], yielding

$$\text{MSE} \geq \frac{1}{N_u} \text{tr}(\kappa\,(\mathbf{x})) \approx \frac{2}{\pi^2} \frac{\sigma_\Delta^2}{\Psi(\alpha)} N_u, \tag{4.135}$$

where $\sigma_\Delta^2 = \sigma_0^2 (\Delta/d_0)^\alpha$ and $\Psi(\alpha) = \sum_{m=1}^{N_n} \frac{1}{m^{\alpha-2}}$. This bound is validated in Fig. 4.17 by comparing the true CRB with the approximated bound (4.135) for $r = \Delta$ (left) and $r \to \infty$ (right), and for different values of α. The approximation (4.135) was derived under the assumption of limited coverage and is accurate only for small r or large α. It is important to observe that the term $\Psi(\alpha)$ gives the *connectivity* of the WSN: the higher is the number of reliable connections for each node (i.e., the effective number of cooperating neighbors), the larger is $\Psi(\alpha)$. In the case of $r \to \infty$, we have $\Psi(\alpha) = (N-1)/2 \approx N_u/2$ for $\alpha = 2$, and the MSE is approximately constant for increasing N_u; on the other hand, for $\alpha > 2$ the effective number of connections decreases rapidly and tends to 1. The same holds for a coverage radius equal to Δ, which leads to $\Psi(\alpha) = 1\ \forall\alpha$.

4.4.3 Bounds for TOA Estimation in the Presence of Interference

In many cases, wireless systems are supposed to operate in the presence of interference (e.g., UWB), then the knowledge of the impact of the interference on the positioning accuracy may be of fundamental

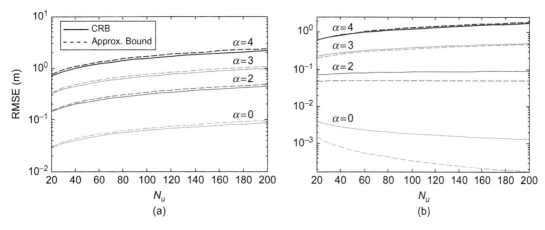

FIGURE 4.17

Average CRB along the array versus the number of UNs for different values of path-loss exponent: true CRB (solid line) and approximation (4.135) (dashed line). $\eta_0 = 10$ dB at $d_0 = 1$ m, $\Delta = 5$ m, $r = \Delta$ (left) and $r = \infty$ (right) (from Ref. [45]).

interest. This awareness, for example, can be usefully exploited in cognitive positioning systems (see Section 5.5). For this reason, in this section, we analyze the CRB for TOA estimation in the presence of interference due, for example, to the presence of one or more communication systems sharing the same spectrum. We adopt a multicarrier (MC) signaling scheme and we model the transmitted baseband signal over the symbol period $0 \leq t \leq T_s$ as[9]

$$s(t) = \sum_{k=1}^{K} \sqrt{w_k}\, g(t)\, e^{J\, 2\pi f_k t}, \tag{4.136}$$

over the symbol interval $[-T_s/2, T_s/2]$. In this equation, $f_k = (k - K/2)\Delta$ is the kth subcarrier frequency shift with respect to the center frequency, Δ is the subcarrier spacing, and $g(t)$ is a pulse with duration T_s and energy E_g. The weights $w_k \geq 0$ permit spectrum shaping and

$$P_t = \frac{E_g}{T_s} \sum_{k=1}^{K} w_k \tag{4.137}$$

represents the power of the baseband signal.[10] In practice, the weights w_k are limited by peak power constraints, as is detailed in Section 5.5.1 when considering the optimization of the signal spectrum.

---

[9]A guard interval between symbols is assumed to avoid intersymbol interference at the receiver.

[10]The corresponding RF power is $\frac{E_g}{2T_s} \sum_{k=1}^{K} w_k$.

Assuming that Δ is small compared to the channel coherence bandwidth, the baseband received signal corresponding to (4.136) is

$$r(t) \cong s_r(t - \tau) + v(t), \tag{4.138}$$

with

$$s_r(t) = \sum_{k=1}^{K} \alpha_k \sqrt{w_k}\, g(t)\, e^{j\, 2\pi f_k t}, \tag{4.139}$$

where τ is the propagation delay, $\alpha_k = a_k e^{j\phi_k}$ denotes the complex channel coefficient at frequency f_k, and $v(t)$ is the total disturbance due to thermal noise and interference. In particular, $v(t)$ is the sum of two terms, say $v_N(t)$ and $v_I(t)$, where $v_N(t)$ is complex AWGN with spectral density N_0 for each component, and $v_I(t)$ is a stationary interference term with power spectral density $S_I(f)$ for each component. Thus, the power spectral density of each component of $v(t)$ is $S_v(f) = N_0 + S_I(f)$. In addition, the interference is modeled as a zero-mean complex Gaussian process. As will be detailed in Section 5.5.1, it can be assumed that $S_I(f)$ is known at the receiver [29, 40, 52].

We consider the best achievable accuracy in estimating the TOA parameter τ from the observation of $r(t)$ in the presence of interference.[11]

The Fourier transform of $s_r(t - \tau)$ in (4.139) is

$$S_r(f, \boldsymbol{\theta}) = \sum_{k=1}^{K} \alpha_k \sqrt{w_k}\, G(f - f_k)\, e^{-j\, 2\pi f \tau}, \tag{4.140}$$

where $G(f)$ is the Fourier transform of $g(t)$, and $\boldsymbol{\theta} \triangleq [\tau\ a_1 \cdots a_K\ \phi_1 \cdots \phi_K]$ is a vector collecting all the channel parameters.

As the disturbance $v(t)$ is colored, we assume without loss of generality that the received signal is first passed through a whitening filter with a frequency response [67]

$$|H(f)|^2 = \frac{1}{S_v(f)}. \tag{4.141}$$

Accordingly, the log-likelihood function can be written as

$$\ln \Lambda(\widetilde{\boldsymbol{\theta}}) = \mathfrak{Re}\left\{ \int_{-\infty}^{\infty} x(t) u^*(t, \widetilde{\boldsymbol{\theta}})\, dt \right\} - \frac{1}{2} \int_{-\infty}^{\infty} |u(t, \widetilde{\boldsymbol{\theta}})|^2\, dt, \tag{4.142}$$

where $\widetilde{\boldsymbol{\theta}}$ is a possible value of $\boldsymbol{\theta}$, $x(t) = r(t) \otimes h(t)$ is the convolution of the received waveform $r(t)$ with the impulse response of the whitening filter $h(t)$, $u(t, \widetilde{\boldsymbol{\theta}}) = \tilde{s}_r(t - \tilde{\tau}) \otimes h(t)$, and

$$\tilde{s}_r(t) = \sum_{k=1}^{K} \tilde{\alpha}_k \sqrt{w_k}\, g(t)\, e^{j2\pi f_k t}. \tag{4.143}$$

---

[11] The observation interval is assumed sufficiently long to comprise the whole received signal not withstanding the a priori uncertainty on the actual value of τ.

Equivalently, the whitening operation can be performed by correlating $r(t)$ with a pulse $q(t, \widetilde{\boldsymbol{\theta}})$ with the following Fourier transform [67]:

$$Q(f, \widetilde{\boldsymbol{\theta}}) \propto \widetilde{S}_r(f, \widetilde{\boldsymbol{\theta}}) / \sqrt{S_v(f)} \tag{4.144}$$

and the log-likelihood function is obtained as [67]

$$\ln \Lambda(\widetilde{\boldsymbol{\theta}}) = \Re \left\{ \int_{-\infty}^{\infty} r(t) q^*(t, \widetilde{\boldsymbol{\theta}}) \, dt \right\} - \frac{1}{2} \int_{-\infty}^{\infty} \widetilde{s}_r(t - \widetilde{\tau}) q^*(t, \widetilde{\boldsymbol{\theta}}) \, dt. \tag{4.145}$$

The FIM **J** for the unknown vector $\boldsymbol{\theta}$ has elements [67]

$$[\mathbf{J}]_{m,n} = \Re \left\{ \int_{-\infty}^{\infty} \frac{\partial \widetilde{S}_r^*(f, \boldsymbol{\theta})}{\partial \widetilde{\theta}_m} S_v^{-1}(f) \frac{\partial \widetilde{S}_r(f, \boldsymbol{\theta})}{\partial \widetilde{\theta}_n} \, df \right\}. \tag{4.146}$$

In (4.146), $\widetilde{\theta}_n$ is the nth element of $\widetilde{\boldsymbol{\theta}}$, and with a slight abuse of notation, $\partial \widetilde{S}_r(f, \boldsymbol{\theta}) / \partial \widetilde{\theta}_n$ denotes the partial derivative of $\widetilde{S}_r(f, \widetilde{\boldsymbol{\theta}})$ with respect to $\widetilde{\theta}_n$ computed for $\widetilde{\boldsymbol{\theta}} = \boldsymbol{\theta}$.

After some manipulations from (4.140) and (4.146), it is found that

$$\mathbf{J} = \begin{bmatrix} \mathbf{J}_{\tau\tau} & \mathbf{J}_{\tau a} & \mathbf{J}_{\tau\phi} \\ \mathbf{J}_{\tau a}^T & \mathbf{J}_{aa} & \mathbf{J}_{a\phi} \\ \mathbf{J}_{\tau\phi}^T & \mathbf{J}_{a\phi}^T & \mathbf{J}_{\phi\phi} \end{bmatrix}, \tag{4.147}$$

where the elements of **J** are expressed as follows:

$$\mathbf{J}_{\tau\tau} = 4\pi^2 \Re \left\{ \sum_{k=1}^{K} \sum_{l=1}^{K} \alpha_k^* \alpha_l \sqrt{w_k w_l} \, y_{k,l}(2) \right\}, \tag{4.148}$$

$$[\mathbf{J}_{\tau a}]_m = -2\pi \sqrt{w_m} \, \Im \left\{ e^{j\phi_m} \sum_{k=1}^{K} \alpha_k^* \sqrt{w_k} \, y_{k,m}(1) \right\}, \tag{4.149}$$

$$[\mathbf{J}_{\tau\phi}]_m = -2\pi \sqrt{w_m} \, \Re \left\{ \alpha_m \sum_{k=1}^{K} \alpha_k^* \sqrt{w_k} \, y_{k,m}(1) \right\}, \tag{4.150}$$

$$[\mathbf{J}_{aa}]_{m,n} = \sqrt{w_m w_n} \, \Re \left\{ e^{j(\phi_n - \phi_m)} y_{m,n}(0) \right\}, \tag{4.151}$$

$$[\mathbf{J}_{a\phi}]_{m,n} = -\sqrt{w_m w_n} \, \Im \left\{ e^{-j\phi_m} \alpha_n y_{m,n}(0) \right\}, \tag{4.152}$$

$$[\mathbf{J}_{\phi\phi}]_{m,n} = \sqrt{w_m w_n} \, \Re \left\{ \alpha_m^* \alpha_n y_{m,n}(0) \right\}, \tag{4.153}$$

with

$$y_{m,n}(i) \triangleq \int_{-\infty}^{\infty} f^i S_v^{-1}(f) P^*(f - f_m) G(f - f_n) \, df, \tag{4.154}$$

for $i = 0, 1, 2$ and $m, n = 1, 2, \ldots, K$.

Inspection of (4.147) reveals that the FIM can be put in the form of

$$\mathbf{J} = \begin{bmatrix} \mathbf{J}_{\tau\tau} & \mathbf{B} \\ \mathbf{B}^T & \mathbf{C} \end{bmatrix}, \tag{4.155}$$

with $\mathbf{B} \triangleq \begin{bmatrix} \mathbf{J}_{\tau a} & \mathbf{J}_{\tau\phi} \end{bmatrix}$ and $\mathbf{C} \triangleq \begin{bmatrix} \mathbf{J}_{aa} & \mathbf{J}_{a\phi} \\ \mathbf{J}_{a\phi}^T & \mathbf{J}_{\phi\phi} \end{bmatrix}$. Thus, the CRB for parameter τ can be evaluated as

$$\kappa(\tau) = \left(\mathbf{J}_{\tau\tau} - \mathbf{B}\mathbf{C}^{-1}\mathbf{B}^T \right)^{-1}. \tag{4.156}$$

Equation (4.156) takes simpler forms in the following special cases.

Disjoint Spectra

If $|G(f)|$ is approximately zero outside $-\Delta/2 \leq f \leq \Delta/2$, from (4.154) we have $y_{m,n}(i) = 0$ for $m \neq n$ and (4.148)–(4.153) become

$$\mathbf{J}_{\tau\tau} = 4\pi^2 \sum_{k=1}^{K} |\alpha_k|^2 w_k \gamma_k(2), \tag{4.157}$$

$$\mathbf{J}_{\tau a} = \mathbf{0}, \tag{4.158}$$

$$\left[\mathbf{J}_{\tau\phi}\right]_m = -2\pi w_m |\alpha_m|^2 \gamma_m(1), \tag{4.159}$$

$$\mathbf{J}_{aa} = \text{diag}\{w_1\gamma_1(0), w_2\gamma_2(0), \ldots, w_K\gamma_K(0)\}, \tag{4.160}$$

$$\mathbf{J}_{a\phi} = \mathbf{0}, \tag{4.161}$$

$$\mathbf{J}_{\phi\phi} = \text{diag}\left\{ w_1|\alpha_1|^2\gamma_1(0), \ldots, w_K|\alpha_K|^2\gamma_K(0) \right\}, \tag{4.162}$$

with

$$\gamma_k(i) \triangleq \int_{-\infty}^{\infty} f^i S_v^{-1}(f) |G(f - f_k)|^2 df, \quad i = 0, 1, 2. \tag{4.163}$$

Thus, substituting (4.157)–(4.162) into (4.156) yields

$$\kappa(\tau)|_{\text{disjoint spectra}} = \left(\mathbf{J}_{\tau\tau} - \mathbf{J}_{\tau\phi}\mathbf{J}_{\phi\phi}^{-1}\mathbf{J}_{\tau\phi}^T \right)^{-1} = \left(\sum_{k=1}^{K} w_k \lambda_k \right)^{-1}, \tag{4.164}$$

with

$$\lambda_k = 4\pi^2 |\alpha_k|^2 \left(\gamma_k(2) - \frac{\gamma_k(1)}{\gamma_k(0)} \right). \tag{4.165}$$

We see that the contribution of each subcarrier to the CRB is determined by the corresponding weight w_k, the squared channel gain $|\alpha_k|^2$, the spectrum of pulse $g(t)$, and the power spectral density $S_I(f)$ of the interference around f_k.

Slowly-Varying $S_v(f)$

The coefficient λ_k in (4.165) can be further simplified assuming $S_v(f) \cong S_v(f_k) = N_0 + S_I(f_k)$ for $|f - f_k| \le \Delta/2 \,\forall k$. Correspondingly (4.163) becomes

$$\gamma_k(i) \cong \frac{1}{S_v(f_k)} \int_{-\infty}^{\infty} f^i |G(f - f_k)|^2 df$$

$$= \frac{1}{S_v(f_k)} \int_{-\infty}^{\infty} (f + f_k)^i |G(f)|^2 df. \tag{4.166}$$

Then, defining

$$\beta_i \triangleq \frac{1}{E_g} \int_{-\infty}^{\infty} f^i |G(f)|^2 df \qquad i = 1,2 \tag{4.167}$$

and bearing in mind that

$$\int_{-\infty}^{\infty} |G(f)|^2 df = E_g, \tag{4.168}$$

we obtain

$$\gamma_k(2) = \frac{E_g}{S_v(f_k)} (\beta_2 + 2f_k \beta_1 + f_k^2), \tag{4.169}$$

$$\gamma_k(1) = \frac{E_g}{S_v(f_k)} (\beta_1 + f_k), \tag{4.170}$$

$$\gamma_k(0) = \frac{E_g}{S_v(f_k)}. \tag{4.171}$$

Finally, substituting (4.169)–(4.171) into (4.165) produces

$$\lambda_k = \frac{4\pi 2 E_g |\alpha_k|^2 (\beta_2 - \beta_1^2)}{N_0 + S_I(f_k)}. \tag{4.172}$$

The physical meanings of β_2 and β_1 are as follows. From (4.167), we recognize that β_2 gives the *mean-squared bandwidth* (the *effective bandwidth*) of $g(t)$ while β_1 represents the *skewness* of the spectrum $|G(f)|^2$. Note that if $g(t)$ is real valued, $|G(f)|$ is an even function and β_1 is zero. Equation (4.172) indicates that the contribution of the kth subcarrier is proportional to $|\alpha_k|^2/(N_0 + S_I(f_k))$. Thus, λ_k gets larger and the CRB reduces as the channel gain increases and/or the interference spectral density around f_k decreases.

The CRB in (4.164) using (4.165) or (4.172) will be adopted in Section 5.5.1 to optimize the ranging signal structure in the presence of interference through the modification of weights $\{w_k\}$.

References

[1] IEEE standard for information technology—telecommunications and information exchange between systems—local and metropolitan area networks—specific requirement part 15.4: Wireless medium access control (MAC) and physical layer (PHY) specifications for low-rate wireless personal area networks (WPANs). IEEE Std 802.15.4a-2007 (Amendment to IEEE Std 802.15.4-2006), 2007, pp. 1–203.

[2] A. Amar, A.J. Weiss, Direct position determination of multiple radio signals, EURASIP J. Appl. Signal Process. 2005 (1) (2005) 37–49.

[3] A. Amar, A.J. Weiss, Localization of narrowband radio emitters based on Doppler frequency shifts, IEEE Trans. Signal Process. 56 (11) (2008) 5500–5508.

[4] H. Anouar, A.M. Hayar, R. Knopp, C. Bonnet, Ziv–Zakai lower bound on the time delay estimation of UWB signals, in: Int. Symp. on Commun., Control, and Signal Processing (ISCCSP), Marrakech, Morocco, Sovisoft Oy, Tampere, Finland, 2006.

[5] H. Anouar, A.M. Hayar, R. Knopp, C. Bonnet, Lower bound on time-delay estimation error of UWB signals, in: Proc. Asilomar Conference on Signals, Systems and Computers (ACSSC 2007), IEEE, Pacific Grove, CA, 2007, pp. 1350–1354.

[6] M.J. Appel, R. LaBarre, D. Radulovic, On accelerated random search, SIAM J. Optim. 14 (3) (2004) 708–731.

[7] J.N. Ash, R.L. Moses, On the relative and absolute positioning errors in self-localization systems, IEEE Trans. Signal Process. 56 (11) (2008) 5668–5679.

[8] Y. Bar-Shalom, X.R. Li, T. Kirubarajan, Estimation with Applications to Tracking and Navigation: Theory Algorithms and Software, John Wiley & Sons, New York, 2001.

[9] K.L. Bell, Y. Steinberg, Y. Ephraim, H.L.V. Trees, Extended Ziv–Zakai lower bound for vector parameter estimation, IEEE Trans. Inform. Theory 43 (2) (1997) 624–637.

[10] S. Bellini, G. Tartara, Bounds on error in signal parameter estimation, IEEE Trans. Commun. 22 (3) (1974) 340–342.

[11] M.G.D. Benedetto, Understanding Ultra Wide Band Radio Fundamentals, Prentice Hall PTR, Upper Saddle River, NJ, 2004.

[12] K. Borre, D. Akos, N. Bertelsen, P. Rinder, S. Jensen, A Software-Defined GPS and Galileo Receiver. A Single-Frequency Approach. Applied and Numerical Harmonic Analysis, Birkhäuser, Boston, 2007.

[13] D. Cassioli, M.Z. Win, A.F. Molisch, The ultra-wide bandwidth indoor channel: from statistical model to simulations. IEEE J. Sel. Areas Commun. 20 (6) (2002) 1247–1257.

[14] D. Chazan, M. Zakai, J. Ziv, Improved lower bounds on signal parameter estimation, IEEE Trans. Inf. Theory 21 (1) (1975) 90–93.

[15] P. Closas, C. Fernández-Prades, Bayesian nonlinear filters for direct position estimation, in: Proc. of the IEEE Aerospace Conference, Big Sky, MT, IEEE, 2010, pp. 1–12.

[16] P. Closas, C. Fernández-Prades, D. Bernal, J.A. Fernández-Rubio, Bayesian direct position estimation, in: Proceedings of the ION GNSS 2008, Savannah, GA, 2008, pp. 183–190.

[17] P. Closas, C. Fernández-Prades, J.A. Fernández-Rubio, Maximum likelihood estimation of position in GNSS, IEEE Signal Process. Lett. 14 (5) (2007) 359–362.

[18] P. Closas, C. Fernández-Prades, J.A. Fernández-Rubio, Direct position estimation approach outperforms conventional two-steps positioning, in: Proceedings of the XVII European Signal Processing Conference, EUSIPCO, Glasgow, Scotland, 2009, pp. 1958–1962.

[19] P. Closas, C. Fernández-Prades, J.A. Fernández-Rubio, A. Ramírez-González, On the maximum likelihood estimation of position, in: Proceedings of the ION GNSS 2006, Fort Worth, TX, 2006, pp. 1800–1810.

[20] D. Dardari, C.-C. Chong, M.Z. Win, Improved lower bounds on time-of-arrival estimation error in realistic UWB channels, in: IEEE International Conference on Ultra-Wideband, ICUWB 2006, Waltham, MA, USA, IEEE, 2006, pp. 531–537.

[21] D. Dardari, C.-C. Chong, M.Z. Win, Threshold-based time-of-arrival estimators in UWB dense multipath channels, IEEE Trans. Commun. 56 (8) (2008) 1366–1378.

[22] D. Dardari, A. Conti, U. Ferner, A. Giorgetti, M.Z. Win, Ranging with ultrawide bandwidth signals in multipath environments, Proc. of IEEE, Spec. Issue UWB Technol. & Emerg. Appl. 97 (2) (2009) 404–426.

[23] D. Dardari, M.Z. Win, Ziv–Zakai bound of time-of-arrival estimation with statistical channel knowledge at the receiver, in: IEEE International Conference on Ultra-Wideband, ICUWB 2009, Vancouver, Canada, 2009, pp. 624–629.

[24] C. Fernández-Prades, Advanced Signal Processing Techniques for GNSS Receivers, PhD thesis, Dept. of Signal Theory and Communications, Universitat Politècnica de Catalunya (UPC), Barcelona, Spain, 2006.

[25] S. Gezici, Z. Tian, G.B. Giannakis, H. Kobayashi, A.F. Molisch, H.V. Poor, and Z. Sahinoglu, Localization via ultra-wideband radios: A look at positioning aspects for future sensor networks, IEEE Signal Process. Mag. 22 (2005) 70–84.

[26] W. Gifford, D. Dardari, M.Z. Win, Improved lower bounds on time-of-arrival estimation error for wideband signals in the presence of multipath, IEEE Trans. Commun. 2011. In preparation.

[27] M.S. Grewal, L.R. Weill, A.P. Andrews, Global Positioning Systems, Inertial Navigation, and Integration, John Wiley & Sons, New York, 2001.

[28] F. Gustafsson, F. Gunnarsson, N. Bergman, U. Forssell, J. Jansson, R. Karlsson, P.J. Nordlund, Particle filters for positioning, navigation and tracking. IEEE Trans. Signal Process. 50 (2) (2002) 425–437.

[29] S. Haykin, Cognitive radio: Brain-empowered wireless communications, IEEE J. Sel. Areas Commun. 23 (2) (2005) 201–220.

[30] L. Huang, Z. Xu, B.M. Sadler, Simplified Ziv–Zakai time delay estimation bound for ultra-wideband signals, in: Proc. Workshop on Short Range Ultra-Wideband Radio Syst., Santa Monica, CA, 2006.

[31] J.P. Iannello, Large and small error performance limits for multipath time delay estimation, IEEE Trans. Acoust. Speech Signal Process. ASSP-34 (2) (1986) 245–251.

[32] D.H. Johnson, D.E. Dudgeon, Array Signal Processing: Concepts and Techniques, Prentice-Hall, Englewood Cliffs, NJ, 1993.

[33] D. Jourdan, D. Dardari, M.Z. Win, Position error bound for UWB localization in dense cluttered environments, IEEE Trans. Aerosp. Electron. Syst. 44 (2) (2008) 613–628.

[34] E. Kaplan, Understanding GPS. Principles and Applications, Artech House Publishers, London, 1996.

[35] S.M. Kay, Fundamentals of Statistical Signal Processing: Estimation Theory, Prentice-Hall, Englewood Cliffs, NJ, 1993.

[36] L.l. Scharf, Statistical Signal Processing. Detection, Estimation and Time Series Analysis, Addison-Wesley, Reading, MA, 1991.

[37] E. Larsson, Cramer–Rao bound analysis of distributed positioning in sensor networks, IEEE Signal Process. Lett. 11 (3) (2004) 334–337.

[38] J. Li, R. Wu, An efficient algorithm for time delay estimation, IEEE Trans. Sig. Process. 46 (8) (1998) 2231–2235.

[39] T.G. Manickam, R.J. Vaccaro, D.W. Tufts, A least-squares algorithm for multipath time-delay estimation, IEEE Trans. Sig. Process. 42 (11) (1994) 3229–3233.

[40] J. Mitola, G.Q. Maguire, Cognitive radio: Making software radios more personal, IEEE Personal Commun. Mag. 6 (4) (1999) 13–18.

[41] A.F. Molisch, Ultrawideband propagation channels – Theory, measurement, and modeling, IEEE Trans. Veh. Technol. 54 (5) (2005) 1528–1545.

[42] A.F. Molisch, D. Cassioli, C.-C. Chong, S. Emami, A. Fort, B. Kannan, et al., A comprehensive standardized model for ultrawideband propagation channels, IEEE Antennas Propagat. Mag. 54 (11) (2006) 3151–3166. Special Issue on Wireless Communications.

[43] R.A. Monzingo, T.W. Miller, Introduction to Adaptive Arrays, John Wiley & Sons, New York, 1980.

[44] R.L. Moses, D. Krishnamurthy, R. Patterson, A self-localization method for wireless sensor networks, EURASIP J. Appl. Signal Process. 2003 (4) (2003) 348–358.

[45] M. Nicoli, D. Fontanella, Fundamental performance limits of TOA-based cooperative localization, in: Proc. IEEE Int. Conf. on Communications (ICC '09), SyCoLo Workshop, IEEE, Dresden, 2009.

[46] B. Parkinson, J. Spilker (Eds.), Global Positioning System: Theory and Applications, vol. I, II, vol. 163–164 of Progress in Astronautics and Aeronautics, American Institute of Aeronautics, Inc., Washington DC, 1996.

[47] N. Patwari, J.N. Ash, S. Kyperountas, A.O. Hero, R.L. Moses, N.S. Correal, Locating the nodes: Cooperative localization in wireless sensor networks, IEEE Signal Process. Mag. 22 (4) (2005) 54–69.

[48] N. Patwari, A.O. Hero, M. Perkins, N.S. Correal, R.J. O'Dea, Relative location estimation in wireless sensor networks, IEEE Trans. Signal Process. 8 (51) (2003) 2137–2148.

[49] R. Plackett, Some theorems in least squares. Biometrika 37 (1/2) (1950) 149–157.

[50] J.G. Proakis, Digital Communications, fourth ed., McGraw-Hill, Inc., New York, 10020, 2001.

[51] Y. Qi, H. Kobayashi, H. Suda, On time-of-arrival positioning in a multipath environment, IEEE Trans. Veh. Technol. 55 (5) (2006) 1516–1526.

[52] Z. Quan, S. Cui, H.V. Poor, A.H. Sayed, Collaborative wideband sensing for cognitive radios, IEEE Signal Process. Mag. 25 (6) (2008) 60–73.

[53] M. Rydström, A. Urruela, E.G. Ström, A. Svensson, A low complexity algorithm for distributed sensor positioning, in: Proceedings of the European Wireless Conference, Nicosia, Cyprus, vol. 2, 2005, pp. 714–718.

[54] B.M. Sadler, L. Huang, Z. Xu, Ziv–Zakai time delay estimation bound for ultra-wideband signals, in: Proc. IEEE Int. Conf. Acoustics, Speech, and Signal Processing, vol. 3, Honolulu, HI, IEEE, 2007, pp. 549–552.

[55] G. Seco-Granados, Antenna Arrays for Multipath and Interference Mitigation in GNSS Receivers, PhD thesis, Dept. of Signal Theory and Communications, Universitat Politècnica de Catalunya (UPC), Barcelona, Spain, 2000.

[56] G. Seco-Granados, J.A. Fernández-Rubio, C. Fernández-Prades, ML estimator and hybrid beamformer for multipath and interference mitigation in GNSS receivers, IEEE Trans. Signal Process. 53 (3) (2005) 1194–1208.

[57] Y. Shen, M. Win, Fundamental limits of wideband localization—part I: A general framework. IEEE Trans. Inf. Theory 56 (10) 4956–4980.

[58] Y. Shen, H. Wymeersch, M. Win, Fundamental limits of wideband cooperative localization via Fisher information, in: Proc. IEEE WCNC'07, IEEE, Hong Kong, 2007, pp. 3951–3955.

[59] Y. Shen, H. Wymeersch, M. Win, Fundamental limits of wideband localization—part II: Cooperative networks, IEEE Trans. Inf. Theory 56 (10) (2010) 4981–5000.

[60] H. Sheng, P. Orlik, A.M. Haimovich, L.J. Cimini Jr., J. Zhang, On the spectral and power requirements for ultra-wideband transmission, in: Proc. IEEE Int. Conf. Commun. (ICC), vol. 1, IEEE, Anchorage, AK, 2003, pp. 738–742.

[61] J. Soubielle, I. Fijalkow, P. Duvaut, A. Bibaut, GPS positioning in a multipath environment, IEEE Trans. Signal Process. 50 (1) (2002) 141–150.

[62] P. Stoica, R. Moses, Introduction to Spectral Analysis, Prentice-Hall, Upper Saddler River, NJ, 1997.

[63] P. Stoica, A. Nehorai, MUSIC maximum likelihood and Cramer–Rao bound, IEEE Trans. Acoust. Speech and Signal Process. 37 (5) (1989) 720–741.

[64] G. Strang, K. Borre, Linear Algebra, Geodesy, and GPS, Cambridge Press, Wellesley, MA, 1997.

[65] H.V. Trees, Optimum Array Processing. Detection, Estimation and Modulation Theory, Part IV, Wiley Interscience, New York, 2002.

[66] J.B.-Y. Tsui, Fundamentals of Global Positioning System Receivers. A Software Approach, John Willey & Sons, Inc., New York, 2000.

[67] H.L. Van Trees, Detection, Estimation, and Modulation Theory, first ed., John Wiley & Sons, Inc., New York, 10158-0012, 1968.

[68] A.J. Viterbi, CDMA: Principles of Spread Spectrum Communication, Addison-Wesley Wireless Communication Series, Addison-Wesley, Reading, MA, 1995.

[69] E. Weinstein, A.J. Weiss, Fundamental limitations in passive time delay estimation—part II: Wide-band systems, IEEE Trans. Acoust. Speech Signal Process. ASSP-32 (5) (1984) 1064–1078.

[70] E. Weinstein, A.J. Weiss, A general class of lower bounds in parameter estimation, IEEE Trans. Inform. Theory 34 (2) (1988) 338–342.

[71] A.J. Weiss, Direct position determination of narrowband radio frequency transmitters, IEEE Signal Process. Lett. 11 (5) (2004) 513–516.

[72] A.J. Weiss, E. Weinstein, Fundamental limitations in passive time delay estimation—part I: Narrow-band systems, IEEE Trans. Acoust. Speech Signal Process. 31(2) (1983) 472–486.

[73] M.Z. Win, A. Conti, S. Mazuelas, Y. Shen, W.M. Gifford, D. Dardari, Network localization and navigation via cooperation, IEEE Commun. Mag. (2011).

[74] M.Z. Win, R.A. Scholtz, Impulse radio: How it works, IEEE Commun. Lett. 2 (2) (1998) 36–38.

[75] M.Z. Win, R.A. Scholtz, Characterization of ultra -wide bandwidth wireless indoor communications channel: A communication theoretic view, IEEE J. Sel. Areas Commun. 20 (9) (2002) 1613–1627.

[76] R. Wooding, The multivariate distribution of complex normal variables, Biometrika 43 (1956) 212–215.

[77] Z. Xu, B. Sadler, Time delay estimation bounds in convolutive random channels, IEEE J. Sel. Top. Signal Process. 1 (3) (2007) 418–430.

[78] A. Zeira, P.M. Schultheiss, Realizable lower bounds for time delay estimation, IEEE Trans. Signal. Process. 41 (11) (1993) 3102–3113.

[79] A. Zeira, P.M. Schultheiss, Realizable lower bounds for time delay estimation: Part 2—Threshold phenomena, IEEE Trans. Signal Process. 32 (5) (1994) 1001–1007.

[80] A. Mallat, C. Oestges, L. Vandendorpe, CRBs for UWB multipath channel estimation: Impact of the overlapping between the MPCs on MPC gain and TOA estimation, in: Proc. IEEE Int. Conf. on Communication (ICC '09), 2009.

[81] P. Closas, C. Fernández-Prades, J.A. Fernández-Rubio, Cramér–Rao bound analysis of positioning approaches in GNSS receivers, IEEE Trans. Signal Process. 57 (10) (2009) 3775–3786.

Innovative Signal Processing Techniques for Wireless Positioning

Davide Dardari, Mario Di Dio, Andrea Emmanuele, Diana Fontanella, Sinan Gezici, Mohammad Reza Gholami, Michel Kieffer, Eva Lagunas, Jérôme Louveaux, Achraf Mallat, Montse Nájar, Monica Navarro, Monica Nicoli, Luca Reggiani, Mats Rydström, Erik G. Ström, Luc Vandendorpe, Francesca Zanier

This chapter presents a collection of the latest research results in the field of wireless positioning carried out within the European Network of Excellence NEWCOM++. They represent a necessarily partial panorama of the "hottest topics" in advanced wireless positioning.

In Section 5.1, some recent UWB TOA/AOA estimation techniques operating in the frequency domain are presented. In addition, the effect of interference is investigated, and some solutions to mitigate the performance degradation are described.

It is well known that the adoption of multiple antennas at both the transmitter and the receiver in MIMO configuration leads to improvement in performance of communication systems in terms of diversity gain or higher capacity. MIMO systems can also be successfully employed in positioning systems. Some preliminary considerations regarding the theoretical performance limits and practical TOA estimation schemes are reported in Section 5.2.

To counteract the effects of measurement errors, especially when working in harsh environments with severe NLOS configurations, some advanced geometric positioning approaches are described in Section 5.3.

In Section 5.4, the cooperative positioning topic is addressed by showing how cooperation among agents can in general improve the positioning accuracy by extending classical localization algorithms to the cooperative scenario or by introducing new approaches such as those based on belief propagation.

One of the most recent promising topics is cognitive positioning (CP) in which the main idea is to design the positioning system according to the cognitive radio (CR) paradigm with the purpose to use efficiently the radio spectrum and to optimize the system performance in the presence of interference. Some aspects relating to optimal ranging signal design and practical TOA estimators in CP systems are discussed in Section 5.5.

5.1 ADVANCED UWB POSITIONING TECHNIQUES

Most localization techniques for UWB signals are based on time-domain approaches. ML-based TOA estimators have been illustrated in Chapter 3, but they suffer from high sampling constraints. We have seen that to cope with these sampling requirements a viable alternative is given by energy-based TOA estimators developed using sub-Nyquist sampling rates (e.g., Refs. [24, 25]). However, TOA estimation can also be tackled by processing the signal in the frequency domain. Section 5.1.1 is devoted to the discussion of two approaches (an optimal and a suboptimal one) for UWB TOA estimation, based on the phase estimate of the signal Fourier transform, instead of on the classic time-domain techniques. A technique for the joint estimation of range (TOA) and angle (AOA) is analyzed in Section 5.1.2, whereas solutions to deal with the presence of interference are illustrated in Section 5.1.3.

The dramatic loss of accuracy caused by NLOS propagation makes one look for ranging techniques that are robust to impairments generated by multipath and path obstructions. The frequency-domain high-resolution TOA estimation techniques described in Section 5.1.4 are an example of ranging methods suitable for NLOS conditions. On the other hand, the presence of multipath can be detected and mitigated through techniques that can estimate the bias on the TOA estimate due to NLOS propagation.

5.1.1 **TOA Estimators Operating in the Frequency Domain**

In this section, we consider the TOA estimation in UWB systems by exploiting the discrete Fourier transform (DFT) of the received signal. In particular, by analyzing the DFT of the received signal, it is possible to derive the MLE of the instantaneous (with respect to the frequency) phase of the useful received signal. Two instantaneous ranging techniques based on TOA and the instantaneous phase will be illustrated. The first one is the MLE and the second one is based on the slope characterization of the instantaneous phase. Based on the instantaneous TOA estimates obtained at different frequencies, for each technique (ML-based and slope-based) the optimal estimator by maximum ratio combination is computed.

System Model

Let $s(t)$ be the real band-pass transmitted signal. The signal received through an AWGN channel can be written as

$$r(t) = \alpha s(t - \tau) + n(t), \tag{5.1}$$

where α and τ are the gain and the time delay of the channel, respectively, and $n(t)$ is the AWGN with monolateral power spectral density N_0. We will show later that the estimators can also be used in multipath UWB channels and approximately the same performance can be achieved in the case of nonoverlapping multipath components.

Let $u(t)$, $v(t)$, and $w(t)$ be the complex envelopes of $s(t)$, $r(t)$, and $n(t)$ with respect to the central frequency f_c of $s(t)$. Then, we can write (5.1) as

$$v(t) = \alpha e^{-j2\pi f_c \tau} u(t - \tau) + w(t),$$

where $w(t)$ is a complex AWGN with unilateral PSD equal to $\gamma_w = 2N_0$ (N_0 for each of in-phase and in-quadrature components). Note that the energy of $s(t)$ is half the energy of $u(t)$ ($E_s = E_u/2$). The frequency f_c is defined as

$$f_c = \frac{\int_0^{+\infty} f |S(f)|^2 df}{\int_0^{+\infty} |S(f)|^2 df}, \tag{5.2}$$

where $S(f)$ is the Fourier transform of $s(t)$. Note that when f_c is defined as in (5.2), we can write the mean quadratic bandwidth of $s(t)$ as

$$\beta_s^2 = \frac{\int_{-\infty}^{+\infty} 4\pi^2 f^2 |S(f)|^2 df}{\int_{-\infty}^{+\infty} |S(f)|^2 df} = \frac{\int_0^{+\infty} 4\pi^2 (f + f_c)^2 |U(f)|^2 df}{\int_0^{+\infty} |U(f)|^2 df}. \tag{5.3}$$

Assume that $v(t)$ passes through an ideal low-pass filter with cutoff frequency equal to $B_W/2$, where B_W is the bandwidth of $s(t)$. The signal at the output of the low-pass filter can be written as

$$\tilde{v}(t) = \alpha e^{-j2\pi f_c \tau} u(t - \tau) + \tilde{w}(t) \tag{5.4}$$

$$= \alpha e^{-j2\pi f_c \tau} u_\tau(t) + \tilde{w}(t), \tag{5.5}$$

where $\tilde{w}(t)$ is a complex zero-mean wide-sense stationary Gaussian process. The covariance of $\tilde{w}(t)$ is given by

$$C_{\tilde{w}}(t) = 2N_0 B_W \frac{\sin(\pi Bt)}{\pi Bt} = \sigma_{\tilde{w}}^2 \frac{\sin(\pi Bt)}{\pi Bt}, \tag{5.6}$$

where $\sigma_{\tilde{w}}^2 = 2N_0 B_W = C_{\tilde{w}}(0)$ is the variance of $\tilde{w}(t)$.

The result obtained in (5.4) follows from low-pass filtering the complex envelope of the real band-pass received signal. However, note that the same result can be obtained if the real band-pass received signal passes first through a band-pass filter and then we consider its complex envelope.

Assume now that $\tilde{v}(t)$ is sampled at the rate $f_s = 1/T_s = B_W$. Equations (5.4) and (5.5) lead to

$$\tilde{v}[n] = \alpha e^{-j2\pi f_c \tau} u(nT_s - \tau) + \tilde{w}[n]$$

$$= \alpha e^{-j2\pi f_c \tau} u_\tau[n] + \tilde{w}[n],$$

where $\tilde{v}[n] = \tilde{v}(nT_s)$, $n = 0, \ldots, N-1$, N is the number of samples and $T = NT_s$ is the observation time. The component $\tilde{w}[n]$ is a complex zero-mean Gaussian sequence. The covariance of $\tilde{w}[n]$ can be obtained from (5.6):

$$C_{\tilde{w}}[n, n'] = C_{\tilde{w}}((n' - n)T_s) = \sigma_{\tilde{w}}^2 \frac{\sin(\pi(n' - n))}{\pi(n' - n)} \tag{5.7}$$

$$= \begin{cases} \sigma_{\tilde{w}}^2 = 2N_0 B_W & n' = n \\ 0 & n \neq n', \end{cases} \tag{5.8}$$

where $\sigma_{\tilde{w}}^2$ is the variance of $\tilde{w}[n]$. It can be inferred from (5.8) that $\{\tilde{w}[n]\}$ is a white sequence (hence, $\tilde{w}[n]$, $n = 0, \ldots, N-1$, are independent, and identically distributed (i.i.d.)).

The DFT $\tilde{V}[k]$ of $\tilde{v}[n]$, with $k = -N/2, \ldots, N/2 - 1$, can be written as

$$\tilde{V}[k] = \sum_{n=0}^{N-1} \tilde{v}[n] e^{-j2\pi \frac{nk}{N}} = \alpha e^{-j2\pi f_c \tau} U_\tau[k] + \tilde{W}[k], \tag{5.9}$$

where $\tilde{W}[k]$ is a complex zero-mean white Gaussian sequence with variance equal to $\sigma_{\tilde{W}}^2 = N\sigma_{\tilde{w}}^2$ [64]. Given that the Nyquist sampling theorem is respected, we can write

$$\tilde{V}[k]T_s = \tilde{V}(f_k), \tag{5.10}$$

where $\tilde{V}(f)$ is the Fourier transform of $\tilde{v}(t)$ and $f_k = k/(NT_s) = k\Delta f$. If $\tilde{v}[k]$ is real, we have $\tilde{V}[-k] = \tilde{V}^*[k]$, where $*$ denotes the complex conjugate. Taking into account that $U_\tau[k] = e^{-j2\pi f_k \tau} U[k]$, $\tilde{V}[k]$ can be written as

$$\tilde{V}_k = \alpha e^{-j2\pi(f_c + f_k)\tau} U_k + \tilde{W}_k$$

$$= \alpha e^{-j\theta_k} U_k + \tilde{W}_k$$

$$= G_k + \tilde{W}_k,$$

where $V_k = V[k]$ and $G_k = \alpha e^{j\theta_k} U_k$, and θ_k can be written as

$$\theta_k = 2\pi(f_c + f_k)\tau = \theta_{U_k} - \theta_{G_k}. \tag{5.11}$$

Statistics of the Observation

In this part, we give some statistics related to the observation $\tilde{\mathbf{V}} = \left[\tilde{V}_{-N/2}, \ldots \tilde{V}_{N/2-1}\right]^T$. Given that $\{\tilde{W}_k\}$ is a complex zero-mean white Gaussian sequence, denoting by ρ_Z, θ_Z, X_Z, and Y_Z the modulus, phase, real part, and imaginary part of the complex number Z, we can write the pdf of $\tilde{V}_k$ and $\tilde{\mathbf{V}}$ (with respect to cartesian and polar coordinates) as

$$T_{\tilde{V}_k}(X_{\tilde{V}_k}, Y_{\tilde{V}_k}) = \frac{1}{2\pi\sigma^2} e^{-\frac{(X_{\tilde{V}_k} - X_{G_k})^2 + (Y_{\tilde{V}_k} - Y_{G_k})^2}{2\sigma^2}}$$

$$T_{\tilde{V}_k}(\rho_{\tilde{V}_k}, \theta_{\tilde{V}_k}) = \frac{\rho_{\tilde{V}_k}}{2\pi\sigma^2} e^{-\frac{\rho_{\tilde{V}_k}^2 + \rho_{G_k}^2 - 2\rho_{\tilde{V}_k}\rho_{G_k}\cos(\theta_{\tilde{V}_k} - \theta_{G_k})}{2\sigma^2}} \tag{5.12}$$

$$T_{\tilde{\mathbf{V}}} = \sum_{k=-N/2}^{N/2-1} T_{\tilde{V}_k},$$

where $\sigma^2 = \sigma_{\tilde{W}}^2/2 = NN_0 B_W$ is the variance of each of $X_{\tilde{W}_k}$ and $Y_{\tilde{W}_k}$. $T_{\tilde{V}_k}(\rho_{\tilde{V}_k}, \theta_{\tilde{V}_k})$ is obtained from $T_{\tilde{V}_k}(X_{\tilde{V}_k}, Y_{\tilde{V}_k})$ using the following transformation formula:

$$T_{\tilde{V}_k}(\rho_{\tilde{V}_k}, \theta_{\tilde{V}_k}) = \left|\frac{\partial(X_{\tilde{V}_k}, Y_{\tilde{V}_k})}{\partial(\rho_{\tilde{V}_k}, \theta_{\tilde{V}_k})}\right| T_{\tilde{V}_k}(X_{\tilde{V}_k}, Y_{\tilde{V}_k}), \tag{5.13}$$

where $|\partial(X_{\tilde{V}_k}, Y_{\tilde{V}_k})/\partial(\rho_{\tilde{V}_k}, \theta_{\tilde{V}_k})| = \rho_{\tilde{V}_k}$ is the determinant of the Jacobian matrix.

The pdf of $\theta_{\tilde{V}_k}$ can be obtained from (5.12) by integrating $T_{\tilde{V}_k}(\rho_{\tilde{V}_k}, \theta_{\tilde{V}_k})$ with respect to $\rho_{\tilde{V}_k}$:

$$T_{\theta_{\tilde{V}_k}}(\theta_{\tilde{V}_k}) = \int_0^{+\infty} T_{\tilde{V}_k}(\rho_{\tilde{V}_k}, \theta_{\tilde{V}_k}) d\rho_{\tilde{V}_k}$$

$$= \frac{e^{-\frac{\eta_k}{2}}}{2\pi} + \frac{\sqrt{\eta_k}\cos(\theta_{\tilde{V}_k} - \theta_{G_k})}{2\sqrt{2\pi}} e^{-\frac{\eta_k}{2}\sin^2(\theta_{\tilde{V}_k} - \theta_{G_k})}$$

$$\times \mathrm{erfc}\left(\sqrt{\frac{\eta_k}{2}}\cos(\theta_{\tilde{V}_k} - \theta_{G_k})\right), \tag{5.14}$$

where $\eta_k = |G_k|^2/\sigma^2 = \alpha^2|U_k|^2/\sigma^2$ is the instantaneous SNR, obtained at the frequency f_k, and $\mathrm{erfc}(z) = (2/\sqrt{\pi})\int_z^{+\infty} e^{-\xi^2} d\xi$ denotes the complementary error function. Note that the total SNR defined below is independent of N since

$$\eta = \sum_{k=0}^{N-1} \frac{\alpha^2|U_k|^2}{\sigma^2} = \sum_{k=0}^{N-1} \frac{\alpha^2|U(f_k)/T_s|^2 N T_s \Delta f}{N N_0 B_W}$$

$$= \sum_{k=0}^{N-1} \frac{\alpha^2|U(f_k)|^2 \Delta f}{N_0} \approx \frac{\alpha^2 E_u}{N_0} = \frac{2\alpha^2 E_s}{N_0}, \tag{5.15}$$

where it has been taken into account that $U_k = U(f_k)/T_s$ (5.10), $\Delta f = 1/(NT_s)$, $B_W = 1/T_s$, and $E_u = 2E_s$.

In Fig. 5.1(a) we show $|G_k|^2$, $|\tilde{V}_k|^2$, and $|\tilde{W}_k|^2$, $k = 0, N/2 - 1$ with respect to f_k for $\rho = 20\,\mathrm{dB}$ and $N = 128$ when a Gaussian pulse $p(t) = \exp(-2\pi t^2/T_p^2)$ with $T_p = 1\,\mathrm{ns}$ is transmitted. We take $T_s = 1/B_W = T_p/4$. In Fig. 5.1(b) we show η_k in dB with respect to f_k for $N = 32$ and 128 ($T = 8$ and $32\,\mathrm{ns}$). Note that we have considered only the positive frequencies because the considered signal is real ($\tilde{V}_{-k} = \tilde{V}_k^*$).

5.1.1.1 *MLE of the Instantaneous Phase*

In this section, we derive the MLE of the instantaneous phase θ_k (5.11) and compare its variance with the CRB of θ_k. The part of interest of the log-likelihood function of θ_k can be obtained by substituting θ_{G_k} by $\theta_{U_k} - \theta_k$ (Jacobian equal to one) in (5.12) and computing the logarithm

$$\Lambda^{\theta_k} = \frac{\rho_{\tilde{V}_k}^2 + \rho_{G_k}^2 - 2\rho_{\tilde{V}_k}\rho_{G_k}\cos(\theta_{\tilde{V}_k} - \theta_{U_k} + \theta_k)}{2\sigma^2}.$$

The MLE of θ_k is given by

$$\hat{\theta}_k^{ml} = \theta_k; \quad \frac{\partial \Lambda^{\theta_k, \rho_{G_k}}}{\partial \theta_k} = 0 \Rightarrow \hat{\theta}_k^{ml} = \theta_{U_k} - \theta_{\tilde{V}_k}, \tag{5.16}$$

where $\{\hat{\theta}_k^{ml}\}$ are independent because $\{\tilde{W}_k\}$ are i.i.d.

The pdf of $\hat{\theta}_k^{ml}$ can be obtained by substituting $\theta_{\tilde{V}_k}$ by $\theta_{U_k} - \hat{\theta}_k^{ml}$ in (5.14):

$$T_{\hat{\theta}_k^{ml}}(\hat{\theta}_k^{ml}) = \frac{e^{-\frac{\eta_k}{2}}}{2\pi} + \frac{\sqrt{\eta_k}\cos(\hat{\theta}_k^{ml} - \theta_k)}{2\sqrt{2\pi}} e^{-\frac{\eta_k}{2}\sin^2(\hat{\theta}_k^{ml} - \theta_k)}$$

$$\times \mathrm{erfc}\left(\sqrt{\frac{\eta_k}{2}}\cos(\hat{\theta}_k^{ml} - \theta_k)\right).$$

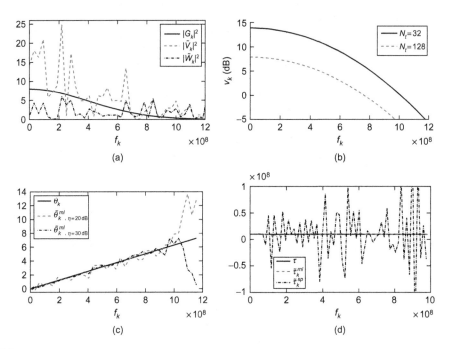

FIGURE 5.1

(a) $|G_k|^2$, $|\tilde{V}_k|^2$, and $|\tilde{W}_k|^2$ with respect to f_k, $\eta = 20$, $N = 128$. (b) η_k in dB with respect to f_k, $\eta = 20$ dB, $N = 32$ and 128 ($T = 8$ and 32 ns). (c) θ_k and $\tilde{\theta}_k$ with respect to f_k, $\tau = 1$ ns, $N = 128$, $\eta = 20$ and 30 dB. (d) τ, $\tilde{\tau}_k^{ml}$ and $\tilde{\tau}_k^{sp}$ with respect to f_k, $\tau = 1$ ns, $N = 128$, $\eta = 20$ dB.

The mean and variance of $\hat{\theta}_k^{ml}$ can be written as

$$\mu_{\hat{\theta}_k^{ml}} = \int_{\theta_k - \pi}^{\theta_k + \pi} \hat{\theta}_k^{ml} T_{\hat{\theta}_k^{ml}}(\hat{\theta}_k^{ml}) d\hat{\theta}_k^{ml} = \theta_k \tag{5.17}$$

$$\sigma_{\hat{\theta}_k^{ml}}^2 = \int_{\theta_k - \pi}^{\theta_k + \pi} \left(\hat{\theta}_k^{ml} - \theta_k\right)^2 T_{\hat{\theta}_k^{ml}}(\hat{\theta}_k^{ml}) d\hat{\theta}_k^{ml}. \tag{5.18}$$

Equation (5.17) shows that $\hat{\theta}_k^{ml}$ is unbiased. A closed-form expression of $\sigma_{\hat{\theta}_k^{ml}}^2$ is not easy to find, but it can be calculated numerically. The maximum value ($\pi^2/3$) of $\sigma_{\hat{\theta}_k^{ml}}^2$ is obtained for $\eta_k = 0$, that is, when $\hat{\theta}_k^{ml}$ is uniformly distributed in $[\theta_k - \pi, \theta_k + \pi]$.

Equations (5.17) and (5.18) are true only if $\hat{\theta}_k^{ml}$ falls in the interval $I_k = [\theta_k - \pi, \theta_k + \pi]$. Given that θ_k is unknown, $\hat{\theta}_k^{ml}$ can only be found in an interval of width 2π independent of θ_k, where the computed values $\hat{\theta}_k^{2\pi}$ and $\hat{\theta}_k^{ml}$ are congruent modulo 2π ($\hat{\theta}_k^{ml} = \hat{\theta}_k^{2\pi} + 2l_k\pi$, $l_k \in \mathbb{Z}$). $\hat{\theta}_k^{2\pi}$ falls completely out of I_k if $l_k \neq 0$.

Moreover, it is possible to approximate $\hat{\boldsymbol{\theta}}^{ml}$ by $\tilde{\boldsymbol{\theta}}^{ml}$ recursively from $\hat{\boldsymbol{\theta}}^{2\pi}$ by adding a multiple $(\hat{l}_k)$ of 2π to $\hat{\theta}_k^{2\pi}$ in order to have $|\hat{\theta}_k^{2\pi} + 2\hat{l}_k\pi - \tilde{\theta}_{k-1}^{ml}| \leq \pi$:

$$\hat{l}_k = \text{round}\left(\frac{\hat{\theta}_k^{2\pi} - \tilde{\theta}_{k-1}^{ml}}{2\pi}\right) \tag{5.19}$$

$$\hat{\theta}_k^{ml} \approx \tilde{\theta}_k^{ml} = \hat{\theta}_k^{2\pi} - 2\hat{l}_k\pi, \tag{5.20}$$

where "round" denotes the round to the nearest integer function. The correction procedure must start from a $\hat{\theta}_k^{2\pi}$ falling in the good interval. For baseband signals we can start from $\theta_0 = 0$. In Fig. 5.1(c), the behavior of θ_k and $\tilde{\theta}_k$ is shown with respect to f_k for $\tau = 1$ ns, $N = 128$ and $\eta = 20$ and 30 dB.

Note that for high SNR, $\hat{\theta}_k^{ml}$ and $\tilde{\theta}_k^{ml}$ are equal while for low SNR, the correction in (5.19) and (5.20) may introduce an error multiple of 2π. Given that $\tilde{\theta}_k^{ml}$ is computed recursively, the estimation error will propagate and will be amplified with frequency. It follows that $\sigma_{\tilde{\theta}_k^{ml}}^2$ (variance of $\tilde{\theta}_k^{ml}$) may depend on multiple parameters (η_k, k, N, etc.), while $\sigma_{\hat{\theta}_k^{ml}}^2$ (5.18) is only a function of η_k. We can expect that for low SNR, $\tilde{\theta}_k^{ml}$ may be biased ($\mu_{\tilde{\theta}_k^{ml}} \neq \theta_k$, where $\mu_{\tilde{\theta}_k^{ml}}$ is the mean value of $\tilde{\theta}_k^{ml}$) and $\sigma_{\tilde{\theta}_k^{ml}}^2$ may be much larger than $\sigma_{\hat{\theta}_k^{ml}}^2$.

The CRB of θ_k is given by

$$\kappa(\theta_k) = \frac{1}{-\mathbb{E}\left\{\frac{\partial^2 \Lambda^{\theta_k, \rho_{G_k}}}{\partial \theta_k^2}\right\}} = \frac{\sigma^2}{\rho_{G_k}^2} = \frac{1}{\eta_k}, \tag{5.21}$$

where $-\mathbb{E}\left\{\frac{\partial^2 \Lambda^{\theta_k, \rho_{G_k}}}{\partial \theta_k^2}\right\}$ is the Fisher information of θ_k.

In Fig. 5.2, we show $\kappa(\theta_k)$, $\sigma_{\hat{\theta}_k^{ml}}^2$, $\tilde{\sigma}_{\tilde{\theta}_k^{ml}}^2$ (obtained via simulations based on 3000 realizations), and $\tilde{\epsilon}_{\tilde{\theta}_k^{ml}}^2$ (mean square error, $\epsilon_{\tilde{\theta}_k^{ml}}^2 = \sigma_{\tilde{\theta}_k^{ml}}^2 + (\mu_{\tilde{\theta}_k} - \theta_k)^2$, obtained also via simulations) with respect to f_k for $\tau = 1$ ns, $N = 128$, and $\eta = 20$ and 30 dB. Recall that η_k is the SNR at the frequency f_k and that $\eta = \sum \eta_k$ (5.15). Note that for a given frequency f_k, η_k is not the same for $\eta = 20$ and 30 dB.

We can see that $\tilde{\epsilon}_{\tilde{\theta}_k^{ml}}^2 \approx \tilde{\sigma}_{\tilde{\theta}_k^{ml}}^2$, which means that $\tilde{\theta}_k^{ml}$ is approximately unbiased. For high η_k, $\sigma_{\hat{\theta}_k^{ml}}^2$ and $\tilde{\sigma}_{\tilde{\theta}_k^{ml}}^2$ closely follow $\kappa(\theta_k)$, while for low η_k, $\tilde{\sigma}_{\tilde{\theta}_k^{ml}}^2$ starts to divert from $\kappa(\theta_k)$ and $\sigma_{\hat{\theta}_k^{ml}}^2$. The latter result was expected. We can also see that $\tilde{\sigma}_{\tilde{\theta}_k^{ml}}^2$ is larger for $\eta = 20$ dB than for $\eta = 30$ dB even when the instantaneous SNR η_k is the same in both cases. In fact, given that $\tilde{\theta}_k^{ml}$ is obtained recursively, the error

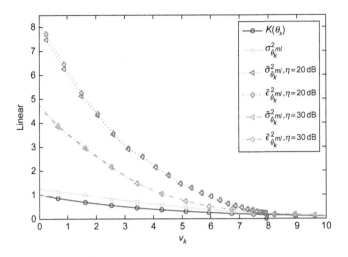

FIGURE 5.2

$\kappa\,(\theta_k)$, $\sigma^2_{\hat\theta^{ml}_k}$, $\tilde\sigma^2_{\tilde\theta^{ml}_k}$ and $\tilde\epsilon^2_{\tilde\theta^{ml}_k}$ with respect to f_k for $\tau = 1$ ns, $N = 128$ and $\eta = 20$ and 30 dB.

will propagate faster with the frequency and will be more amplified for lower SNR ($\eta = 20$ dB) than for higher SNR.

5.1.1.2 *Instantaneous ML and Phase-Slope-Based TOA Estimators*

In this section, we derive the instantaneous MLE ($\hat\tau^{ml}_k$) for TOA and propose another instantaneous esti-mator ($\hat\tau^{sp}_k$) based on the slope of the instantaneous phase. We compare the variances of the considered estimators with the instantaneous CRB for TOA. Note that, here, only the estimator is instantaneous because the TOA is the same for all frequencies. Estimates $\hat\tau^{ml}_k$, for $k = -N/2,\dots,N/2-1$, and $\hat\tau^{sp}_k$, $k = -N/2,\dots,N/2-2$, can be obtained from (5.11) and (5.16):

$$\hat\tau^{ml}_k = \frac{\hat\theta^{ml}_k}{2\pi\,(f_c + f_k)} \approx \frac{\tilde\theta^{ml}_k}{2\pi\,(f_c + f_k)} = \tilde\tau^{ml}_k \tag{5.22}$$

$$\hat\tau^{sp}_k = \frac{\hat\theta^{ml}_{k+1} - \hat\theta^{ml}_k}{2\pi\,\Delta f} \approx \frac{\tilde\theta^{ml}_{k+1} - \tilde\theta^{ml}_{k-1}}{2\pi\,\Delta f} = \tilde\tau^{sp}_k, \tag{5.23}$$

where $\tilde\theta^{ml}_k$ is the approximation of $\hat\theta^{ml}_k$ obtained in (5.19) and (5.20). In Fig. 5.1(d) we show τ, $\tilde\tau^{ml}_k$, and $\tilde\tau^{sp}_k$ with respect to f_k for $\tau = 1$ ns, $N = 128$, and $\eta = 20$ dB.

$\{\hat\tau^{ml}_k\}$ are independent because of the independence of $\{\hat\theta^{ml}_k\}$. In addition, the random variables $\{\hat\tau^{sp}_k\}$ and $\{\hat\tau^{sp}_{k'}\}$ are correlated for $k' - k = \pm 1$ (they both use $\hat\theta^{ml}_{k+1}$ for $k' - k = 1$ and $\hat\theta^{ml}_k$ for $k' - k = -1$) and uncorrelated for $|k' - k| > 1$. We can show that the variance of $\hat\tau^{ml}_k$, variance of $\hat\tau^{sp}_k$, and covariance of

$\hat{\tau}_k^{sp}$ can be written as

$$\sigma_{\hat{\tau}_k^{ml}}^2 = \frac{\sigma_{\hat{\theta}_k^{ml}}^2}{4\pi^2 (f_c + f_k)^2} \tag{5.24}$$

$$\sigma_{\hat{\tau}_k^{sp}}^2 = \frac{\sigma_{\hat{\theta}_k^{ml}}^2 + \sigma_{\hat{\theta}_{k+1}^{ml}}^2}{4\pi^2 \Delta f^2} \tag{5.25}$$

$$C_{\hat{\tau}^{sp}}^{k,k+1} = -\frac{\sigma_{\hat{\theta}_{k+1}^{ml}}^2}{4\pi^2 \Delta f^2}. \tag{5.26}$$

The variance of the MLE is much smaller than the variance of the estimator based on the slope. The ratio of $\sigma_{\hat{\tau}_k^{sp}}^2$ and $\sigma_{\hat{\tau}_k^{ml}}^2$ is given by

$$\frac{\sigma_{\hat{\tau}_k^{sp}}^2}{\sigma_{\hat{\tau}_k^{ml}}^2} = \frac{\sigma_{\hat{\theta}_k^{ml}}^2 + \sigma_{\hat{\theta}_{k+1}^{ml}}^2}{\sigma_{\hat{\theta}_k^{ml}}^2} \frac{(f_c + f_k)^2}{\Delta f^2} \approx 2 \frac{(f_c + k\Delta f)^2}{\Delta f^2},$$

where for $f_c = 0$, we have $\sigma_k^{sp/ml} \approx 2k^2$.

The instantaneous CRB and Fisher information of τ can be obtained from (5.11) and (5.21) using the variable change formula relative to the CRB [64]:

$$\kappa_k(\tau) = \left(\frac{\partial \tau}{\partial \theta_k}\right)^2 \kappa(\theta_k) = \frac{1}{4\pi^2 (f_c + f_k)^2 \eta_k}. \tag{5.27}$$

We can see from (5.21), (5.24), and (5.27) that when $\sigma_{\hat{\theta}_k^{ml}}^2 \to \kappa(\theta_k)$ (for η_k sufficiently high), then $\sigma_{\hat{\tau}_k^{ml}}^2 \to \kappa_k(\tau)$. From (5.21), (5.25), and (5.27), we can see that $\sigma_{\hat{\theta}_k^{sp}}^2$ is much larger than $\kappa_k(\tau)$ even for high SNR ($\sigma_{\hat{\theta}_k^{sp}}^2 / \kappa_k(\tau) \to 2(f_c + f_k)^2 / \Delta f^2$).

In Fig. 5.3, we show $\kappa(\tau_k)$, $\tilde{\epsilon}_{\hat{\tau}_k^{ml}}^2$, and $\tilde{\epsilon}_{\hat{\tau}_k^{sp}}^2$ in dB with respect to f_k for $\tau = 1$ ns, $N = 128$, and $\eta = 25$ dB. We can see that for low f_k (high η_k), $\tilde{\epsilon}_{\hat{\tau}_k^{ml}}^2$ closely follows $\kappa(\tau_k)$.

5.1.1.3 *Optimal ML-Based and Phase-Slope-Based TOA Estimators*

In this section, we derive the global optimal TOA estimators based on the linear combination of the instantaneous TOA-ML and phase-slope-based estimators considered earlier.

Let $\hat{\zeta}_k$ for $k = -N/2, \ldots, N/2 - 1$ be N, unbiased estimators of the same parameter ζ. We can show that the minimum-variance unbiased linear combination of the elements of $\hat{\boldsymbol{\zeta}}$ is given by

$$\hat{\zeta} = \hat{\mathbf{a}}_{\hat{\zeta}}^T \hat{\boldsymbol{\zeta}} \; ; \begin{cases} \hat{\mathbf{a}}_{\hat{\zeta}} = \arg\min \mathbf{a}\{\sigma_{\mathbf{a}^T \hat{\boldsymbol{\zeta}}}^2\} \\ \\ \text{s.t. } \sum \mathbf{a} = 1 \end{cases} \Rightarrow \hat{\mathbf{a}}_{\hat{\zeta}} = \frac{\mathbf{C}_{\hat{\zeta}}^{-1} \mathbf{1}}{\sum (\mathbf{C}_{\hat{\zeta}}^{-1} \mathbf{1})}, \tag{5.28}$$

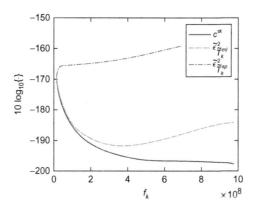

FIGURE 5.3

$\kappa(\tau_k)$, $\tilde{\epsilon}^2_{\hat{\tau}^{ml}_k}$, and $\tilde{\epsilon}^2_{\hat{\tau}^{ml}_k}$ with respect to f_k, $\tau = 1$ ns, $N = 128$, and $\eta = 25$ dB.

where $\sigma^2_{a^T\hat{\zeta}} = \mathbb{E}\left\{(a^T(\hat{\zeta} - \zeta 1))^2\right\}$, $\sum a$ is the sum of the elements of a, $1 = (1 \cdots 1)^T$, and $C_{\hat{\zeta}}$ is the covariance matrix of $\hat{\zeta}$. The variance of $\hat{\zeta}$ is given by

$$\hat{a}^T C_{\hat{\zeta}} \hat{a}. \tag{5.29}$$

Using (5.28) and (5.29), we can obtain the global estimators based on $\hat{\tau}^{ml}$ and $\hat{\tau}^{sp}$ and their variances as

$$\hat{\tau}^{ml} = \frac{\sum_{k=0}^{N-1} \hat{\tau}^{ml}_k / \sigma^2_{\hat{\tau}^{ml}_k}}{\sum_{k=0}^{N-1} 1/\sigma^2_{\hat{\tau}^{ml}_k}} \tag{5.30}$$

$$\hat{\tau}^{sp} = \frac{(C^{-1}_{\hat{\tau}^{sp}} 1)^T}{\sum(C^{-1}_{\hat{\tau}^{sp}} 1)} \hat{\tau}^{sp} \tag{5.31}$$

$$\sigma^2_{\hat{\tau}^{ml}} = \frac{1}{\sum_{k=0}^{N-1} 1/\sigma^2_{\hat{\tau}^{ml}_k}} \tag{5.32}$$

$$\sigma^2_{\hat{\tau}^{sp}} = \hat{a}^T_{\hat{\tau}^{sp}} C_{\hat{\tau}^{sp}} \hat{a}_{\hat{\tau}^{sp}}, \tag{5.33}$$

where the covariance matrix of $\hat{\tau}^{ml}$ is diagonal and $C_{\hat{\tau}^{sp}}$ can be obtained from (5.25) and (5.26). Note that the estimator given in (5.30) is not the MLE. In fact, the MLE aims at maximizing, with respect to τ, the global likelihood function equal to the product of the instantaneous likelihood functions. Here we have maximized the instantaneous likelihood functions, and we have used an optimal linear combination.

Note that in order to compute the estimators given in (5.30) and (5.31), we must know $\sigma^2_{\hat{\tau}^{ml}_k}$ and $C_{\hat{\tau}^{sp}}$. Given that in practice $\sigma^2_{\hat{\tau}^{ml}_k}$ and $C_{\hat{\tau}^{sp}}$ are unknown because the PSD of the noise is unknown, we can substitute $\sigma^2_{\tilde{\tau}^{ml}_k}$ by $1/(f^2_k \rho^2_{U_k})$ and $\mathbf{C}_{\hat{\tau}^{sp}}$ by $(c_{i,j})$, $i,j = -N/2,\ldots,N/2-2$, where $c_{i,i} = 1/\rho^2_{U_k} +$ $1/\rho^2_{U_{k+1}}$, $c_{i,i+1} = -1/\rho^2_{U_{k+1}}$ and $c_{i,i'} = 0$ if $|i'-i| > 1$. The estimators derived using this approximation and $\tilde{\theta}^{ml}_k$ instead of $\hat{\theta}^{ml}_k$ are denoted by $\tilde{\tau}^{ml}$ and $\tilde{\tau}^{sp}$, respectively.

Given that $(\tilde{W}_k)$ (5.9) is a white sequence, the global Fisher information for the estimation of τ is the sum of the instantaneous Fisher information (5.27):

$$J(\tau) = \sum_k f^\tau_k = \sum_k 4\pi^2 \eta_k (f_c + f_k)^2 = \frac{\sum_k 4\pi^2 \rho^2_{G_k}(f_c + f_k)^2}{\sum_k \rho^2_{G_k}}$$

$$\times \frac{\sum_k \rho^2_{G_k}}{\sigma^2} \approx \eta\beta^2_s,$$

where η is the global SNR (5.15) and β^2_s the mean quadratic bandwidth (5.3) of $s(t)$. The global CRB of τ is given by

$$\kappa(\tau) = \frac{1}{J(\tau)} \approx \frac{1}{\eta\beta^2_s}. \tag{5.34}$$

Note that CRB of τ can be obtained directly from (5.1). We can show from (5.32) and (5.34) that $\sigma^2_{\tilde{\tau}^{ml}} \to \kappa(\tau)$ for sufficiently high η. Also from (5.32) and (5.33), $\sigma^2_{\tilde{\tau}^{sp}}/\sigma^2_{\tilde{\tau}^{sp}} > 2\beta^2_s/\Delta f^2$ is obtained.

In Fig. 5.4, we show $\kappa(\tau)$, $\tilde{\epsilon}^2_{\tilde{\tau}^{ml}}$, $\tilde{\epsilon}^2_{\tilde{\tau}^{sp}}$, and $\tilde{\epsilon}^2_{\hat{\tau}}$ ($\hat{\tau}$ is the MLE of τ estimated in time domain) with respect to η for $\tau = 1$ ns and $N = 16$ and 128 when a Gaussian pulse is considered.

We can see that the MSE of the MLE $\hat{\tau}$ is lower than that of the estimates $\tilde{\tau}^{ml}$ and $\tilde{\tau}^{sp}$. In addition, $\tilde{\epsilon}^2_{\tilde{\tau}^{ml}}$ and $\tilde{\epsilon}^2_{\tilde{\tau}^{sp}}$ converge faster to $\kappa(\tau)$ for smaller N even if the total SNR η is the same. The quantities $\tilde{\epsilon}^2_{\tilde{\tau}^{ml}}$ and $\tilde{\epsilon}^2_{\tilde{\tau}^{sp}}$ are very close to each other while we had expected that $\tilde{\epsilon}^2_{\tilde{\tau}^{ml}}$ would be much smaller than $\tilde{\epsilon}^2_{\tilde{\tau}^{sp}}$. In fact, the presence of high-frequency components makes $\tilde{\tau}^{ml}$ a better estimate than $\tilde{\tau}^{sp}$, whereas in our case where a baseband Gaussian pulse is considered, the strong frequency components fall around $f_0 = 0$. We can say that the UWB considered pulse acts as a narrowband pulse. Due to this, we have found that $\tilde{\epsilon}^2_{\tilde{\tau}^{sp}}$ also converges to $\kappa(\tau)$.

In Fig. 5.5, where a baseband cardinal sine pulse is considered, we have obtained the expected results. The quantity $\tilde{\epsilon}^2_{\tilde{\tau}^{ml}}$ converges to $\kappa(\tau)$, and $\tilde{\epsilon}^2_{\tilde{\tau}^{ml}}$ is much smaller than $\tilde{\epsilon}^2_{\tilde{\tau}^{sp}}$. In fact, for a cardinal sine the spectrum is rectangular and the high-frequency components (which improve $\tilde{\tau}^{ml}$) are strong as well as the low-frequency components. We have also found that for a small number of samples ($N = 16$), the estimate $\tilde{\tau}^{ml}$ is even better than the MLE.

5.1.2 Joint Range and Direction of Arrival Estimation

TOA-based approaches can provide very precise location estimates, and therefore they are well suited for UWB systems. Nonetheless, the introduction of spatial information is expected to further increase the accuracy of the estimates. Due to the large bandwidth of the UWB signal, the number of paths may be very large, especially in indoor environments. Therefore, accurate angle estimation becomes very challenging due to scattering from objects in the environments.

Estimation of the angle of arrival (AOA) in UWB systems has been addressed considering subspace techniques [65], but they are of high complexity involving eigenvalue decompositions. In Ref. [92]

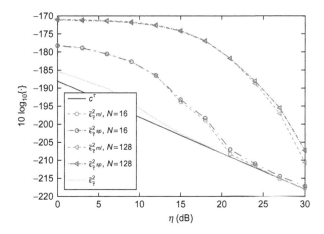

FIGURE 5.4

$\kappa(\tau)$, $\tilde{\epsilon}^2_{\hat{\tau}ml}$, $\tilde{\epsilon}^2_{\hat{\tau}sp}$, and $\tilde{\epsilon}^2_{\hat{\tau}}$ with respect to η; Gaussian pulse, $\tau = 1$ ns, $N = 16$ and 128.

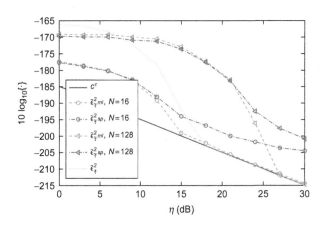

FIGURE 5.5

$\kappa(\tau)$, $\tilde{\epsilon}^2_{\hat{\tau}ml}$, $\tilde{\epsilon}^2_{\hat{\tau}sp}$, and $\tilde{\epsilon}^2_{\hat{\tau}}$ with respect to η; cardinal sine pulse, $\tau = 1$ ns, $N = 16$ and $N = 128$.

the estimation of AOA based on temporal delays is considered but it suffers from constraints associated with sampling requirements. The differences in arrival times of an incoming signal at different antenna elements contain the angle information for a known array geometry [33]. Here a frequency-domain strategy for the joint TOA and AOA estimation compliant with the IEEE 802.15.4a standard is presented, based on time measurements [34, 74].

Working in the frequency domain, any spectral estimation technique can be applied to obtain a power delay profile defined as the signal energy distribution with respect to the propagation delays from which the TOA is estimated. The spectral estimation is based on the periodogram aiming at a

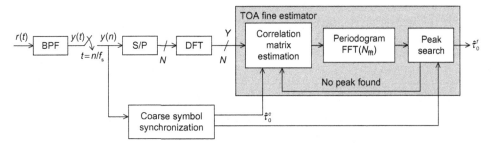

FIGURE 5.6

Block diagram of the noncoherent receiver block for a single antenna.

low-complexity implementation, and it assumes no knowledge of the received pulse waveform. The block diagram of the frequency-domain TOA and DOA estimator is sketched in Fig. 5.6.

Multiantenna Signal Model

The frame structure defined in IEEE 802.15.4a [1] for IR-UWB signals[1] consists of a synchronization header (SHR) preamble, a physical header (PHR), and a data field physical service data unit (PSDU) (see Fig. 5.7). The SHR preamble comprises the ranging synchronization preamble (SYNCH) and the start-of-frame delimiter (SFD). The ranging preamble can consist of 16, 64, 1024, or 4096 symbols. The underlying symbol of the ranging preamble uses length-31 ternary sequences, c_i. Each c_i of length $N_f = 31$ contains 15 zeros and 16 nonzero codes, and has the desired property of perfect periodic autocorrelation.

The mathematical model of the signal transmitted during the SHR is

$$s(t) = \sum_{k=0}^{N_{\text{SYNC}}+N_{\text{SFD}}-1} a_k \psi(t - kT_{\text{sym}}), \tag{5.35}$$

where N_{SYNC} and N_{SFD} are the lengths of the SYNC and the SFD and T_{sym} is the symbol duration. Coefficients a_k are all equal to 1 during the SYNC, whereas they take values $\{-1, 0, +1\}$ during the SFD. Finally $\psi(t)$ is expressed as

$$\psi(t) = \sum_{j=0}^{N_f-1} c_i(j) g_{tx}(t - jT_f). \tag{5.36}$$

In this equation $g_{tx}(t)$ is an ultrashort pulse (monocycle), $T_f = T_{\text{sym}}/N_f$ is the pulse repetition period and $c_i(j)$ denotes the jth element of sequence c_i.

Signal $s(t)$ propagates through an L-path fading channel whose response to $g_{tx}(t)$ is $\sum_{l=0}^{L-1} h_l g_{tx}(t - \tau_l)$. Note that it is assumed that the received pulse from each lth path exhibits the same waveform but

---

[1] The standard specifies two optional signaling formats based on IR-UWB and CSS. The IR-UWB option is mainly for ranging, whereas the CSS signals have better features for data communication. Since this section investigates ranging, we only focus on the IR-UWB option.

SYNCH Preamble (16, 64, 1024, 4096) symbols	Start of frame delimiter, SFD (8, 64) symbols	PHY header (PHR)	Data field (PSDU)

FIGURE 5.7

The IEEE 802.15.4a packet structure.

experiences a different fading coefficient, h_l, and a different delay, τ_l. Without loss of generality we assume $\tau_0 \leq \tau_1 \leq \cdots \leq \tau_{L-1}$, τ_0 being the TOA that is to be estimated.

The received waveform at the qth antenna element for a node equipped with an array of Q antenna elements can be written as

$$r_q(t) = \sum_{k=0}^{N_{\text{SHR}}-1} \sum_{j=0}^{N_f-1} \sum_{l=0}^{L-1} a_k c_i(j) h_{l,q} g_{tx}(t - T_k^j - \tau_{l,q}) + w_q(t), \qquad (5.37)$$

where $w_q(t)$ is thermal noise with two-sided power spectral density $N_0/2$, $T_k^j = jT_f + kT_{\text{sym}}$, $h_{l,q}$ denotes the fading coefficient of the lth path at antenna q, and $N_{\text{SHR}} = N_{\text{SYNC}} + N_{\text{SFD}}$. Given the low duty cycle of UWB signals, it is assumed that the received signal is free of intersymbol interference. The temporal delay $\tau_{l,q}$ at each antenna element depends not only on the propagation delay but also on the direction of arrival. Considering a uniform linear array (ULA), the propagation delay $\tau_{l,q}$ is given by

$$\tau_{l,q} = \tau_l + q\frac{d}{c}\sin(\theta_l) \qquad (5.38)$$

with d being the distance between antenna elements in the array, c the speed of light, and θ_l the angle of arrival of the lth path. Although the scheme is presented assuming a ULA, it can be directly extended to other array configurations.

The signal associated with the jth transmitted pulse corresponding to the kth symbol in the frequency domain yields

$$Y_{jq}^k(w) = \sum_{l=0}^{L-1} a_k c_i(j) h_{l,q} S_j^k(w) e^{-jw\tau_{l,q}} + V_{jq}^k(w) \qquad (5.39)$$

with

$$S_j^k(w) = G(w) e^{-jw((kN_f+j)T_f}, \qquad (5.40)$$

where $G(w)$ denotes the Fourier transform of $g[n]$, which is the sampled received pulse waveform at the output of the band-pass filter (BPF) (see Fig. 5.6) and N_f is the number of frames per symbol. $V_{jq}^k(w)$ is the noise component in the frequency domain associated with the jth frame interval corresponding to the kth symbol. Sampling (5.39) at $w_n = w_0 n$ for $n = 0, 1, \ldots, N-1$, where $w_0 = \frac{2\pi}{NT_s}$, and rearranging the frequency-domain samples $Y_{jq}^k[n]$ into the vector $\mathbf{Y}_{jq}^k \in \mathbb{C}^{N \times 1}$ yields

$$\mathbf{Y}_{jq}^k = \sum_{l=0}^{L-1} a_k c_j h_{l,q} \mathbf{S}_j^k \mathbf{e}_{\tau_{l,q}} + \mathbf{V}_{jq}^k = a_k c_j \mathbf{S}_j^k \mathbf{E}_{l,q} \mathbf{h}_q + \mathbf{V}_{jq}^k, \qquad (5.41)$$

where the matrix $\mathbf{S}_j^k \in \mathbb{C}^{N \times N}$ is a diagonal matrix whose components are the frequency samples of $S_j^k(w)$ and the matrix $\mathbf{E}_{l,q} \in \mathbb{C}^{N \times L}$ contains the delay-signature vectors associated with each arriving delayed signal:

$$\mathbf{E}_{l,q} = \begin{bmatrix} \mathbf{e}_{\tau_{0,q}} & \cdots & \mathbf{e}_{\tau_{j,q}} & \cdots & \mathbf{e}_{\tau_{L-1,q}} \end{bmatrix} \tag{5.42}$$

with $\mathbf{e}_{\tau_{j,q}} = \begin{bmatrix} 1 & e^{-jw_0\tau_{j,q}} & \cdots & e^{-jw_0(N-1)\tau_{j,q}} \end{bmatrix}^T$. The channel fading coefficients are arranged in the vector $\mathbf{h}_q = \begin{bmatrix} h_{0,q} & \cdots & h_{L-1,q} \end{bmatrix}^T \in \mathbb{C}^{L \times 1}$, and the noise samples in vector $\mathbf{V}_{jq}^k \in \mathbb{C}^{N \times 1}$.

5.1.2.1 *TOA Estimation*

The algorithm needs a measure of the TOA at each array element to obtain the final joint TOA and AOA estimation. This section describes how the TOA is estimated from the noisy observations $r_q(t)$ without the knowledge of the pulse shape $g(t)$. For describing the estimation algorithm, the receiver assumes an ideal BPF, the output of which is sampled at Nyquist rate, followed by a DFT module that transforms the signal to the frequency domain (see Fig. 5.6). The frequency samples are processed without the knowledge of the specific pulse shape in the fine TOA estimator stage. The fine estimator requires a previous stage that provides a coarse symbol synchronization. Note that the proposed receiver scheme is a noncoherent receiver and it is implemented in digital form using Nyquist frequency rate.

First, a simple coarse estimation stage that provides the time reference for symbol synchronization and estimates the threshold used in the TOA algorithm is needed. After the synchronization stage, the signal is passed through a fine TOA estimation stage, which provides the final TOA measurement. The TOA defined as τ_0 indicates the beginning of the first complete symbol in the observation interval, being $0 \leq \tau_0 \leq T_{\text{sym}}$.

Coarse TOA Estimation

The coarse estimation consists of an energy estimator and a simple search algorithm that identifies the beginning of the symbol by applying a minimum distance criterion. Since the signal structure in the IEEE 802.15.4.a standard does not include a time-hopping sequence, the minimum distance criterion is applied in this case based on the ternary sequence knowledge, c_i, at symbol level.

It is assumed that the acquisition begins at any point of the SYNCH preamble t_0 and lasts $K_s + 1$ symbols. Note that the acquisition window duration is defined one symbol longer than the number of symbols considered for the fine timing estimation. Hence the minimum acquisition window will be equivalent to two symbols duration to perform the fine estimation over a single symbol $K_s = 1$.

To find the beginning of the next symbol the frame number which the first detected pulse belongs to is needed. Let $y[m] = y(mT_s)$ denote the discrete-time received signal, where T_s is the sampling period. The time-domain samples of the received signal in the ith time interval of duration T_f are defined as

$$y_{\text{frame},i}[n] = y[(i-1)K_f + n] \quad \text{for } n = 1, \ldots, K_f, \tag{5.43}$$

where $i = 1, \ldots, N_f(K_s + 1)$. Rearranging the time-domain samples $y_{\text{frame},i}[n]$ in the vector $\mathbf{y}_{\text{frame},i} \in \mathbb{C}^{K_f \times 1}$, with $K_f = \lfloor T_f/T_s \rfloor$ being the number of samples in a frame interval, the energy at each frame interval in one symbol period is obtained by averaging, for each of the N_f frames, over all K_{s+1} symbols in the acquisition interval. That is,

$$E_{\text{frame},j} = \sum_{k=0}^{K_s} \|\mathbf{y}_{\text{frame},j+kN_f}\|^2 \quad j = 1,\ldots,N_f. \tag{5.44}$$

Then the algorithm searches the 16 maxima corresponding to the 16 frames containing pulses and estimates the ternary sequence $\hat{c}_i$. From the original ternary sequence c_i, the vector $\mathbf{d}$ is defined as a vector comprising the 16 positions of the sequence c_i containing ± 1. Then the circulant matrix $\Delta_{\rho_{c_i}}$ is defined whose rows contain the relative delays in the number of frame intervals between two consecutive pulses within a symbol period. Each row is a shifted version of the previous one. More specifically, defining $\rho_{c_i}(n) = d(n+1) - d(n) - 1$ for $n = 1,\ldots,15$ and $\rho_{c_i}(16) = N_f - d(16) + d(1)$ as the number of frames between two consecutive transmitted pulses, the circulant matrix $\Delta_{\rho_{c_i}}$ is given by

$$\Delta_{\rho_{c_i}} = \begin{bmatrix} \rho_{c_i}(1) & \rho_{c_i}(2) & \cdots & \rho_{c_i}(15) & \rho_{c_i}(16) \\ \rho_{c_i}(2) & \rho_{c_i}(3) & \cdots & \rho_{c_i}(16) & \rho_{c_i}(1) \\ \vdots & \vdots & \ddots & \vdots & \vdots \\ \rho_{c_i}(16) & \rho_{c_i}(1) & \cdots & \rho_{c_i}(14) & \rho_{c_i}(15) \end{bmatrix}. \tag{5.45}$$

Hence, with the estimated ternary sequence $\hat{c}_i$ is conformed the vector $\hat{\mathbf{d}}$, which contains the estimated pulse positions or, in other words, which contains the 16 positions of the estimated sequence $\hat{c}_i$ containing ± 1. Therefore, the relative distance between the 16 estimated positions of the pulses form the vector

$$\Delta\mathbf{d} = \begin{bmatrix} \hat{\rho}_{c_i}(1) & \hat{\rho}_{c_i}(2) & \cdots & \hat{\rho}_{c_i}(15) & \hat{\rho}_{c_i}(16) \end{bmatrix}. \tag{5.46}$$

If we denote the first pulse within the symbol by the number 1, the second by the number 2, and so on until the sixteenth pulse with the number 16, then the estimated number of the first detected pulse $u \in \{1,\ldots,16\}$ is carried out by finding the closest row of $\Delta_{\rho_{c_i}}$, which provides lower MSE with respect to $\Delta\mathbf{d}$:

$$u = \arg\min_{j=1,\ldots,16} \left\| \Delta\mathbf{d} - \Delta_{\rho_{c_i}}|_j \right\|^2, \tag{5.47}$$

where $\Delta_{\rho_{c_i}}|_j$ denotes the jth row of the matrix $\Delta_{\rho_{c_i}}$. From the estimated pulse number u, the frame number $v \in \{1,\ldots,N_f\}$ to which the first detected pulse belongs to is estimated. That is,

$$v = \mathbf{d}(u) - \hat{\mathbf{d}}(1) + 1. \tag{5.48}$$

Then the coarse TOA estimation can be directly identified as

$$\hat{\tau}_0^c = (N_f - v + 1)T_f. \tag{5.49}$$

The temporal resolution of this estimation is a frame period T_f.

Fine TOA Estimation

Once the beginning of the symbol is coarsely estimated, working in the frequency domain any type of spectral estimation can be applied to obtain a power profile defined as the signal energy distribution with respect to the propagation delays.

The fine TOA estimation $\hat{\tau}_0$ is obtained from the TOA coarse estimation $\hat{\tau}_0^c$ and a high-resolution time delay $\tilde{\tau}$ estimate of the first arriving path with respect to the time reference obtained in the coarse estimation stage. The TOA estimation resulting from the fine estimation stage is given by

$$\hat{\tau}_0 = \hat{\tau}_0^c + \tilde{\tau}. \tag{5.50}$$

The TOA estimator consists of finding the first delay, $\tilde{\tau}$, that exceeds a given threshold, P_{th}, in the PDP:

$$\tilde{\tau} = \min \arg_\tau \{P(\tau) > P_{th}\}. \tag{5.51}$$

Given the signal frequency-domain structure obtained in (5.41), the power delay profile can be obtained by estimating the energy of the frequency-domain signal filtered by the delay signature vector, $\mathbf{e}_\tau = \{1 \quad e^{-jw_0\tau} \quad \dots \quad e^{-jw_0(N-1)\tau}\}^T$, at each time delay resulting in the quadratic form

$$P(\tau) = \mathbf{e}_\tau^H \mathbf{R} \mathbf{e}_\tau, \tag{5.52}$$

where $\mathbf{R} \in \mathbb{C}^{N \times N}$ is the frequency-domain signal correlation matrix computed from the received signal at each of the array antennas.

The quadratic form (5.52) allows for a low-complexity implementation by applying the FFT to the following coefficients:

$$\tilde{R}_n = \begin{cases} \sum_{k=n+1}^{N} \mathbf{R}_{k-n,k} & : 0 \leq n \leq N-1 \\ \sum_{k=1}^{N+n} \mathbf{R}_{k-n,k} & : -N+1 \leq n \leq 0 \end{cases}, \tag{5.53}$$

where $\tilde{R}_n$ is the sum of the nth diagonal elements of the correlation matrix $\mathbf{R}$. So, the maximization problem resorts to maximizing the following expression:

$$P(\tau) = \sum_{n=1-N}^{-N+1} \tilde{R}_n e^{-jw_0\tau n}. \tag{5.54}$$

The number of points used in the DFT is equivalent to the number of values of τ to sweep. Therefore, the more points used the more accurate the TOA estimation will be. This number of DFT points will be denoted as M.

A more robust estimation of the correlation matrix $\mathbf{R}$ can be obtained by averaging over K_s symbols. The expression is

$$\mathbf{R} = \frac{1}{N_f K_s} \sum_{k=1}^{K_s} \sum_{j=1}^{N_f} \mathbf{Y}_{jq}^k {\mathbf{Y}_{jq}^k}^H. \tag{5.55}$$

Note that the computation of the correlation matrix also takes advantage of the inherent temporal diversity of the IR-UWB signal, with N_f repeated transmitted pulses for each information symbol, by computing the correlation matrix over the N_f received frames.

5.1.2.2 *Joint TOA and AOA Estimation*

The proposed algorithm considers the use of TOA measurements at each antenna element in an antenna array for a joint estimation of TOA, τ_0, and AOA, θ_0. Once the previous TOA estimator is applied at each array element, the delay measurements are arranged in a vector $\mathbf{m}_\tau$,

$$\mathbf{m}_\tau = \begin{bmatrix} \hat{\tau}_{0,1} & \cdots & \hat{\tau}_{0,Q} \end{bmatrix}^T, \tag{5.56}$$

which depends on the TOA and the AOA by the geometric relation (see (5.38))

$$\mathbf{m}_\tau = \mathbf{1}\tau_0 + \mathbf{x}\sin(\theta_0) + \mathbf{n}, \tag{5.57}$$

where $\mathbf{1} = \begin{bmatrix} 1 & 1 & \cdots & 1 \end{bmatrix}^T$, $\mathbf{x} = \frac{d}{c}\begin{bmatrix} 0 & 1 & \cdots & Q-1 \end{bmatrix}^T$ contains the ordered indices of the array antenna elements, and $\mathbf{n}$ denotes the measurement noise vector. The measurement vector $\mathbf{m}_\tau$ can also be written as

$$\mathbf{m}_\tau = \mathbf{Z}\mathbf{b} + \mathbf{n}, \tag{5.58}$$

where $\mathbf{Z} = \begin{bmatrix} \mathbf{1} & \mathbf{x} \end{bmatrix}^T$ and $\mathbf{b} = \begin{bmatrix} \tau_0 & \psi \end{bmatrix}^T$ is the vector of parameters to be estimated, with $\psi = \sin(\theta_0)$. Then, the best linear unbiased estimator [64] is given by

$$\hat{\mathbf{b}} = \left(\mathbf{Z}^T\mathbf{Z}\right)^{-1}\mathbf{Z}^T\mathbf{m}_\tau. \tag{5.59}$$

Here, it is assumed that the noise associated with the measurements is white and Gaussian and has the same variance for all array elements (which is reasonable since the elements in the array experience similar propagation conditions).

Numerical Results

For numerical evaluation of the algorithm, we consider the channel models developed within the framework of IEEE 802.15.4a. In particular, the "CM1 residential" LOS channel model is used. The spatial dependency on the angle of arrival is introduced as in Refs. [33, 74]. All simulations are given for 100 independent channel realizations. The main simulation parameters are shown in Table 5.1. The pulse $g_{tx}(t) = c(t)\cos(2\pi f_0 t)$ is a modulated eighth-order Butterworth pulse with a 3 dB bandwidth of 1.3 GHz and a center frequency of $f_0 = 4.5$ GHz according to the European UWB regulations. The pulse repetition period is $T_f = T_{sym}/N_f = 128$ ns. The threshold level P_{th} used in the simulation is defined as the noise power level, which is estimated during the coarse estimation process.

Considering an array equipped with two antennas in a scenario with a single source, Fig. 5.8 depicts the periodograms computed at each array element from which the TOA measurements are obtained. The estimated TOA at each antenna element is given by the first peak that exceeds a given threshold corresponding to the first ray that gets to the array. These measurements yield the vector $\mathbf{m}_\tau$ used in the joint TOA and AOA estimator to obtain both estimated parameters.

One important step to get to the final result is the coarse symbol synchronization stage. Figure 5.9 shows the probability of failure of the synchronization stage with respect to the SNR for an array equipped with different numbers of antennas. One may observe that increasing the number of antennas Q makes the coarse stage more robust due to the larger amount of information collected.

Table 5.1 Joint TOA and AOA Estimation Algorithm: Simulation Parameters

Parameter	Value
Pulse duration $g_{tx}(t)$, T_{p}	0.77 ns
Bandwidth B_{W}	1.3 GHz
Number of frames per symbol, PRF and N_{f}	31
Number of symbols in the acquisition interval, K_{s}	47 symbols
Symbol duration, T_{sym}	3974.4 ns
Sampling rate, $1/T_{\mathrm{s}}$	3 GHz
Number of points of the DFT, N	128
Number of τ values to be swept, M	1024

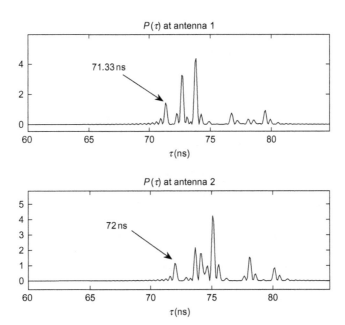

FIGURE 5.8

Power delay profiles for an array equipped with two antennas with antenna elements separation of 36 cm (from Ref. [127]).

Figure 5.10 depicts the RMSE of the estimated TOA, expressed in meters, for arrays equipped with two and three antennas and an antenna spacing of 36 cm. The results show that an estimation accuracy of a few centimeters (6 cm) is asymptotically achieved for high values of SNR in both cases. The results are compared with the CRB derived for the joint TOA–AOA estimate [74]. The asymptotic

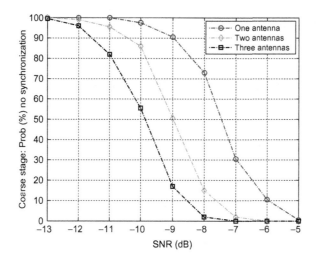

FIGURE 5.9

Coarse stage evaluation with respect to array size (from Ref. [127]).

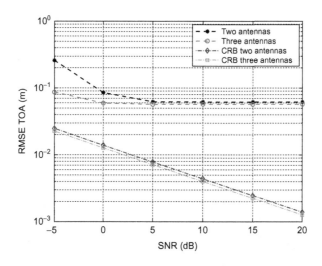

FIGURE 5.10

Time of arrival τ_0 estimation performance for two and three array elements with antenna elements separation of 36 cm (from Ref. [127]).

error floor for high values of SNR arises because the algorithm operates without the knowledge of the pulse waveform and also because of the sampling frequency limitations.

Likewise, Fig. 5.11 shows the estimated sine of the AOA for the same scenario also compared with the CRB. The AOA is randomly generated at each channel realization. We shall remark that the results

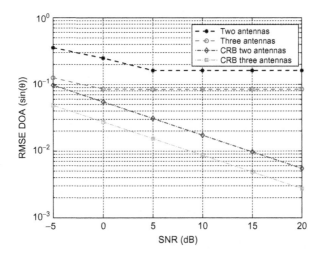

FIGURE 5.11

Direction of arrival estimation performance for two and three array elements with antenna elements separation of 36 cm (from Ref. [127]).

depict directly the output of the estimator, that is, the value of the sine of the AOA, rather than the angle itself. There is no need to transform to the angle domain for later application in positioning algorithms. One may observe that the error tends to increase while the SNR decreases.

5.1.3 TOA Estimation in the Presence of Interference

UWB is intended to operate as an underlay technology, and hence it has to coexist with other communication systems. Therefore, location-aware systems based on UWB technology are expected to experience both narrowband interference (NBI) and multiuser interference (MUI), which can degrade TOA estimation performance [70]. Despite this, relatively few publications address such effects on TOA estimation. A technique based on nonlinear filtering of received samples is proposed in Ref. [100] to mitigate the effect of MUI on ranging. The joint effect of NBI and MUI is considered in Ref. [27], in which a technique to mitigate both of these effects is proposed.

With reference to the TOA ED estimator in Fig. 3.35 and the signal structure described in (3.60), we now illustrate a possible way to mitigate the effects of both NBI and wideband interference (WBI). It has been shown in Ref. [46] that a single NBI can be well approximated by a tone as

$$d(t) = \sqrt{2I}\alpha_{\rm I} \cos(2\pi f_{\rm I}t + \phi_{\rm I}), \tag{5.60}$$

where I is the average received power, $f_{\rm I}$ is the center frequency of the NBI spectrum, and $\alpha_{\rm I}$ and $\phi_{\rm I}$ are the amplitude and phase, respectively, of the fading associated with the NBI. In this case, we define the interference to noise power ratio (INR) as $\mathrm{INR} \triangleq IT_{\rm s}/N_0$. On the other hand, the expression for the MUI depends on the structure of transmitted signals as well as on the channel characteristics. For example, when a number U of UWB interferers are present, the MUI contribution to $d(t)$ in (3.60) can be written as

$$d(t) = \sum_{u=2}^{U} \sum_{n=0}^{N_t-1} w^{(u)}(t - c_n^{(u)} T_c - n T_f). \tag{5.61}$$

Typical energy matrix $\mathbf{V}$ in (3.63) is plotted in Fig. 5.12(a), which shows the energy collected at the output of the ED in the presence of both NBI and MUI (remember that N_t is the number of pulses in the preamble and K is the number of time slots in each observation subinterval). It can be seen that the energy samples corresponding to the desired signal are time aligned and partially buried under a floor caused by NBI and noise. This floor may increase the probability of early detection and thus degrades the TOA estimator performance. Note that, unlike the desired signal, the energy samples corresponding to MUI are not time aligned, resulting in impulsive behavior, due to the use of the TH technique.

Figure 5.12(a) suggests the use of filtering techniques to reduce the effects of interference on TOA estimation. To reduce the impulse behavior of MUI, nonlinear processing can be applied to each column of $\mathbf{V}$. In particular, the use of a *min filter* gives

$$z_k = \sum_{n=0}^{N_t-H} \min\{v_{n,k}, v_{n+1,k}, \ldots, v_{n+H-1,k}\}, \tag{5.62}$$

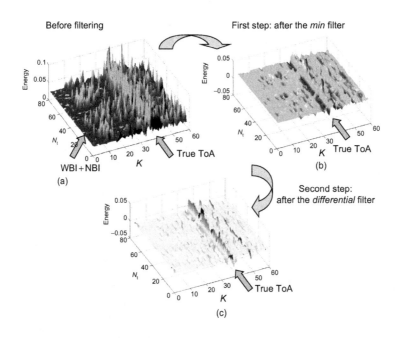

FIGURE 5.12

Energy matrix before filtering (a), after *min* filter (b), and after the *min* + *differential* filter (c) for a system with $N_{sym} = 20$, $N_s = 4$, $T_{ob} = 120$ ns and $T_{int} = 2$ ns, in an IEEE 802.15.4a CM4 channel and affected by MUI and NBI.

where $k = 0, \ldots, K - 1$, and H is the length of the filter [100]. The energy matrix at the output of the *min filter* is shown in Fig. 5.12(b). In scenarios where both NBI and MUI are present, Ref. [27] proposes to adopt a double-filtering scheme. In particular, *min filter* is applied to each matrix column as in (5.62), whose outputs $\{\tilde{z}_k\}$ are then further processed by a *differential filter* yielding

$$z_k = \tilde{z}_k - \tilde{z}_{k+1}, \quad k = 0, \ldots, K - 1 \tag{5.63}$$

with $\tilde{z}_K = 0$. The purpose of (5.63) is to reduce the presence of floors, typically caused by NBI, and to emphasize the beginning of the multipath cluster.[2] The energy matrix at the output of the differential filter is shown in Fig. 5.12(c). The ability of this scheme to mitigate both NBI and MUI is demonstrated by comparing Figs 5.12(a) and 5.12(c).

Various filtering techniques are compared in Fig. 5.13 in terms of the RMSE of the ST estimator described in Section 3.4.2.2 in the absence and presence of NBI and MUI using the IEEE 802.15.4a CM4 channel [80]. In particular, the NBI is modeled as a tone with frequency $f_I = 3.5\,\text{GHz}$ subject to Rayleigh fading. To model MUI, we consider an additional user and define the signal-to-interference ratio as $\text{SIR} \triangleq E_s^{(1)}/E_s^{(2)}$. The performance of single *averaging*, *min*, and *differential filters* drastically

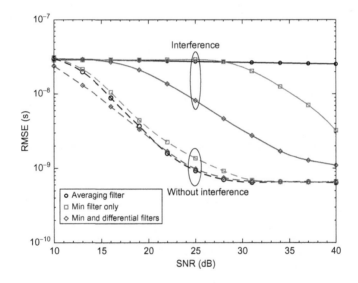

FIGURE 5.13

The RMSE as a function of SNR in the presence and absence of interference. The IEEE 802.15.4a CM4 channel characterizing NLOS indoor propagation in large office environments is used. ED ST-based estimator is considered. The signal is TH with $T_f = 128$ ns. The band-pass pulse has RRC envelope, $f_0 = 4$ GHz, roll-off $\nu = 0.6$, $T_p = 1$ ns. Parameters $T_a = 100$ ns, $T_{ob} = 120$ ns, $T_{int} = 2$ ns, $N_{sym} = 400$, and $N_s = 4$ are also considered, together with different 2D filtering techniques, $H = 5$, in the presence of both NBI, $\text{INR} = 35$ dB, and MUI, $\text{SIR} = -15$ dB (from Ref. [25]).

---

[2]Differential filtering is typically used in image processing to emphasize the presence of edges.

degrades in the presence of NBI and MUI. On the contrary, the cascaded *min and differential filter* provides substantial improvements in performance with respect to other schemes due to its capability to mitigate the combination of both types of interference.

5.1.4 Robust Approaches for TOA Estimation in NLOS Conditions

Location systems based on TOA require an accurate estimation of the first path delay. Unfortunately, in wireless channels multipath propagation and NLOS situations bias the estimation of the first signal arrival. In this context, detectors based on the maximum of the impulse response [111], despite their low computational burden, are not suitable because the delayed paths displace the maximum of the response. Analogously, conventional correlators cannot separate the direct path from the echoes if they are close to the direct signal or the direct propagation path is highly attenuated.

On the other hand, conventional methods based on the ML criterion provide a complete channel description at the expense of a heavy computational burden (see Chapter 3). In general, ML techniques lead to an expensive multidimensional search [13, 98, 115] and in urban environments, where the delay spread of the channel increases, the computational cost of the ML solutions can be extremely high.

As is well known, time delay estimation in the frequency domain and spectral (or spatial) analysis are closely connected [87]. In this context, *high-resolution parametric spectral techniques* such as MUSIC and root-MUSIC can be used as TOA estimators [32, 54, 59]. These methods are commonly referred to in the literature as subspace or singular value decomposition (SVD) methods and are based on the separation of the signal and the noise subspaces. This characterization is costly and not always reliable. Also, the number of propagation paths must be estimated, which is a difficult task when SNR is low or moderate. Another well-known SVD technique is total least-squares estimation of signal parameters via rotational invariance techniques (TLS-ESPRIT), and it is applied to time delay estimation in Ref. [99]. In spite of avoiding the traditional scanning of the MUSIC algorithm, its main drawbacks are that it requires an eigendecomposition and the number of incoming rays must be determined a priori. To overcome these problems, and bearing in mind the analogy between time delay estimation and spectral analysis, high-resolution methods based on the *minimum variance (MV)* and on the *normalized minimum variance (NMV)* are successfully developed and applied to location systems in Refs. [120] and [84]. These techniques can provide an accurate estimation of the first signal arrival at an affordable cost in high-multipath environments, even when the LOS signal is strongly attenuated.

On the other hand, a great improvement on the ranging accuracy could be obtained if a reliable estimate of the range bias due to NLOS propagation were available. Simplified models based on geometric and experimental characterization have been proposed for this purpose.

In the following sections two approaches to tackle the NLOS impairment to TOA estimation are presented. The first approach (Section 5.1.4.1) is based on spectral techniques and can provide high-resolution estimates of TOA parameters in the frequency domain. The second one (Section 5.1.4.2) develops models of NLOS propagation which are used to predict and correct the bias due to the NLOS propagation when using UWB signals.

5.1.4.1 *Frequency-Domain High-Resolution TOA Estimation*

Herein efficient implementations of the grid search of the *minimum variance* (MV) and *normalized minimum variance* (NMV) TOA estimators based on FFT are described, allowing a significant reduction on the computational burden. Furthermore, polynomial approaches that transform the grid search of the minimum variance criterion into a polynomial rooting procedure are exposed. Based on the power delay profile (PDP), the *root derivative minimum variance* (RDMV) algorithm is obtained. Another polynomial method closely related to the latter is the *root minimum variance* (RMV). This technique was briefly introduced in Ref. [5] and later on analyzed in detail in Ref. [118] in the context of AOA estimation. These algorithms are quite general and not necessarily confined to UWB systems.

Minimum Variance (MV) and Normalized Minimum Variance (NMV) Criteria

The minimum variance criterion can be applied to the frequency model exposed in (4.23) and reused in (5.39). The criterion consists of finding the filter $\mathbf{w}(\tau)$ that maximizes the output SNR and satisfies the constraint $\mathbf{w}^H(\tau)\mathbf{Se}_\tau = 1$, yielding

$$\mathbf{w}(\tau) = \frac{\mathbf{R}_y^{-1}\mathbf{Se}_\tau}{\mathbf{e}_\tau^H \mathbf{S}^H \mathbf{R}_y^{-1}\mathbf{Se}_\tau}, \tag{5.64}$$

where $\mathbf{R}_y$ is the estimated correlation matrix obtained from a collection of N observations within the coherence time of the delays, $\mathbf{S} \in \mathbb{C}^{N \times N}$ is a diagonal matrix whose components are the frequency samples of the spectrum of the transmitted signal (5.40), and $\mathbf{e}_\tau$ is the vector containing the harmonic components sampled at the time delay τ.

Taking into account (5.64), the PDP is given by

$$G(\tau) = \mathbf{w}(\tau)^H \mathbf{R}_y \mathbf{w}(\tau) = \frac{1}{\mathbf{e}_\tau^H \mathbf{S}^H \mathbf{R}_y^{-1}\mathbf{Se}_\tau}. \tag{5.65}$$

The expression (5.65) constitutes the MV-TOA estimator and provides the well-known matched filter bank point of view of the minimum variance criterion [113]. In the traditional approach, a grid search is performed and the LOS delay is estimated as the first maximum of the PDP above a threshold level related to the noise power. Nevertheless, the grid search of the maxima of the PDP can be efficiently implemented using an FFT. As demonstrated in Refs. [113] and [83], the expression (5.65) can be reformulated as a DFT:

$$G(\tau) = \frac{1}{\sum\limits_{n=1-N}^{N-1} D_n \mathrm{e}^{-jn\omega}}, \tag{5.66}$$

D_n being the sum of the nth diagonal elements of the matrix $\mathbf{S}^H \mathbf{R}_y^{-1}\mathbf{S}$. Obviously, time delay information is now wrapped in the phase $\omega = \omega_0 \tau$.

A fast implementation of the grid search can be achieved by computing the expression (5.66) using an FFT of length N_{FFT}, where N_{FFT} is the number of desired grid points.

In practice, the MV power delay spectrum is strongly dependent on the filter bandwidth, and the presence of the shaping filter causes significant side lobes, which can be misinterpreted as arrivals. Normalizing the PDP (5.65) by the equivalent noise bandwidth, the power delay spectrum density is obtained as follows:

$$S(\tau) = \frac{P(\tau)}{\mathbf{w}(\tau)^H \mathbf{w}(\tau)} = \frac{\mathbf{e}_\tau^H \mathbf{S}^H \mathbf{R}_y^{-1}\mathbf{Se}_\tau}{\mathbf{e}_\tau^H \mathbf{S}^H \mathbf{R}_y^{-2}\mathbf{Se}_\tau}, \tag{5.67}$$

where $P(\tau)$ is the power delay profile defined as in (5.52). From this expression, the NMV-TOA estimator is performed, and the first arrival is determined as the first power delay spectrum density maximum above a threshold level. As is well known, NMV solutions present better resolution properties than the MV estimators.

The FFT approach of the NMV-TOA estimators based on the work of Ref. [83] was derived in Ref. [8]. The computational burden is dramatically reduced by computing the grid search of the NMV estimator as a quotient of two N_{FFT}-point-FFT, where N_{FFT} is the number of desired grid points.

Root Derivative Minimum Variance (RDMV) Criterion

In Ref. [8] the authors propose a new approach to the MV criterion, which transforms the traditional grid search of the maxima into a polynomial rooting procedure. The maximization of the PDP to find the delays corresponding to each channel path is equivalent to minimizing the denominator of the expression (5.66):

$$\max P(\tau) \Leftrightarrow \min e_\tau^H S^H R_y^{-1} S e_\tau \Leftrightarrow \min \sum_{n=1-N}^{N-1} D_n e^{-jn\omega}. \tag{5.68}$$

Writing $P(\tau)$ in terms of $\rho = e^{-j\omega}$ as in (5.54), and deriving we obtain the following:

$$\frac{\partial}{\partial \omega} P(\tau) = -j \sum_{n=1-N}^{N-1} n D_n \rho^n = 0. \tag{5.69}$$

Then the TOA can be obtained by identifying the root that corresponds to the first peak (relative maximum), evaluated through the PDP second derivative. Note that the second derivative is easily calculated from the polynomial itself:

$$r'_{root}(\rho) = - \sum_{n=1-N}^{N-1} n^2 D_n \rho^n. \tag{5.70}$$

Root Minimum Variance (RMV) Criterion

The denominator of the expression (5.65) is a quadratic form. Thus, finding the peaks of the PDP is equivalent to finding the roots (minima) of the denominator polynomial:

$$\sum_{n=1-N}^{N-1} D_n \rho^n = 0. \tag{5.71}$$

By evaluating the polynomial (5.71) in the unit circle, we obtain the PDP.

Performance Comparison

In Fig. 5.14 the *ranging RMSEs* obtained with the different high-resolution TOA estimators are compared with the CRB. A mean number of 10 incoming rays and a delay spread between $3 \cdot 10^{-7}$ and $6 \cdot 10^{-7}$ seconds, typical values in urban environments, have been considered. The spectral versions of the MV- and NMV-TOA estimators have been computed using a 4096-point FFT implementation.

5.1.4.2 *Characterization of the Range Bias*

The potentialities of high-accuracy positioning systems, such as those based on the UWB technology, might vanish in the presence of NLOS channel conditions due to the ranging errors (typically positively

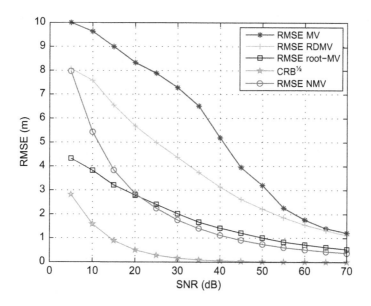

FIGURE 5.14

RMSE on ranging versus SNR, for different frequency-domain TOA estimation algorithms (from Ref. [128]).

biased) caused by obstacles. In many cases, the DP is not completely obstructed but it experiences an extra propagation delay due to the different materials encountered by the electromagnetic wave [62].

The measured range to the ith anchor can then be modeled as

$$r_i = d_i + b_i + \epsilon_i, \tag{5.72}$$

where d_i is the true distance between the agent and the ith anchor, b_i is the bias, and ϵ_i is Gaussian noise, independent of b_i, with zero mean and variance σ_i^2.

The bias b_i can be treated either as a random variable, if a statistical characterization is available, or as a deterministic quantity if it is somehow known. Here we introduce a simple deterministic model characterizing the bias introduced by NLOS conditions in indoor scenarios when using UWB signaling. Examples of statistical model can be found in Refs. [2, 26, 61, 123].

Deterministic Model for the Bias: The Wall Extra Delay (WED) Model

It is shown in Ref. [26] that the bias depends primarily on the walls obstructing the DP signal. The bias to the ith anchor, b_i, can therefore be modeled as

$$b_i = E_i \cdot c$$

$$E_i = \sum_{k=1}^{N_e^{(i)}} W_k^{(i)} \cdot \Delta_k, \tag{5.73}$$

where c is the speed of light, E_i is the total time delay caused by NLOS conditions, $W_k^{(i)}$ is the number of walls introducing the same extra delay value Δ_k (i.e., the number of walls of the same material and thickness), and $N_e^{(i)}$ is the number of different extra delay values. The total number of walls separating the ranging devices is $W^{(i)} = \sum_{k=1}^{N_e^{(i)}} W_k^{(i)}$. This model is named the *wall extra delay (WED) model*. When every wall in the scenario has the same thickness and composition (i.e., $\Delta_k = \Delta$ for each k), (5.73) simplifies to

$$b_i = W^{(i)} \Delta \cdot c. \tag{5.74}$$

Values for Δ_k and $W_k^{(i)}$ can be obtained once the environment topology is known by using experimental data or using ray-tracing tools. As will be demonstrated later, a priori knowledge of the bias can sometimes be obtained using the WED model if an initial rough estimate of the agent position is available. In that case, the approximate bias value can be simply subtracted from the range measurements.

5.1.4.3 *Positioning Bias Correction*
We have already anticipated that in general NLOS configurations produce a ranging bias, which is often the major source of positioning errors. By analyzing experimental data using the UWB technology, in Refs. [22, 26] it has been shown that the bias is strictly related to the number of walls encountered by the signal. Assuming that prior knowledge of the environment topology is available, it is possible to refine the agent's position estimate once an initial rough estimate is obtained. In many cases, knowledge of the room in which the agent is located will suffice as an initial estimate. These considerations suggest the following *two-step positioning algorithm* when prior information is available [26]:

- *First estimate*—An initial rough position estimate $\hat{\mathbf{p}}^{(1)}$ is obtained using the LS method (3.19), by setting $\hat{d}_i = r_i$, or other localization algorithms.
- *Range correction*—Biases due to propagation through walls are subtracted from range measurements according to the WED model for b_i in (5.73), where the number of walls separating the agent and each anchor is calculated using the first position estimate and the topology information.
- *Refinement*—A second LS position estimate $\hat{\mathbf{p}}^{(2)}$ is calculated with the corrected (unbiased) range values.

We present a numerical analysis of positioning performance based on experimental data used in Section 3.1.2.1 and refer to the scenario shown in Fig. 3.13. The effect of prior information and extra delay correction on positioning using the two-step algorithm previously presented can be observed in Fig. 5.15. By comparing Figs 3.14 and 5.15, we can conclude that the correction of the range measurements using the WED model and knowledge of the environment topology leads to a significant performance improvement for many locations. Positioning errors less than 1 m are achieved in most locations.

5.2 MIMO POSITIONING SYSTEMS
In an environment characterized by rich multipath, the presence of multiple antennas in both transmitter and receiver (multiple-input multiple-output (MIMO)) allows the exploitation of the spatial diversity in the channel, and it has a twofold impact on the network link: Transmission performance can have a considerable improvement with a consequent positive impact on the noisy measured parameters

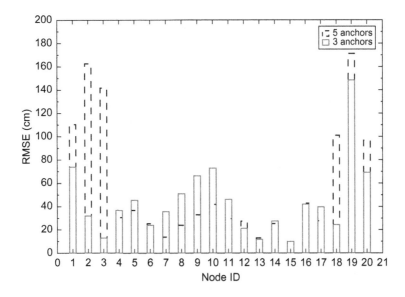

FIGURE 5.15

RMSE as a function of agent position in the presence of prior information (two-step algorithm). $N = 3$ (tx1, tx3, tx5) and $N = 5$ anchors are considered.

(or reduction in the transmitted power) and multiple hybrid parameters (time of arrival, angles of arrival, and departure of the strongest paths) can be estimated simultaneously at the same node [73, 119]. In recent years, considerable work has also concerned the MIMO radar application [7, 19, 36, 37, 48, 72]. An interesting extension of the MIMO concept is the case of different nodes in a subnetwork that cooperate to form improved nodes with multiple antennas: This form of cooperation, often referred to as *virtual MIMO*, is realized at the physical layer without involving the positioning algorithm itself, but it can improve channel quality and the number of measured parameters [58].

In this section, space diversity provided by multiple antennas is studied for accurate TOA estimation. In other words, the effects of space diversity on positioning systems are investigated.

5.2.1 CRB for the Joint Estimation of TOA and AOA in MIMO Systems

For the sake of completeness, we provide hereafter a theoretical evaluation of the advantages of MIMO systems, deriving the CRB for the joint estimation of TOA and AOA and comparing them with single-input single-output (SISO), single-input multiple-output (SIMO), and multiple-input single-output (MISO) systems. In our case we have to estimate two parameters: τ, the time delay between the transmitter and the receiver, and θ, the AOA.

With reference to Fig. 5.16, we consider an antenna array TX of N elements for transmission and an array RX of M elements for reception. Let (ρ_n, θ_n) be the polar coordinates of TX_n (nth element of TX) with respect to G_T (geometric center of TX) and (r_m, ϕ_m) the polar coordinates of RX_m (mth element of RX) with respect to G_R (geometric center of RX). Let $i_T(\theta) = \sum_{n=1}^{N} \rho_n^2 \sin^2(\theta - \theta_n)$ and

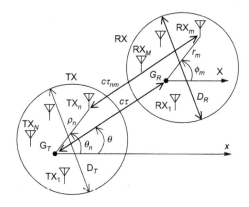

FIGURE 5.16

MIMO ranging system configuration (from Ref. [129]).

$i_R(\theta) = \sum_{m=1}^{M} r_m^2 \sin^2(\theta - \phi_m)$ denote the *inertias* of the arrays TX and RX in the direction θ.[3] Notice that ρ_n is the distance between the geometric center of the transmission array and the nth element of the array.

We assume that the communication channel between TX and RX is AWGN. Even if this assumption is not realistic, it allows us to obtain the lower bounds of the MIMO system and better understand the basic parameters that affect the performance. We also assume that TX and RX are sufficiently far with respect to TX antennas spacing so that

$$\begin{cases} \alpha_{nm} \approx \alpha \\ \tau_{nm} \approx \tau - \frac{\rho_n}{c} \cos(\theta - \theta_n) + \frac{r_m}{c} \cos(\theta - \phi_m), \end{cases} \quad \forall(n,m)$$

where α_{nm} and τ_{nm} are the gain and the time delay of the channel between TX$_n$ and RX$_m$ and c the speed of light.

Let $r_m(t)$ be the complex envelope of the signal received at RX$_m$. We can express it as

$$r_m(t) = \alpha \sum_{n=1}^{N} e^{-j2\pi f_c \tau_{nm}} s_n(t - \tau_{nm}) + n_m(t) = \sum_{n=1}^{N} v_{nm}(t) + n_m(t), \tag{5.75}$$

where f_c is the frequency of the carrier, $s_n(t)$ the complex envelope of the signal transmitted by TX$_n$, $n_m(t)$ the complex envelope of the thermal noise at RX$_m$, and $v_{nm}(t)$ the useful signal received by RX$_m$ from TX$_n$. The PSD of $n_m(t)$ is equal to $2N_0$ ($N_0/2$ being the PSD of the real noise). We assume that $n_1(t), \dots, n_M(t)$ are i.i.d.

---

[3]The inertia of an array in a direction θ is the sum of the squares of the array elements in a Cartesian coordinate system that is centered in the geometric center of the array and whose x-axis is in the θ direction. Thus, the inertia (without direction) of a set of points with respect to a reference point denotes here the sum of the distance squares between the reference point and the set of points.

FIM and CRB for the Joint Estimation of τ and θ

In this section, we derive the FIM and the CRBs for the joint estimation of τ and θ. We assume that the unknown parameters are only τ and θ; all other parameters are supposed to be known.

The CRBs of τ and θ (κ^τ and κ^θ respectively) are the diagonal elements of the CRB matrix given by

$$\kappa^{\tau,\theta} = (\mathbf{J}(\tau,\theta))^{-1}, \quad \mathbf{J}(\tau,\theta) = -\mathbb{E}\left\{\begin{bmatrix} \frac{\partial^2 \Lambda(\tau,\theta)}{\partial \tau^2} & \frac{\partial^2 \Lambda(\tau,\theta)}{\partial \tau \partial \theta} \\ \frac{\partial^2 \Lambda(\tau,\theta)}{\partial \theta \partial \tau} & \frac{\partial^2 \Lambda(\tau,\theta)}{\partial \theta^2} \end{bmatrix}\right\} \triangleq \begin{bmatrix} \mathsf{J}(\tau,\tau) & \mathsf{J}(\tau,\theta) \\ \mathsf{J}(\theta,\tau) & \mathsf{J}(\theta,\theta) \end{bmatrix}, \tag{5.76}$$

where $\mathbf{J}(\tau,\theta)$ is the Fisher information matrix (FIM), and $\Lambda(\tau,\theta)$ is the log-likelihood function of τ and θ. If $\mathbf{C}^{\hat{\tau},\hat{\theta}}$ denotes the covariance matrix of the unbiased estimator $(\hat{\tau},\hat{\theta})$, the matrix $(\kappa^{\tau,\theta} - \mathbf{C}^{\hat{\tau},\hat{\theta}})$ is semidefinite negative $\forall(\hat{\tau},\hat{\theta})$.

The part of interest of the log-likelihood function can be written as

$$\Lambda(\tau,\theta) = \frac{-1}{2N_0} \sum_{m=1}^{M} \int_{-\infty}^{+\infty} \left| r_m(t) - \alpha \sum_{n=1}^{N} e^{-j2\pi f_c \tau_{nm}} s_n(t - \tau_{nm}) \right|^2 dt. \tag{5.77}$$

Assume now that some *orthogonal codes* are adopted to make $s_1(t),\dots,s_N(t)$ orthogonal. Depending on the receiver antenna position, $s_1(t),\dots,s_N(t)$ are not necessarily orthogonal at the receiver. In narrowband and wideband systems the distance between two neighboring elements is the order of $\lambda_c/2$, λ_c being the carrier wavelength. The maximum value of $\Delta\tau_{n',n}$ is equal to D/c, D being the diameter of the transmitting array. Since in narrowband and wideband systems the symbol period is much larger than the carrier period, $\Delta\tau_{n',n}$ will not have any significant impact on the CRBs, which can be approximated as

$$\kappa_0^\tau = \frac{1}{M N \eta (2\pi)^2 \beta_s^2} \tag{5.78}$$

$$\kappa_0^\theta = \frac{c^2}{\eta (2\pi)^2 \beta_s^2 [N i_R(\theta) + M i_T(\theta)]},$$

where $\eta \triangleq \alpha^2 E_s/N_0$ is the signal-to-noise ratio, β_s is the effective bandwidth of $s(t)$ as defined in (4.4), and the subscript o indicates *orthogonality of the received signals*. For simplicity, we have assumed $s_n(t)$ to have the same energy E_s and the same effective bandwidth β_s.

Comparison between CRBs in SISO, SIMO, MISO, and MIMO Systems

Let us first give the CRBs of τ and θ in SISO, SIMO, and MISO systems:

$$\kappa_{SISO}^\tau = \frac{1}{\eta(2\pi)^2\beta_s^2} \tag{5.79}$$

$$\kappa_{SIMO}^\tau = \frac{1}{M \eta(2\pi)^2\beta_s^2}, \quad \kappa_{SIMO}^\theta = \frac{c^2}{i_R(\theta) \eta(2\pi)^2\beta_s^2} \tag{5.80}$$

$$\kappa_{MISO,o}^\tau = \frac{1}{N \eta(2\pi)^2\beta_s^2}, \quad \kappa_{MISO,o}^\theta = \frac{c^2}{i_T(\theta) \eta(2\pi)^2\beta_s^2}, \tag{5.81}$$

where for MISO we have assumed that the received signals are orthogonal. Note that in the SISO case the CRB coincides with that given in (4.3).

We can see that κ_{SIMO}^{τ} is M times smaller than c_{SISO}^{τ}; in fact, the signal containing the information on τ is received M times by the receiver. $\kappa_{MISO,o}^{\tau}$ is N times smaller than κ_{SISO}^{τ} because receiving N orthogonal signals is equivalent to receiving one transmitted signal N times. κ_o^{τ} for MIMO in (5.78) is MN times smaller than c_{SISO}^{τ} because the N orthogonal signals are received M times by the receiver.

The expressions of κ_{SIMO}^{θ} and $\kappa_{MISO,o}^{\theta}$ are similar because MISO becomes identical to SIMO when the received signals are orthogonal. In the expression of κ_o^{θ} for MIMO (5.78), we can see the term $N i_R(\theta) + M i_T(\theta)$ in the denominator. $N i_R(\theta)$ states that each of the N transmitted signals carries the same information on θ to the reception array while $M i_R(\theta)$ states that the information on θ carried by the set of transmitted signals is received M times at the receiver.

5.2.2 A Practical Range Estimator for SIMO Systems

A SIMO system is considered to be a first step, and the benefits of diversity for ranging is quantified by CRBs. In addition, a practical range estimator with low computational complexity is proposed, and its performance is investigated via theoretical and numerical calculations. It is shown that the proposed estimator approximately achieves the CRB at high SNRs.

Consider an SIMO system with N receive antenna elements, and suppose that the maximum distance between the antenna pairs divided by the speed of light is considerably smaller than the symbol duration. Then, the baseband received signal at the ith antenna can be expressed as [42]

$$r_i(t) = \alpha_i s(t - \tau) + n_i(t), \quad t \in [0, T], \tag{5.82}$$

for $i = 1, \ldots, N$, where $s(t)$ is the baseband representation of the transmitted signal, α_i is the channel coefficient of the received signal at the ith antenna, τ is the TOA, and $n_i(t)$ is a complex-valued white Gaussian noise process with zero mean and single-side spectral density N_0. It is assumed that noise processes at different receiver branches are independent and that there is sufficient separation (comparable to the signal wavelength) between all antenna pairs so that different channel coefficients can be observed at different antennas. This is unlike a phased array structure in which $\alpha_i = \alpha \; \forall i$ [42].

The ranging problem in an SIMO system involves the estimation of the TOA τ from the received signals at N receiver antennas. In addition, the channel coefficients $\boldsymbol{\alpha} = [\alpha_1 \cdots \alpha_N]$ are also unknown and need to be considered in the estimation problem in general. If the complex channel coefficients are represented as $\alpha_i = a_i e^{j\phi_i}$ for $i = 1, \ldots, N$, the vector of unknown signal parameters can be expressed as $\boldsymbol{\lambda} = [\tau \; \boldsymbol{a} \; \boldsymbol{\phi}]$, where $\boldsymbol{a} = [a_1 \cdots a_N]$ and $\boldsymbol{\phi} = [\phi_1 \cdots \phi_N]$.

From (5.82), the log-likelihood function for $\boldsymbol{\lambda}$ can be expressed as [93]

$$\Lambda(\boldsymbol{\lambda}) = k - \sum_{i=1}^{N} \frac{1}{2N_0} \int_0^T |r_i(t) - \alpha_i s(t - \tau)|^2 \, dt, \tag{5.83}$$

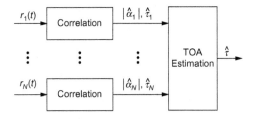

FIGURE 5.17

An asymptotically optimal algorithm for joint TOA and range estimation (from Ref. [42]).

where k represents a term that is independent of λ. Then, the ML estimate for λ can be obtained from (5.83) as

$$\hat{\lambda}_{\text{ML}} = \arg\max_{\lambda} \sum_{i=1}^{N} \frac{1}{N_0} \int_0^T \mathcal{R}\left\{\alpha_i^* r_i(t)s^*(t-\tau)\right\} \mathrm{d}t - \frac{E_s|\alpha_i|^2}{2N_0}, \tag{5.84}$$

where $E_s = \int_{-\infty}^{\infty} |s(t)|^2 \mathrm{d}t$ is the signal energy. Here, for a complex number z, $\mathcal{R}\{z\}$ and $\mathcal{I}\{z\}$ represent its real and imaginary parts, respectively.

In general, the ML solution in (5.84) requires optimization over an $(N+1)$-dimensional space, which can have prohibitive complexity in scenarios with a large number of receive antennas. Here, a two-step suboptimal estimator, as shown in Fig. 5.17, is proposed, which performs joint channel and delay estimation at each output branch in the first step, and implements a simple delay (range) estimator in the second step. Note that the algorithm exploits the multiple-output structure of an SIMO system, which facilitates individual signal processing, such as correlation or matched filter-based channel coefficient and delay estimation, at each receiver branch [42].

In the first step of the estimator, each branch processes its received signal individually and provides estimates of the channel coefficient and the delay, based on an ML approach [42]. For the ith branch, the ML estimates of $\alpha_i \ (= a_i e^{j\phi_i})$ and τ can be obtained from $r_i(t)$ in (5.82) as [42]

$$\left(\hat{\tau}_i, \hat{\phi}_i\right) = \arg\max_{\tau, \phi_i} \mathcal{R}\left\{e^{-j\phi_i} \int_0^T r_i(t)s^*(t-\tau)\mathrm{d}t\right\}, \tag{5.85}$$

$$\hat{a}_i = |\hat{\alpha}_i| = \frac{1}{E_s} \mathcal{R}\left\{e^{-j\hat{\phi}_i} \int_0^T r_i(t)s^*(t-\hat{\tau}_i)\mathrm{d}t\right\}, \tag{5.86}$$

for $i = 1, \ldots, N$. Note that the ML estimation results in a correlator, as in (5.85), which provides the delay and phase estimates, and the channel amplitude can be directly estimated from those estimates as in (5.86).

In the second step, the estimates for the channel amplitudes and the delays are used to estimate the TOA as follows:

$$\hat{\tau} = \frac{\sum_{i=1}^{N} \widehat{\eta}_i \hat{\tau}_i}{\sum_{i=1}^{N} \widehat{\eta}_i}, \tag{5.87}$$

where $\widehat{\eta}_i = E_s |\hat{\alpha}_i|^2 / \sigma_i^2$. In other words, the TOA is estimated as a weighted average of the delay estimates obtained at the N receiver branches, where the weights are proportional to the SNR estimates at the respective branches [42].

The computational complexity of the two-step estimator in Fig. 5.17 is dominated by the optimization operations in (5.85). In other words, the estimator requires the solution of N optimization problems, each over a two-dimensional space. On the other hand, the optimal ML solution in (5.84) requires optimization over an $(N+1)$-dimensional space, which is computationally more complex than the proposed algorithm. In fact, as N increases, the optimal solution becomes quite impractical.

The reduction in the computational complexity of the two-step algorithm results in its suboptimality in general compared to the ML algorithm in (5.84). However, under certain circumstances, it can be shown that the two-step scheme performs very closely to the optimal solution; that is, it approximately achieves the CRB of the original problem.

To this end, first consider the following lemma, which provides an approximate model for the estimates in (5.85) and (5.86) under certain conditions [42].

Lemma 1: *For the signal model in (5.82) when s(t) has symmetric spectrum, the delay estimate in (5.85) and the channel amplitude estimate in (5.86) can be modeled, at high SNR, as*

$$\hat{\tau}_i = \tau + v_i, \tag{5.88}$$

$$|\hat{\alpha}_i| = |\alpha_i| + \eta_i, \tag{5.89}$$

for i = 1,...,N, where v_i and η_i are independent zero-mean Gaussian random variables with variances $N_0/(\tilde{E}|\alpha_i|^2)$ and N_0/E_s, respectively, where $\tilde{E}$ is the energy of the first derivative of s(t). In addition, v_i and v_j (η_i and η_j) are independent for $i \neq j$ [42].

Lemma 1 establishes the approximate unbiasedness and efficiency of the two-step estimator, as implied by the following proposition [42].

Proposition 1: *For the delay and channel amplitude estimates as modeled in Lemma 1, the TOA estimator in (5.87) is an unbiased estimator of τ with the following variance:*

$$\text{Var}\{\hat{\tau}\} = \frac{1}{\tilde{E}} \mathbb{E} \left\{ \sum_{i=1}^{N} \frac{|\hat{\alpha}_i|^4}{N_0 |\alpha_i|^2} \left(\sum_{i=1}^{N} \frac{|\hat{\alpha}_i|^2}{N_0} \right)^{-2} \right\}, \tag{5.90}$$

where the expectation is over $|\hat{\alpha}_i|$ values modeled by (5.89).

Note that the variance of the two-step estimator in (5.90) is always larger than the CRB in (5.80). However, as E_s/N_0 gets higher, $|\hat{\alpha}_i|$ gets closer to $|\alpha_i|$ (Lemma 1), and the variance in (5.90) becomes approximately equal to the CRB.

Numerical Results
In order to compare CRBs for generic SIMO systems and phased arrays, and to analyze performance of the proposed two-step algorithm, a ULA structure with $N = 5$ antennas is considered for a narrowband

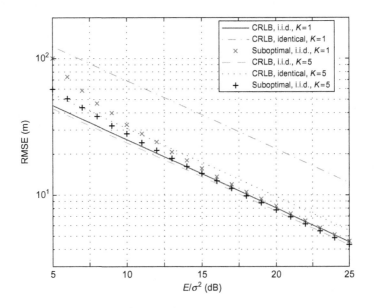

FIGURE 5.18

The RMSE of the two-step algorithm and the CRBs (from Ref. [42]).

signal with 1-MHz bandwidth and 3-GHz carrier frequency. The channel is modeled to be Ricean fading with a K-factor of K.

In Fig. 5.18, the RMSEs of the two-step algorithm ("suboptimal") are plotted for $K = 1$ and $K = 5$, together with the CRBs for the case of i.i.d. fading channel coefficients at different receiver branches.[4] Also shown in the figure are the CRBs for the phased array case, in which the antenna elements are closely spaced together so that the channel coefficients are identical at all the antennas.

It is observed from the figure that the accuracy is better for the i.i.d. fading scenarios, especially for the cases without strong line-of-sight components (i.e., for small Ks). In addition, the two-step algorithm converges to the CRB at high SNRs as expected, although it does not perform that well at low SNRs [42].

5.3 ADVANCED GEOMETRIC LOCALIZATION APPROACHES

NLOS propagation can dramatically impair the accuracy performance also for cooperative positioning algorithms. For this reason, recent advances on algorithm design focus on robust strategies to mitigate NLOS effects. Section 5.3.1 presents some results on *distributed bounded-error estimation* in a network of wireless nodes. Measurement noise is assumed to be bounded with known limits, and one

---

[4]The antennas are spaced 10 cm apart in the i.i.d. fading case.

aims at evaluating the *set* of all values of the parameter vector, which is consistent with the measurements and the bounds on the measurement noise. It provides a guaranteed and robust estimator in a distributed context using interval analysis. Here, *guaranteed* means that no parameter value consistent with a fixed number of measurements is missed. Section 5.3.2, on the other hand, considers distributed positioning rendered into a feasibility problem solved using projections onto convex sets (POCS). The positioning problem is formulated as finding a point in the intersection of a number of convex sets that are found from the measurements. To find a feasible point, POCS can be employed.

5.3.1 Bounded-Error Distributed Estimation

Challenging problems arise when considering parameter or state estimation using measurements provided by a WSN. Two main types of estimation techniques may be considered. In centralized approaches, all measurements obtained by the sensors are transmitted to a central processing unit; see, for example, Ref. [110]. Many data have then to be sent to a given point of the network. Moreover, this solution is not robust to a failure of the central processing unit, since the estimate is only available at that point of the network. Alternative distributed estimation techniques for constant [29] and time-varying parameters [14] have been provided. In this case, each sensor is responsible for the processing of its measurement and of data provided by neighboring sensors. An increased robustness to failure of the central processing unit is thus obtained.

Nevertheless, distributed solutions may not be very robust against erroneous measurements provided by some defective sensors. Despite robust estimators having been proposed in a centralized context, using bounded-error estimation [57] or linear programming [91, 105, 106], the extension of these techniques to a distributed context is far from being trivial.

This section presents some results on distributed bounded-error estimation in a WSN [66]. Measurement noise is assumed to be bounded with known bounds, and one aims at evaluating the *set* of all values of the parameter vector, which is consistent with the measurements and the bounds on the measurement noise. It provides a guaranteed and robust estimator in a distributed context using interval analysis [55]. By guaranteed, it is meant that no parameter value consistent with a fixed number of measurements is missed.

The robust bounded-error parameter estimation problem in a WSN is recalled first. Then, a centralized approach is described to provide a reference to the distributed approach presented a second time. Implementation issues and simulation results are finally provided.

Problem Formulation

Consider a network of N sensors spread in an environment. The aim is to provide the estimation of an unknown parameter vector $\mathbf{p}^* \in \mathbb{P}$ containing the position of the sensors, denoting by $\mathbb{P}$ the set of all possible positions, using measurements $\mathbf{y}_i$, $i \in [\![1, N]\!]$ provided by each sensor of the network. The measurement $\mathbf{y}_i$ is assumed to be linked to $\mathbf{p}^*$ via the measurement model

$$\mathbf{y}_i = \mathbf{f}_i(\mathbf{p}^*) + \mathbf{e}_i \quad i \in [\![1, N]\!], \tag{5.91}$$

where $\mathbf{e}_i$ is assumed to remain in some known interval vector (box) $[\underline{\mathbf{e}}, \overline{\mathbf{e}}]$. Introducing $[\mathbf{y}_i] = \mathbf{y}_i - [\underline{\mathbf{e}}, \overline{\mathbf{e}}]$, one has

$$\mathbf{f}_i(\mathbf{p}^*) \in [\mathbf{y}_i] \quad i \in [\![1, N]\!]. \tag{5.92}$$

Bounded-error parameter estimation [78, 121] aims at characterizing the *set* $\mathbb{S}_0$ of all parameter values, which are consistent with the measurements, the measurement model, and the bounds on the measurement noise, that is,

$$\mathbb{S}_0 = \{\mathbf{p} \in \mathbb{P} | \forall i \in [\![1,N]\!], \mathbf{f}_i(\mathbf{p}) \in [\mathbf{y}_i]\}. \tag{5.93}$$

When there are some outliers, for example in the case of defective sensors, for some measurements, the noise may not remain in its bounds, $\mathbb{S}_0$ may then be empty. In such situations, one may define a set estimator for $\mathbf{p}$ robust to q outliers as follows:

$$\mathbb{S}_q = \{\mathbf{p} \in \mathbb{P} | \lambda(\mathbf{p}) \geq N - q\}, \tag{5.94}$$

where

$$\lambda(\mathbf{p}) = \sum_{i=1}^{N} \mathbb{I}_{[\mathbf{y}_i]}(\mathbf{f}_i(\mathbf{p})) \tag{5.95}$$

and

$$\mathbb{I}_A(x) = \begin{cases} 1 & \text{if } x \in A \\ 0 & \text{else.} \end{cases} \tag{5.96}$$

Guaranteed inner and outer approximations $\underline{\mathbb{S}}_q$ and $\overline{\mathbb{S}}_q$ of $\mathbb{S}_q$ may be obtained for any value of q using interval analysis [55], provided that all measurements are collected at a CPU, to which all measurements $\mathbf{y}_i$ and models $\mathbf{f}_i$ have been transmitted. By guaranteed, it is meant that one is able to *numerically prove* that in $\mathbb{P} \setminus \overline{\mathbb{S}}_q$, all values of the parameter vector are inconsistent with at least $q+1$ measurements, and that in $\underline{\mathbb{S}}_q$, all values of the parameter vector are consistent with at least $N - q$ measurements.

This section proposes a *distributed* robust bounded-error estimator, that is, to provide an estimation algorithm that is able to evaluate at *each* sensor i of the network an outer approximation $\overline{\mathbb{S}}_{q,i}$ of $\mathbb{S}_q$ using only a subset of the measurements available in the network. The aim is to be robust to a failure of the CPU, to compute at each sensor of the network partial estimates with only a subset of the measurements, and if possible, to reduce the amount of data exchanged within the network. In what follows, the network is assumed to be entirely connected, that is, any sensor of the WSN is able to exchange information with any other sensor, in one or several hops.

5.3.1.1 *Robust Centralized Approach*

In this approach, all sensors send their measurements and measurement functions to a central processing unit. The robust bounded-error approach presented in Ref. [57] is briefly recalled here to serve as reference for the distributed approach detailed in the next paragraphs.

The robust estimator is based on the notion of *inclusion function*, introduced by interval analysis [55, 81]. Consider a function $\mathbf{f} : \mathcal{D} \subset \mathbb{R}^\alpha \longrightarrow \mathbb{R}^\beta$; an inclusion function $[\mathbf{f}]$ for $\mathbf{f}$ has to be such that

$$\forall [\mathbf{x}] \subset \mathcal{D} \quad [\mathbf{f}]([\mathbf{x}]) \supset \mathbf{f}([\mathbf{x}]). \tag{5.97}$$

The *natural inclusion function* is an *inclusion function* obtained by replacing all occurrences of the variable x in the formal expression of $f(x)$ by the interval counterpart $[x]$. It allows one to compute an outer approximation of the range of f over any interval $[x] \subset \mathcal{D}$. For more details, see Refs. [55, 81].

Assuming that an inclusion function $[\lambda]$ for λ in (5.95) is available, one may use the SIVIA algorithm [56] to evaluate an inner approximation $\underline{\mathbb{S}}_q$ and an outer approximation $\overline{\mathbb{S}}_q$ of $\mathbb{S}_q$ comprising unions of nonoverlapping boxes of $\mathbb{P}$. $\underline{\mathbb{S}}_q$ and $\overline{\mathbb{S}}_q$ are initialized as $\emptyset$.

Starting with a working last-in first-out (LIFO) list $\mathcal{W}$ of boxes to be processed, initially containing the box $[\mathbf{p}]_0 = \mathbb{P}$, SIVIA extracts a box $[\mathbf{p}]$ from $\mathcal{W}$ and applies the following tests:

- If $[\lambda]([\mathbf{p}]) \subset [N-q, N]$, then all parameters in $[\mathbf{p}]$ are consistent with at least $N-q$ measurements or more and $[\mathbf{p}]$ is stored in $\underline{\mathbb{S}}_q$ and $\overline{\mathbb{S}}_q$.
- If $[\lambda]([\mathbf{p}]) \subset [0, N-q[$, then all parameters in $[\mathbf{p}]$ are *not* consistent with $q+1$ measurements or more, and $[\mathbf{p}]$ is dropped.
- If the size of $[\mathbf{p}]$ is larger than some parameter ε, $[\mathbf{p}]$ is bisected into two subboxes $[\mathbf{p}]'$ and $[\mathbf{p}]''$, which are stored in $\mathcal{W}$.
- If the size of $[\mathbf{p}]$ is smaller than ε, it is stored into $\overline{\mathbb{S}}_q$.

One of the interesting features of this approach is that it is not necessary to specify a priori the sensors that are defective. Only the number q of erroneous data the estimator has to be robust to has to be specified. The approach considered in the algorithms GOMNE described in Ref. [57] consists of starting with $q = 0$ and increasing q until a nonempty solution set $\underline{\mathbb{S}}_q$ is obtained. Note that with this approach, the solution set $\underline{\mathbb{S}}_q$ is only guaranteed to contain the true parameter value $\mathbf{p}^*$ if the number of outliers is actually less than q.

5.3.1.2 *Idealized Robust Distributed Approach*

In this context, each sensor has to process its own measurement and information transmitted by neighboring sensors. One aims at characterizing $\mathbb{S}_q$ in a guaranteed way, as in the centralized approach.

Consider the subset of measurement indexes $J \subset [\![1, N]\!]$, and define the set

$$\mathbb{S}_q^J = \bigcup_{I \subset J, \text{card}(I) = \text{card}(J) - q} \left(\bigcap_{i \in I} \mathbb{P}_i \right), \tag{5.98}$$

of all parameters consistent with $\text{card}(J) - q$ or more measurements provided by sensors with index J, with $\mathbb{P}_i = \{\mathbf{p} \in \mathbb{P} | \mathbf{f}_i(p) \in [\mathbf{y}_i]\}$, the set of parameters consistent with the measurement provided the sensor i and $\text{card}(A)$, the cardinal number of the set A. One may easily verify that $\mathbb{S}_q = \mathbb{S}_q^{[\![1,N]\!]}$ and

$$\forall J_1 \subset J_2 \subset [\![1, N]\!] \quad \mathbb{S}_q^{J_1} \supset \mathbb{S}_q^{J_2} \supset \mathbb{S}_q. \tag{5.99}$$

Assume that a sensor has evaluated $\mathbb{S}_q^{J_1}$ and that $\mathbb{S}_q^{J_2}$ has been provided by one of its neighbors. According to (5.99), to obtain a better outer approximation of $\mathbb{S}_q$, the sensor has to compute $\mathbb{S}_q^{J_1 \cup J_2}$. If $J_1 \cap J_2 \neq \emptyset$, there is no simple relation between $\mathbb{S}_q^{J_1} \cap \mathbb{S}_q^{J_2}$ and $\mathbb{S}_q^{J_1 \cup J_2}$. Now, if $J_1 \cap J_2 = \emptyset$, one may easily prove that $\mathbb{S}_q^{J_1} \cap \mathbb{S}_q^{J_2} \supset \mathbb{S}_q^{J_1 \cup J_2}$, but both sets are not equal in general. In fact, to compute $\mathbb{S}_q^{J_1 \cup J_2}$, all $\mathbb{S}_0^{J_1}, \ldots, \mathbb{S}_q^{J_1}$ and $\mathbb{S}_0^{J_2}, \ldots, \mathbb{S}_q^{J_2}$ are needed (see Ref. [71]). Thus, each sensor has to transmit $\mathbb{S}_0^J, \ldots, \mathbb{S}_q^J$ in place of only $\mathbb{S}_q^J$.

For any subset of indexes $J \subset [\![1, N]\!]$, consider the set $\Gamma_q^J = \left\{ \mathbb{S}_0^J, \mathbb{S}_1^J, \ldots, \mathbb{S}_q^J \right\}$; see Fig. 5.19. In what follows, sensors send and receive such sets and try to compute $\Gamma_q^{[\![1,N]\!]}$ to obtain $\mathbb{S}_q^{[\![1,N]\!]} = \mathbb{S}_q$. The number of tolerated outliers q is assumed to be fixed a priori.

FIGURE 5.19

Representation of some set $\Gamma_2^J = \{\mathbb{S}_0^J, \mathbb{S}_1^J, \mathbb{S}_2^J\}$ (from Ref. [71]).

Initially, each sensor i processes its own measurement to get $\mathbb{P}_i$ and $\Gamma_q^{\{i\}} = (\mathbb{P}_i, \mathbb{P}, \ldots, \mathbb{P})$. Then, it broadcasts this first estimate to its neighboring sensors and receives similar structures. After a first round of communication, the ith sensor is able to improve its estimates as follows. If $J_1 \cap J_2 = \emptyset$ (*combination constraint*), then $\Gamma_q^{J_1}$ and $\Gamma_q^{J_2}$ can be used to get $\Gamma_q^{J_1 \cup J_2}$ by computing each $\mathbb{S}_{q'}^{J_1 \cup J_2}$ as [71]

$$\forall q' \in [\![0, q]\!] \quad \mathbb{S}_{q'}^{J_1 \cup J_2} = \bigcup_{q_1 + q_2 = q'} \mathbb{S}_{q_1}^{J_1} \cap \mathbb{S}_{q_2}^{J_2}. \tag{5.100}$$

For the next round of communication, each sensor broadcasts the best Γ_q^J (with the largest card(J)). Once all sensors have exchanged improved estimates, new improvements are possible. The two phases (estimation and communication) may be performed until convergence, that is, until all sensors have obtained $\mathbb{S}_q^{[\![1,N]\!]} = \mathbb{S}_q$, which occurs in finite time [71]. Computations may also be stopped at any time, each sensor of the network having an outer approximation of $\mathbb{S}_q$, which improves when more data are exchanged.

5.3.1.3 *Implementation Issues and Results*

In Section 5.3.1.2, sets such as $\mathbb{S}_q^J$ are assumed to be transmitted. This is not possible in general, since the shape of such sets may be quite complex. Here, external approximations $\overline{\mathbb{S}}_q^J$ are considered, in order to be able to determine a guaranteed outer approximation of the solution $\overline{\mathbb{S}}_q$. Such outer approximations may consist of any simple geometric shape, such as ellipsoids, polytopes, or unions of nonoverlapping boxes or subpavings [55], which are considered here.

A single subpaving of $\mathbb{P}$ can be used to represent $\overline{\Gamma}_q^J$. A subpaving may be easily described by a binary tree. Each leaf of the tree has to be labeled with ℓ to indicate that the corresponding box is a subset of $\overline{\mathbb{S}}_\ell^J$. This subpaving implementation of $\overline{\Gamma}$ as labeled binary trees allows computation for each sensor to become only unions and intersections of subpavings.

Each node stores intermediate results with $\overline{\Gamma}_q^J = (\overline{\mathbb{S}}_0^J, \overline{\mathbb{S}}_1^J, \ldots, \overline{\mathbb{S}}_q^J)$ in place of Γ_q^J and transmitted to neighboring nodes. Efficient routing protocols, such as *Optimized Link State Routing Protocol* [20], may be used to satisfy the recombination constraint more easily. The nodes of the network use *multipoint distribution relays* for transmission. For a given node, only a subset of its neighbors relays its message. The selection of multipoint distribution relays is adaptive and done in real time. For more details, see Ref. [20]. We impose here a dynamic hierarchical structure where a sensor selects its multipoint distribution relays and sends its sets $\overline{\Gamma}$ only to its multipoint distribution relays.

A simple single source localization in a 2D-environment with a WSN is considered; see Fig. 5.20. A network of nine regularly spaced nodes is considered. Each sensor measures the power it receives

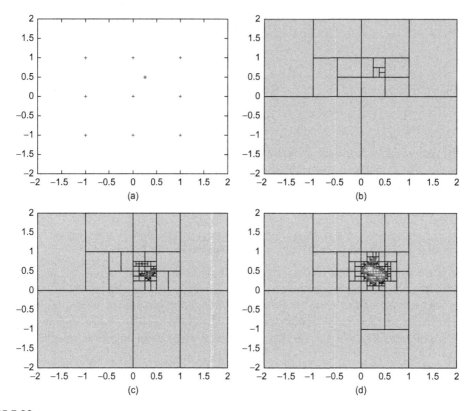

FIGURE 5.20

Considered network of nine sensors (blue) and one source (red) (a); solution obtained using a centralized estimation technique for $q = 0$ (b), $q = 1$ (c), and $q = 2$ (d); all plots are in $[-2, 2]^2$ (from Ref. [71]).

from the source. All measurement errors are bounded: for a received power y_i by the ith sensor, the noise-free measurement is assumed to belong to the interval $[y] = \left[\frac{y}{w}, yw\right]$ with $w = 1.7$ (this corresponds to a quite large relative measurement noise). Two outliers are introduced by hand, concerning sensors 4 and 6. The location of the source $\mathbf{p}^* = (\theta_1, \theta_2)$ has then to be estimated.

The following measurement model is considered for the ith sensor:

$$y_{m,i} = \frac{P_0}{d((\theta_1, \theta_2), (\theta_{1i}, \theta_{2i}))^\eta}, \tag{5.101}$$

where $(\theta_{1i}, \theta_{2i})$ is the location of the sensor i, and where $d(P_1, P_2)$ is the distance between P_1 and P_2. Moreover, $P_0 = 1$ and $\eta = 2$ are assumed to be known.

For a number of outliers $q \in \{0, 1, 2\}$, the centralized robust bounded-error estimator for (θ_1, θ_2) provides the results represented in Fig. 5.20. In distributed approach, Fig. 5.21 shows the estimates obtained by sensor 4. The sets of $\overline{\Gamma}_q$ are outer approximations of sets of Γ_q.

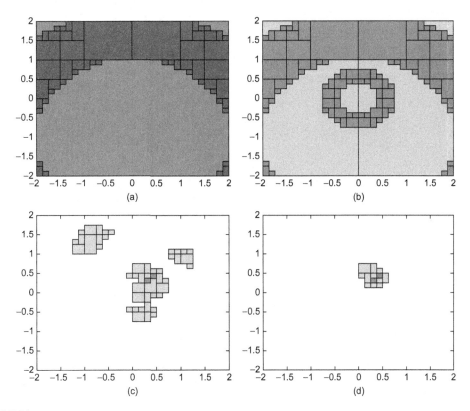

FIGURE 5.21

Estimates $\overline{\Gamma}_2^J$ available at sensor 4 using the described distributed estimator, with $\overline{\mathbb{S}}_0^J$ in dark grey, $\overline{\mathbb{S}}_1^J$ in grey, and $\overline{\mathbb{S}}_2^J$ in light grey, all represented in the box $[-2, 2]^2$; initial estimate $\overline{\Gamma}_2^{\{4\}}$ (a), $\overline{\Gamma}_2^{\{4,5\}}$ (b), $\overline{\Gamma}_2^{\{2,4,5,6,8\}}$ (c), and final estimate $\overline{\Gamma}_2^{[1,N]} = \Gamma_2$ (d) (from Ref. [71]).

With reference to Fig. 5.22, by increasing the number of nodes of the network, one may evaluate the number of iterations required until convergence. The complexity appears linear with the number of sensors. The described guaranteed robust bounded-error distributed estimation algorithm is robust to any number q of outliers. It is able to provide at each sensor of the WSN an outer approximation of the set of all values of the parameter vector, which are consistent with all except q measurements, or more, the model structure and the noise bounds. It is not necessary to specify a priori measurements that are deemed as outliers. The maximum number of outliers allowed by the estimator has to be specified a priori. If labeled trees are used to represent subpavings, themselves used to describe $\overline{\Gamma}$, exchanged between sensors, the complexity of these structures is not affected by the value of q. The complexity of the algorithm has to be evaluated more carefully. The sets $\overline{\Gamma}$ are quite complex, and their transmission may require some resources. One could imagine an alternative way to provide a robust estimator by exchanging measurements within clusters and exchange estimates between cluster heads to optimize the amount of data to be exchanged within the network.

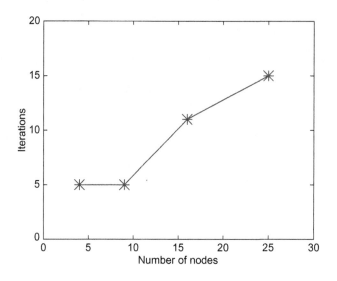

FIGURE 5.22

Evolution of convergence time with number of sensors (from Ref. [71]).

5.3.2 Projections onto Convex Sets (POCS) Algorithms

In localization systems, a characteristic signal is detected and measured by a set of sensors with fixed or known positions, and the positioning problem is then often formulated as a weighted least-squares (WLS) optimization problem. Local minima and saddle points in the WLS objective function complicate the optimization procedure, since straight forward gradient-descent-based methods tend to converge erroneously, occasionally producing estimates with large errors.

To tackle the problem with cost function local minima, a positioning algorithm based on POCS [18] was proposed some years ago by Blatt and Hero [10, 11]. As anticipated in Section 3.3, the algorithm is both robust to local minima in the objective function of low complexity, and has a computation which is possible to distribute over the nodes in, for example, a sensor network. In Refs. [10, 11], distance estimates based on received signal strength (RSS) measurements were considered, but the algorithm is applicable in any positioning application where sensor–source distance estimates are available. This method, henceforth referred to as circular POCS, was shown to perform well in the case of a source node located inside the sensor convex hull, defined by the outer perimeter nodes in the network, but experiences problems in locating sources outside this convex hull and requires knowledge of channel parameters such as path loss and transmission power. In this section, we give a brief overview of the classical circular POCS algorithm and then present two recent variants or augmentations of the POCS method proposed by Blatt and Hero. One is called hyperbolic POCS and the other is called line POCS.

Hyperbolic POCS augments the classical circular POCS method since it is also capable of locating sources outside the sensor convex hull. It therefore presents a strong complement to the method presented in Refs. [10, 11]. In addition, hyperbolic POCS allows for distance estimates with a constant unknown bias, for instance the class of pseudo time-of-arrival (pTOA) measurements [4].

In general, positioning algorithms that are based on distance estimates derived from RSS measurements need to know or somehow estimate channel parameters (such as path-loss exponent and transmission power). In practice, however, the transmission power and path-loss exponent may be unknown or difficult to estimate accurately. Line POCS relaxes assumptions on a priori known path-loss exponent and transmission power to only assume that they are similar for all links (but unknown). Line POCS is therefore able to operate in a wider range of applications and scenarios than circular and hyperbolic POCS (including scenarios where transmission power and path loss are unknown and difficult to estimate accurately). Numerical simulations presented at the end of this section show that line POCS is able to achieve quite accurate positioning despite these relaxed assumptions.

Let N sensors with known coordinates $\left\{\mathbf{z}_i = [z_{i,1}, z_{i,2}]^T \in \mathbb{R}^2\right\}_{i=1}^N$, called anchor nodes or reference node, be located in a region. Assume a vector $\mathbf{m}$ of M noisy measurements is made by these sensors. Measurements are in some way a function of the distances $\left\{d_i(\mathbf{x}_0) = \|\mathbf{x}_0 - \mathbf{z}_i\|\right\}_{i=1}^N$, $\|\cdot\|$ denoting Euclidean norm, between sensors and the source node, located at the unknown coordinates $\mathbf{x}_0 = [x_{0,1}, x_{0,2}]^T \in \mathbb{R}^2$. For instance, RSS measurements r_i are often modeled as being inversely proportional to the distance between the source and sensor, raised to a path-loss exponent β [95], that is, $r_i \propto d_i^{-\beta}(\mathbf{x}_0)$, while pTOA measurements t_i are often modeled as directly proportional to the distance, but include an unknown clock-offset Δ [97], that is, $t_i \propto d_i(\mathbf{x}_0) + \Delta$. Let the positive definite covariance matrix of $\mathbf{m}$ be $\mathbf{V}$, and its expected value $\mathbb{E}\{\mathbf{m}\} = \mu(\mathbf{x}_0)$. A weighted least-squares estimator of $\mathbf{x}_0$ from $\mathbf{m}$ is

$$\hat{\mathbf{x}}_0 = \arg\min_{\mathbf{x}} \left[\mathbf{m} - \mu(\mathbf{x})\right]^T \mathbf{V}^{-1} \left[\mathbf{m} - \mu(\mathbf{x})\right]. \tag{5.102}$$

As noted above, the objective function in (5.102) often contains many local minima and saddle points. Such local minima may in some cases be a limiting factor in trying to reach the performance bounds discussed in Section 5.4.1.

In Ref. [11], the circular POCS method is used to derive an approximation to (5.102), that is, an ML approximation in the case of zero-mean Gaussian measurement noise. Circular POCS is shown to produce promising results in terms of robustness, convergence, and performance, while offering low complexity. However, circular POCS, as described in Ref. [11], only yields good performance in terms of robustness and positioning error if the source is located inside the sensor convex hull $\mathcal{H}$, defined by

$$\mathcal{H} = \left\{\mathbf{x} \in \mathbb{R}^2 : \mathbf{x} = \sum_{i=1}^N \alpha_i \mathbf{z}_i, \alpha_i \geq 0, \sum_{i=1}^N \alpha_i = 1\right\}. \tag{5.103}$$

Also, circular POCS cannot incorporate an unknown measurement bias, as would be present if, for example, pTOA measurements were available. Further, if POCS is based on distance estimates derived from RSS measurements, then some a priori information about propagation parameters is needed.

We first give a brief overview of the circular POCS method [10, 11, 18] and then present the hyperbolic POCS and line POCS variants that make POCS-based positioning applicable over a wider range of scenarios and assumptions.

Let $\mathcal{N}_r$ be the set of sensors that are capable of measuring RSS, synchronized source time of arrival, or round-trip time measurements, from which distances $\{d_i(\mathbf{x}_0)\}_{i \in \mathcal{N}_r}$ may be directly estimated. If measurement errors are assumed to be zero mean and uncorrelated between measurements, the objective function in (5.102) is a weighted sum of squares. Each term in this sum attains its minimum on the circumference of a disc $\mathcal{D}_i$, defined by

$$\mathcal{D}_i = \left\{ \mathbf{x} \in \mathbb{R}^2 : d_i(\mathbf{x}) \leq \hat{d}_i \right\},$$

where $\hat{d}_i$ is an estimate of $d_i(\mathbf{x}_0)$, derived from the signal model. An estimator of $\mathbf{x}_0$, as proposed in Ref. [11] and derived in Ref. [18], is any point in the intersection $\mathcal{D}$ of the convex sets $\mathcal{D}_i$,

$$\hat{\mathbf{x}}_0 \in \mathcal{D} = \bigcap_{i \in \mathcal{N}_r} \mathcal{D}_i, \tag{5.104}$$

or, if $\mathcal{D}$ is empty, which might be the case due to measurement noise,

$$\hat{\mathbf{x}}_0 = \arg\min_{\mathbf{x}} \sum_{i \in \mathcal{N}_r} \| \mathbf{x} - \mathcal{P}_{\mathcal{D}_i}(\mathbf{x}) \|, \tag{5.105}$$

where $\mathcal{P}_{\mathcal{D}_i}(\mathbf{x})$ is the unique orthogonal projection of $\mathbf{x}$ onto the convex set $\mathcal{D}_i$,

$$\mathcal{P}_{\mathcal{D}_i}(\mathbf{x}) = \arg\min_{\mathbf{y} \in \mathcal{D}_i} \| \mathbf{x} - \mathbf{y} \|.$$

The POCS method for distance estimates, as derived in Ref. [18, ch. 5] is, as follows:

1. Initialization $\mathbf{x}^0$ is arbitrary.
2. $\mathbf{x}^{\nu+1} = \mathbf{x}^\nu + \lambda_\nu \left[\mathcal{P}_{\mathcal{D}_{\iota(\nu)}}(\mathbf{x}^\nu) - \mathbf{x}^\nu \right]$,
   where $\{\lambda_\nu\}_{\nu=0}^\infty$ are relaxation parameters and $\iota(\nu) = \nu \mod |\mathcal{N}_r|$, where $|\mathcal{N}_r|$ denotes the cardinality of set $\mathcal{N}_r$.

The relaxation parameters $\{\lambda_\nu\}_{\nu=0}^\infty$ are first set equal to one, and, after a given number ν_0 of iterations, are decreased as

$$\lambda_\nu = \left\lceil \frac{\nu - \nu_0 + 1}{|\mathcal{N}_r|} \right\rceil^{-1}, \tag{5.106}$$

where $\lceil a \rceil$ denotes the smallest integer greater than or equal to a. See Ref. [10] for a discussion on these relaxation parameters. Further, the projection function is

$$\mathcal{P}_{\mathcal{D}_i}(\mathbf{x}) = \begin{cases} \mathbf{x}, & \text{if } d_i(\mathbf{x}) \leq \hat{d}_i \\ \mathbf{z}_i + \left[\hat{d}_i \cos\phi \quad \hat{d}_i \sin\phi \right]^T, & \text{otherwise,} \end{cases}$$

where ϕ is the argument of the complex number $c = (x_1 - z_{i,1}) + j(x_2 - z_{i,2})$, which is denoted by $\phi = \angle c$.

Let $\mathcal{N}_\tau$ be the set of sensors that are capable of exchanging distance estimates and forming differences $\tau_{i,j} = \hat{d}_i - \hat{d}_j$. Such differences are often formed to cancel an unknown but constant bias in distance estimates [97]. It is assumed that $|\tau_{i,j}| < \| \mathbf{z}_i - \mathbf{z}_j \|$, that is, that the difference satisfies the triangle inequality. Each difference equation that satisfies the triangle inequality defines a hyperbola in the plane with foci in $\mathbf{z}_i$ and $\mathbf{z}_j$. To describe this hyperbola in standard form, we define an invertible rotation and translation transform $T_{i,j}(\mathbf{x})$ of a point $\mathbf{x}$ relative to sensors i and j as

$$T_{i,j}(\mathbf{x}) = \mathbf{G}_{i,j}(\mathbf{x} - \mathbf{b}_{i,j}),$$

where

$$G_{i,j} = \begin{bmatrix} \cos\theta & \sin\theta \\ -\sin\theta & \cos\theta \end{bmatrix}, \ \theta = \angle c_{i,j},$$

$c_{i,j} = z_{j,1} - z_{i,1} + j(z_{j,2} - z_{i,2})$, and $\mathbf{b}_{i,j} = (\mathbf{z}_i + \mathbf{z}_j)/2$. We denote the inverse of $T_{i,j}(\cdot)$ by $T_{i,j}^{-1}(\cdot)$. The hyperbola divides the plane into two disjoint sets, of which one set is convex (Fig. 5.23). Let $\mathcal{Y}_{i,j}$ be the unbounded convex set defined by the hyperbola corresponding to the difference $\tau_{i,j}$ of, possibly biased, distances estimated by sensors i and j,

$$\mathcal{Y}_{i,j} = \left\{ \mathbf{x} \in \mathbb{R}^2 : T_{i,j}^T(\mathbf{x}) \begin{bmatrix} \frac{1}{a^2} & 0 \\ 0 & -\frac{1}{b^2} \end{bmatrix} T_{i,j}(\mathbf{x}) \leq 1 \right\},$$

where $a = \tau_{i,j}/2$ and

$$b = \sqrt{\frac{\|\mathbf{z}_i - \mathbf{z}_j\|^2}{4} - a^2} \tag{5.107}$$

define the semimajor and semiminor axes of the rotated and translated hyperbola. Points on the hyperbola itself, where equality holds in (5.107), may be parameterized by

$$\mathbf{h}(t) = T_{i,j}^{-1}\left(\begin{bmatrix} a\cosh t & b\sinh t \end{bmatrix}^T \right), \ t \in \mathbb{R}.$$

The projection of $\mathbf{x}$ onto $\mathcal{Y}_{i,j}$, illustrated in Fig. 5.23, is given by

$$\mathcal{P}_{\mathcal{Y}_{i,j}}(\mathbf{x}) = \begin{cases} \mathbf{x}, & \text{if } \mathbf{x} \in \mathcal{Y}_{i,j} \\ \mathbf{h}(t_p), & \text{otherwise,} \end{cases}$$

where

$$t_p = \arg\min_t \left\| \mathbf{x} - \mathbf{h}(t) \right\|^2. \tag{5.108}$$

The solution to (5.108) is found in closed form from the natural logarithm of the real and non-negative roots[5] of the polynomial

$$g(\ell) = (a^2 + b^2)\ell^4 - (2a[T_{i,j}(\mathbf{x})]_1 + 2b[T_{i,j}(\mathbf{x})]_2)\ell^3$$
$$+ (2a[T_{i,j}(\mathbf{x})]_1 - 2b[T_{i,j}(\mathbf{x})]_2)\ell - a^2 - b^2,$$

where $[\mathbf{x}]_i$ denotes the ith element of vector $\mathbf{x}$. Hyperbolic POCS now follows the same iteration, using the same set of relaxation parameters, as that of circular POCS described above.

Circular POCS has, as mentioned above, been shown to produce good results when the source is located inside the sensor convex hull $\mathcal{H}$ and somewhat poorer results for sources outside $\mathcal{H}$, as exemplified in Figs 5.24 and 5.25. The reason for this behavior is that the area of $\mathcal{D}$, denoted $A(\mathcal{D})$, becomes large if the source is located far from $\mathcal{H}$.

---

[5] If more than one real non-negative root is found, we select the root that gives the smallest norm in (5.108).

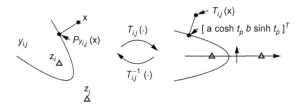

FIGURE 5.23

Projection of $\mathbf{x}$ onto $\mathcal{Y}_{i,j}$ (from Ref. [130]).

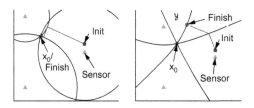

FIGURE 5.24

Sources inside convex hull, circular and hyperbolic POCS (from Ref. [130]).

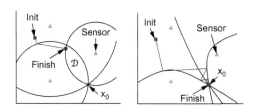

FIGURE 5.25

Sources outside convex hull, circular and hyperbolic POCS (from Ref. [130]).

Numerical simulations show that hyperbolic POCS, while occasionally exhibiting poor performance for sources inside the convex hull, is often capable of locating sources outside $\mathcal{H}$, as exemplified in Fig. 5.25. The reason is, again, the area and shape of $\mathcal{Y} = \bigcap_{(i,j) \in \mathcal{N}_\tau} \mathcal{Y}_{i,j}$, which does not grow in the same way $\mathcal{D}$ does as sources move away from $\mathcal{H}$. From the properties of the intersection operator, we have

$$A(\mathcal{D} \cap \mathcal{Y}) \leqslant A(\mathcal{D}).$$

Hyperbolic POCS should therefore be a good complement to the circular POCS method, when sources outside the convex hull can be expected. An iterative method that forms and projects onto both bounded circular and unbounded hyperbolic sets will henceforth be denoted hybrid POCS (HPOCS).

Both the circular and hyperbolic POCS methods described above require distance estimates (possibly biased in the case of hyperbolic POCS) to define convex sets, that is, hyperbolas or discs. When the only measurements available are RSS measurements, and the path-loss exponent and transmission powers are unknown, these POCS methods are clearly not suitable for positioning purposes. We can, however, relax the assumption on known values for both β and P_0 and instead only assume that they are similar for all links. Instead of estimating distances from RSS measurements, for example using the estimator in (5.113), we base the positioning algorithm on the assumption that RSS on average decreases with distance. Based on this assumption, we can compare two received powers and deduce that the source node is more likely to be closer to the anchor node with the greater received power. As an example, let p_k for $k = 1, 2, \ldots, N(N-1)/2$ denote all possible anchor node pairs $(\mathbf{z}_i, \mathbf{z}_j)$ such that $1 \le i < j \le N$. Given a pair $p_k = (\mathbf{z}_i, \mathbf{z}_j)$, we can divide $\mathbb{R}^2$ into two half-planes whose boundary is a bisector that is perpendicular to the line that joins the nodes $\mathbf{z}_i$ and $\mathbf{z}_j$. We are primarily interested in the half-plane that includes the anchor node with the highest RSS measurement. Formally, we define

$$
\mathcal{H}_k =
\begin{cases}
\left\{ \mathbf{x} \in \mathbb{R}^2 : \frac{(\mathbf{x}-\mathbf{z}_i)^T (\mathbf{z}_j-\mathbf{z}_i)}{\|\mathbf{z}_j-\mathbf{z}_i\|} \ge \frac{\|\mathbf{z}_j-\mathbf{z}_i\|}{2} \right\}, & P_i \le P_j \\
\left\{ \mathbf{x} \in \mathbb{R}^2 : \frac{(\mathbf{x}-\mathbf{z}_i)^T (\mathbf{z}_j-\mathbf{z}_i)}{\|\mathbf{z}_j-\mathbf{z}_i\|} < \frac{\|\mathbf{z}_j-\mathbf{z}_i\|}{2} \right\}, & P_j < P_i.
\end{cases}
\tag{5.109}
$$

From (5.109), one can deduce that the source node with unknown position most likely is to be found in the intersection of sets $\mathcal{H}_k, k = 1, 2, \ldots, N(N-1)/2$, that is,

$$
\hat{\mathbf{x}}_0 \in \mathcal{A} = \bigcap_{k=1}^{N(N-1)/2} \mathcal{H}_k.
\tag{5.110}
$$

Assuming that $\mathcal{H}_k$ is nonempty, we note similarities to (5.104) and conclude that a POCS method can be used to find a point in the intersection (which in this case is a convex polygonal region [94]). The point found by POCS can then be taken as an estimate of the unknown node position. This concept makes up the core of the line POCS positioning algorithm. Figure 5.26 shows the intersection of sets $\mathcal{H}_k$ for three anchor nodes that contain the position of the source node.

In a practical implementation of the line POCS method, we have considered the regular half-planes described above and also a variant based on a shifted bisector. In the first approach, we try to find

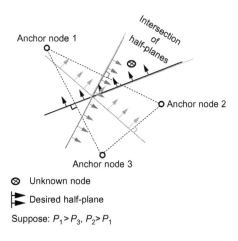

FIGURE 5.26

Intersection of three half-planes that contain the unknown source node (from Ref. [131]).

a point on the perimeter of the intersection of half-planes. To do this, we use projections onto the bisector that is perpendicular to the line joining the anchor nodes of a given pair. The result of this projection is a point that has the same distance to each of the two anchor nodes in the pair, even if these nodes have measured widely different RSSs. Since the main assumption made about RSS is that it decreases in some fashion with distance, the anchor node with a greater received power should be located closer to the source than the other anchor, not at the same distance. We therefore slightly alter the first method to comply better with the assumptions made about RSS measurements. Effectively, this means that we should look for some area different from the half-plane defined by the bisector line. Since the source node is more likely to be located somewhere inside the half-plane corresponding to greater RSS, performance could be improved if we try to search for a point inside that half-plane. In the second variant of line POCS, we therefore define half-planes $\mathcal{H}'_k$ similar to $\mathcal{H}_k$, but where the boundary now goes through the node with the strongest received power, still perpendicular to the line joining the two nodes of the pair:

$$\mathcal{H}'_k = \begin{cases} \left\{ \mathbf{x} \in \mathbb{R}^2 : \frac{(\mathbf{x}-\mathbf{z}_j)^T(\mathbf{z}_i-\mathbf{z}_j)}{\|\mathbf{z}_j-\mathbf{z}_i\|} \leq 0 \right\}, & P_i \leq P_j \\ \left\{ \mathbf{x} \in \mathbb{R}^2 : \frac{(\mathbf{x}-\mathbf{z}_i)^T(\mathbf{z}_j-\mathbf{z}_i)}{\|\mathbf{z}_j-\mathbf{z}_i\|} \leq 0 \right\}, & P_j < P_i. \end{cases} \tag{5.111}$$

In the second approach, because we project onto the half-plane that is likely to contain the source node, we can use relaxation parameters similar to those used in circular POCS. In the first approach, using relaxation parameters might bring a point into the "wrong" half-plane, which is not desired. Figure 5.27 shows the basic concepts of two methods of projection.

The projections needed for line POCS 1 and 2 are shown in Fig. 5.28. Here it is assumed that $P_i \geq P_j$. Note that $\mathbf{z}_m = (\mathbf{z}_i + \mathbf{z}_j)/2$ for line POCS 1 and $\mathbf{z}_m = \mathbf{z}_i$ for line POCS 2.

Simulation results (presented in part below) show that both algorithms converge to a point on the intersection of half-planes in just a few iterations.

To provide some numerical examples, we let a set of N sensors be located in a region. Assume a subset $\mathcal{N}_r$ of these sensors are capable of estimating distance by measuring the power of signals received from a source [23, 90, 95]. The RSS measurement P_i at the ith sensor, measured in decibels,

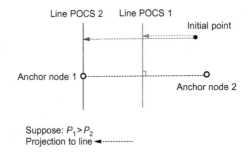

FIGURE 5.27

Projection onto half-planes for the two variants of line POCS (from Ref. [131]).

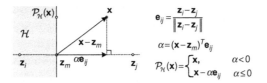

FIGURE 5.28

Projection onto the half-plane $\mathcal{H}$.

is given by

$$P_i = P_0 - 10\beta \log \frac{d_i(\mathbf{x}_0)}{d_0} + w_{P_i}, \tag{5.112}$$

where P_0 is the received signal power in decibels at calibration distance d_0, and β is a path-loss exponent [95], which is assumed known to all algorithms except line POCS. Measurements are affected by log-normal shadowing, that is, $w_{P_i} \sim \mathcal{N}(0, \sigma_{P_i}^2)$. Any measurement noise is assumed insignificant in comparison to the log-normal shadowing and is therefore incorporated in the shadowing term. Throughout this section, $d_0 = 1$ m.

The ML estimate of distance $d_i(\mathbf{x}_0)$, given P_i, is

$$\hat{d}_i = d_0 10^{(P_0 - P_i)/10\beta}. \tag{5.113}$$

This estimate is readily shown to be slightly biased [90]. However, its squared bias is small in comparison to its variance.

Further, let $\mathcal{N}_\tau$ be the subset of sensors that are capable of forming differences between distance estimates. The difference between distance estimates at sensors i and j is

$$\tau_{i,j} = \hat{d}_i - \hat{d}_j, \tag{5.114}$$

with $\hat{d}_i$ given by (5.113). If the same measurement P_i is used in many difference equations, then these differences will be correlated in the obvious way.

It should be noted that the difference operation in (5.114) cancels any constant bias in distance estimates. The hyperbolic POCS method may therefore also be implemented in a positioning system based on pseudo time-of-arrival measurements [97].

A network layout with N sensors, uniformly distributed over a square area with side S, denoted $\mathbf{z}_i \sim \mathcal{U}([-S/2, S/2], [-S/2, S/2])$, was generated. One source node was randomly placed in the area, and a set of RSS measurements was generated according to (5.112). For each set of simulation parameters, 4000 networks were generated. In the HPOCS method, the projection order was such that projections were first made onto each hyperbolic set, then onto each circular set, before starting over and projecting onto hyperbolic sets again.

Two scenarios are investigated: in scenario A, measurements are generated according to (5.112); in scenario B, measurements are, with probability P_0, affected by a calibration error $\eta < 0$, in addition to the log-normal shadowing, as

$$P_i = P_0 - 10\beta \log \frac{d_i(\mathbf{x}_0)}{d_0} + \eta + w_{P_i}, \tag{5.115}$$

causing a positive bias in distance estimates that varies in magnitude with distance. This type of calibration error could arise, for example, if the path between the source and sensor contains unforeseen obstacles, or if there is an unexpected decrease in transmitted power.

The results of the circular, hyperbolic, and hybrid POCS algorithms are compared to those of a weighted least-squares optimization method (5.102), approximated by Matlab's gradient-based lsqnonlin routine, both randomly initialized and initialized by the HPOCS output. Results are plotted as the estimated cumulative density function (CDF) of the absolute positioning error $\|\hat{x}_0 - x_0\|$. Relaxation parameters for all POCS methods are first set equal to one for v_0 iterations and are then decreased as in (5.106). All difference estimates $\tau_{i,j}$ are tested with respect to the triangle inequality, with an added margin K; for example, if $|\tau_{i,j}| + K > \|z_i - z_j\|$, difference $\tau_{i,j}$ is not used in iterations. This margin was chosen empirically and may be further optimized. Simulation parameters common to scenarios A and B are given in Table 5.2. It should be noted that no major effort has been put into optimizing the performance of the gradient-based method. However, it is our opinion that obtained results illustrate well the behavior of gradient-based optimization methods in the presence of multiple local minima. Networks with more sensors would contain more local minima, causing randomly initialized gradient-based methods to converge erroneously more often. In fact, a randomly initialized WLS approach may not even be feasible in a large network.

In this first example, all sensors measure RSS and can communicate their measurements to all other sensors, that is, $|\mathcal{N}_r| = |\mathcal{N}_\tau| = N$. The variance of log-normal shadowing is $\sigma_{P_i}^2 = 4$ dB^2 $\forall i$, $K = 2$, and the source has random coordinates $x_0 \sim \mathcal{U}([-10, 10], [-10, 10])$.

The simulation results, plotted in Fig. 5.29, appear typical for a scenario in which the source may appear outside $\mathcal{H}$. The lsqnonlin routine is a closer approximation to the ML solution when convergence to a global minimum occurs, but suffers from outliers due to convergence to local minima. From the simulation results, we note that randomly initialized WLS converges to local minima in approximately 30% of cases, but this fraction would increase significantly if more sensors were added to the network. In this scenario, sensors do not always surround the source node, explaining the difference in performance between circular and hyperbolic POCS. It is interesting to note the performance gain when the two POCS methods are combined into the HPOCS method. It is also deduced that performance may be increased if a gradient-descent-based method is initialized by the HPOCS position estimate. However, this gain in performance comes at the cost of a significant increase in complexity.

In Fig. 5.30, we use the same simulation parameters, except now the source node is more often located closer to the origin, $x_0 \sim \mathcal{U}([-3, 3], [-3, 3])$. Two effects are immediately noted. First, as expected, the circular POCS method now outperforms the hyperbolic POCS method. Second, overall

---

Table 5.2 Common Simulation Parameters (scenarios A and B)

$N = 10$, $P_0 = 30$ dB, $\beta = 4$, $z_i \sim \mathcal{U}([-10, 10], [-10, 10])$
$v_0 = 10 N$ unrelaxed iterations
$20 N$ relaxed iterations, $\lambda_v = \lceil \frac{v - v_0 + 1}{N} \rceil^{-1}$
Random initialization $x_{\text{init}} \sim \mathcal{U}([-10, 10], [-10, 10])$

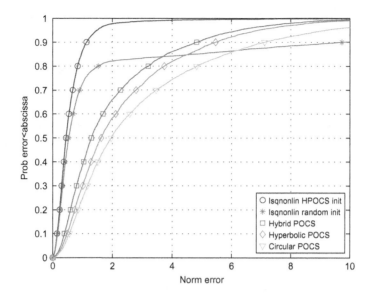

FIGURE 5.29

CDF of $\|\hat{\mathbf{x}}_0 - \mathbf{x}_0\|$, $\mathbf{x}_0 \sim \mathcal{U}([-10, 10], [-10, 10])$ (from Ref. [130]).

performance is significantly improved. This is due to the favorable geometrical conditions, allowing for a low geometric dilution of precision (GDOP) [4].

The effect of calibration errors on the POCS methods and on the lsqnonlin method is exemplified in Fig. 5.31, where $\sigma_{P_i}^2 = 3$ dB^2 $\forall i$, $\eta = -8$ dB with $P_0 = 0.7$ and $\eta = 0$ dB with probability $1 - P_0$, while $\mathbf{x}_0 \sim \mathcal{U}([-10, 10], [-10, 10])$. Margins for HPOCS and hyperbolic POCS were set at $K = 7$ and $K = 2$, respectively.

It is seen that a large negative error in RSS measurements does not penalize the POCS methods as much as the WLS approximation. It is also interesting to note that circular POCS performs well in approximately the first 50% of the error probability density function mass, where enough sensors surround the source node, because distance estimates with a large positive bias are implicitly discarded in the circular POCS method. If some distance estimates have a large bias, while the source node is located well inside $\mathcal{H}$, the convex sets associated with outlier measurements will be large, and therefore not affect the POCS iteration. Unfortunately, the same cannot be said for distance estimates with a negative bias, which will have a degrading effect on the performance of circular POCS methods. Compare Fig. 5.24 and (5.105).

Also, if a large bias, regardless of sign, is present in some distance estimates, difference estimates will often not fulfill the triangle inequality and are therefore discarded. If biases of similar magnitude and equal sign are present in a majority of distance estimates (not the case shown in Fig. 5.31), biases are often reduced in the difference operation. Indeed, if all distance estimates suffer from a constant bias or distance calibration error, hyperbolic POCS would remain unaffected due to a total cancelation of this bias. It should be noted that the WLS approximation is naive and that a fairer comparison would be to treat $\{\eta_i\}$ as nuisance parameters in the WLS estimator.

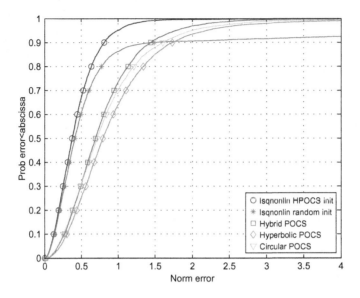

FIGURE 5.30

CDF of $\|\hat{\mathbf{x}}_0 - \mathbf{x}_0\|$, $\mathbf{x}_0 \sim \mathcal{U}([-3,3],[-3,3])$ (from Ref. [130]).

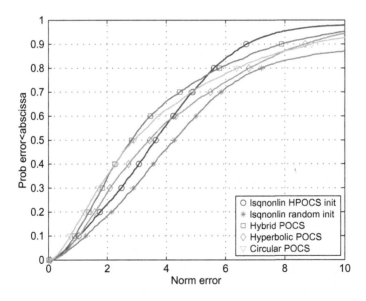

FIGURE 5.31

CDF of $\|\hat{\mathbf{x}}_0 - \mathbf{x}_0\|$ for scenario B (calibration error) (from Ref. [130]).

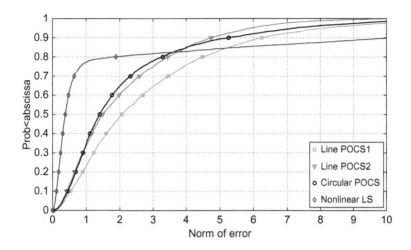

FIGURE 5.32

CDF of line POCS and circular POCS $z \sim \mathcal{U}([-10, 10], [-10, 10])$, $\sigma^2 = 4$ dB, $\beta = 4$, $P_0 = 30$ dB.

In the rest of this section, we assume that all algorithms except line POCS know the path-loss exponent and the transmission power and compare line POCS to circular POCS using the signal model in (5.115). Simulation parameters are the same as those given in Table 5.2, except for the number of sensors that was set to $N = 15$.

In Fig. 5.32, the CDF of the positioning error for WLS, circular POCS and line POCS is plotted. Like in previous simulations, WLS has superior performance as long as it converges to the global minimum, which in this case occurs in approximately 80% of cases. When WLS converges to a local minimum the estimation error is significant, and the position estimates are not reliable. It can be seen that the performance of line POCS (second approach) is quite close to that of circular POCS, while regular line POCS has the worst performance in this example.

As discussed above, when the source node is outside the sensor perimeter (or convex hull), circular POCS is not able to locate the unknown node correctly. We can make a similar statement for line POCS: If the intersection of half-planes is often large (and in some cases even unbounded), the performance of the algorithm decreases considerably.

One important performance metric in iterative algorithms is the rate of convergence. In some applications using POCS for positioning, it can be crucial that the number of iterations is small, in order not to expend too much energy. This can, for instance, be the case if an iterative algorithm is used for positioning in a sensor network where energy is a scarce resource that must be conserved. In the next simulation, we compare the positioning MSE of line POCS and circular POCS versus the number of projections. In this simulation, the parameters are the same as in previous simulations except for $N = 6$. As can be seen in Fig. 5.33, circular POCS converges somewhat faster than line POCS, while line POCS2 shows a lower residual MSE after convergence. It is again confirmed that line POCS2 outperforms line POCS1 on all accounts. For further comparison, we show the results from another

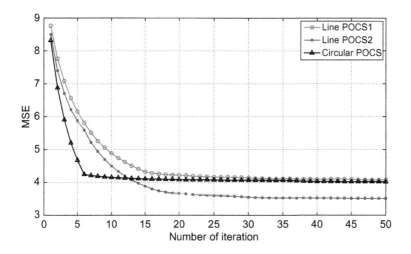

FIGURE 5.33

Convergence rate of line POCS and circular POCS ($\mathbf{z} \sim \mathcal{U}([-10, 10], [-10, 10])$, $\sigma^2 = 4$ dB, $\beta = 4$, $P_0 = 30$ dB).

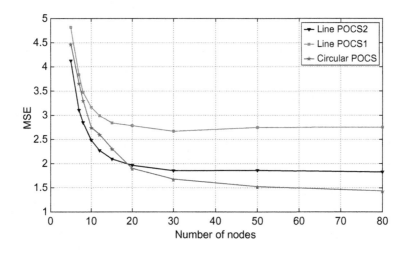

FIGURE 5.34

MSE versus number of nodes ($\mathbf{z} \sim \mathcal{U}([-10, 10], [-10, 10])$, $\sigma^2 = 4$ dB, $\beta = 4$, $P_0 = 30$ dB).

simulation that investigates the impact of node density, that is, the number of anchor nodes per unit area. The results of the simulation for line POCS and circular POCS are shown in Fig. 5.34. For low anchor node density networks, the performance of line POCS2 is in fact better than circular POCS, while in moderate to high density networks circular POCS outperforms line POCS.

5.4 **COOPERATIVE POSITIONING**

As already discussed in previous chapters, one common denominator in most WSN applications is the need for position or range data to accompany sensor readings. The positioning system must be energy efficient and robust to imperfections in hardware, varying multipath and fading propagation conditions, NLOS channels, and possibly also strong interference. However, in WSNs usually only few nodes are placed in known positions to reduce the installation costs, thus leaving several unknown-location devices out of the range of any reference node (due to the short transmission range). This is the case for ad hoc network scenarios as well. Thereby, conventional positioning techniques employed in cellular mobile systems or wireless local area networks (WLANs) are not appropriate, and localization needs to be carried out cooperatively by the nodes, that is, by allowing nodes in unknown locations to exchange measurements on a peer-to-peer basis (Fig. 5.35).

Several motivations support the adoption of cooperative approaches with respect to conventional ones: Large networks with large numbers of nodes may have a great potential advantage from the use of an extended number of measures, possibly in a distributed way; ad hoc wireless networks with no or few fixed nodes need cooperation for computing coordinates of all the nodes; the integration of nodes with different technological platforms (e.g., heterogeneous networks with WiMAX/4G systems, global navigation satellite system (GNSS), and wireless personal area network (WPAN) solutions) may have a substantial benefit over a cooperative approach. For example, one of the main applications of cooperative localization is in vehicular ad hoc networks (VANETs) where anticollision and warning applications are of extreme importance for decreasing one of the leading causes of death on the roads. Considerable effort has been put into this area, in research or industrial projects. Cooperation is a necessity because of the dynamic nature of vehicular ad hoc networks and the severe demand of fast response from the network and high reliability [35, 53, 88, 104].

In general, cooperative techniques may rely on the extension of the conventional approaches to the cooperative one. The problem formulation is similar to that of the noncooperative approaches. It is based on the definition of cost functions involving the estimated distances/angles among nodes. Considerable work has been done in this area (see Refs. [89, 122, 123] and references cited therein).

● Node with unknown location
△ Access point or node with known location

FIGURE 5.35

Principle of cooperative localization: Nodes with unknown positions also cooperate with the positioning algorithm.

In network-based positioning (e.g., for 2G-3G networks) or large ad hoc networks, the location of either fixed or moving nodes can be estimated by algorithms processed in reference stations or locally in all the nodes. The algorithms may be solved in a centralized way, or by iterating the exchange of measurements among nodes till convergence (distributed localization methods).

Of course, various application requirements (such as scalability, energy efficiency, and accuracy) will influence the design of cooperative localization systems. An open direction for research is the design of cooperative distributed algorithms that take into account energy, complexity, and/or rate constraints, especially for low-complexity wireless sensor networks. Real-time localization of moving agents is another important aspect of many positioning applications. A primary aspect is the adaptation of Bayesian tracking (e.g., particle filters) to distributed processing in which the analysis of the internode signaling overhead needed for message exchanging is crucial. Other crucial issues are the analysis of distributed algorithm performance in terms of convergence and accuracy and data fusion approaches combining different positioning techniques/systems. This last wide topic will be covered in Chapter 6.

Section 5.4.1 is dedicated to the description of the main design strategies for cooperative positioning algorithms, centralized and distributed.

The extension of LS-based positioning schemes is presented in Section 5.4.2, whereas Section 5.4.3 considers cooperative and distributed positioning rendered into a feasibility problem solved using projections onto convex sets (POCS). The application of different methods for cooperative positioning is analyzed.

Section 5.4.4 considers positioning using *active* and *passive* anchor nodes: A subset of anchor nodes (active) initiates the range estimation procedure based on two-way-TOA (TW-TOA), whereas the passive anchors, which can listen to both the signals from the active ones and the acknowledgments from the agent nodes, help the active nodes to enhance the target positioning accuracy due to TDOA measurements.

Finally, Section 5.4.5 introduces a different approach for distributed positioning, based on belief propagation (BP). First, the basic BP algorithm is developed, and then a more efficient implementation based on a distributed particle filter structure is presented.

5.4.1 Introduction to Cooperative Localization

This section offers an overview of the main issues for the design of cooperative positioning algorithms.

In *centralized solutions*, the measures provided by the nodes in known or unknown locations are transmitted toward a unique point that processes all the data and achieves an estimate of nodes deployment in the network. This processing node may result from the network organization (e.g. a star network) or from a particular node technology. The advantage of this solution is its simple organization (e.g. in small networks) and performance if the processing node has a large computational capability. The main drawback is the amount of transmissions to the unique processing point, which is expressed in terms of consumed energy and increased traffic load in the network.

On the other hand, in *distributed solutions* the algorithm execution occupies all the nodes, and the coordinate result is achieved locally. Distributed solutions are more attractive than conventional centralized ones, as they avoid forwarding measurements to a central processor, thus reducing the communication energy costs. In distributed approaches, sensors exchange information only with their neighbours, and the location of the unknown nodes is obtained iteratively by successive refinements.

Each node exchanges information with its neighbors and refines its estimate based on the neighbors' information, iterating the process till convergence. The advantage of this approach with respect to the centralized one is communication and energy cost saving [96]: Delivering data from each node to the central unit, processing the whole collected data, and disseminating location information back to all nodes may lead to unfeasible energy consumption. This is true especially in large networks in which many hops are needed to reach the central unit. On the other hand, distributed processing is based on (repeated) one-hop only transmissions and thus can provide significant cost reduction (clearly if the number of iterations required to obtain the desired location accuracy is not too large). A possible way to extend conventional localization methods to distributed processing is *network multilateration* or *iterative multilateration* as described in Chapter 3 [68, 89, 102].

A promising solution for distributed localization is *message-passing algorithms*, such as the well-known *belief propagation* (BP) method [50, 126]. Localization in this case is formulated as an inference problem on a graphical model that can be solved by iterative message passing [31, 101, 103, 123]. The communication network used for collaboration among the nodes is described by a connectivity graph and modeled as a Bayesian network. Each node computes a local belief of its position based on its own measurements and the marginal beliefs provided by the linked nodes. The location is estimated by refining the belief computation through iterated message exchanges. However, efficient representations of the probability densities are needed to avoid huge communication overhead for belief exchange. Neither grid-based nor Gaussian approaches are well suited to solve this problem, since the location space is too large for efficient discretization and the localization model is usually nonlinear non-Gaussian. Analytical approximations and sampled representation of the probability densities have been proposed based on the use of Gaussian mixture [50], particle filters (PFs), or Monte Carlo methods [3]. By PF, the location density can be efficiently represented by a set of nonuniform samples (particles) weighted according to their likelihood.

Distributed mobile positioning is also an open direction for research. In dynamic networks where the nodes to be localized are moving, Bayesian tracking algorithms have been used to reduce false localizations due to multipath and NLOS conditions [82]. Distributed PF algorithms have been proposed in Refs. [21, 52, 109], whereas a factor graph approach is considered in Ref. [123] . The theory of Bayesian filtering and PFs, as well as their use in positioning problems, will be addressed in Chapter 6.

The choice between centralized and distributed solutions depends primarily on the node/network technology (e.g., a processing node may not be available in ad hoc networks or WSN), the application (e.g., a relative coordinate system may be sufficient in several situations), the response latency (e.g., in large wireless networks), and the energy budget. These considerations are summarized in Table 5.3. We may observe that the design choice is driven by a trade-off between the amount of data to be transmitted in the network and the computational load that nodes technology can guarantee. Distributed or centralized localization algorithms are generally suboptimal with respect to the minimization of a global cost function, and their optimization degree is obviously related to their complexity, cost, and energy consumption.

One of the design key points for an effective use of cooperative approaches is the *energy budget* of the network. This is a crucial aspect for WSN where nodes and battery lives coincide, and it is especially true for the cooperative approach since the final budget will be affected by transmission rate and processing of data among the nodes. Transmission is usually much more expensive than data processing, and this is important in the evaluation of distributed cooperative algorithms if data traffic can be reduced.

Table 5.3 Main Factors Driving the Choice of a Centralized or Distributed Network Algorithm

Factor	Centralized Solution	Distributed Solution
Processing	Only at central unit (CU)	At all nodes
Transmission	Multihop delivery of data to and from the CU	Repeated single-hop transmission among neighbors
Response latency	Depending mainly on multihop delivery time	Depending mainly on the number of iterations
Energy budget	High consumption at and near the CU	Small consumption at each node

So the key advantage of cooperative localization strategies is related to the energy budget. Message exchange among cooperating nodes has to be evaluated with respect to the final performance advantage, and in distributed solutions, local processing and multihop routing of messages could mean less network traffic (especially less long-distance traffic) with respect to centralized solutions.

The Cramér–Rao lower bound (CRB) provides a useful means for the analysis of positioning accuracy also in cooperative networks. A CRB analysis for distributed positioning based on TOA ranging is also given in Ref. [69], for both conventional and cooperative positioning, showing the effects of clock bias on TOA measurements. In Ref. [108], the analysis of node positions error bound in TOA-based cooperative localization shows that anchors and unknown nodes are essentially equivalent in the cooperative approach: Anchors are just special nodes with infinite accuracy in the localization process.

5.4.2 Cooperative LS

In this section, we study two statistical positioning algorithms based on the LS approach to obtain a distributed version of them. Let us consider a two-dimensional cooperative network with $N + M$ sensor nodes. Suppose M agent nodes with unknown positions are randomly placed at $\mathbf{z}_i = [x_i, y_i]^T \in \mathbb{R}^2$, $i = 1, \ldots, M$, and the remaining N reference nodes are located at known positions, $\mathbf{z}_j = [x_j, y_j]^T \in \mathbb{R}^2$, $j = M+1, \ldots, N+M$. Every agent can connect to nearby reference nodes and some other agents. Let us define $\mathcal{A}_i = \{(i,j)|, \text{ agent } i \text{ can communicate with reference node } j\}$ and $\mathcal{B}_i = \{(i,j)| \ i \neq j, \text{ agent } i \text{ can communicate with agent } j\}$ as the sets of all reference and agent nodes that are connected to agent i. Suppose that sensor nodes are able to estimate distances to nodes with which they are connected, giving rise to the following observation:

$$\hat{d}_{i,j} = d(\mathbf{z}_i, \mathbf{z}_j) + \epsilon_{i,j} , \ i = 1, \ldots, M, \ j \in \mathcal{A}_i \cup \mathcal{B}_i, \tag{5.116}$$

where $d(\mathbf{z}, \mathbf{z}_j) = \|\mathbf{z} - \mathbf{z}_j\|$ is the actual distance from sensor node j to the point $\mathbf{z}$ and $\epsilon_{i,j}$ is the measurement error. We assume that measurement errors are i.i.d. [75]. Figure 5.36 shows a cooperative network comprising two agent nodes and four reference nodes.

In nonlinear least squares (NLS) position estimates, based on the ranging measurement (5.116), can be obtained as the solution to the following nonlinear nonconvex minimization problem:

$$\hat{\mathbf{Z}} = \arg \min_{\substack{\mathbf{z}_i \in \mathbb{R}^2 \\ i=1,\ldots,M}} \sum_{i=1}^{M} \sum_{j \in \mathcal{A}_i \cup \mathcal{B}_i} \left(\hat{d}_{i,j} - d(\mathbf{z}_i, \mathbf{z}_j)\right)^2, \tag{5.117}$$

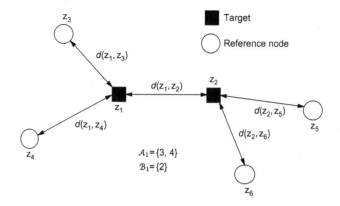

FIGURE 5.36

A typical cooperative network with two agents and four reference nodes (from Ref. [132]).

where $\hat{\mathbf{Z}} = [\hat{\mathbf{z}}_1, \dots, \hat{\mathbf{z}}_M]$. There is, in general, no analytical solution to (5.117) and we resort to numerical methods. For the rest of this section, let

$$\tilde{\mathbf{z}}_j = \begin{cases} \hat{\mathbf{z}}_j, & j = 1, \dots, M \\ \mathbf{z}_j, & j = M+1, \dots, M+N. \end{cases} \tag{5.118}$$

A distributed solution of NLS can be obtained for cooperative networks as follows:

- $\mathcal{C}_i = \emptyset$, $i = 1, \dots, M$
- For $k = 0$ until convergence or predefined number K
- For $i = 1, \dots, M$

$$\hat{\mathbf{z}}_i = \arg \min_{\mathbf{z}_i \in \mathbb{R}^2} \sum_{j \in \mathcal{A}_i \cup \mathcal{C}_i} \left(\hat{d}_{i,j} - d(\mathbf{z}_i, \tilde{\mathbf{z}}_j) \right)^2,$$

$$\text{if } i \notin \mathcal{C}_m \text{ then } \mathcal{C}_m = \mathcal{C}_m \cup \{i \cap \mathcal{B}_m\}, \quad m = 1, \dots, M. \tag{5.119}$$

In practice, this algorithm is a possible implementation of incremental positioning schemes introduced in Section 3.4.3, where at each iteration an NLS approach is applied considering the agents localized in the previous iteration as additional "virtual" anchor nodes. It is clear that the performance of the algorithm strongly depends on initial position estimates. If it is not well initialized, it might converge to local minima, resulting in large errors.

For positioning, it is possible to have a linear closed-form algorithm. One possibility is that investigated in Ref. [26], where the linearization (3.19) is considered leading to a linear LS approach. In the sequel of this section, we illustrate a two-step linear estimator proposed in Ref. [49] and similar to that presented in Ref. [26]. In the first step we obtain a coarse estimate, and it is then refined in the second step. Let us define $\mathcal{E}_i \subseteq \mathcal{B}_i$ as the set of all agent nodes, which have already been localized connected to the agent i.

Suppose that there are at least three noncollinear nodes with known positions around an agent and that noise is small compared to distances. For the first step of updating for agent i, squaring both sides of (5.116) yields, after neglecting small terms, and some manipulations,

$$\tilde{d}_{i,j} = \hat{d}_{i,j}^2 - \|\tilde{z}_j\|^2 \approx \left[-2\tilde{z}_j^T \ \ 1 \right] \psi_i + 2d(z_i, \tilde{z}_j)\epsilon_{i,j} \ , \ j \in \mathcal{A}_i \cup \mathcal{E}_i, \tag{5.120}$$

where $\psi_i = \left[z_i^T \ \|z_i\|^2 \right]^T$. Now a set of linear equations can be written as

$$d_i = A_i\psi_i + v_i, \tag{5.121}$$

where

$$d_i = \left[\tilde{d}_{i,j_1}, \tilde{d}_{i,j_2}, \dots, \tilde{d}_{i,j_L} \right]^T$$
$$A_i = \left[a_{j_1}^T, \dots, a_{j_L}^T \right]^T,$$
$$v_i = \left[2d(z_i, \tilde{z}_{j_1})\epsilon_{i,j_1}, \dots, 2d(z_i, \tilde{z}_{j_L})\epsilon_{i,j_L} \right]^T$$
$$\{j_1, \dots, j_L\} = \mathcal{A}_i \cup \mathcal{E}_i,$$

and where $L = |\mathcal{A}_i \cup \mathcal{E}_i|$ is the cardinality of set $\mathcal{A}_i \cup \mathcal{E}_i$, and $a_{j_p} = \left[-2\tilde{z}_{j_p}^T \ \ 1 \right]$. Suppose that the matrix A_i is full rank, then the unknown parameter ψ_i can be estimated by [64]

$$\hat{\psi}_i = (A_i^T C_{v_i}^{-1} A_i)^{-1} A_i^T C_{v_i}^{-1} d_i, \tag{5.122}$$

where the weighting matrix C_{v_i}, for i.i.d. measurement noise, is given as [49]

$$C_{v_i} = \text{diag}\{4d^2(z_i, \tilde{z}_{j_1}), \dots, 4d^2(z_i, \tilde{z}_{j_L})\}.$$

The covariance matrix of $\hat{\psi}_i$ is $\text{cov}(\hat{\psi}_i) = (A_i^T C_{v_i}^{-1} A_i)^{-1}$ [64].

To compute the weighting matrix C_{v_i}, we use the measured distances in (5.116), since the real distances are not available. It has been shown in Ref. [49] that the degradation of replacing the real distances with estimated distances is negligible.

Now we update the set $\mathcal{E}_m$ and the ith agent's position as

$$\mathcal{E}_m = \mathcal{E}_m \cup \{i \cap \mathcal{B}_m\}, \quad m = 1, \dots, M,$$
$$\hat{z}_i = [\hat{\psi}_i]_{2 \times 1}, \tag{5.123}$$

where, in general, $A_{n \times m}$ denotes the upper left $n \times m$ part of matrix A. The first step of updating for agent i, which gives a coarse estimate, is just run once. In subsequent steps, the previous estimate is refined using a new estimator. Let us apply a first-order Taylor-series expansion around $\hat{z}_i$ for the measurement in (5.116) to get

$$\hat{d}_{i,j} = d(z_i, \tilde{z}_j) + \nabla d^T(z_i, \tilde{z}_j)|_{\hat{z}_i}(z_i - \hat{z}_i) + \epsilon_{i,j}, \quad j \in \mathcal{A}_i \cup \mathcal{E}_i,$$

where the gradient is found as $\nabla d^T(x, \tilde{x})|_{\hat{x}} = \hat{x}^T/d(\hat{x}, \tilde{x})$. The new linear set of measurements can be written in matrix form as

$$\tilde{d}_i = G_i \Delta z_i + \epsilon_i, \tag{5.124}$$

where $\Delta \mathbf{z}_i = \mathbf{z}_i - \hat{\mathbf{z}}_i$ and vectors $\tilde{\mathbf{d}}_i$, $\boldsymbol{\varepsilon}_i$, and matrix $\mathbf{G}_i$ are obtained as follows:

$$\tilde{\mathbf{d}}_i = \left[\hat{d}_{i,j_1} - d(\mathbf{z}_i, \tilde{\mathbf{z}}_{j_1}), \dots, \hat{d}_{i,j_P} - d(\mathbf{z}_i, \tilde{\mathbf{z}}_{j_P}) \right]^T,$$

$$\mathbf{G}_i = \begin{bmatrix} \nabla d^T(\mathbf{z}_i, \tilde{\mathbf{z}}_{j_1})|_{\hat{\mathbf{z}}_i} \\ \vdots \\ \nabla d^T(\mathbf{z}_i, \tilde{\mathbf{z}}_{j_P})|_{\hat{\mathbf{z}}_i} \end{bmatrix}, \quad \boldsymbol{\epsilon}_i = \left[\epsilon_{i,j_1}, \dots, \epsilon_{i,j_P} \right]^T,$$

where $\{j_1, \dots, j_P\} = \mathcal{A}_i \cup \mathcal{E}_i$ and $P = |\mathcal{A}_i \cup \mathcal{E}_i|$.

To solve the linear equation in (5.124), we add a constraint on the unknown parameter $\Delta \mathbf{z}_i$ to be small (if possible). Therefore, we can consider the following optimization problem:

$$\underset{\Delta \mathbf{z}_i}{\text{minimize}} \, \| \tilde{\mathbf{d}}_i - \mathbf{G}_i \Delta \mathbf{z}_i \|^2 + \gamma \, \| \Delta \mathbf{z}_i \|^2, \tag{5.125}$$

where the regularized parameter $\gamma > 0$ determines the trade-off between $\| \tilde{\mathbf{d}}_i - \mathbf{G}_i \Delta \mathbf{z}_i \|^2$ and $\| \Delta \mathbf{z}_i \|^2$. The solution to (5.125) can be obtained as [12]

$$\widehat{\Delta \mathbf{z}_i} = \left(\mathbf{G}_i^T \mathbf{G}_i + \gamma \mathbf{I}_2 \right)^{-1} \mathbf{G}_i^T \tilde{\mathbf{d}}_i, \tag{5.126}$$

where $\mathbf{I}_n$ denotes an $n \times n$ identity matrix. The updated estimate is

$$\hat{\hat{\mathbf{z}}}_i = \hat{\mathbf{z}}_i + \widehat{\Delta \mathbf{z}_i}. \tag{5.127}$$

The linear estimator obtained in this section works well when the level of noise is small. When there are not enough nodes with known or estimated position around an agent, this estimator cannot locate that agent. In fact, it is required to have at least three known noncollinear nodes around an agent to uniquely estimate that agent's position.

Using the same experimental data adopted for results in Fig. 5.15, the impact of cooperation on positioning is investigated in Figs 5.37 and 5.38 by considering the algorithm in (5.119) and the linearization approach (3.19) proposed in Ref. [26]. Figure 5.37 presents the RMSE as a function of the number of iterations K of the iterative LS algorithm. We assume $N = 3$ anchors (tx1, tx3, tx5) and two agents with the capability of performing interagent ranging. Agent 1 is located in position 8, and the cooperating node (agent 2) is located in position 10 (LOS condition) or 18 (NLOS condition). Anchor tx5 is assumed as reference for the linear LS algorithm in (3.19). These configurations are chosen because they lead to two distinct interesting situations. When the two agents are located in LOS, they can attain a highly accurate interagent range estimate. When the agents are located in NLOS (different rooms), the interagent range estimate is expected to be worse.

Figure 5.37 shows that cooperation in LOS can strongly improve the RMSE and that the incremental LS algorithm converges after few iterations. Also note that the resulting RMSE for cooperation with four iterations and $N = 3$ anchors is better than the case of $N = 5$ anchors without cooperation (Fig. 5.15).

In Fig. 5.38, the RMSE is shown for $N = 3$ (tx1,tx3,tx5) and $K = 4$ iterations when agent 1, located in position 8, cooperates with agent 2, located in different positions. As can be noted, cooperation is not always advantageous with this approach. In fact, the best position for the cooperating agent node is location 10, which is in LOS. NLOS positions 11 and 12 lead to a significant performance improvement, whereas LOS positions 7 and 9 do not lead to any performance gain. The reason is that,

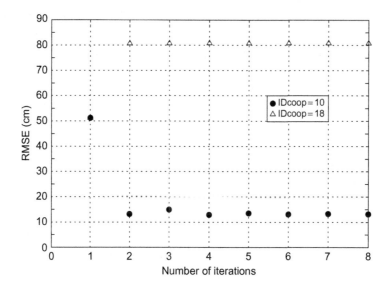

FIGURE 5.37

RMSE as a function of number of iterations when agent 1, located in position 8, cooperates with agent 2, in position 10 or 18. $N = 3$ (tx1,tx3,tx5) anchors are considered. Tx5 is taken as reference for the linearized LS algorithm (from Ref. [26]).

in addition to interagent link reliability, the geometric configuration of the nodes plays a fundamental role in determining the position estimation accuracy.

5.4.3 Cooperative POCS

In this section, we extend the geometric positioning algorithm based on the projections onto convex sets (POCS) approach to the cooperative scenario. For the cooperative positioning, applying POCS is tricky, and it needs some considerations. Assuming positive distance measurements errors, the intersection is nonempty and we can apply the so-called outer approximation (OA) approach to solve the positioning problem. In OA, the intersection is approximated by a regular convex shape, for example a polytope or an ellipse. The application of OA for cooperative networks is also tricky. For cooperative networks, it is easier to use a simple convex shape for OA, for example a disc. Unfortunately, the OA problem is not easy to solve in an efficient way, and the final shape may not be the optimum shape. This section studies application of different methods for cooperative positioning. It also shows that a hybrid method can improve the performance of available approaches. For instance, distributed constrained NLS (constrained NLS (CNLS)), which combines OA and NLS, performs well in some situations.

The estimators introduced in previous sections fail to work in NLOS or when anchor density is low. Geometric positioning is another technique that can be applied to the problem. In this section, we study robust techniques based on a geometric interpretation of the positioning problem that avoid both drawbacks.

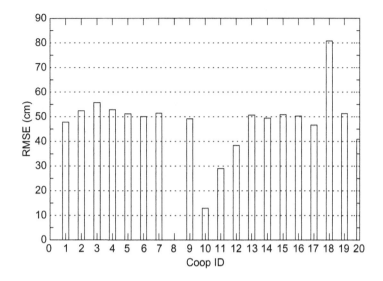

FIGURE 5.38

RMSE when agent 1, located in position 8, cooperates with agent 2 as a function of agent 2 position. $N = 3$ (tx1,tx3,tx5) anchors considered, $K = 4$ (from Ref. [26]).

For agent i, from (5.117) it is clear that the minimum of each term is obtained when $d(\mathbf{z}_i, \mathbf{z}_j) = \hat{d}_{i,j}$. Now, suppose that we define the disc $\mathcal{D}(\mathbf{z}_j, \hat{d}_{i,j})$ centered at $\mathbf{z}_j$ as

$$\mathcal{D}(\mathbf{z}_j, \hat{d}_{i,j}) = \left\{ \mathbf{z} \in \mathbb{R}^2 : \|\mathbf{z} - \mathbf{z}_j\| \le \hat{d}_{i,j} \right\}, \quad j \in \mathcal{A}_i \cup \mathcal{B}_i; \tag{5.128}$$

it is then reasonable to define an estimate of $\mathbf{z}_i$ as a point in the intersection, that is, $\mathcal{D}_i$, of the discs $\mathcal{D}(\mathbf{z}_j, \hat{d}_{i,j})$

$$\hat{\mathbf{z}}_i \in \mathcal{D}_i = \bigcap_{j \in \mathcal{A}_i \cup \mathcal{B}_i} \mathcal{D}(\mathbf{z}_j, \hat{d}_{i,j}). \tag{5.129}$$

Geometric positioning algorithms solve the following feasibility problem:

$$\text{find } \mathbf{Z} = [\mathbf{z}_1, \dots, \mathbf{z}_M] \text{ such that } \mathbf{z}_i \in \mathcal{D}_i, \ i = 1, \dots, M.$$

In the sequel, we study this class of estimators. The first estimator that we study uses the projection onto convex set method to solve the positioning problem. To apply POCS in cooperative positioning, we must unambiguously define all the discs, $\mathcal{D}(\mathbf{z}_j, \hat{d}_{i,j})$, for every agent i. From (5.128), it is clear that some discs, that is, discs centered around a reference node, can be defined without any ambiguity, that is, $\mathcal{D}(\mathbf{z}_j, \hat{d}_{i,j})$, $j \in \mathcal{A}_i$. Other discs derived from measurements between agents, that is, discs centered around an unknown agent, have an unknown center, that is, $\mathcal{D}(\mathbf{z}_j, \hat{d}_{i,j})$, $j \in \mathcal{B}_i$.

Let us consider Fig. 5.39 where for agent node 1, we want to involve the measurement between agent 2 and agent 1. Since there is no prior knowledge about the position of agent 2, the disc centered around agent 2 cannot be involved in the localization process for agent 1. Suppose based on applying POCS to the discs defined by reference nodes $\mathbf{z}_5$ and $\mathbf{z}_6$, we obtain an initial estimate $\hat{\mathbf{z}}_2$ for agent 2.

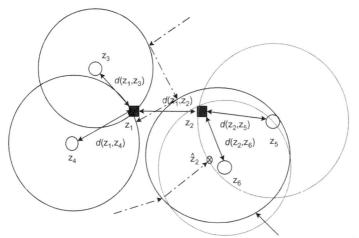

Estimated disc form agent 2 (radius $\hat{d}_{1,2}$)

FIGURE 5.39

Applying POCS for agent 1 considering initial estimate of agent 2, $\hat{\mathbf{z}}_2$ (from Ref. [132]).

Now, based on distance estimate $\hat{d}_{1,2}$, we can define a new disc centered around $\hat{\mathbf{z}}_2$. This new disc can be combined with the two other discs defined by reference nodes $\mathbf{z}_3$ and $\mathbf{z}_4$. Figure 5.39 shows the process for localizing agent 1. For agent 2, the same procedure is followed.

Now we can implement cooperative POCS (Coop-POCS) as follows:

- $\mathcal{C}_i = \emptyset, \quad i = 1, \ldots, M$
- For $k = 0$ until convergence or predefined number K
- For $i = 1, \ldots, M$, find $\hat{\mathbf{z}}_i$ with POCS such that

$$\hat{\mathbf{z}}_i \in \mathcal{D}_i = \bigcap_{j \in \mathcal{A}_i \cup \mathcal{C}_i} \mathcal{D}(\tilde{\mathbf{z}}_j, \hat{d}_{i,j}),$$

if $i \notin \mathcal{C}_m$ then $\mathcal{C}_m = \mathcal{C}_m \cup \{i \cap \mathcal{B}_m\}, \quad m = 1, \ldots, M.$

Note that POCS also handles an empty intersection [9]. For inconsistent problems for the agent i, that is, empty intersection, POCS minimizes the sum of squared distances to the sets $\mathcal{D}(\tilde{\mathbf{z}}_j, \hat{d}_{i,j}), j \in \mathcal{A}_i \cup \mathcal{C}_i$.

As discussed earlier, the position of an unknown agent can be found in the intersection of a number of discs. The intersection in general may have any convex shape. Based on the assumption of positive measurement noise in Section 5.4.4, the agent i definitely lies inside the intersection

$$\bigcap_{j \in \mathcal{A}_i} \mathcal{D}(\tilde{\mathbf{z}}_j, \hat{d}_{i,j}).$$

In contrast to POCS, which tries to find a point in the intersection as an estimate, the outer approximation (OA) method determines an outer approximation of the intersection and then takes one point inside as an estimate. The main problem is how to accurately approximate the intersection.

There is some work in the positioning literature to approximate the intersection by convex regions such as polytopes or ellipsoids [30].

To apply OA for cooperative networks, we first see how to approximate the intersection by a simple region that can be exchanged easily between agents. We consider a disc approximation of the intersection since it can be easily obtained and exchanged between agents.

Using simple geometry, we are able to find all intersection points between different discs and finally find a minimum disc that passes through them and covers the intersection. Let $\mathbf{z}_k^l$, $k = 1,\ldots,L$ be the set of intersection points. Among all intersection points, some of them are redundant and will be discarded. Therefore, the common points that belong to the intersection are selected as $\mathcal{S}_{int} = \{\mathbf{z}_k^l | \mathbf{z}_k^l \in \mathcal{D}_i\}$. The problem therefore renders to finding a disc that contains $\mathcal{S}_{int}$ and cover the intersection, which is a well-known convex optimization problem [6, 12].

Now cooperative OA (Coop-OA) can be implemented as

- $\mathcal{C}_i = \emptyset$, $i = 1,\ldots,M$
- For $k = 0$ until convergence or predefined number K
- For $i = 1,\ldots,M$

$$\mathcal{D}(\hat{\mathbf{z}}_i, \hat{R}_i) = \text{OA}\left\{\bigcap_{j \in \mathcal{A}_i \cup \mathcal{C}_i} \mathcal{D}(\tilde{\mathbf{z}}_j, \tilde{R}_j)\right\},$$

$$(\tilde{\mathbf{z}}_j, \tilde{R}_j) = \begin{cases} (\mathbf{z}_j, \hat{d}_{i,j}), & j \in \mathcal{A}_i \\ (\hat{\mathbf{z}}_j, \hat{d}_{i,j} + \hat{R}_j) & j \in \mathcal{C}_i \end{cases},$$

$$\text{if } i \notin \mathcal{C}_m \text{ then } \mathcal{C}_m = \mathcal{C}_m \cup \{i \cap \mathcal{B}_m\}, \quad m = 1,\ldots,M.$$

To see how this method works, consider Fig. 5.40 where agent 2 helps agent 1 to improve its positioning. Target 2 can be found in the intersection derived from two discs centered around $\mathbf{z}_5$ and $\mathbf{z}_6$ in noncooperative mode (semioval shape). Suppose that we approximate it by a disc (small dashed circle). To help agent 1 to approximate its intersection in cooperative mode, this region should be involved in finding the intersection for agent 1. We can easily extend every point of this disc by $\hat{d}_{1,2}$ to come up with a bigger disc (big dashed circle). It is seen that the approximated intersection for agent 1 is smaller than that for the noncooperative case. Note if we use the exact intersection extended, we end up with a better approximation for the intersection of agent 1.

We now consider the intersection obtained in Coop-OA as a constraint for NLS methods (CNLS) to improve the performance of the algorithm mentioned in (5.119). Suppose that for agent i we obtain a final disc as $\mathcal{D}_i(\hat{\mathbf{z}}_i, \hat{R}_i)$, where $\hat{\mathbf{z}}_i$ and $\hat{R}_i$ are the center and the radius of the disc, respectively. It is clear that we can define $\|\mathbf{z}_i - \hat{\mathbf{z}}_i\| \leq \hat{R}_i$ as a constraint for the ith agent in the optimization problem (5.119).

Simulation Results

In this section, computer simulations are performed to evaluate the performance of different algorithms. To compare different methods, we consider the cumulative density function (CDF) of the positioning error. The simulation area, a $100\,\text{m} \times 100\,\text{m}$ square area, consists of a number of reference nodes placed at fixed positions and also 100 agents randomly distributed over the area. We assume a pair of nodes can connect and estimate the distance between each other if that distance is less than

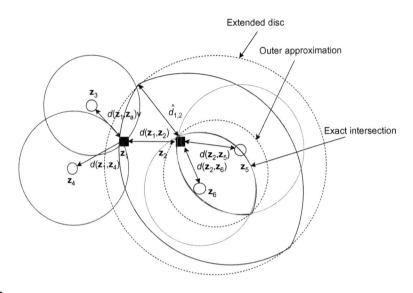

FIGURE 5.40

Extending the convex region involving agent 2 to help agent 1 to find smaller intersection (from Ref. [132]).

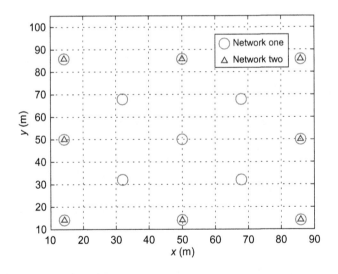

FIGURE 5.41

Simulation environment for networks 1 and 2; 100 agent nodes are randomly placed inside the area (from Ref. [132]).

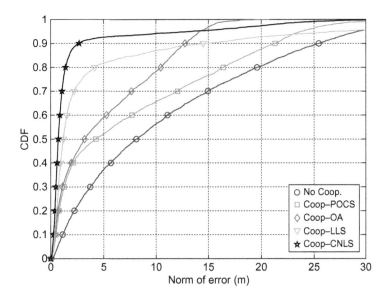

FIGURE 5.42

CDFs of different algorithms for network 1, LOS situation (from Ref. [132]).

20 m. The measurement noise is assumed to have an exponential distribution with mean (and standard deviation) equal to 50 cm. We consider two different networks based on the number of reference nodes shown in Fig. 5.41. To implement the CNLS estimator, we use the MATLAB routine `fmincon` [116] initialized and constrained with Coop-OA. Without any attempt to optimize the regularization parameter, we simply set $\gamma = 0.5$. To make a fair comparison between different methods, we consider two scenarios for these network deployments, LOS and NLOS situations. In cooperative mode, the estimated parameters, points or discs, are exchanged among different agents 10 times. All CDFs were obtained for 30 networks. Note that for the noncooperative case we used POCS to find an estimate.

The CDFs of errors for the different methods in an LOS scenario are shown in Figs 5.42 and 5.43. There are some interesting results to be pointed out. As can be seen, CNLS outperforms the other methods in most of the cases. The linear estimator for network 1 works well for most of the agents since there are enough known reference nodes around them. In network 2, this method fails since there are less than three known reference nodes for an agent. It is seen that the geometric approach can improve the performance of cooperative positioning compared to the noncooperative case. The geometric methods also work well when the number of anchors is decreased. It is also seen that Coop-POCS shows good performance for small errors compared to the Coop-OA especially when the number of reference nodes is decreased.

To see the performance of different methods in an NLOS situation, we run another simulation for network 1 and network 2 where 20% of measurements suffer from an NLOS situation. To simulate an NLOS measurement, we add a uniform random number between 0 and 20 m to the LOS measurement. Figures 5.44 and 5.45 show the error CDFs of different methods. As can be seen in comparison to the

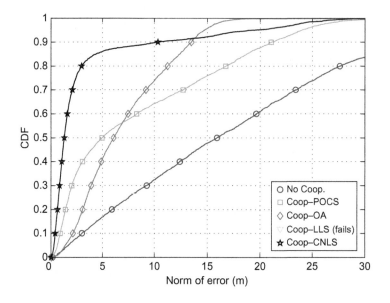

FIGURE 5.43

CDFs of different algorithms for network 2, LOS situation (from Ref. [132]).

LOS case (Figs 5.42 and 5.43), the geometric approaches are more robust compared to the statistical approaches. Figure 5.44 also shows that the linear approach performs poorly and it cannot locate agents well even if there are enough reference nodes in network 1. Since the outer approximation method, as well as POCS, is highly robust and can easily handle NLOS measurements, CNLS still performs well.

Regarding convergence, we have observed that all algorithms converge after a few iterations; for example, five iterations are enough for convergence.

5.4.4 Positioning Using Active and Passive Anchors

Among the various position-dependent parameter measurement approaches, as described in Chapter 3, TW-TOA has been considered an effective approach in the literature (e.g., Ref. [43]), especially when using UWB signals in WSNs, mainly because of its relatively high accuracy and lack of synchronization requirements. In this approach, a reference node sends a signal to a target node and waits for a response from it. The round-trip time delay between the reference node and the target node gives an estimate of the distance between them. As the number of reference nodes in a WSN increases, the position of the target node can be estimated more accurately via TW-TOA estimation. Since, in practice, there are some limitations on increasing the number of reference nodes due to power and complexity constraints, the idea of cooperation between reference nodes is proposed in Ref. [40] to decrease the number of transmissions, and its theoretical analysis is presented in Ref. [43]. In this method, some reference nodes, called primary reference nodes (PRNs), initiate range estimation by sending a signal. The target replies to received signals by sending an acknowledgment. Suppose that there are some other

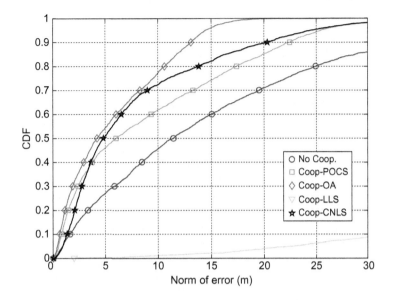

FIGURE 5.44

CDFs of different algorithms for network 1, NLOS situation (from Ref. [132]).

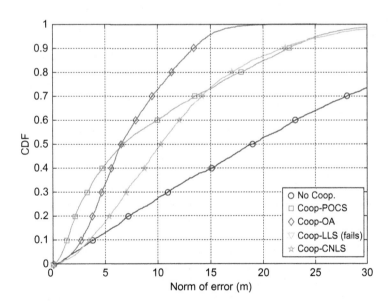

FIGURE 5.45

CDF of different algorithms for network 2, NLOS situation (from Ref. [132]).

reference nodes, which can listen to both signals, and are called secondary reference nodes (SRNs). It has been shown that the SRNs can help the PRNs to estimate the target position more accurately [43]. In fact, it is possible to get the same performance with fewer PRNs when measurements from the SRNs are involved in the positioning process. The model considered here is based on cooperation between reference nodes, which is different from targets' cooperation in a cooperative network [123].

In this approach, it is assumed that the SRNs are able to receive signals from both the target and the PRNs. Therefore, the SRNs are able to measure the TDOA between the target signal and the signals of the PRNs. In this case, an MLE derived in Ref. [43] can be employed to improve the positioning accuracy compared to the noncooperative approach. However, due to the nonlinear nature of the cost function in the MLE, a numerical method, for example an iterative search algorithm with a good initial point, should be used to obtain an accurate estimate of the target's position. There might be some drawbacks in using a numerical method in practice. For example, high complexity or convergence problems can limit the use of numerical methods. To avoid drawbacks of numerical methods, a two-step linear estimator using two different linearization techniques is presented. In the first step, using a noncooperative preprocessing similar to previous works for conventional networks, a coarse estimator is obtained. In the second step, using the first step estimation, another linear set of measurements different from the first step is derived. The second-step estimator is a regularized least squares, which refines the first estimate.

System Model

Let us consider a two-dimensional network[6] with $N+M$ reference nodes located at known positions, $\mathbf{x}_i = [x_{i,1}, x_{i,2}]^T \in \mathbb{R}^2$, $i = 1, \ldots, N+M$. Suppose that N PRNs are used to measure the TW-TOA between the PRNs and the target to be located and that M SRNs are able to listen and measure signals transmitted by the PRNs and the target. For simplicity, we assume that the first N sensors are the primary nodes and the remaining M sensors are the secondary nodes.

Let $\mathcal{C} = \{(i,j) \mid \text{PRN } i \text{ and SRN } j \text{ are connected}\}$ be the set of all pairs with one primary node and one secondary node, which are connected. In this discussion, for simplicity, we assume a fully connected network. The TW-TOA measurement between primary node i and the target, located at coordinates $\boldsymbol{\theta} = [\theta_1, \theta_2]^T \in \mathbb{R}^2$, can be written as [43]

$$\hat{t}_i = \frac{r_i}{c} + \frac{\tilde{n}_{T,i}}{2} + \frac{\tilde{n}_{i,T}}{2}, \quad i = 1, \ldots, N, \tag{5.130}$$

where c is the speed of propagation, $r_i = \|\mathbf{x}_i - \boldsymbol{\theta}\|$ is the distance between the ith PRN and the point $\boldsymbol{\theta}$, $\tilde{n}_{i,T}$ is the TOA estimation error at the target node for the signal transmitted from the ith PRN, and $\tilde{n}_{T,i}$ is the TOA estimation at the ith PRN for the signal transmitted from the target node. The estimation errors are modeled as zero-mean Gaussian random variables with variances $\sigma_{T,i}^2/c^2$ and $\sigma_{i,T}^2/c^2$, that is, $\tilde{n}_{T,i} \sim \mathcal{N}(0, \sigma_{T,i}^2/c^2)$ and $\tilde{n}_{i,T} \sim \mathcal{N}(0, \sigma_{i,T}^2/c^2)$ [43].

Suppose that the SRNs are able to measure the TOA based on the received signal from the target and the PRNs. The TOA estimate of the ith PRN in the jth SRN is

$$\hat{t}_{i,j} = T_{o_i} + \frac{r_{i,j}}{c} + \tilde{n}_{i,j}, \quad (i,j) \in \mathcal{C}, \tag{5.131}$$

---

[6]The generalization to a three-dimensional scenario is straightforward, but is not explored here.

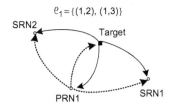

FIGURE 5.46

A primary node initiates positioning by transmitting a signal to the target, whereupon the target, replies to the received signal. Both signals are received in the secondary node (from Ref. [133]).

where the ith PRN sends its signal at time instant T_{o_i}, which is unknown to the SRN, $r_{i,j} = \|\mathbf{z}_i - \mathbf{z}_j\|$ is the distance between primary node i and secondary node j, and $\tilde{n}_{i,j}$ is modeled as a Gaussian random variable $\tilde{n}_{i,j} \sim \mathcal{N}(0, \sigma_{i,j}^2/c^2)$. Suppose that the response signal from the target to this signal is also received by the jth SRN. The TOA estimate for this signal is given by

$$\hat{t}_{i,T,j} = T_{o_i} + \frac{r_i}{c} + \frac{r_j}{c} + \tilde{n}_{i,T} + \tilde{n}_{T,j}, \quad (i,j) \in \mathcal{C}. \tag{5.132}$$

Having these two measurements in the SRN, namely, measurement in (5.131) and in (5.132), we are able to measure the TDOA between the ith PRN and the target, which corresponds to the distance from the ith PRN to the target plus the distance from the target to the jth SRN.

Positioning Algorithms

To gain some insight into the problem, let us consider Fig. 5.46, where one PRN performs TW-TOA estimation with the target. Namely, the PRN sends a signal to the target, and the target replies to this signal. Here, we assume that either an estimate of the turnaround time is available [43] or the turn-around time is extremely small such that it can be neglected. Suppose that two other nodes (SRN1 and SRN2) listen to both signals. Since the distances between the reference nodes are known, it is possible in the secondary node to estimate the time reference from (5.131); hence, the SRNs are able to estimate the overall distance from the PRN to the target and the target to the SRN as follows (see Fig. 5.47):

$$z_i^j = c(\hat{t}_{i,T,j} - \hat{T}_{o_i}) = r_i + r_j + n_{T,j} + n_{i,T} - n_{i,j}, \quad (i,j) \in \mathcal{C}, \tag{5.133}$$

where $\hat{T}_{o_i}$ is an estimate of T_{o_i}, for example $\hat{T}_{o_i} = \hat{t}_{i,j} - r_{i,j}/c = T_{o_i} + \tilde{n}_{i,j}$, and $n_{i,T} = c\tilde{n}_{i,T}, n_{i,T} = c\tilde{n}_{i,T}$, and $n_{i,j} = c\tilde{n}_{i,j}$.

From (5.130), the distance estimate to the target in the ith PRN is expressed as

$$z_i = c\hat{t}_i = r_i + \frac{n_{i,T}}{2} + \frac{n_{T,i}}{2}, \quad i = 1, \dots, N. \tag{5.134}$$

In the sequel, we consider the vector of measurements $\mathbf{z}$ as follows:

$$\mathbf{z} = [z_1, \dots, z_N, z_1^1, \dots, z_1^M, \dots, z_N^1, \dots, z_N^M]^T. \tag{5.135}$$

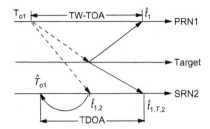

FIGURE 5.47

Computing TW-TOA and TDOA for PRN1 and SRN2 (from Ref. [133]).

It is clear that the vector $\mathbf{z}$ can be modeled as Gaussian random vector $\mathbf{z} \sim \mathcal{N}(\boldsymbol{\mu}, \mathbf{C})$, where mean $\boldsymbol{\mu} = [\mu_1, \ldots, \mu_N, \mu_1^1, \ldots, \mu_1^M, \ldots, \mu_N^1, \ldots, \mu_N^M]$ and covariance matrix $\mathbf{C}$ are computed as follows [43]:[7]

$$\mu_i = r_i, \quad \mu_i^j = r_i + r_{N+j},$$

$$\mathbf{C} = E\{(\mathbf{z} - \boldsymbol{\mu})(\mathbf{z} - \boldsymbol{\mu})^T\} = \begin{bmatrix} \mathbf{C}_{11} & \mathbf{C}_{12} \\ \mathbf{C}_{21} & \mathbf{C}_{22} \end{bmatrix}, \tag{5.136}$$

where matrices $\mathbf{C}_{11} \in \mathbb{R}^{N \times N}$, $\mathbf{C}_{12} = \mathbf{C}_{21}^T \in \mathbb{R}^{N \times NM}$, and $\mathbf{C}_{22} \in \mathbb{R}^{NM \times NM}$ can be obtained as follows:

$$\mathbf{C}_{11} = \text{diag}\left\{\left(\frac{\tilde{\sigma}_1^2}{4} + \frac{\sigma_1^2}{4}\right), \ldots, \left(\frac{\tilde{\sigma}_N^2}{4} + \frac{\sigma_N^2}{4}\right)\right\},$$

$$\mathbf{C}_{12} = \begin{bmatrix} \mathbf{v}_1^T & \cdots & \mathbf{0} \\ \vdots & \ddots & \vdots \\ \mathbf{0} & \cdots & \mathbf{v}_N^T \end{bmatrix}, \quad \mathbf{v}_i = \frac{\tilde{\sigma}_i^2}{2}\mathbf{1}_M,$$

$$\mathbf{C}_{22} = \text{blkdiag}\{\mathbf{W}_1, \ldots, \mathbf{W}_N\}, \mathbf{1}_M = [1, \ldots, 1]^T$$

$$\mathbf{W}_i = \tilde{\sigma}_i^2 \mathbf{1}_M \mathbf{1}_M^T + \text{diag}\{\sigma_{N+1}^2 + \sigma_{i,N+1}^2, \ldots, \sigma_{N+M}^2 + \sigma_{i,N+M}^2\},$$

where (blk)diag$(X_1, \ldots, X_N)$ is a (block) diagonal matrix with diagonal element $X_1, \ldots, X_N$.

The MLE is obtained by the following optimization problem [64]:

$$\hat{\boldsymbol{\theta}} = \underset{\boldsymbol{\theta} \in \mathbb{R}^2}{\text{argmin}}(\mathbf{z} - \boldsymbol{\mu})^T \mathbf{C}^{-1}(\mathbf{z} - \boldsymbol{\mu}). \tag{5.137}$$

---

[7]For simplicity of notation, we assume $\sigma_{T,i} = \sigma_i$ and $\sigma_{i,T} = \tilde{\sigma}_i$.

With some manipulations, (5.137) can be expressed as [43]

$$\hat{\theta} = \underset{\theta \in \mathbb{R}^2}{\text{argmin}} \sum_{i=1}^{N} \left\{ \left(\frac{2}{\sigma_i^2} - \frac{1}{s_i \sigma_i^4} \right) (z_i - r_i)^2 - \frac{1}{s_i} \left(\sum_{j=N+1}^{N+M} \frac{z_i^j - r_i - r_j}{4\sigma_j^2} \right)^2 \right.$$

$$\left. - \frac{z_i - r_i}{s_i \sigma_i^2} \sum_{j=N+1}^{N+M} \frac{(z_i^j - r_i - r_j)}{2\sigma_j^2} + \sum_{j=N+1}^{N+M} \frac{(z_i^j - r_i - r_j)^2}{2\sigma_j^2} \right\}, \tag{5.138}$$

where $s_i \triangleq 1/(2\tilde{\sigma}_i^2) + 1/(2\sigma_i^2) + \sum_{j=N+1}^{N+M} 1/(4\sigma_j^2)$. As can be seen, there is no closed-form solution for the ML estimate, and we are forced to use numerical methods, for example an iterative search with good initial point. Note that for the conventional network, that is, a network comprising only primary nodes, MLE changes to [43]

$$\hat{\theta} = \underset{\theta \in \mathbb{R}^2}{\text{argmin}} \sum_{i=1}^{N} \frac{2}{\sigma_i^2} (z_i - r_i)^2. \tag{5.139}$$

To obtain a closed-form solution to the positioning problem, which avoids MLE's drawback, we apply linearization techniques for the measurements in both PRNs and SRNs. The estimator is implemented in two steps: coarse and fine. Note that it is required to have at least three noncollinear reference nodes for this estimator.

One way to obtain a linear model versus the target position is to apply the method described in Section 5.4.2.

Simulation Results

In this section, computer simulations are performed to evaluate the performance of the described estimator. To compare different methods, we consider RMSE and CDF of positioning errors. The CRB for the cooperative case is computed in Ref. [43]. Figure 5.48 shows the simulation environment where three PRNs and one SRN are located on the corner of a 100 m × 100 m square area. The target is randomly placed inside the area over a grid of 19 × 19 points. We run simulations for 4000 realizations of measurement noise. In simulations, we assume $\tilde{\sigma}_i = \sigma_i = \sigma_{i,j} = \sigma$. For the implementation of the MLE, we used grid search inside the area. For the fine estimator without any attempt to optimize γ, we simply set it equal to 0.01 for both conventional and cooperative networks. Figure 5.49 shows the RMSE of the CRB and the proposed estimator for both cooperative and noncooperative networks. As can be seen, for both networks, the linear estimator, as well as MLE, attains the CRB for high SNR.

As the variance of noise increases, the gap between the proposed estimator and the CRB also increases. It also shows that the cooperation can improve the performance for low SNRs. It is seen that the fine estimator outperforms the coarse estimator for both networks. For this deployment, the coarse estimator for the cooperative case outperforms the optimal unbiased estimator in the noncooperative case. For further evaluation, we consider the position error CDF for the linear estimators in both cooperative and noncooperative modes. In this simulation, we consider the network of Fig. 5.48 and set $\sigma_T = \sigma_i = 10$ m. Figure 5.50 shows the position error CDF for both cases. It can be seen that in the cooperative mode the estimator outperforms the one in the noncooperative mode. It also shows that the fine estimator improves the estimation accuracy for both networks compared to the coarse estimator. Figure 5.51 shows the RMSE of estimators for both conventional and cooperative networks. As can be seen in the cooperative mode, the bias value is extremely small compared to CRB. Hence, the fine estimator for cooperative mode can be considered an unbiased estimator.

5.4.5 Distributed Positioning Based on Belief Propagation

In this section, we consider the issue of estimating cooperatively the locations of nodes in a UWB ad hoc network, where nodes are able to communicate *directly* (peer-to-peer) with one another (rather than through a coordinating base station in between). In this way, the agent can estimate the position of the nodes cooperatively by exploiting all the available internode range measurements, not only those with respect to the anchors. Cooperation can dramatically increase the localization performance with respect to conventional techniques, especially in ad hoc networks with low anchor density and/or low transmission coverage [85, 89, 123].

We are interested in a *distributed* approach in which each node estimates *locally* its own position based on range measurements obtained from neighbors and some prior information, if available. Our investigation focuses on stochastic networks; that is, we assume that nodes are randomly deployed over a finite one-/two-dimensional (1D/2D) space, according to a prior distribution. For the solution of the distributed Bayesian inference problem, we employ the *belief propagation* (BP) algorithm [50]. This is an iterative procedure that allows each node to compute the a posteriori pdf (belief) of its own position by a repeated exchange of pdfs (messages) with nearby nodes.

In Section 5.4.5.2, we investigate a distributed localization approach based on BP, which extends the results of Ref. [50]. However, a main problem in BP is the efficient representation of beliefs and messages that have to be evaluated and exchanged by nodes. As a matter of fact, the computation and transmission of these functions can have an unacceptably high cost for practical purposes, depending on the dimension of the deployment. Here, we adopt a particle filter (PF) BP approach (PF-BP) in which each node represents its beliefs and messages through random sets of particles, updating them

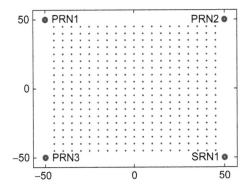

FIGURE 5.48

Simulation environment with three primary nodes and one secondary node. The coordinates are in meters (from Ref. [133]).

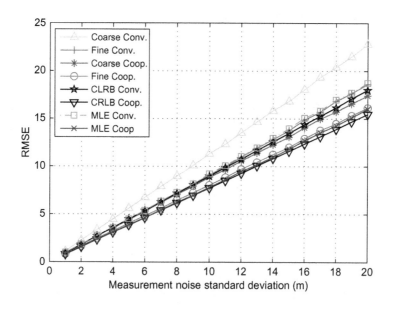

FIGURE 5.49

RMSE of the CRB, MLE, and the linear estimators for both cooperative and noncooperative cases (from Ref. [133]).

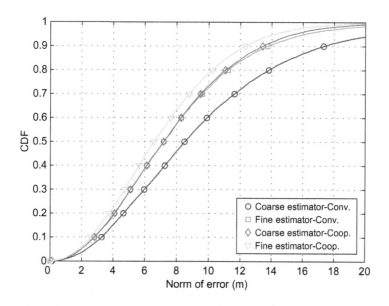

FIGURE 5.50

CDF of the linear estimators for both cooperative and noncooperative networks ($\sigma = 10$ m) (from Ref. [133]).

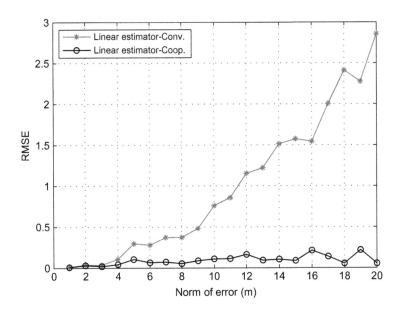

FIGURE 5.51

RMSE of the linear estimators for both cooperative and noncooperative networks (from Ref. [133]).

by importance resampling and weighting [38]. Fusion of messages from different neighbors (defined over different sets of particles) is obtained by an efficient resampling over an auxiliary common set of particles. The fusion method is less complex than techniques usually adopted for combining PF and BP, based on Gibbs sampling [114], as it is linear in the number of particles (while Gibbs sampling is quadratic).

In Section 5.4.5.3, we analyze the performance of the PF-BP method on simulated 1D/2D random networks and experimental UWB measurements. We also compare the BP-PF performances with fundamental performance limits based on the CRB. While the deterministic (or local) CRB has been widely investigated for localization of a fixed configuration of nodes (see, e.g., Refs. [44, 85, 89] and references therein), we propose a novel analysis based on Bayesian (or global) bounds that are valid for any set of unknown node positions drawn from the a priori distribution. We derive the Bayesian CRB [117] and other types of global bounds for the considered localization scenario. Finally, we evaluate the performance of the distributed PF-BP algorithm on both simulated and experimental TOA measurements, comparing the results with fundamental bounds.

5.4.5.1 Network Model

We consider a stochastic network of N nodes deployed over a finite space. The generic kth node position, $k = 1, \ldots, N$, is denoted by the vector $x_k \in \mathcal{X}$ that takes values over the nD finite space $\mathcal{X} \subset \mathbb{R}^n$, with $n \in \{1, 2\}$ (for 1D or 2D localization scenarios). For convenience of notation, we assume that the first N_u nodes have unknown positions $\mathbf{x} = [x_1 \ldots x_{N_u}]^T$, while the remaining $N_a = N - N_u$ nodes are anchors with known positions $\mathbf{x}_a = [x_{N_u+1} \ldots x_N]^T$. We also assume that $N_u \gg N_a$. The agent positions $\mathbf{x}$ are modeled as independent RVs distributed within the nD finite space $\mathcal{X}$ with a given pdf

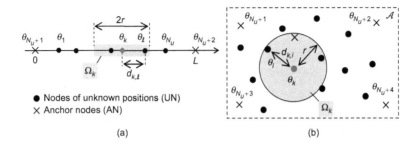

FIGURE 5.52

Examples of localization scenarios: (a) 1D scenario, $\theta_k = x_k$; (b) 2D scenario, $\theta_k = x_k \equiv (x_k, y_k)$, $\mathcal{A} = \mathcal{X} \in \mathbb{R}^2$.

$p(\mathbf{x}) = \prod_{k=1}^{N_u} p_k(x_k)$, whereas anchors are assumed to be placed in fixed positions. Examples of 1D and 2D deployment are shown in Fig. 5.52.

Cooperative estimation of the locations $\mathbf{x}$ is obtained from pairwise distance measurements $z_{k,\ell}$ made between any pair of (either anchor or agent) nodes k and ℓ, for $k, \ell = 1, \ldots, N$. To account for energy constraints, we assume a limited transmission coverage, which results in the following definition of the set of neighbors for the generic node k: $\Omega_k = \{x_\ell : \|x_k - x_\ell\| \leq r\}$, where r is the coverage radius. For $\ell \in \Omega_k$, the measurement from node k to node ℓ is modeled as

$$z_{k,\ell} = d_{k,\ell} + e_{k,\ell}, \tag{5.140}$$

where $d_{k,\ell} = \|x_k - x_\ell\|$ is the distance between the two nodes and $e_{k,\ell} \sim \mathcal{N}(0, \sigma_{k,\ell}^2)$ denotes the Gaussian measurement uncertainty with zero mean and variance $\sigma_{k,\ell}^2$. Errors associated with different links, which are made between different pairs of nodes, are assumed to be independent of one another.

5.4.5.2 *A Distributed BP-Based Positioning Algorithm*

We consider in this section a possible approach for the solution of the localization problem in a wireless network that cannot rely on a centralized infrastructure. In this condition, each node must estimate its own position locally, using only noisy measurements obtained from neighbors and some a priori information, if available. Starting from the results in Ref. [50], we investigate a distributed localization approach based on BP. This is an iterative local message-passing procedure that allows each node to compute, through successive refinements, the marginal a posteriori distribution of its own position. We apply the algorithm to a simple 1D network modeled as described in Section 5.4.5.1, comparing its performance with those of the centralized MLE and the CRB.

We assume that the agent locations $\mathbf{x}$ are randomly distributed within the 1D space with known pdf $p(\mathbf{x})$. More specifically, the wireless network is modeled as an ordered set of N_u nodes that are uniformly distributed in the finite spatial interval $(0, L)$ with length L. We further assume that two anchors are placed at the two ends, that is, $x_{N_u+1} = 0$ and $x_{N_u+2} = L$. According to the considerations in Refs. [47, 112], the nodes form a binomial point process whose distance distributions from a common reference point (in this case, the coordinate of the first anchor) are shown in Fig. 5.53(a).

An exhaustive discussion on the BP method can be found in Ref. [50] and in Chapter 6. Here we recall only the two main equations that characterize the BP procedure. At the iteration t, node k receives from each neighbor $\ell \in \Omega_k$ a message, $m_{\ell \to k}^t(x_k)$, which represents the belief about node-k position

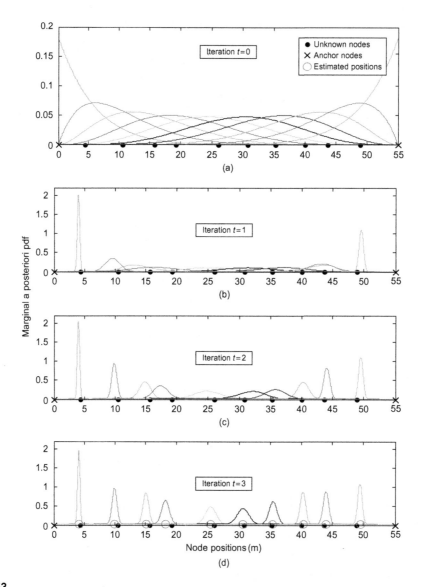

FIGURE 5.53

Marginal posterior distributions estimated by nodes (BP algorithm) after each iteration versus node positions. The simulation is performed on a noisy regular deployment with $SNR_0 = 15$ dB at $d_0 = 1$ m, $\alpha = 4$ and $r = 12$ m.

based on node-ℓ observations up to the current iteration. Node k approximates its a posteriori pdf as

$$\hat{p}^t(x_k|\mathbf{z}) \propto p(x_k) \prod_{\ell \in \Omega_k} m^t_{\ell \to k}(x_k), \tag{5.141}$$

where $\propto$ stands for proportional. The new location estimate $\hat{x}^t_k$ is drawn from (5.141) using the MAP criterion:

$$\hat{x}^t_k = \arg\max_{x_k} \hat{p}^t(x_k|\mathbf{z}),$$

where $\mathbf{z} = \{z_{k,\ell}\}$ is the set of noisy observations. The a posteriori pdf is then used to compute the message $m^{t+1}_{k \to \ell}(x_k)$ to be transmitted from node k to node ℓ for the next iteration according to

$$m^t_{k \to \ell}(x_\ell) \propto \int \psi_{k,\ell}(x_k,x_\ell)\, \psi_k(x_k) \prod_{i \in \Omega_k, i \neq \ell} m^{t-1}_{i,k}(x_k)\, dx_k, \tag{5.142}$$

where the *pairwise potential* $\psi_{k,\ell}(x_k,x_\ell)$ represents the measurement likelihood for the link (k,ℓ):

$$\psi_{k,\ell}(x_k,x_\ell) = \mathcal{P}\big(I_{\Omega_k}(\ell) = 1\big) p\big(z_{k,\ell}|x_k,x_\ell\big), \tag{5.143}$$

where $\mathcal{P}(\ell \in \Omega_k) = \mathcal{P}(d_{k,\ell} \leq r)$ is the *probability of visibility* that is calculated from the a priori distributions (see Fig. 5.53(a)), while $p(z_{k,\ell}|x_k,x_\ell)$ can be easily obtained by recalling that $z_{k,\ell}$ conditioned to (x_k,x_ℓ) has a Gaussian distribution $\mathcal{N}\big(\|x_k - x_\ell\|, \sigma^2_{k,\ell}\big)$, as assumed in (5.140). Then

$$m^{t+1}_{k \to \ell}(x_\ell) \propto \int p(z_{k,\ell}|x_k,x_\ell) \frac{\hat{p}^t(x_k|\mathbf{z})}{m^t_{\ell \to k}(x_k)} dx_k, \tag{5.144}$$

where the measurement likelihood $p(z_{k,\ell}|x_k,x_\ell)$ is, from (5.140), Gaussian with mean $\|x_k - x_\ell\|$ and variance $\sigma^2_{k,\ell}$. Steps involving (5.141)–(5.144) are repeated till convergence. The algorithm is initialized with $\hat{p}^0(x_k|\mathbf{z}) = p_k(x_k)$. The complete algorithm is summarized in Algorithm 5.1.

In Figs 5.53 and 5.54, the performance of the BP algorithm is evaluated for a binomial point process sensor network realization, with $N_u = 10$ agents distributed over a 1D space of length $L = 55$ m, with $N_r = 2$ anchors at the ends of the segment. Beliefs are uniformly sampled with sampling interval of 0.2 m; the tolerance parameter for convergence is $\varepsilon = 0.05$.

Figure 5.53 shows the marginal a posteriori beliefs for all nodes at iterations $t = 0,1,2,3$ and the corresponding MAP estimates. The localization accuracy obtained by BP for $t = 3$ is compared with the MLE performance and the CRB in Fig. 5.54(a). The RMSE for the whole array is $\text{RMSE}_{\text{BP}} = 0.576$ m, $\text{RMSE}_{\text{MLE}} = 0.339$ m, $\text{RMSE}_{\text{CRB}} = 0.307$ m.

5.4.5.3 *BP-Based Cooperative Positioning Using Particle Filters*

We now perform distributed Bayesian estimation of the random vector $\mathbf{x}$, given the set of noisy observations $\mathbf{z} = \{z_{k,\ell}\}$ and the a priori information $\{p(x_k)\}^{N_u}_{k=1}$, by the BP procedure. The self-localization problem is formulated as an undirected graphical model (or Markov random field) in which the vertices are the nodes, with associated random positions x_k, and the edges connect nodes that exchange measurements with one another. Each node computes the marginal a posteriori pdf of its own location, $p(x_k|\mathbf{z})$, by successive refinements, and it uses this belief to localize itself through MAP estimation.

1: **Initialization** $(t = 0)$:
2: **for** each node $k = 1$ to N_u **do**
3:     the node obtains local information and computes its own potential $\psi_k(x_k)$;
4:     the node collects the range estimates $\{z_{k,\ell}\}$ from neighbors, by broadcasting its node ID (k) and listening for other node broadcasts $(\ell \in \Omega_k)$;
5:     the node computes the pairwise potential $\{\psi_{k,\ell}(x_k, x_\ell)\}$ for each neighbor $\ell \in \Omega_k$;
6:     the node initializes the received messages to $m_{\ell,k}^0(x_k) = 1$ and the marginal posterior distribution as to
$$\hat{p}^0\left(x_k | \{z_{k,\ell}\}\right) = \psi_k(x_k).$$
7: **end for**
8: **Iterations** $(t = 1, 2, \ldots)$:
9: **repeat**
10·   **for** each node $k = 1$ to N_u **do**
11:         the node computes the message $m_{k,\ell}^t(x_\ell)$ to send to node ℓ, $\forall \ell \in \Omega_k$, according to (5.142);
12:         the node sends $m_{k,\ell}^t(x_\ell)$ to its neighbors and receives the neighbors' messages $m_{\ell,k}^t(x_k)$, $\forall \in \Omega_k$;
13:         the node updates the marginal a posteriori pdf $\hat{p}^t\left(x_k | \{z_{k,\ell}\}\right)$ according to (5.141);
14:         the node computes the MAP estimate of its own position as $\hat{x}_k^t = \arg\max\left\{\hat{p}^t\left(x_k | \{z_{k,\ell}\}\right)\right\}$ (interpolation is implemented before the maximization);
15:     **end for**
16: **until** $\dfrac{\|\hat{x}_n^t - \hat{x}_n^{t-1}\|}{\hat{x}_n^{t-1}} > \varepsilon$ (a given threshold)

Algorithm 5.1

Distributed positioning algorithm based on BP

The main issue for practical implementation of the BP algorithm is the efficient representation of beliefs/messages to be calculated/exchanged among nodes. A grid-based approach relying on a regular sampling of the location space has an unacceptably high cost for both computation and transmission of beliefs. A more efficient method has been proposed in Refs. [50, 114] based on Gaussian mixture models for the representation of beliefs/messages and Gibbs sampling or mixture importance sampling for the fusion of messages from different neighbors. An alternative technique using PF has recently been proposed in Ref. [51] for a generic Bayesian estimation problem, with messages from neighbors sharing the same set of particles. Here, we consider a modified PF approach, called the PF-BP method (see Ref. [38] for details), in which each node represents its beliefs (5.141) and messages (5.144) through random sets of particles as follows:

$$\hat{p}^t(x_k|\mathbf{z}) \approx \sum_{n=1}^{N_p} \omega_{k,n}^t \delta\left(x_k - x_{k,n}^t\right), \tag{5.145}$$

$$m_{k \rightarrow \ell}^{t+1}(x_\ell) \approx \sum_{n=1}^{N_p} \gamma_{k \rightarrow \ell,n}^{t+1} \delta\left(x_\ell - x_{k \rightarrow \ell,n}^{t+1}\right), \tag{5.146}$$

where $\{x_{k,n}^t, \omega_{k,n}^t\}_{n=1}^{N_p}$ denote the particles and weights for the a posteriori pdf of the kth node location, while $\{x_{k \rightarrow \ell,n}^{t+1}, \gamma_{k \rightarrow \ell,n}^{t+1}\}_{n=1}^{N_p}$ are those of the messages for the node ℓ position. Particles are updated locally at each node by importance resampling and weighting [3, 86]. A general discussion on Bayesian estimation theory and particle filters can be found in Chapter 6.

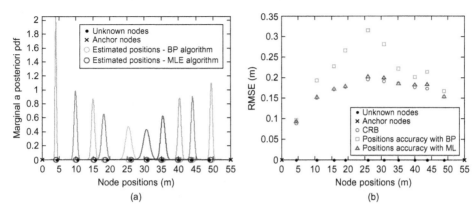

FIGURE 5.54

(a) Marginal a posteriori pdfs at the last BP iteration and positions estimated by BP and ML. (b) Location accuracy obtained with the BP algorithm, MLE algorithm, and the CRB versus node positions; the MSEs are averaged over 100 experiments. This simulation is performed with a higher $SNR_0 = 20$ dB and a smaller sampling interval (0.1 m).

The main issue in PF-BP localization is the fusion, through the product in (5.141), of the messages obtained from different neighbors and thus sampled over different sets of particles. A possibility to approximate the message product is to smooth and resample the messages over an auxiliary common set of particles. Here, we propose to join all neighbors' particles in the common set of $|\Omega_k| \cdot N_p$ particles (a smaller set could be obtained by selecting a reduced number of significant samples from each message). The computation of messages over the new particle set is obtained by smoothing the messages (5.146) with a kernel function $w(\mathbf{x})$ (see Ref. [38] for investigation of different types of kernel). The resampled messages are finally multiplied according to (5.141) to get the updated a posteriori belief. Then, each node k has to evaluate particles and weights $\{x_{k \to \ell,n}^{t+1}, \gamma_{k \to \ell,n}^{t+1}\}_{n=1}^{N_p}$ for the message to be sent to each neighbor ℓ. This is carried out by applying importance sampling to (5.144), using as importance density the belief for the node-ℓ position approximated at node k (see Ref. [38] for details).

Fundamental Performance Limits for Stochastic Networks

For a given configuration of nodes $\mathbf{x}$, the deterministic (or local) CRB, denoted by $\kappa_{CRB}(\mathbf{x})$, provides a lower bound on the covariance matrix $\mathbf{C}(\mathbf{x})$ of any unbiased estimate $\hat{\mathbf{x}}(\mathbf{z})$ drawn from $\mathbf{z}$ [117]:

$$\mathbf{C}(\mathbf{x}) \triangleq \mathbb{E}_{\mathbf{z}}[(\hat{\mathbf{x}}(\mathbf{z}) - \mathbf{x})(\hat{\mathbf{x}}(\mathbf{z}) - \mathbf{x})^T] \geq \mathbf{J}^{-1}(\mathbf{x}) \triangleq \kappa_{CRB}(\mathbf{x}). \tag{5.147}$$

The term $\mathbf{J}(\mathbf{x})$ denotes the FIM, whose expression for 1D/2D localization problems can be found in Refs. [85, 89, 107, 108] and in Section 4.4.2.2.

The deterministic CRB depends on the actual value of $\mathbf{x}$, that is, on a specific network deployment. When considering random deployments, on the other hand, it is of interest to evaluate a Bayesian (or global) bound that should be valid for any value of $\mathbf{x}$ drawn from $p(\mathbf{x})$. The Bayesian CRB (BCRB) [117], denoted by κ_{BCRB}, belongs to this class of bounds. It is a lower bound for the covariance matrix

C of any Bayesian estimate $\hat{\mathbf{x}}(\mathbf{z})$, based on the measurements $\mathbf{z}$ *and* the a priori information $p(\mathbf{x})$:

$$\mathbf{C} \triangleq \mathbb{E}_{\mathbf{x}} \{\mathbf{C}(\mathbf{x})\} \geq \mathbf{J}_{\mathrm{B}}^{-1} \triangleq \boldsymbol{\kappa}_{\mathrm{BCRB}}, \tag{5.148}$$

with $\mathbf{J}_{\mathrm{B}}$ denoting the Bayesian information matrix (BIM),

$$\mathbf{J}_{\mathrm{B}} = \underbrace{\mathbb{E}_{\mathbf{x}} \{\mathbf{J}(\mathbf{x})\}}_{\mathbf{J}_{\mathrm{D}}} + \underbrace{\mathbb{E}_{\mathbf{x}} \left\{ \frac{\partial \ln p(\mathbf{x})}{\partial \mathbf{x}} \frac{\partial^T \ln p(\mathbf{x})}{\partial \mathbf{x}} \right\}}_{\mathbf{J}_{\mathrm{P}}}. \tag{5.149}$$

The matrices $\mathbf{J}_{\mathrm{D}}$ and $\mathbf{J}_{\mathrm{P}}$ represent the information obtained from the data and the a priori pdf, respectively. We recall that in the case of uniform a priori pdf (with finite support) the BCRB is not defined as the weak unbiasedness condition [117] is not fulfilled (the a priori pdf is not twice differentiable); still the BCRB holds for $N \to \infty$ and/or large SNR as the measurements become more informative and the a priori information loses importance. In these asymptotic conditions (high SNR bound), $\mathbf{J}_{\mathrm{B}} \approx \mathbf{J}_{\mathrm{P}}$ and the BCRB reduces to the inverse of the average FIM, that is, to the modified CRB (MCRB):

$$\boldsymbol{\kappa}_{\mathrm{MCRB}} \triangleq (\mathbb{E}_{\mathbf{x}} \{\mathbf{J}(\mathbf{x})\})^{-1}. \tag{5.150}$$

If we consider a non-Bayesian estimate that does not exploit any a priori information and treats $\mathbf{x}$ as a deterministic parameter, performances are expected to be worse than for the Bayesian estimate. In this case, a tighter bound for the error covariance $\mathbf{C}$ is given by the average of the local CRBs with respect to $\mathbf{x}$, here called the average CRB (ACRB):

$$\boldsymbol{\kappa}_{\mathrm{ACRB}} \triangleq \mathbb{E}_{\mathbf{x}} \{\boldsymbol{\kappa}_{\mathrm{CRB}}(\mathbf{x})\} = \mathbb{E}_{\mathbf{x}} \left\{ \mathbf{J}(\mathbf{x})^{-1} \right\}. \tag{5.151}$$

From Jensen's inequality, we have

$$(\mathbb{E}_{\mathbf{x}} \{\mathbf{J}(\mathbf{x})\} + \mathbf{J}_{\mathrm{P}})^{-1} \leq (\mathbb{E}_{\mathbf{x}} \{\mathbf{J}(\mathbf{x})\})^{-1} \leq \mathbb{E}_{\mathbf{x}} \left\{ \mathbf{J}(\mathbf{x})^{-1} \right\}, \tag{5.152}$$

and thus holds

$$\boldsymbol{\kappa}_{\mathrm{BCRB}} \leq \boldsymbol{\kappa}_{\mathrm{MCRB}} \leq \boldsymbol{\kappa}_{\mathrm{ACRB}}. \tag{5.153}$$

In the next paragraph, all the above CRB bounds will be analyzed for the considered localization scenario.

Performance Analysis

We consider the deployment in Fig. 5.52(a) with $N_{\mathrm{a}} = 2$ anchors at 0 m and $L = 45$ m, and $N_{\mathrm{u}} = 8$ agents. The coverage radius for cooperation is $r = 10$ m. In practical systems, a priori information about the node positions can be obtained during the installation phase, based on some knowledge of the spatial configuration of the environment or the wireless network structure (e.g., unknown nodes are known to be clustered around a number of anchors); the a priori pdf $p(\mathbf{x})$ should be defined so as to account for this information. For simulation here we consider a simplified scenario where agent positions are independent random variables distributed over the interval $[0, L]$ according to the a priori pdfs $\{p_k(x_k)\}_{k=1}^{8}$ shown in Fig. 5.55 (top figure): $p_k(x_k)$ is a raised cosine of width 2Δ centered around the position $k\Delta$, where $\Delta = L/(N_{\mathrm{u}} + 1) = 5$ m is the mean internode distance. Measurements are generated according to (5.140) with $\sigma_{k,\ell} = \sigma \approx 1.8$ m.

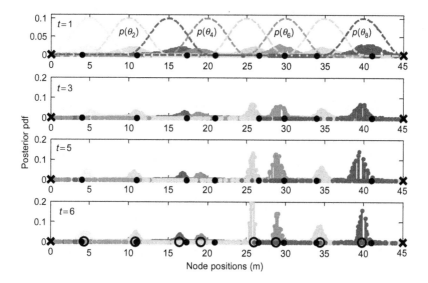

FIGURE 5.55

Marginal posterior pdf estimated by nodes at iterations $t = 1, 3, 5, 6$. Anchors are represented with "×" and agents with "●."

An example of node beliefs calculated by the BP procedure is shown in Fig. 5.55. The algorithm is implemented using a triangular window $w(\mathbf{x})$ of support $L_w = 1$ m and $N_p = 100$ particles. Figure 5.55 shows the marginal a posteriori distributions approximated by nodes at some iterations ($t = 1, 3, 5, 6$) and the estimated positions (open circles) at the last iteration. Looking at the evolution of the beliefs over the iterations, it can be seen that the accuracy and the convergence rate are worse for central nodes which are far away from anchors. After six iterations the RMSE for the PF-BP algorithm, averaged over the nodes of the fixed array, is $\mathsf{RMSE} \approx 1$ m.

We compare now the performance of the PF-BP algorithm with the lower bounds discussed in the previous paragraph. The network model is simulated according to the a priori distribution in Fig. 5.55 (top figure) with $L = 55$ m, $N_u = 10$, and mean internode distance $\Delta = 5$ m. Network configurations for which the localization problem is not solvable (i.e., the FIM is singular) are not considered. The left plot in Fig. 5.56 shows the RMSE of the location estimate versus the inverse range error $1/\bar{\sigma}$ computed over 100 realizations of the network layout, each of them characterized by 100 sets of measurements. The range error is normalized with respect to the coverage radius $r = 10$ m so that $\bar{\sigma} = \sigma/r$. The accuracy of the PF-BP algorithm is shown to approach the BCRB while it is upper bounded by the ACRB. As expected, the BCRB is a more realistic bound with respect to the ACRB for Bayesian estimation as it accounts for the information provided by the a priori distribution $p(\mathbf{x})$. For small range error σ (high SNR bound), we observe that the BCRB tends to the MCRB as the information drawn from measurements ($\mathbf{J}_D$) becomes dominant with respect to the a priori information ($\mathbf{J}_P$). For $\sigma = 4.5$ m, the right plot in Fig. 5.56 shows the localization accuracy as a function of the node index k. Since nodes are randomly scattered around a regular grid with internode interval Δ, the performances along the array are similar to those observed for regular networks (see Ref. [85]), with accuracy that decreases toward

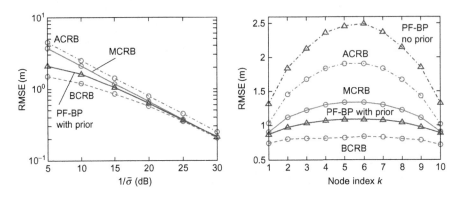

FIGURE 5.56

Accuracy of PF-BP compared to the lower bounds versus the inverse normalized range accuracy (left) and the nodes along the 1D array (right).

the center of the array. In addition to PF-BP with known $p_k(x_k)$, the figure also shows the performance of the algorithm when a priori information is not available, that is, for $p_k(x_k)$ uniform in $[0, L]$. In this case, ACRB is a tighter bound for distributed localization.

Finally, the PF-BP algorithm is applied to the database of range measurements collected during the measurement campaign described in Chapter 7. The sensor nodes, numbered from 1 to 20, were placed in fixed positions (static scenario) in a typical office environment, with walls and obstacles (NLOS conditions). Figure 5.57 presents the localization obtained by the application of PF-BP to a randomly chosen set of range measurements. We assume that only four nodes know their location $(2, 6, 13, 17)$, whereas the other 16 nodes have to be localized. The algorithm is implemented using $N_p = 100$ particles; the product of all incoming messages is computed over a collection of $|\Omega_k| \cdot N_p$ particles for each node. The windowing function $w(\mathbf{x})$ is a Gaussian kernel with effective width σ_w equal to the error standard deviation characterizing the radio link; this quantity is computed as a sample variance over the set of measurements $\{z_{k,\ell}^{(m)}\}$: $\sigma_{k,\ell}^2 = \frac{1}{N_m} \sum_{n=1}^{N_m} \left(z_{k,\ell}^{(m)} - d_{k,\ell}\right)^2$, where $N_m = 1000$ is the number of range estimates per link. Note that negative and zero distance measurements have been removed from the data set in the preprocessing phase. The a priori distributions are chosen as Gaussian with mean equal to the real positions and variance 10 m.

Figure 5.57(a) shows the agent position estimates obtained by the PF-BP algorithm and the 3-σ uncertainty ellipses computed according to (5.147), based on the hypothesis of unbiased Gaussian range errors. The RMSE averaged over all nodes is 1.5 m. Note however that the RMSE fluctuates and the reachable accuracy is not uniform along the deployment, as shown by the CRB ellipses: nodes 1, 19, and 20 seem to be in bad positions for self-localization. Figure 5.57(b–d) shows the a posteriori distributions for nodes 1 (orange), 12 (green), and 19 (blue) at iterations $t = 0$ (initialization), $t = 1$, and $t = 5$ (end). As already observed for a 1D geometry in Fig. 5.55, particles representing the resampled beliefs tend to concentrate in the neighborhood of the true positions. Due to the large number of neighbors for each node, the convergence is fast and the algorithm reaches its best performance in few iterations (3–5).

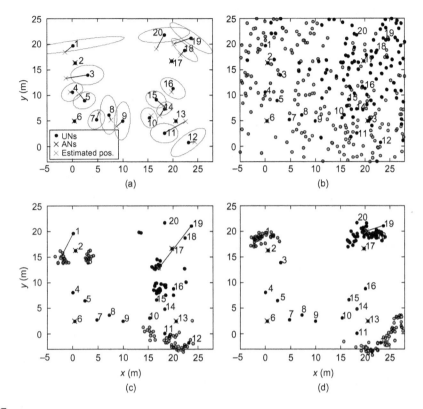

FIGURE 5.57

Distributed localization from real UWB-TOA measurements: PF-BP position estimates and 3-σ uncertainty ellipses computed by the local CRB, in the hypothesis of unbiased Gaussian range measurements (a); particle-based representation of the posterior distributions for nodes 1, 12 and 19 at iterations $t = 0$, $t = 1$ and $t = 5$ (b–d).

5.5 COGNITIVE POSITIONING FOR COGNITIVE RADIO TERMINALS

The formidable development of wireless communication systems in recent years has led to a progressive definition of rules for accessing the radio spectrum, posing several tasks in the organization and in the sharing strategies for such a precious resource. Traditionally, the radio spectrum has been divided into two categories: licensed and unlicensed bands. In unlicensed bands, users can freely transmit. However, the amount of interference introduced by a transmission system operating in those bands must be limited to avoid impairments to other users' communications. Thus, some limitations on the power spectral density are usually introduced. Nevertheless, too large a number of users transmitting on the same unlicensed band (in the same area) can be responsible for a high overall aggregated interference, sufficient to prevent communications.

There is an emerging concept in the spectrum usage, known as *cognitive radio* (CR) [79]. CR aims at providing a more efficient and flexible usage of the radio spectrum. The licensed spectrum is not

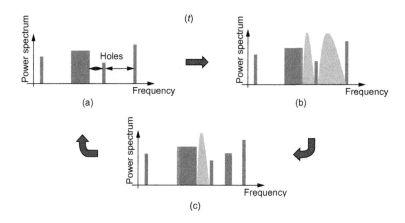

FIGURE 5.58

The cognitive radio cycle: spectrum sensing (a), agile spectrum generation (b), continuously sense and adapt (c).

used, for instance, in some geographic areas; there are also many applications for emergency where the use of the spectrum has a very low duty cycle (e.g., services that use the spectrum only occasionally, but with high priority). Moreover, it has been observed, by investigations on the radio spectrum usage including some revenue-rich urban areas, that some frequency bands are largely unoccupied most of the time or are only partially occupied and that the remaining frequency bands are heavily used.

CR permits one in principle to satisfy the necessity of greater efficiency in the radio spectrum occupation. The basic idea is that a CR terminal can sense its environment (see Fig. 5.58) and location and then adapt some of its features (power, frequency, modulation, etc.) allowing the dynamic reuse of the available spectrum. This could, in theory, lead to a multidimensional reuse of spectrum in space, frequency, and time, exceeding the severe limitations in the spectrum and bandwidth allocations that have slowed down broadband wireless communications development.

Modern wireless networks expect more and more availability of location information about the wireless terminals, driven by requirements coming from applications, or just for better network resources allocation [16, 60]. Thus, signal-intrinsic capability for accurate localization is a goal of fourth generation (4G) as well as beyond-4G (B4G) networks [39, 76]. Therefore, the CR concept can also be applied in the context of high-definition location systems where the accuracy in ranging (hence, in localization) can be varied according to the available bandwidth, thus leading to the new concept of *cognitive positioning* (CP) [15–17, 45, 63].

In the following, we will see that a multicarrier (MC) signal format also gives new opportunities in terms of enhanced accuracy time-delay estimation that ultimately translates into enhanced accuracy positioning. We will first derive some results for a generic multicarrier signal, and then we will specialize and analyze in detail the special case of a band-limited filter-bank multitone modulated signal.

5.5.1 Cognitive TOA Estimation

In Section 4.4.3, CRBs are derived for TOA estimation in systems that employ an OFDM type signal in the presence of interference [28, 63]. Here we show how these results can be exploited to design the optimal waveform shape that minimizes the CRB given a sensed interference spectrum and power spectral emission constraints. In addition, an ML-TOA estimation algorithm is derived and its performance is evaluated via simulations to assess the impact of the proposed optimal signal shape on a practical estimator.

The transmitted baseband signal (4.136) considered in Section 4.4.3 relating to an OFDM-like system is here reported for convenience [28]:

$$s(t) = \sum_{k=1}^{K} \sqrt{w_k} \, g(t) \, e^{j2\pi f_k t} \,, \tag{5.154}$$

for $t \in [-T_s/2, T_s/2]$, where $f_k = (k - K/2)\Delta$ is the kth subcarrier frequency shift with respect to the center frequency, Δ is the subcarrier spacing, $g(t)$ is a pulse with duration T_s and energy E_g. The weights $w_k \geq 0$ facilitate spectrum shaping and

$$P_t = \frac{E_g}{T_s} \sum_{k=1}^{K} w_k \tag{5.155}$$

denotes the power of the baseband signal. Recall that in general the weights w_k are limited by peak power constraints when considering the optimal signal spectrum [28].

Optimal Weights Design

We concentrate on the weight assignment that minimizes the CRB. It is assumed that the interference spectral density $S_I(f)$ is known or estimated, which is commonly possible in a system employing CR techniques with a spectrum sensing unit [79].

The optimal weights must satisfy constraints on the emitted signal spectrum imposed by regulatory masks (e.g., the FCC mask for ultrawide bandwidth signals). Let $B(f)$ denote the equivalent baseband version of the power spectral density mask. Then, defining $\boldsymbol{w} \triangleq (w_1, w_2, \ldots, w_K)^T$ and $\boldsymbol{\lambda} \triangleq (\lambda_1, \lambda_2, \ldots \lambda_K)^T$ (cf. (4.164)), the optimal weights are found as the solution of the following problem:

$$\underset{\boldsymbol{w}}{\text{maximize}} \, \boldsymbol{\lambda}^T \boldsymbol{w} \tag{5.156}$$

subject to
$$\mathbf{1}^T \boldsymbol{w} \leq 1 \tag{5.157}$$
$$\boldsymbol{w} \geq \mathbf{0} \tag{5.158}$$
$$\boldsymbol{w} \leq \boldsymbol{b}, \tag{5.159}$$

where $\boldsymbol{x} \preceq \boldsymbol{y}$ means that each element of $\boldsymbol{x}$ is smaller than or equal to the corresponding element of $\boldsymbol{y}$, $\mathbf{1}$ is the vector of all ones, $\boldsymbol{b} \triangleq [b_1 \ b_2 \ \ldots \ b_K]^T$, and $b_k \triangleq B(f_k) \Delta / P_t$ is the normalized emission power constraint on the kth subcarrier.

This is a classical linear programming problem and its solution can be obtained in closed form as follows: Without loss of generality, assume that the λ_k are in a decreasing order, that is, $\lambda_1 > \lambda_2 > \cdots > \lambda_K$. Then, the optimal weights are recursively computed as

$$
\begin{aligned}
w_1^{(\text{opt})} &= \min\{1, b_1\}, \\
w_2^{(\text{opt})} &= \min\{1 - w_1^{(\text{opt})}, b_2\}, \\
w_3^{(\text{opt})} &= \min\{1 - w_1^{(\text{opt})} - w_2^{(\text{opt})}, b_3\}, \\
&\vdots
\end{aligned}
\tag{5.160}
$$

and so on. This result has the following intuitive interpretation. We start by selecting the *best* subcarrier (the one associated with the largest component of λ) and we assign to it the maximum allowed power, which is $\min\{1, b_1\}$. Next, we select the best of the remaining subcarriers and again assign to it the maximum allowed power (which is the minimum between b_2 and the residual power $1 - w_1^{(\text{opt})}$). We proceed in this way until all the available power is used or no other subcarriers are available (which happens if $\sum_{i=1}^{K} b_i \leq 1$).

Practical TOA Estimator

Without loss of generality we assume that $\{1, 2, \ldots, \bar{K}\}$, with $\bar{K} \leq K$, is the subset of indices k corresponding to $w_k > 0$. Then, (4.139) can be written in the equivalent form:

$$
s_r(t) = \sum_{k=1}^{\bar{K}} \alpha_k \sqrt{w_k}\, g(t)\, e^{j\, 2\pi f_k t}.
\tag{5.161}
$$

Letting $\tilde{g}_k(t) = \sqrt{w_k}\, g(t) e^{j 2\pi f_k t} \otimes h(t)$ and taking (5.161) into account, the log-likelihood function reads

$$
\ln \Lambda(\tilde{\boldsymbol{\theta}}) = \mathfrak{Re}\left\{ \sum_{k=1}^{\bar{K}} \tilde{\alpha}_k^* x_k(\tilde{\tau}) \right\} - \frac{1}{2} \sum_{k=1}^{\bar{K}} \sum_{l=1}^{\bar{K}} \tilde{\alpha}_k^* \tilde{\alpha}_l \rho_{k,l}
\tag{5.162}
$$

with

$$
x_k(\tilde{\tau}) = \int_{-\infty}^{\infty} x(t) \tilde{g}_k^*(t - \tilde{\tau})\, dt
\tag{5.163}
$$

and

$$
\rho_{k,l} = \int_{-\infty}^{\infty} \tilde{g}_k^*(t - \tilde{\tau}) \tilde{g}_l(t - \tilde{\tau})\, dt = \int_{-\infty}^{\infty} \tilde{g}_k^*(t) \tilde{g}_l(t)\, dt
$$

$$
= \int_{-\infty}^{\infty} G_k^*(f) G_l(f)\, df = \sqrt{w_k}\sqrt{w_l} \int_{-\infty}^{\infty} P^*(f - f_k) P(f - f_l) |H(f)|^2 df.
\tag{5.164}
$$

Using a matrix notation, (5.162) can be written as

$$
\ln \Lambda(\tilde{\boldsymbol{\theta}}) = \mathfrak{Re}\left\{ \tilde{\boldsymbol{\alpha}}^H \mathbf{x}(\tilde{\tau}) \right\} - \frac{1}{2} \tilde{\boldsymbol{\alpha}}^H \mathbf{R} \tilde{\boldsymbol{\alpha}},
\tag{5.165}
$$

where $\tilde{\boldsymbol{\alpha}} = [\tilde{\alpha}_1, \tilde{\alpha}_2, \ldots, \tilde{\alpha}_{\bar{K}}]^T$, $\mathbf{x}(\tilde{\tau}) = [x_1(\tilde{\tau}), x_2(\tilde{\tau}), \ldots, x_{\bar{K}}(\tilde{\tau})]^T$, and $\mathbf{R}$ is the Hermitian $\bar{K} \times \bar{K}$ correlation matrix:

$$\mathbf{R} = \begin{bmatrix} \rho_{1,1} & \rho_{1,2} & \cdots & \rho_{1,\bar{K}} \\ \rho_{2,1} & \rho_{2,2} & \cdots & \rho_{2,\bar{K}} \\ \vdots & \vdots & \ddots & \vdots \\ \rho_{\bar{K},1} & \rho_{\bar{K},2} & \cdots & \rho_{\bar{K},\bar{K}} \end{bmatrix}. \tag{5.166}$$

Our goal is to maximize (5.165) with respect to $\tilde{\tau}$ and $\tilde{\boldsymbol{\alpha}}$. For this purpose, taking $\tilde{\tau}$ fixed and letting $\tilde{\boldsymbol{\alpha}}$ vary, the maximum of (5.165) is achieved for

$$\widehat{\boldsymbol{\alpha}} = \mathbf{R}^{-1}\mathbf{x}(\tilde{\tau}). \tag{5.167}$$

Next, substituting (5.167) in (5.165) and maximizing with respect to $\tilde{\tau}$ produces

$$\hat{\tau} = \underset{\tilde{\tau}}{\operatorname{argmax}} \left\{ \mathbf{x}^H(\tilde{\tau}) \mathbf{R}^{-1} \mathbf{x}(\tilde{\tau}) \right\}. \tag{5.168}$$

The last equation gives the desired ML estimate of τ. The ML estimate of $\boldsymbol{\alpha}$ is obtained from (5.167) by replacing $\tilde{\tau}$ with $\hat{\tau}$ (however, we are not interested in estimating $\boldsymbol{\alpha}$).

From (5.164), it can be seen that, based on the disjoint spectrum assumption made in Section 4.4.3, we have $\rho_{k,l} = 0$ for $k \neq l$. Accordingly, (5.168) becomes

$$\hat{\tau} = \underset{\tilde{\tau}}{\operatorname{argmax}} \left\{ \sum_{k=1}^{\bar{K}} \rho_{k,k}^{-1} |x_k(\tilde{\tau})|^2 \right\}. \tag{5.169}$$

It should be noted that the time-delay estimators in (5.168) and (5.169) require a simple one-dimensional search, whose complexity increases only linearly with the duration of the search interval.

Numerical Examples

A scenario with a subcarrier spacing $\Delta = 1$ MHz and $K = 128$ subcarriers is considered. The channel coefficients α_k are modeled as independent complex-valued Gaussian random variables with unit average power. The results are obtained by averaging over 500 independent channel realizations. Pulse $g(t)$ in (5.154) is modeled as a sinc pulse, namely, $g(t) = \sqrt{E_p \Delta} \sin(\pi t \Delta)/(\pi t \Delta)$. Parameters β_1 and β_2 in (4.167) are set to 0 and $\Delta^2/12$, respectively. The results are expressed in terms of the square root of the CRB on the *ranging error*, which is computed as the product of the square root of the CRB on TOA error multiplied by the speed of light.

The interference spectral density $S_I(f)$ takes a constant value of $N_I = 2N_0$ for the subcarrier indices from 23 to 106 while it is zero elsewhere. In Fig. 5.59, the performance of the TOA estimator and the CRB are illustrated with optimal and conventional (uniform) weights in the case of an interference avoidance strategy. This means that the transmitted power is set to zero at the subcarriers with interference (i.e., $w_k = 0$ for $23 \leq k \leq 106$) while uniform or optimal power allocation is used with the remaining subcarriers. It is seen that the optimal and uniform allocation strategies provide the same TOA estimation accuracy in this case. In addition, it is observed that the estimation errors increase significantly in the presence of interference when the subcarriers with interference are not utilized. In

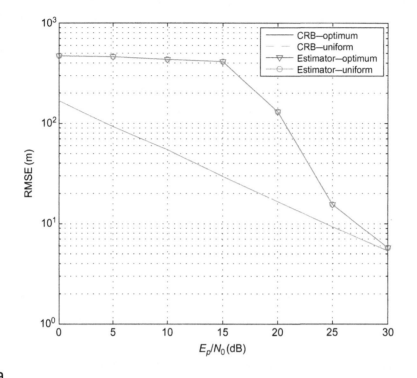

FIGURE 5.59

RMSE versus E_p/N_0 for the optimal and conventional (uniform) algorithms in the presence of interference with a flat spectral density in the interval $23 \leq k \leq 106$. In this scenario, the subcarriers with interference are not used (interference avoidance).

Fig. 5.60, the same scenario is considered except that all the subcarriers can now be employed. In this case, it is observed that the optimal algorithm improves both the CRB and the TOA estimation accuracy of the ML algorithm compared to the conventional (uniform) algorithm. In addition, the mean error values are smaller than those in the interference avoidance case (see Fig. 5.59), as expected. We conclude that subcarriers with interference should be employed to better use the frequency diversity and enhance TOA estimation performance.

5.5.2 Filter-Bank Multicarrier Ranging Signals

After the general development of the previous subsection, we will now examine in detail the specific format of a band-limited multicarrier signal with a binary ranging code to perform (cognitive) positioning. Although the results that we show here are to a large extent a direct consequence of the general framework above, we would like to develop in detail this "case study" to further highlight the potential of a multicarrier signal for ranging and positioning. We will assume that the chip rate of the ranging code is $R_c = 1/T_c$ and that the code repetition length L (the code period LT_c) is very large.

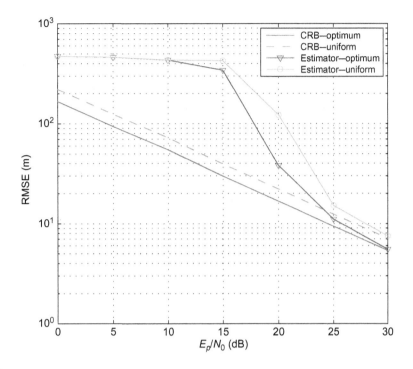

FIGURE 5.60

RMSE versus E_p/N_0 for the optimal and conventional (uniform) algorithms in the presence of interference with a flat spectral density in the interval $23 \leq k \leq 106$. In this scenario, subcarriers with interference are also used.

A basic ranging MC signal can be constructed following the general arrangement of multicarrier modulation: The input chip stream $c[i]$ of the ranging code is parallelized into N substreams with an MC symbol rate $R_s = R_c/N = 1/(NT_c) = 1/T_s$, where T_s is the time duration of the "slow-motion" ranging chips in the N parallel substreams. We can use a *polyphase* notation for the kth ranging subcode $(k = 0, 1, N-1)$ in the kth substream (subcarrier) as $c^{(k)}[n] \triangleq c[nN + k]$ where k, $0 \leqslant k \leqslant N-1$, is the subcode identifier and/or the subcarrier index, while n is a time index that addresses the nth MC symbol (block) of time length $T_s = NT_c$. The substreams are then modulated onto a raster of evenly spaced subcarriers with frequency spacing f_{sc} and the resulting modulated signals are added to give the (baseband equivalent of the) overall ranging signal. In filter-bank multicarrier modulation (FB-MCM) (also called *filtered multitone* (FMT)) the spectrum on all subcarriers is strictly band-limited and nonoverlapping, akin to conventional single channel per carrier (SCPC). The resulting signal is

$$x(t) = \sqrt{\frac{2 \cdot P_T}{N}} \sum_{n} \sum_{k=0}^{N-1} c^{(k)}[n] g(t - nT_s) \, e^{j2\pi k(1+\alpha)t/T_s}, \tag{5.170}$$

where P_T is the signal power and $g(t)$ is a band-limited pulse, for instance a square-root raised cosine pulse with roll-off factor α—in this case the subcarrier spacing is $(1+\alpha)/T_s$. Figure 5.61 shows the

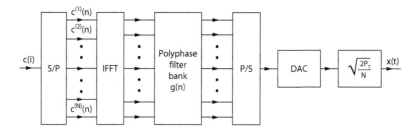

FIGURE 5.61

Filter-bank multicarrier modulator.

arrangement of a multicarrier modulator implementing (5.170). It is well known that this arrangement has an efficient realization based on the usual IFFT processing of multicarrier modulators, followed by a suited polyphase filter bank based on the prototype filter $g(t)$ [41].

Unlike OFDM, subcarrier orthogonality is attained in the frequency domain and holds irrespective of the signal observation time.

The bandwidth occupancy of an FB-MCM signal can be easily calculated based on the hypothesis that the chips of the ranging code are i.i.d., as is the case for a very long PN code. The PSD $S_x(f)$ of $x(t)$ is

$$S_x(f) = \frac{2 \cdot P_T}{N} \sum_{k=0}^{N-1} S\left(f - k\frac{(1+\alpha)}{T_s}\right),$$ (5.171)

where

$$S(f) = \frac{\mathbb{E}\{|c^{(k)}[n]|^2\}}{T_s}|G(f)|^2 = G_N(f)$$ (5.172)

is the PSD of each substream with chip rate $1/T_s$, in which $G(f)$ is the Fourier transform of the pulse $g(t)$, and $G_N(f)$ is a Nyquist frequency-raised-cosine function with roll-off factor α, and with $G_N(0) = T_s$. Figure 5.62 shows a sample FB-MCM spectrum to show its band-limitation feature. Unlike OFDM, FB-MCM does not need virtual carriers, since the spectrum is strictly band-limited and the sampling frequency in the modulators/demodulators obey Nyquist's rule.

When T_{obs} is sufficiently large ($T_{obs} = N_m T_s$, $N_m \gg 1$), the MCRB for such a signal can be computed as

$$\tilde{k}(\tau) = \frac{T_c^2}{8\pi^2 \frac{E_c}{N_0} \frac{N_m}{N}\left[\xi_g + \frac{(1+\alpha)^2}{N}\sum_k k^2\right]},$$ (5.173)

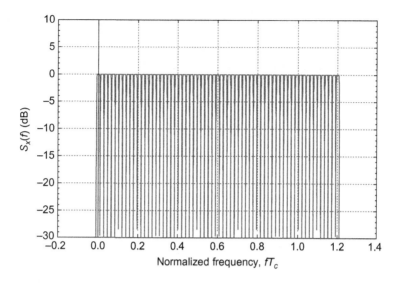

FIGURE 5.62

PSD of an FB-MCM signal with $N = 64$ carriers and a square root raised cosine pulse with $\alpha = 0.2$.

where $E_c = P_T \cdot T_c$ and ξ_g is the so-called pulse shape factor (PSF),[8] a normalized version of the Gabor bandwidth of pulse $g(t)$ [77]:

$$
\xi_g \triangleq \frac{T_s^2 \cdot \int\limits_{-\infty}^{\infty} f^2 \cdot |G(f)|^2 df}{\int\limits_{-\infty}^{\infty} |G(f)|^2 df}. \tag{5.174}
$$

5.5.3 Cognitive Bounds and Algorithms with Multicarrier Signals

As shown in the previous subsection, a multicarrier signal naturally allows for selective allocation of signal power across a wide bandwidth. To accommodate this feature, we only need to further introduce an amplitude coefficient on each subcarrier in (5.170):

$$
x(t) = \sqrt{\frac{2P_T}{N}} \sum_{n} \sum_{k=0}^{N-1} p_k\, c^{(k)}[n] g(t - nT_s)\, e^{j2\pi k(1+\alpha)t/T_s}, \tag{5.175}
$$

where p_k^2 is the relative power weight of carrier k ($p_k \geq 0$), which satisfies

$$
\sum_k p_k^2 = N \tag{5.176}
$$

---

[8] For a square-root raised cosine pulse with roll-off factor α, $\xi_g = 1/12 + \alpha^2(1/4 - 2/\pi^2)$.

to give the nominal total transmitted power P_T. Some p_ks can also be zero, indicating that the relevant subcarrier or even a whole subband is not being used. The relevant MCRB is found to be

$$\tilde{\kappa}(\tau) = \frac{T_c^2}{8\pi^2 \frac{E_c}{N_0} \frac{N_m}{N} \left[\xi_g + \frac{(1+\alpha)^2}{N} \sum_k k^2 p_k^2 \right]}. \tag{5.177}$$

A nice problem is now finding the *power distribution* that provides the best timing estimation accuracy, that is, that minimizes (5.177) through maximizing

$$\sum_k k^2 p_k^2 \tag{5.178}$$

with the constraint (5.176). When N is fixed and the spectrum is symmetric ($f_G = 0$), (5.178) is maximized for the optimal power scheme

$$\begin{cases} p_k = \sqrt{\frac{N}{2}} & k = \pm \frac{N-1}{2} \\ p_k = 0 & k \neq \pm \frac{N-1}{2} \end{cases}, \tag{5.179}$$

indicating a configuration in which the power of the signal is concentrated at the edge of the bandwidth. This is a well-known result of the Gabor bandwidth theory. When comparing the optimal distribution with what we get with the uniform (flat) power allocation $p_k \equiv 1$, we see for very large N that

$$\tilde{\kappa}(\tau)|_{\text{optimal}} \cong \frac{\tilde{\kappa}(\tau)|_{\text{flat}}}{3}. \tag{5.180}$$

An MC signal with uneven, adaptive power distribution can be adopted to implement a *cognitive positioning system* (CPS) [15]. In our envisioned FB-MCM scheme for positioning, the proper power allocation allows us to achieve the desired positioning accuracy, not only in an AWGN channel, but also in an additive colored Gaussian noise (ACGN) channel. Colored noise arises from variable levels of interference produced by coexisting (possibly primary) systems on different frequency bands. The key assumption is that such interference can be modeled as a Gaussian process. This is certainly justified in wireless networks with unregulated multiple access techniques such as CDMA and/or UWB.

A simple case study for a CPS may start from the following assumption: We use only M out of the N available subcarriers the total assigned bandwidth can be partitioned in, and we use them all at the same power level. The signal format is (5.175) with $p_k = \sqrt{N/M}$ for M subcarriers, and $p_k = 0$ for the remaining $N - M$ elements. The corresponding MCRB (symmetric spectrum) is minimized by a configuration in which the M subcarriers are split into two groups of $M/2$ subcarriers each and placed at the edges of the bandwidth. Figure 5.63 depicts the shape of the PSD of such a signal with $N = 64$ and $M = 16$. The (optimal) MCRB can be easily computed and turns out to be

$$\tilde{\kappa}(\tau) = \frac{T_c^2}{8\pi^2 \frac{E_c}{N_0} \frac{N_m}{N}} \cdot \frac{1}{\left[\xi_g + \frac{(1+\alpha)^2}{M} \frac{(M+2)(M^2+4M-3MN-6N+3N^2+3)}{12} \right]}. \tag{5.181}$$

Let us compare this result with the MCRB that applies to M contiguous carriers (symmetric around $f = 0$), that is, for the same total net spectral occupancy. We only need to replace N with M in (5.173)

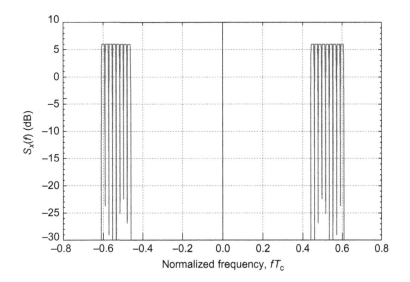

FIGURE 5.63

PSD of a multicarrier signal with even power allocation on two noncontiguous bands.

and do the summation:

$$\tilde{\kappa}(\tau)' = \frac{T_c^2}{8\pi^2 \frac{E_c}{N_0} \frac{N_m}{M} \left[\xi_g + \frac{(1+\alpha)^2}{12} \left(M^2 - 1 \right) \right]}$$

$$\cong \frac{T_c^2}{2\pi^2 \frac{E_c}{N_0} \frac{(1+\alpha)^2}{3} N_m \cdot M} \qquad \text{(large } M\text{)}. \qquad (5.182)$$

If we assume $M \ll N$, (5.181) reads

$$\tilde{\kappa}(\tau) \cong \frac{T_c^2}{2\pi^2 \frac{E_c}{N_0} (1+\alpha)^2 N_m \cdot N} \ll \tilde{\kappa}(\tau)' \qquad \text{(large } M\text{)}, \qquad (5.183)$$

that is considerably smaller than (5.182). The effect of having two bands for positioning far apart in the frequency domain is apparent.

5.5.3.1 *Bounds for CP in the ACGN Channel*

If we drop the assumption of white background noise, we can find the power allocation scheme that gives the minimum MCRB [124] for time-delay estimation in a Gaussian channel whose (additive) noise has a variable PSD $S_n(f)$. A further nice approach to solve the problem, which leads to exactly the same results as in (5.160), is the following. Our starting point is the computation of $\kappa(\tau)|_{\text{ACGN}}$ as a function of the "partial" CRBs $\kappa_k(\tau)$ we would get observing each subcarrier separately from the others on its own kth subband only. By virtue of signal orthogonality, we get

$$\kappa(\tau)|_{\text{ACGN}} = \frac{1}{\sum_k \kappa_k^{-1}(\tau)}. \tag{5.184}$$

After straightforward computations, we have

$$\kappa(\tau)|_{\text{ACGN}} = \frac{1}{4\pi^2 T_{\text{obs}}(\Delta f)^3} \cdot \left(\sum_k k^2 \frac{S_x(k\Delta f)}{S_n(k\Delta f)} \right)^{-1}, \tag{5.185}$$

where $\Delta f = f_{\text{sc}}$ is the subcarrier spacing, and where the PSD of the transmitted signal and of the noise are considered constant across each subband. A fundamental result is obtained if we let $N \to \infty$ (thus $\Delta f \to df$ and $\sum \to \int$) [125] :

$$\kappa(\tau)|_{\text{ACGN}} = \frac{1}{4\pi^2 T_{\text{obs}}} \cdot \left(\int_B f^2 \frac{S_x(f)}{S_n(f)} df \right)^{-1}. \tag{5.186}$$

5.5.3.2 *Optimum Signal Design for Cognitive Positioning in the ACGN Channel*

Coming back to the problem of enhancing time-delay estimation accuracy, and sticking for simplicity to the finite subcarriers version of the problem, we have to minimize the MCRB (5.185) with the constraint (5.176) on the total signal power. Considering that $S_x(k\Delta f)$ is proportional to p_k^2, we can find the power distribution p_k^2 that maximizes

$$\sum_k \frac{k^2 p_k^2}{S_n(k\Delta f)} \tag{5.187}$$

subject to the constraints $\sum p_k^2 = N$ and, of course, $p_k \geq 0$. The optimal distribution is easily found to be

$$\begin{cases} p_k = 0 & k \neq k_M \\ p_{k_M} = \sqrt{N} & k = k_M \end{cases}, \tag{5.188}$$

where

$$k_M = \underset{k}{\text{argmax}} \frac{k^2}{S_n(k\Delta f)}, \tag{5.189}$$

that corresponds to placing all the power onto the subband for which the *squared frequency-to-noise ratio* (SFNR) $k^2/S_n(k\Delta f)$ is maximum.

A more realistic case study for CP in ACGN also considers possible power limitations on each subcarrier that prevents us from concentrating all the signal power onto the edge subcarriers (for AWGN) or on the subcarriers with the best SFNR as above. We have thus the further constraint

$$0 \leq p_k^2 \leq P_{\text{max}} < N. \tag{5.190}$$

The solution to this new power allocation problem can be easily found via linear programming:

1. Order the square frequency-to-noise-ratios $SFNR_k$ from the highest to the lowest; set the currently allocated power to zero; mark all carriers available.
2. Find the available power as the difference between the total power N and the currently allocated power. If it is null, then STOP; if it is larger than P_{max}, then put the maximum power P_{max} on the available carrier with the highest SFNR, else put on the same carrier the (residual) available power.
3. Update the currently allocated power by adding the one just allocated, and remove the just allocated carrier from the list of available carriers. If the list is empty, then STOP, else go to (2).

This results in a set of bounded-power subcarriers that gives the optimum power allocation (minimum MCRB for time-delay estimation) with ACGN, the same criterion as in (5.160).

5.5.3.3 *Algorithms for Cognitive Positioning*

How can the opportunities for CP be exploited? Conventional delay estimators are not directly applicable to a multicarrier signal whose subcarriers are scattered over noncontiguous bands [67]. In addition to the general ML-inspired algorithm (5.169), let us examine a simple case study of an FB-MCM signal in AWGN, where the active subcarriers are concentrated at the two band edges, and all have the same power, as in Fig. 5.63, but with an asymmetric spectrum on positive frequencies only for simplicity. After baseband conversion, matched filtering on each subcarrier, sampling at the multicarrier symbol rate $1/T_s = 1/NT_c$ on each subcarrier, and removal of the ranging chip $c^{(k)}[n]$ on each subcarrier (despreading), the signal model is

$$r^{(k)}[n] = \sqrt{\frac{2P_T}{M}} g_N(\tau) \exp\{j2\pi k(1+\alpha)\tau/T_s\} + IChI^{(k)}(\mathbf{c},\tau;n) + w^{(k)}[n], \qquad (5.191)$$

where $w^{(k)}[n]$ is the nth noise component on the kth subcarrier, whose I/Q (mutually independent) components both have variance $N_0/T_s = N_0/(NT_c)$, and where $IChI^{(k)}(\mathbf{c},\tau;n)$ is the interchip interference term arising on subcarrier k due to the sampling offset τ. Assuming that coarse delay acquisition has already taken place, so that the (residual) timing offset is comparable to the chip time T_c, we can assume in (5.191) $g_N(\tau) \simeq 1$ and $IChI^{(k)}(\mathbf{c},\tau;n) \simeq 0$, so that the approximated, simplified signal model turns out to be

$$r^{(k)}[n] = \sqrt{\frac{2P_T}{M}} \exp\{j2\pi k(1+\alpha)\tau/T_s\} + w^{(k)}[n]. \qquad (5.192)$$

The subcarrier index k runs across the union of the two disjoint bands $B_L = [-N/2, -N/2+M/2-1]$ and $B_R = [N/2-M/2, N/2-1]$, so that the relevant CRB is (5.181).

A sample "cognitive" delay estimator for this signal structure is really simple: We start by computing two subcarrier *phase* estimates, the one on the left-edge section B_L and the other on the right-edge section B_R, using the standard (low complexity) maximum-likelihood (ML) algorithm:

$$\hat{\theta}_L = \angle \sum_{n=0}^{N_m-1} \sum_{k \in B_L} r^{(k)}[n]$$

$$\hat{\theta}_R = \angle \sum_{n=0}^{N_m-1} \sum_{k \in B_R} r^{(k)}[n], \qquad (5.193)$$

where the operator $\angle\{\cdot\}$ denotes the phase of its complex-valued argument. Each estimate is derived after observing $M/2$ values in the frequency domain and repeating such observations for N_m multi-carrier symbol periods in time. We also conventionally associate the two phase estimates with the two *center frequencies* of the left-edge and right-edge sections, respectively, whose mutual frequency distance is equal to $(N - M/2)(1 + \alpha)/T_s$. After this is done, we derive the delay estimate as the *slope* of the line that connects the two points $(-N/2 + M/4 - 1/2, \hat{\theta}_L)$ and $(N/2 - M/4 - 1/2, \hat{\theta}_R)$ on the (*frequency, phase*) plane:

$$\hat{\tau} = \frac{T_s}{1 + \alpha} \frac{\left| \left| \hat{\theta}_R \right|_{2\pi} - \left| \hat{\theta}_L \right|_{2\pi} \right|_{2\pi}}{2\pi (N - M/2)} \tag{5.194}$$

or, in vector form,

$$\exp\left\{ j\, 2\pi \frac{\hat{\tau}(1 + \alpha)(N - M/2)}{T_s} \right\} = \exp\left\{ j\, \hat{\theta}_R \right\} \cdot \exp\left\{ -j\, \hat{\theta}_L \right\}. \tag{5.195}$$

We term this simple algorithm described in Fig. 5.64 as *delay estimation through phase estimation* (DEPE). The operator $|x|_{2\pi}$ returns the value of x modulo 2π to avoid phase ambiguities and is trivial to implement when operating with fixed-point arithmetic on a digital hardware.

It is clear that the operating range of the estimator is quite narrow. In order not to have estimation ambiguities, we have to make sure that

$$-\pi \le |\hat{\theta}_R|_{2\pi} - |\hat{\theta}_L|_{2\pi} < \pi. \tag{5.196}$$

Therefore, the range is bounded to

$$\frac{|\tau|}{T_c} \le \frac{N}{2(1 + \alpha)(N - M/2)}. \tag{5.197}$$

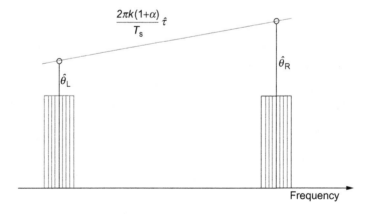

FIGURE 5.64

Pictorial representation of the DEPE algorithm.

When $N \gg M \gg 1$, this interval is basically equal to a fraction $1/(1+\alpha)$ of the chip time, so that the use of the DEPE algorithm is restricted to fine estimation of the residual time offset after coarse acquisition is over.

The DEPE algorithm can be shown to be unbiased, apart from the usual issue of the discontinuity of the $\angle$ function across $\pm\pi$, that may become annoying for low values of C/N_0. Figure 5.65 depicts the normalized mean estimated value (MEV) curves of the DEPE algorithm (i.e., the average estimated value $E\{\hat{\tau}\}$ as a function of the true delay τ for different values of E_c/N_0) as derived by simulation. The parameter N is equal to 2048, while $M = 128$. As mentioned above, such curves show that the algorithm is unbiased in a broad range around the true value.

It is also easy to evaluate the estimation error variance of the DEPE estimator. It is known that $\hat{\theta}_L$ and $\hat{\theta}_R$ in (5.194) have an estimation variance $\sigma_{\hat{\theta}}^2$ that (for sufficiently high E_s/N_0) achieves its own CRB. Considering the signal expression in (5.192) and that the estimation window for each estimate is equal to $M/2$, the variance is given by

$$\sigma_{\hat{\theta}}^2 = \frac{1}{N_m}\kappa(\theta) = \frac{1}{N \cdot N_m}\frac{1}{E_c/N_0}. \tag{5.198}$$

Since the two phase estimates in (5.194) are independent, we get

$$\sigma_{\text{DEPE}}^2(\hat{\tau}) = \left(\frac{NT_c}{1+\alpha}\right)^2 \frac{2 \cdot \sigma_{\hat{\theta}}^2}{4\pi^2(N-M/2)^2}$$

$$= \frac{T_c^2}{2\pi^2[(N-M/2)(1+\alpha)]^2}\frac{N}{N_m \cdot E_c/N_0}. \tag{5.199}$$

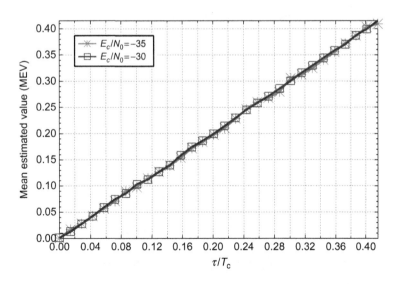

FIGURE 5.65

Bias curves of the DEPE algorithm.

Assuming $N \gg M \gg 1$, this variance is approximated by

$$\sigma^2_{\text{DEPE}}(\hat{\tau}) = \frac{T_c^2}{2\pi^2 N \cdot N_m (1+\alpha)^2} \frac{1}{E_c/N_0}, \tag{5.200}$$

that turns out to be exactly equal to (5.183). This shows that DEPE attains its CRB in most practical situations. We report anyway in Fig. 5.66 the ratio between the estimation variance (5.199) and the actual MCRB (5.181). We can see that this ratio is close to 1 for any practical values of N and M.

Simulation Results

The performance of the DEPE algorithm has been evaluated by simulation in terms of its *mean square estimation error* (MSEE). We use the simplified signal model (5.192) in which the time delay τ satisfies the limitation (5.197). Figure 5.67 shows our simulation results for $N = 1024$ or 2048 and with $M = 128$. As predicted by Fig. 5.66, the DEPE algorithm attains its own MCRB (5.181).

The basic strategy of DEPE can be generalized to a ranging signal that spans more than two sub-bands. The starting point is still the derivation of a single-phase estimate for each subband, followed by linear regression to fit a line across the phase values. The estimated signal delay $\hat{\tau}$ is trivially the slope of the regression line. We will not enter here into the details of such computation, but we wish to give a few numerical results for another simple test case of a multicarrier ranging signal of $M = 128$ active subcarriers split into four equi-spaced and equi-span subbands. The signal format is (5.175) with $p_k = \sqrt{N/M}$ for M active subcarriers and $p_k = 0$ for the remaining $N - M$ elements, so that the relevant MCRB is easily computed using (5.177). The MSEE performance of the algorithm, which is

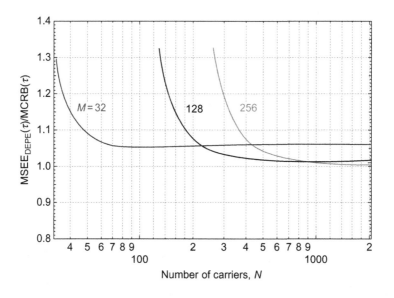

FIGURE 5.66

Comparison between the RMSEE and the MCRB of the DEPE algorithm—SRRC pulse with $\alpha = 0.2$.

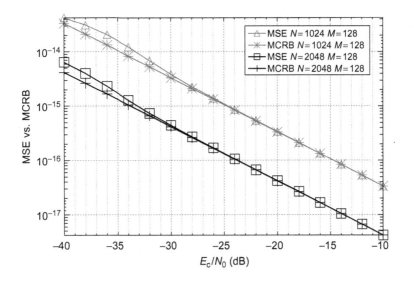

FIGURE 5.67

Comparison between MCRB and measured MSE of the DEPE algortihm—SRRC pulse with $\alpha = 0.2$.

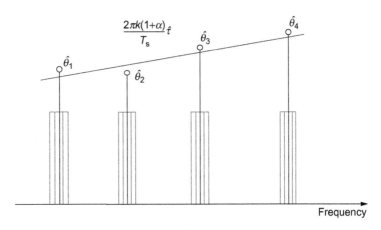

FIGURE 5.68

Pictorial representation of the extended DEPE algorithm applied on a signal divided into four subbands.

again sketched in Fig. 5.68, is shown in Fig. 5.69 for $N = 1024$ and 2048 and $M = 128$. We notice that the extended DEPE algorithm still attains its own MCRB.

Concluding Remarks

A multicarrier signal can be considered "highly modular" in the sense that playing with the subcarrier amplitudes and/or switching on/off some subcarriers (or subbands altogether), we can improve time-delay estimation accuracy. A good candidate for an MC ranging signal is *filtered multitone* (filter-bank

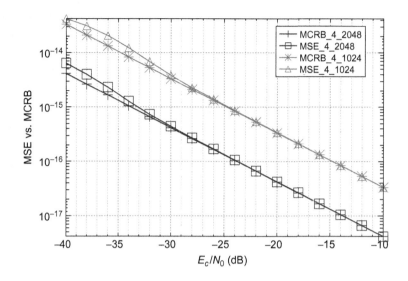

FIGURE 5.69

Comparison between the MCRB and the simulated MSEE of the extended DEPE algortihm—SRRC pulse with $\alpha = 0.2$.

multicarrier modulation) with variable-amplitude subcarriers. In the presence of strong interference on the FB-MCM subbands, which we can model as an ACGN channel, it is easy to find the optimum power allocation of the FB-MCM radio signal on the different subbands to attain the minimum MCRB. "Cognitive" algorithms for delay estimation on noncontiguous bands can be devised and compare favorably with their own MCRBs. Further research is needed to devise simple cognitive algorithms operating with unequal subband spacing and/or frequency span. The estimators are also to be adapted to the case of a frequency-selective wireless multipath channel, provided that the multipath components of the channel can be resolved (i.e., their relative delay is larger than the chip time).

References

[1] 802.15.4a-2007: IEEE Standard for Information Technology – Telecommunications and information exchange between systems – Local and metropolitan area networks – specific requirement. Part 15.4: Wireless Medium Access Control (MAC) and Physical Layer (PHY) Specifications for Low-Rate Wireless Personal Area Networks (WPANs), 2007.

[2] B. Alavi, K. Pahlavan, Modeling of the TOA-based distance measurement error using UWB indoor radio measurements, IEEE Commun. Lett. 10 (4) (2006) 275–277.

[3] M.S. Arulampalam, S. Maskell, N. Gordon, T. Clapp, A tutorial on particle filters for online nonlinear/non-Gaussian Bayesian tracking, IEEE Trans. Signal Process. 50 (2) (2002) 174–188.

[4] Y. Bar-Shalom, X.R. Li, T. Kirubarajan, Estimation with Applications to Tracking and Navigation, John Wiley and Sons, New York, 2001.

[5] A.J. Barabell, Improving the resolution performance of eigenstructure-based direction-finding algorithms, in: Proceedings of the IEEE International Conference on Acoustics, Speech and Signal Processing, ICASSP 1993, Minneapolis, MN, 27–30, April 1983, pp. 336–339.

[6] A. Beck, On the convexity of a class of quadratic mappings and its application to the problem of finding the smallest ball enclosing a given intersection of ball. J. Global Optim. 39 (2007) 113–126.

[7] I. Bekkerman, J. Tabrikian, Target detection and localization using MIMO radars and sonars, IEEE Trans. Signal Process. 10 (2006) 3873–3883.

[8] L. Blanco, J. Serra, M.Nájar, Low complexity TOA estimation for wireless location, in: Proceedings of the IEEE Global Telecommunication Conference, GLOBECOM 06, San Francisco, CA, Nov.–Dec. 2006.

[9] D. Blatt, A. Hero, Energy-based sensor network source localization via projection onto convex sets, IEEE Trans. Signal Process. 54 (9) (2006) 3614–3619.

[10] D. Blatt, A.O. Hero, APOCS: A rapidly convergent source localization algorithm for sensor networks, in: Proceedings IEEE/SP 13th Workshop on Statistical Signal Processing, Bordeaux, France 2005, pp. 1214–1219.

[11] D. Blatt, A.O. Hero, Sensor network source localization via projection onto convex sets (POCS), in: Proceedings IEEE International Conference on Acoustics, Speech, and Signal Processing, Philadelphia, USA, 2005, pp. 1065–1068.

[12] S. Boyd, L. Vandenberghe, Convex Optimization, Cambridge University Press, 2004.

[13] Y. Bresler, A. Macovski, Exact maximum likelihood parameter estimation of superimposed exponentials in noise, IEEE Trans. Acoust. Speech Signal Process. 34 (5) (1986) 1081–1089.

[14] F. Cattivelli, C.G. Lopes, A.H. Sayed, Diffusion recursive least-squares for distributed estimation over adaptative networks, IEEE Trans. Signal Process. 56 (5) (2008) 1865–1877.

[15] H. Celebi, H. Arslan, Cognitive positioning systems, IEEE Trans. Commun. 6 (12) (2007), 4475–4483.

[16] H. Celebi, H. Arslan, Utilization of location information in cognitive wireless networks, IEEE Wireless Comm. 14 (4) (2007) 6–13.

[17] H. Celebi, H. Arslan, Enabling location and environment awareness in cognitive radios, Elsevier Comput. Commun. (Special Issue on Advanced Location-Based Services) 31 (6) (2008) 1114–1125.

[18] Y. Censor, S.A. Zenios, Parallel Optimization, Theory, Algorithms, and Applications, Oxford University Press, 1997.

[19] C.-Y. Chen, P. Vaidyanathan, MIMO radar space-time adaptive processing using prolate spheroidal wave functions, IEEE Trans. Signal Process. 56 (2008) 623–635.

[20] T. Clausen, P. Jacquet, A. Laouiti, P. Muhlethaler, A. Qayyum, L. Viennot, Optimized link state routing protocol for ad hoc networks, in: Proc. IEEE Int. Multi Topic Conference, 2001. IEEE INMIC 2001. Technology for the 21st Century, 2001, pp. 62–68.

[21] M. Coates, Distributed particle filters for sensor networks, in: Proceedings IEEE Information Processing in Sensor Networks (IPSN'04), Berkeley, CA, 2004, pp. 99–107.

[22] A. Conti, D. Dardari, L. Zuari, Cooperative UWB based positioning systems: CDAP algorithm and experimental results, in: International Symposium on Spread Spectrum Techniques and Applications, ISSSTA 2008, Bologna, Italy, 2008, pp. 1–6.

[23] J.A. Costa, N. Patwari, A.O. Hero, Achieving high-accuracy distributed localization in sensor networks, in: Proceedings IEEE International Conference on Acoustics, Speech and Signal Processing, vol. 3, Philadelphia, PA, 2005, pp. 641–644.

[24] A.A. D'Amico, U. Mengali, L. Taponecco, Energy-based TOA estimation, IEEE Trans. Wireless Commun. 7 (3) (2008) 1–10.

[25] D. Dardari, A. Conti, U. Ferner, A. Giorgetti, M.Z. Win, Ranging with ultrawide bandwidth signals in multipath environments, Proc. IEEE, Spec. Issue UWB Tech. Emerg. Appl. 97 (2) (2009) 404–426.

[26] D. Dardari, A. Conti, J. Lien, M.Z. Win The effect of cooperation on localization systems using UWB experimental data, EURASIP J. Adv. Signal Process. Special Issue on Cooperative Localization in Wireless Ad Hoc and Sensor Networks 2008 (2008) Article ID 513873, pp. 1–11.

[27] D. Dardari, A. Giorgetti, M.Z. Win, Time-of-arrival estimation of UWB signals in the presence of narrowband and wideband interference, in: Proceedings IEEE International Conference on Ultra Wideband (ICUWB), Singapore, Sept. 2007, pp. 1–6.

[28] D. Dardari, Y. Karisan, S. Gezici, A.A. D'Amico, U. Mengali, Performance limits on ranging with cognitive radio, in: IEEE International Conference on Communications (ICC) Workshops, Dresden, Germany, 2009, pp. 1–5.

[29] M.H. DeGroot, Reaching a consensus, J. Am. Stat. Assoc. 69 (345) (1974) 118–121.

[30] L. Doherty, K.S.J. Pister, L. El Ghaoui, Convex position estimation in wireless sensor networks, in: Proc. IEEE INFOCOM 2001, Anchorage, AK, vol 3, 2001, pp. 1655–1663.

[31] L. Doyle, A. Kokaram, S. Doyle, T. Forde, Ad hoc networking, Markov random fields, and decision making, IEEE Signal Process. Mag. 23 (5) (2006) 63–73.

[32] L. Dumont, M. Fattouche, G. Morrison, Super-resolution of multipath channels in a spread spectrum location system, IEEE Electron. Lett. 30 (19) (1994) 1583–1584.

[33] E. Lagunas, M. Najar, M. Navarro, UWB joint TOA and DOA estimation, in: Proceedings of IEEE International Conference on Ultra-Wideband, Vancouver, Canada, Sept. 2009, pp. 839–843.

[34] M. Najar, E. Lagunas, L. Taponecco, A. D'Amico, TOA estimation in UWB: Comparison between time and frequency domain processing, in: Proceedings of the 2nd International Conference on Mobile Lightweight Wireless Systems, Barcelona, Spain, 2010.

[35] Federal Communications Commission, Dedicated short range communications (DSRC) service, report and order 03-324. http://www.federalregister.gov/articles/2004/08/03/04-16087/dedicated-short-range-communication-services-and-mobile-service-for-dedicated-short-range, 2004 (last accessed: July 2011).

[36] E. Fishler, A. Haimovich, R. Blum, L. Cimini, D. Chizhik, R. Valenzuela, MIMO radar: An idea whose time has come, in: Proceedings IEEE International Conference on Radar, Philadelphia, PA, 2004.

[37] E. Fishler, A. Haimovich, R. Blum, L. Cimini, D. Chizhik, R. Valenzuela, Spatial diversity in radars—models and detection performance, IEEE Trans. Signal Process. 54 (2006) 823–838.

[38] D. Fontanella, M. Nicoli, L. Vandendorpe, Bayesian localization in sensor networks: Distributed algorithm and fundamental limits, in: 2010 IEEE International Conference on Communications (ICC), Cape Town, 2010, pp. 1–5.

[39] S. Frattasi, M. Monti, R. Prasad, A cooperative localization scheme for 4G wireless communications, in: Radio and Wireless Symposium, 2006 IEEE, San Diego, CA, 17–19, January 2006, pp. 287–290.

[40] R. Fujiwara, K. Mizugaki, T. Nakagawa, D. Maeda, M. Miyazaki, TOA/TDOA hybrid relative positioning system using UWB-IR, in: IEEE Radio and Wireless Week, 2009, pp. 679–682.

[41] Y. Gao, Z. Gao, W. Zhu, X. Yang, The research on the design of filter banks in filtered multitone modulation, in: Wireless Communications and Networking Conference, 2005 IEEE, New Orleans, LA, vol. 1, March 2005, pp. 584–588.

[42] S. Gezici, Z. Sahinoglu, Ranging in a single-input multiple-output (SIMO) system, IEEE Commun. Lett. 12 (3) (2008) 197–199.

[43] S. Gezici, Z. Sahinoglu, Enhanced position estimation via node cooperation, in: Proceedings IEEE International Conference on Communications (ICC), Cape Town, South Africa, 2010, pp. 1–6.

[44] S. Gezici, Z. Tian, G.B. Giannakis, H. Kobayashi, A.F. Molisch, H.V. Poor, Z. Sahinoglu, Localization via ultra-wideband radios, IEEE Signal Process. Mag. 22 (4) (2005) 70–84.

[45] A. Giorgetti, M. Chiani, D. Dardari, R. Minutolo, M. Montanari, Cognitive radio with ultra-wide bandwidth location-capable nodes, in: IEEE Military Communications Conference, 2007 (MILCOM 2007), Orlando, Florida, USA, 2007, pp. 1–7.

[46] A. Giorgetti, M. Chiani, M. Z. Win, The effect of narrowband interference on wideband wireless communication systems, IEEE Trans. Commun. 53 (12) (2005) 2139–2149.

[47] M. Haenggi, On distances in uniformly random networks, IEEE Trans. Inf. Theory 51 (10) (2005) 3584–3586.

[48] A. Haimovich, R. Blum, L. Cimini, MIMO radar with widely separated antennas, IEEE Signal Process. Mag. 25 (2008) 116–129.

[49] K.C. Ho, X. Lu, L. Kovavisaruch, Source localization using TDOA and FDOA measurements in the presence of receiver location errors: Analysis and solution, IEEE Trans. Signal Process. 55 (2) (2007) 684–696.

[50] A. Ihler, J. Fisher, R.L. Moses, A. Willsky, Nonparametric belief propagation for self-localization of sensor networks, IEEE J. Sel. Areas Commun. 23 (4) (2005) 809–819.

[51] A. Ihler, D. McAllester, Particle belief propagation, in: D. van Dyk, M. Welling (Eds.), Proceedings of the Twelfth International Conference on Artificial Intelligence and Statistics (AISTATS) 2009, J. Machine Learning Res., Workshop and Conf. Proc., vol. 5, Clearwater Beach, Florida, 2009, pp. 256–263.

[52] G. Ing, M.J. Coates, Parallel particle filters for tracking in wireless sensor networks, in: Proceedings IEEE 6th Workshop on Signal Processing Advances in Wireless Communications, New York, 5–8 June 2005, pp. 935–939.

[53] International Organization for Standardization (ISO), TC204 CALM. http://www.isotc204wg16.org.

[54] A. Jakobsson, A.L. Swindlehurst, P. Stoica, Subspace-based estimation of time delays and Doppler shifts, IEEE Trans. Signal Process. 46 (9) (1998), 2472–2483.

[55] L. Jaulin, M. Kieffer, O. Didrit, E. Walter, Applied Interval Analysis. Springer-Verlag, London, 2001.

[56] L. Jaulin, E. Walter, Set inversion via interval analysis for nonlinear bounded-error estimation, Automatica 29 (4) (1993) 1053–1064.

[57] L. Jaulin, E. Walter, O. Didrit, Guaranteed robust nonlinear parameter bounding, in: Proceedings of CESA'96 IMACS Multiconference (Symposium on Modelling, Analysis and Simulation), Lille, France, 1996, pp. 1156–1161.

[58] S. Jayaweera, Virtual MIMO-based cooperative communication for energy-constrained wireless sensor networks, IEEE Trans. Wireless Commun. 5 (2006) 984–989.

[59] Y. Jeong, H. Ryou, C. Lee, A high resolution time delay estimation technique in frequency-domain for positioning systems, in: Proceedings of the IEEE Vehicular Technology Conference 2002, Fall, Vancouver, BC, vol. 4, 2002, pp. 2318–2321.

[60] P. Jia, M. Vu, T. Le-Ngoc, Capacity impact of location-aware cognitive sensing, in: Proceedings IEEE Global Telecommunications Conference (GLOBECOM), Honolulu, HI, 2009, pp. 1–6.

[61] D. Jourdan, D. Dardari, M.Z. Win, Position error bound for UWB localization in dense cluttered environments, IEEE Trans. Aerosp. Electron. Syst. 44 (2) (2008) 613–628.

[62] D.B. Jourdan, J.J. Deyst, M.Z. Win, N. Roy, Monte-Carlo localization in dense multipath environments using UWB ranging, in: IEEE International Conference on Ultra-Wideband (ICU 2005), ZüRICH, Switzerland, 2005, pp. 314–319.

[63] Y. Karisan, D. Dardari, S. Gezici, A.A. D'Amico, U. Mengali, Range estimation in multicarrier systems in the presence of interference: Performance limits and optimal signal design, IEEE Trans. Wirel. Commun. 2011. Submitted.

[64] S. Kay, Fundamentals of Statistical Signal Processing: Estimation Theory, Prentice-Hall, Upper Saddle River, NJ, 1993.

[65] H. Keshavarz, Weighted signal-subspace direction-finding of ultra-wideband sources, in: IEEE International Conference on Wireless and Mobile Computing, Networking and Communications, 2005. (WiMob'2005), Montreal, Canada, 2005, pp. 23–29.

[66] M. Kieffer, E. Walter, Centralized and distributed source localization by a network of sensors using guaranteed set estimation, in: Proceedings of ICASSP, vol. IV, Toulouse, France, 15–19 May 2006, pp. 977–980.

[67] F. Kocak, H. Celebi, S. Gezici, K.A. Qaraqe, H. Arslan, H.V. Poor, Time-delay estimation in dispersed spectrum cognitive radio systems, EURASIP J. Adv. Signal Process. 2010 (Special Issue on Adv Signal Proc . Cognitive Radio Netw.) Article ID 675959, 10 pages. doi: 10.1155/2010/675959.

[68] K. Langendoen, N. Reijers, Distributed Localization in Wireless Sensor Networks: A Quantitative Comparison, Elsevier Science, New York, 2003.

[69] E. Larsson, Cramer–Rao bound analysis of distributed positioning in sensor networks, IEEE Signal Process. Lett. 11 (3) (2004) 334–337.

[70] J.Y. Lee, S. Yoo, Large error performance of UWB ranging in multipath and multiuser environments, IEEE Trans. Microw. Theory Tech. 54 (4) (2006) 1887–1895.

[71] J.B. Leger, M. Kieffer, Guaranteed robust distributed estimation in a network of sensors, in: Proceedings International Conference on Acoustics, Speech, and Signal Processing, Dallas, TX, 14–19 March 2010, pp. 3378–3381.

[72] J. Li, P. Stoica, L. Xu, W. Roberts, On parameter identifiability of MIMO radar, IEEE Signal Process. Lett. 14 (2007) 968–971.

[73] X. Li, Collaborative localization with received-signal strength in wireless sensor networks, IEEE Trans. Veh. Technol. 56 (2007) 3807–3817.

[74] M. Navarro, IEEE 8th Workshop on Signal Processing Advances in Wireless Communications, SPAWC 2007, Marina Congress Center, Helsink, Finland, June 2007, pp. 1–5.

[75] S. Marano, W.M. Gifford, H. Wymeersch, M.Z. Win, Nonparametric obstruction detection for UWB localization, in: Proceedings IEEE Global Telecommunications Conference, Honolulu, HI, 2009, pp. 1–6.

[76] C. Mayorga, F. della Rosa, S. Wardana, G. Simone, M. Raynal, J. Figueiras, S. Frattasi, Cooperative positioning techniques for mobile localization in 4G cellular networks, in: IEEE Int. Conf., Pervasive Services, Istanbul, Turkey, 15–20 July 2007, pp. 39–44.

[77] U. Mengali, A.N. D'Andrea, Synchronization Techniques for Digital Receivers, Plenum Press, New York, 1997.

[78] M. Milanese, J. Norton, H. Piet-Lahanier, E. Walter (Eds), Bounding Approaches to System Identification, Plenum Press, New York, NY, 1996.

[79] J. Mitola, G.Q. Maguire, Cognitive radio: Making software radios more personal, IEEE Personal Commun. Mag. 6 (4) (1999) 13–18.

[80] A.F. Molisch, D. Cassioli, C.-C. Chong, S. Emami, A. Fort, B. Kannan, et al., A comprehensive standardized model for ultrawideband propagation channels, IEEE Trans. Antennas Propag. 54 (11) (2006) 3151–3166.

[81] R.E. Moore, Methods and Applications of Interval Analysis, SIAM, Philadelphia, PA, 1979.

[82] C. Morelli, M. Nicoli, V. Rampa, U. Spagnolini, Hidden Markov models for radio localization in mixed LOS/NLOS conditions, IEEE Trans. Signal Process. 55 (4) (2007) 1525–1542.

[83] B. Musicus, Fast MLM power spectrum estimation from uniformly spaced correlations, IEEE Trans. Acoust. Speech Signal Process. 33 (1985) 1333–1335.

[84] M. Nájar, J.M. Huerta, J. Vidal, J.A. Castro, Mobile location with bias tracking in non-line-of-sight, in: Proceedings of the IEEE International Conference on Acoustics, Speech and Signal Processing, Montreal, Quebec, 17–21 May 2004, pp. 956–959.

[85] M. Nicoli, D. Fontanella, Fundamental performance limits of TOA-based cooperative localization, in: Proceedings IEEE International Conference on Communications (ICC '09), SyCoLo Workshop, Dresden, 18 June 2009, pp. 1–5.

[86] M. Nicoli, C. Morelli, V. Rampa, A jump Markov particle filter for localization of moving terminals in dense multipath indoor scenarios, IEEE Trans. Signal Process. 56 (8) (2008) 3801–3809.

[87] M. A. Pallas, G. Jourdain, Active high resolution time delay estimation for large BT signals, IEEE Trans. Signal Process. 39 (4) (1991), 781–788.

[88] R. Parker, S. Valaee, Vehicular node localization using received-signal-strength indicator, IEEE Trans. Veh. Technol. 56 (2007) 3371–3380.

[89] N. Patwari, J. Ash, S. Kyperountas, A.H. III, R. Moses, N. Correal, Locating the nodes, IEEE Signal Process. Mag. 22 (4) (2005) 54–69.

[90] N. Patwari, A.O. Hero, M. Perkins, N.S. Correal, R.J. O'Dea, Relative location estimation in wireless sensor networks, IEEE Trans. Signal Process. 8 (51) (2003) 2137–2148.

[91] J. Picard, A. Weiss, Accurate geolocalization in the presence of outliers using linear programming, in: Proceedings European Signal Processing Conference (EUSIPCO), 2009.

[92] L. Pierucci, P.J. Roig, UWB localization on indoor MIMO channels, in: IEEE International Conference on Wireless and Mobile Computing, Networking and Communications, Glasgow, 24–28 Aug., 2005. (WiMob'2005), IEEE, Montreal, Canada, 2005, pp. 890–894.

[93] H.V. Poor, An Introduction to Signal Detection and Estimation, Springer-Verlag, New York, 1994.

[94] P. Preparata, M. Shamos, Computational Geometry, Springer-Verlag, 1985.

[95] J.G. Proakis, Digital Communications, fourth ed., McGraw-Hill, 2001.

[96] M. Rabbat, R. Nowak, Distributed optimization in sensor networks, in: Proceedings of the Information Processing in Sensor Networks Conf., Berkeley, CA, 26–27 April 2004, pp. 20–27.

[97] M. Rydström, A. Urruela, E.G. Ström, A. Svensson, A low complexity algorithm for distributed sensor positioning, in: Proceedings of the European Wireless Conference, vol. 2, Nicosia, Cyprus, 10–13 April 2005, pp. 714–718.

[98] H. Saarnisaari, ML time delay estimation in a multipath channel, in: Proceedings Spread Spectrum Techniques ans Applications Conference 1996, vol 3, 1996, pp. 1007–1011.

[99] H. Saarnisaari, TLS-ESPRIT in a time delay estimation, in: Proceedings of the IEEE Vehicular Technology Conference 1997, vol 3, 1997, pp. 1619–1623.

[100] Z. Sahinoglu, I. Guvenc, Multiuser interference mitigation in noncoherent UWB ranging via nonlinear filtering, EURASIP J. Wireless Commun. Netw. Issue 2, April 2006.

[101] V. Saligrama, M. Alanyali, O. Savas, Distributed detection in sensor networks with packet losses and finite capacity links, IEEE Trans. Signal Process. 54 (11) (2006) 4118–4132.

[102] A. Savvides, C. Han, M. Srivastava, Dynamic fine-grained localization in ad hoc networks of sensors, in: Proceedings IEEE Mobicom, Rome, 2001, pp. 166–179.

[103] J. Schiff, D. Antonelli, A. Dimakis, D. Chu, M. Wainright, Robust message-passing for statistical inference in sensor networks, in: Proceedings IEEE Information Processing in Sensor Networks (IPSN'07), Cambridge, MA, 25–27, April 2007.

[104] R. Schubert, M. Schlingelhof, H. Cramer, G. Wanielik, Accurate positioning for vehicular safety applications—the SAFESPOT approach, in: Proceedings IEEE VTC-Spring 2007, Dublin, 22–25 April 2007, pp. 2506–2510.

[105] S. Severi, G. Abreu, D. Dardari, A quantitative comparison on multihop algorithms, in: Proceedings Workshop on Positioning, Navigation and Communication (WPNC 2010), Dresden, Germany, March 2010, pp. 1–6.

[106] S. Severi, G. Abreu, G. Destino, D. Dardari, Understanding and solving flip-ambiguity in network localization via semidefinite programming, in: IEEE Global Communications Conference (GLOBECOM 2009), Honolulu, Hawaii, USA, 2009, pp. 1–6.

[107] Y. Shen, M. Win, Fundamental limits of wideband localization—part I: A general framework, IEEE Trans. Inf. Theory 56 (10) (2010) 4956–4980.

[108] Y. Shen, H. Wymeersch, M. Win, Fundamental limits of wideband localization—part II: Cooperative networks, IEEE Trans. Inf. Theory 56 (10) (2010) 4981–5000.

[109] X. Sheng, Y. Hu, P. Ramanathan, Distributed particle filter with GMM approximation for multiple targets localization and tracking in wireless sensor network, in: Proceedings of the 4th Int. Symp. on Information Processing in Sensor Networks (IPSN '05), UCLA, Los Angeles, CA, 25–27 April 2005, pp. 181–188.

[110] X. Sheng, Y.H. Hu, Maximum likelihood multiple-source localization using acoustic energy measurements with wireless sensor networks, IEEE Trans. Signal Process. 53 (1) (2005) 44–53.

[111] J. Soubielle, I. Fijalkow, P. Duvant, GPS positioning in a multipath environment, IEEE Trans. Signal Process. 50 (1) (2002), 141–150.

[112] S. Srinivasa, M. Haenggi, Distance distributions in finite uniformly random networks: Theory and applications, IEEE Trans. Veh. Technol. 59 (2) (2009), 940–949.

[113] P. Stoica, R. Moses, Spectral Analysis of Signals, Prentice-Hall, New York, 2005.

[114] E. Sudderth, A. Ihler, W. Freeman, A. Willsky, Nonparametric belief propagation, Proc. IEEE CVPR I (2003) 605–612.

[115] A.L. Swindlehurst, Time delay and spatial signature estimation using known asynchronous signals, IEEE Trans. Signal Process. 46 (2) (1998), 449–462.

[116] The Mathworks Inc., Online. http://www.mathworks.com, accessed, 2010.

[117] H.L.V. Trees, Detection, Estimation, and Modulation Theory: Part I, second ed., John Wiley and Sons, Inc., New York, NY, 2001.

[118] H.V. Trees, Optimum Array Processing, first ed., John Wiley and Sons, New York, 2002.

[119] D. Tse, P. Viswanath, Fundamentals of Wireless Communications, Cambridge University Press, Cambridge, UK, 2005.

[120] J. Vidal, M. Nájar, R. Játiva, High resolution time-of-arrival detection for wireless positioning systems, in: IEEE Vehicular Technology Conference, VTC 2002, Fall, vol. 4, Vancouver, Canada, 2002, pp. 2283–2287.

[121] E. Walter (Ed.), Special Issue on Parameter Identification with Error Bounds, vol. 32, 1990. Mathematics and Computers in Simulation.

[122] M.Z. Win, A. Conti, S. Mazuelas, Y. Shen, W.M. Gifford, D. Dardari, Network localization and navigation via cooperation, IEEE Commun. Mag. 49 (5) (2011), 56–62.

[123] H. Wymeersch, J. Lien, M. Win, Cooperative localization in wireless networks, Proc. IEEE 97 (2) (2009) 427–450.

[124] F. Zanier, G. Bacci, M. Luise, Criteria to improve time-delay estimation of spread spectrum signals in satellite positioning, IEEE J. Sel. Top. Signal Process. 3 (5) (2009) 748–763.

[125] A. Zeira, A. Nehorai, Frequency-domain Cramer–Rao bound for Gaussian processes, Acoustics, Speech and Signal Processing, IEEE Transactions, 38 (6) (1990) 1063–1066.

[126] Zhang, Dai, Dynamic self-calibration in collaborative wireless networks using belief propagation with Gaussian particle filtering, in: Proceedings 41st Ann. Conf. on Information Sciences and Systems, CISS '07, Baltimore, MD, 2007, pp. 771–776.

[127] E. Lagunas, M. Nájar, M. Navarro, Joint TOA and DOA estimation compliant with IEEE 802.15.4a standard, Int. Symp. on Wireless Pervasive Computing (ISWPC), Modena, Italy, May 2010, pp. 157–162.

[128] L. Blanco, J. Serra, M. Nájar, Minimum variance time of arrival estimation for positioning, Signal Process. (Eurasip–Elsevier) 90 (8) (2010) 2611–2620.

[129] A. Mallat, J. Louveaux, L. Vandenorpe, CRBS for the joint estimation of TOA and AOA in wideband MISO and MIMO systems: Comparison with SISO and SIMO systems, 2009 IEEE Int. Conf. on Communications (ICC 2009), Dresden, 14–18 June 2009.

[130] M. Rydstrom, E.G. Strom, A. Svensson, Robust sensor network positioning based on projections onto circular and hyperbolic convex sets (POCS), in: IEEE 7th Workshop on Signal Processing Advances in Wireless Communications (SPAWC), 2–5 July 2006, pp. 1–5, doi:10.1109/SPAWC.2006.346399.

[131] M.R. Gholami, E.G. Ström, M. Rydström, Positioning of node using plane projection onto convex sets, in: IEEE Wireless Communications and Networking Conf. ISBN/ISSN: 978-142446398-5.

[132] M.R. Gholami, H. Wymreersch, E.G. Ström, M. Rydström, Robust distributed positioning algorithms for cooperative networks, in: IEEE 12th Workshop on Signal Processing Advances in Wireless Communications (SPAWC), 2011 (to appear).

[133] M.R. Gholami, S. Gezici, E.G. Ström, M. Rydström, Hybrid TW-TOA/TDOA positioning algorithms for cooperative wireless networks, in: IEEE Int. Conf. on Communications (ICC), 2011.

Signal Processing for Hybridization

6

**Mauricio A. Caceres Duran, Pau Closas, Emanuela Falletti,
Carles Fernández-Prades, Montse Nájar, Francesco Sottile**

Our everyday life is surrounded by a continuous set of radiating sources coming from everywhere. The rapid deployment and pervasiveness of wireless communication systems (cellular networks, metropolitan area networks, local area networks, personal area networks, and wireless sensor networks), in combination with the availability of flexible, low-cost commercial off-the-shelf components, offers a bundle of electromagnetic signals that suggests investigating their *joint exploitation* in order to compute user's position in combination with GNSS. Even more, the mobile device could include a low-cost inertial measurement unit, providing complementary information of a completely different nature.

All these considerations point to the need of hybridization techniques to allow a smart combination of observations from multiple sources and gather that information in order to achieve an inference about the user position. These sources have different degrees of accuracy, reliability, and coverage. The final goal of this approach is the achievement of *seamless positioning systems* working in any area and at any time.

At first glance, the simplest way to perform a combination of that kind could consist of the position computation for each different system and then averaging the results using weights for each technology according to its expected accuracy. However, it is likely that this is not the optimum solution: Indeed, problems arising from data of different nature arriving sequentially in time find a natural framework in statistics. Bayesian models allow us to include prior knowledge about the phenomenon being modeled, handle sequential arrival of data, and provide online inference, thus becoming a very well-suited mathematical foundation for hybrid positioning purposes. This chapter offers a presentation of classic and innovative methods of signal processing designed to cope with the hybridization issues in positioning and navigation problems; all these methods are based on the Bayesian estimation theory.

Because of the nonlinearity of the positioning systems, the optimal Bayesian approach is hard to solve in a closed form; thus, different suboptimal solutions based on extended Kalman filters (EKFs), particle filters (PFs), and their derivatives are presented. An introduction to Bayesian filtering theory and methods is given in Section 6.1. Then, some representative examples of application of these methods to different hybrid positioning scenarios are discussed: Combination of different types of terrestrial range measurements (Section 6.2), data fusion of GNSS signals/information and inertial measurements (Section 6.3), and fusion of terrestrial range and GNSS pseudorange measurements in a so-called peer-to-peer scenario (Section 6.4).

Satellite and Terrestrial Radio Positioning Techniques. DOI: 10.1016/B978-0-12-382084-6.00006-4

6.1 AN INTRODUCTION TO BAYESIAN FILTERING FOR LOCALIZATION AND TRACKING

Bayesian filters are a good statistical tool to jointly combine information data measured by hetero-geneous sensors in order to increase the accuracy of the positioning system. Recursive Bayesian estimation is a general probabilistic approach for determining an unknown probability density function recursively over time using incoming measurements and a mathematical process model. Bayesian filters probabilistically estimate a dynamic state of a system from (noisy) observations. In location estimation, the state is the location of a terminal, and location sensors provide observations that are functions of the state. Evolving toward the tracking problem, the state also includes the velocity and, depending on the application, also more complex quantities, such as pitch, roll, yaw, linear and rotational velocities [30].

Given a set of observations $\mathbf{z}$, Bayesian estimators aim to estimate the state vector $\mathbf{x}$ as a realization of an unknown random vector $\mathbf{X}$ with a priori distribution $p_{\mathbf{X}}(\mathbf{x})$. In our case, the state vector $\mathbf{x}$ is the position of the object (and other possible quantities, as said earlier), whereas $\mathbf{z}$ are the measures coming from proper sensors (e.g., a GNSS receiver, a RSS indicator, a TOA indicator, some inertial sensors, etc.). Each element of $\mathbf{x}$ and $\mathbf{z}$ is referred to a particular estimation time t_k, indicated in the discrete-time notation with k. Two consecutive time estimations are spaced by $\Delta t_k = t_{k+1} - t_k$ seconds.

As anticipated in Chapter 1, two classic Bayesian estimators are the minimum mean square error (MMSE) and the maximum a posteriori (MAP). The former estimates the mean of the a posteriori probability density function

$$\hat{\mathbf{x}}_{\mathrm{MMSE}} = \int \mathbf{x} p_{\mathbf{X}|\mathbf{Z}}(\mathbf{x}|\mathbf{z}) \, \mathrm{d}x, \tag{6.1}$$

whereas the latter estimates the mode of the a posteriori probability density function

$$\hat{\mathbf{x}}_{\mathrm{MAP}} = \arg\max_{\mathbf{x}} \left\{ p_{\mathbf{X}|\mathbf{Z}}(\mathbf{x}|\mathbf{z}) \right\}. \tag{6.2}$$

The integration of (6.1) or the maximization of (6.2) can be highly complex if the a posteriori probability density function $p_{\mathbf{X}|\mathbf{Z}}(\mathbf{x}|\mathbf{z})$ is difficult to determine, or if the random variable $\mathbf{X}$ has a high dimension. In this case, the MMSE or MAP estimates can be applied to single components of $\mathbf{X}$ instead of the entire vector; for example, if $\mathbf{X} = [X_1, \ldots, X_N]$, then the MMSE estimate of X_k is given by the mean of the marginal a posteriori distribution $p_{X_k|\mathbf{Z}}(\cdot|\mathbf{z})$ of the variable X_k.

The sequential Bayesian approach gives a recursive estimate of the state of the system by using sequential measurements of some system outputs (i.e., the *observables*). In the rest of this chapter, we give an overview of the Bayesian approach by recalling the concept of Bayesian belief (Section 6.1.1), introducing the dynamic models (Section 6.1.2), discussing the structure of a generic sequential Bayesian filter (Section 6.1.3), and then deriving the two most famous families of sequential Bayesian methods, that is, the Kalman filter (Section 6.1.4) and the particle filter (Section 6.1.5), both presented together with their principal derivations and some application examples.

6.1.1 Bayesian Belief

Bayes filters represent the state at discrete time k by random variables $\mathbf{x}_k$. It is a (possibly nonlinear) function of the earlier states and observations $\mathbf{z}_k$. At each estimation time, a probability distribution over $\mathbf{x}_k$, called *belief*, $Bel(\mathbf{x}_k)$, represents the uncertainty. Bayes filters aim to sequentially estimate such beliefs over the state space conditioned on all information contained in the sensor data. To illustrate, let us assume that the sensor data consists of a sequence of time-indexed sensor observations $\mathbf{z}_1, \mathbf{z}_2, \ldots, \mathbf{z}_k$. The belief $Bel(\mathbf{x}_k)$ is then defined by the posterior density over the random variable $\mathbf{x}_k$ conditioned on all sensor data available at time k:

$$Bel(\mathbf{x}_k) = p(\mathbf{x}_k|\mathbf{z}_{1:k}). \tag{6.3}$$

The *a posteriori density function* $p(\mathbf{x}_k|\mathbf{z}_{1:k})$ allows us to determine the most probable state $\mathbf{x}_k$ at time k, given the history of the observations $\mathbf{z}_{1:k}$.

The complexity grows with the time because the number of measurements of the sensors grows over time. To make the computation tractable, let us assume that the dynamic system is a first-order Markov system, that is, the current state variable $\mathbf{x}_k$ contains all relevant information. For locating objects, the Markov assumption implies that sensor measurements depend only on the current physical location of the moving object and that the location at time k depends only on the earlier state $\mathbf{x}_{k-1}$. States before $\mathbf{x}_{k-1}$ provide no additional information. Under the Markov assumption, we can efficiently compute the belief in (6.3) without losing information.

The desired belief $p(\mathbf{x}_k|\mathbf{z}_{1:k})$ can be computed recursively in two stages: *prediction* and *update*.

Given the a posteriori probability $Bel(\mathbf{x}_{k-1})$ at time $k-1$, the current state at time k is given by the *predictive belief*[1]

Prediction step :

$$Bel^-(\mathbf{x}_k) = \int p(\mathbf{x}_k|\mathbf{x}_{k-1})Bel(\mathbf{x}_{k-1})\,\mathrm{d}\mathbf{x}_{k-1}. \tag{6.4}$$

In location estimation, the *state transition density* $p(\mathbf{x}_k|\mathbf{x}_{k-1})$ represents the *motion model*, that is, where the moving object might be at time k if earlier it was at location $\mathbf{x}_{k-1}$.

Once a new observation (measurement) $\mathbf{z}_k$ is available at time k, the a posteriori probability density function of the state has to be updated through Bayes' rule

$$\begin{aligned} p(\mathbf{x}_k|\mathbf{z}_{1:k}) &= p(\mathbf{x}_k|\mathbf{z}_k, \mathbf{z}_{1:k-1}) \\ &= \frac{p(\mathbf{z}_k|\mathbf{x}_k, \mathbf{z}_{1:k-1})p(\mathbf{x}_k|\mathbf{z}_{1:k-1})}{p(\mathbf{z}_k|\mathbf{z}_{1:k-1})} \\ &= \eta p(\mathbf{z}_k|\mathbf{x}_k)p(\mathbf{x}_k|\mathbf{z}_{1:k-1}), \end{aligned} \tag{6.5}$$

the constant normalizing factor being

$$\eta = \frac{1}{p(\mathbf{z}_k|\mathbf{z}_{1:k-1})} = \frac{1}{\int p(\mathbf{z}_k|\mathbf{x}_k)p(\mathbf{x}_k|\mathbf{z}_{1:k-1})\mathrm{d}\mathbf{x}_k}. \tag{6.6}$$

---

[1]Given that $p(\mathbf{x}_0) = p(\mathbf{x}_0|\mathbf{z}_0)$, where $\mathbf{z}_0$ is the set of no measurements, and $p(\mathbf{x}_0)$ is assumed known, we can assume that the required density at time $k-1$ is available, $p(\mathbf{x}_{k-1}|\mathbf{z}_{1:k-1})$. In the prediction stage, the prediction pdf is obtained by considering that $p(\mathbf{x}_k|\mathbf{x}_{k-1}) = p(\mathbf{x}_k|\mathbf{x}_{k-1}, \mathbf{z}_{1:k-1})$, since a Markov model of order one is assumed.

Thus, we can write the updated belief function

$$Update\ step:$$

$$Bel(\mathbf{x}_k) = \eta p(\mathbf{z}_k|\mathbf{x}_k)Bel^-(\mathbf{x}_k), \tag{6.7}$$

where $p(\mathbf{z}_k|\mathbf{x}_k)$ is the *perception model*, while $Bel(\mathbf{x}_k)$ gives information about the position at time k given the observations up to time k. For location estimation, the perception model, also known as *likelihood function*, is usually considered a property of a given sensor technology. It depends on the types and positions of the sensors and captures their error characteristics.

Bayes filters are an abstract concept in that they only provide a probabilistic framework for recursive state estimation. To implement such filters, it is necessary to specify the likelihood function $p(\mathbf{z}_k|\mathbf{x}_k)$, the dynamics $p(\mathbf{x}_k|\mathbf{x}_{k-1})$, and the representation of the belief $Bel(\mathbf{x}_k)$.

As a first step toward practical implementation of the Bayesian framework, the dynamic model of the problem (system) under consideration must be stated; this will allow us to recognize the likelihood function, the motion model, and the structure of the belief. This is precisely the content of the following section.

Before continuing, it is worth highlighting that, in general, the recursion (6.4)–(6.7), the expression (6.6), and the marginal $p(\mathbf{x}_k|\mathbf{z}_k)$ cannot be solved analytically. There are few cases where the posterior density can be characterized by a sufficient statistic, and this is the case of linear Gaussian models where the Kalman filter (KF) yields the optimal solution. Unfortunately, this is not the case in localization problems, since we will show that measurements depend nonlinearly on parameters. A linearization can be performed in the measurement equation at some point of interest, thus implementing an extended Kalman filter (EKF) algorithm, but this approach is not optimum from a statistical point of view, although it is able to give good results in several cases. Thus, in order to deal with nonlinearities and non-Gaussianities, we must resort to other solutions. Sequential Monte Carlo (SMC) methods and their variants (also known as *particle filters*, *bootstrap filters,* or *survival of the fittest*) constitute a set of simulation-based methods that provide a convenient approach to compute the posterior distribution and its associated features, approaching $p(\mathbf{x}_k|\mathbf{z}_{1:k})$ numerically in a recursive and sequential manner.

6.1.2 Dynamic Models

The problem under study concerns the derivation of efficient methods for online estimation and prediction of the time-varying unknown state of the system, along with the continuous flow of information (observations) from the system. A positioning system is nonlinear and dynamic by nature, and thus, the objective is to recursively compute estimates of states $\mathbf{x}_k$, given measurements $\mathbf{z}_k$ and, possibly, a set of known *system inputs* $\mathbf{u}_k$, at time index k. The system inputs, also called *forcing functions*, are external deterministic functions (signals) that enter the system affecting its states (but are not affected in turn by the states) and that can be assumed known at every instant.

The *state equation* models the evolution of target states (on which measurements depend) as a discrete-time stochastic model. We consider the problem of estimating the sequence of hidden states $\mathbf{X}_k = \{\mathbf{x}_0,\ldots,\mathbf{x}_k\}$, with initial distribution $p(\mathbf{x}_0)$, given the observation $\mathbf{Z}_k = \{\mathbf{z}_1,\ldots,\mathbf{z}_k\}$, which in some way is related to the hidden states. Following a first-order Markov chain approach, this *discrete-time*

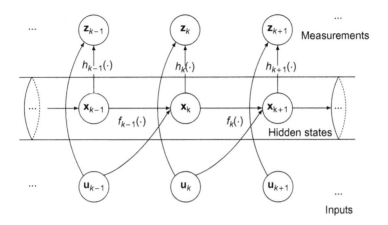

FIGURE 6.1

Conceptual representation of a generic hidden Markov state-space model.

state-space model can be expressed with the following set of equations:

$$\mathbf{x}_k = \mathbf{f}_{k-1}(\mathbf{x}_{k-1}, \mathbf{u}_{k-1}, \mathbf{v}_{k-1}), \tag{6.8}$$

$$\mathbf{z}_k = \mathbf{h}_k(\mathbf{x}_k, \mathbf{u}_k, \mathbf{w}_k), \tag{6.9}$$

where $\mathbf{x}_k \in \mathbb{R}^{N_x}$ is the target *state vector* of order N_x at time k, and $\mathbf{v}_{k-1}$ is the *process* (or *model*) *noise* sequence, which gathers any mismodeling effect or disturbances in the state characterization. Similarly, $\mathbf{z}_k \in \mathbb{R}^{N_z}$ is the *observation* (or *measurement*) *vector* of order N_z at time k, and $\mathbf{w}_k$ is the *observation* (or *measurement*) *noise*. $\mathbf{f}_{k-1}$ and $\mathbf{h}_k$ are *state transition function* and *measurement or observation function*, respectively, and they can be linear or nonlinear. Equation (6.8) is called the *process (or state) equation*, whereas (6.9) models the relation between the measurements and the state and is known as the *measurement (or observation) equation*. Both process and measurement noise ($\mathbf{v}_k$ and $\mathbf{w}_k$) are assumed with known statistics and mutually independent. The combination of (6.8) and (6.9) leads to a Markov chain whose state is not directly observable but indirectly inferred through measurements, an approach known as the *hidden Markov model* (Fig. 6.1). The initial a priori distribution of the state vector is assumed to be known, $p(\mathbf{x}_0)$.

Recalling the recursion enclosed by prediction-update Eqs (6.4) and (6.7), the motion model $p(\mathbf{x}_k|\mathbf{x}_{k-1})$ is described by the state equation (6.8), and the likelihood function $p(\mathbf{z}_k|\mathbf{x}_k)$ is modeled by the measurement equation (6.9). In the rest of this section, we give some examples of dynamic models that are common in localization problems.

Examples of Dynamic Models

A common hypothesis referred to the state equation (6.8) is that the model noise $\mathbf{v}_k$ is additive, white, and Gaussian, that is,

$$p(\mathbf{x}_k|\mathbf{x}_{k-1}) = \frac{1}{\sqrt{2\pi}\,\sigma_m} e^{\frac{||\mathbf{x}_k - \mu_{k-1}||^2}{2\sigma_m^2}}, \tag{6.10}$$

where the variance σ_m^2 accounts for the uncertainty in the movements, and μ_{k-1} depends on $\mathbf{x}_{k-1}$ as will be shown in the following. We see now three examples of mobility models that will be adopted in Chapter 7 to asses the performance of some tracking schemes using experimental data [17]:

[M1] Mobility without speed information. In this case, the pdf (6.10) corresponds to a Gaussian distribution centered in the earlier estimated position, that is, $\mu_{k-1} = \mathbf{x}_{k-1}$.

[M2] Mobility with speed information. This model assumes that the speed and the direction of the motion are known, for example because they are estimated from an inertial device by integrating acceleration measurements. In this case, we have

$$\mu_{k-1} = \mathbf{x}_{k-1} + \upsilon_{k-1}\,\Delta t_k \tag{6.11}$$

with υ_{k-1} being the speed estimated at time $k-1$ and Δt_k the time interval of observation. The standard deviation σ_m depends on the accuracy of the inertial device.

[M3] Mobility with speed learning. Similar to model $M2$, here the speed is evaluated from earlier estimated positions in a sliding window of length N_υ:

$$\upsilon_{k-1} = \frac{1}{N_\upsilon\,\Delta t_k}\sum_{i=1}^{N_\upsilon}(\mathbf{x}_{k-i}-\mathbf{x}_{k-i-1}). \tag{6.12}$$

Advanced mobility models for human movements measured through IMUs can be found in Ref. [77].

Perception Model

The perception model represents the conditioned probability $p(\mathbf{z}_k|\mathbf{x}_k)$. Assuming the measures are i.i.d., the perception model can be defined as

$$p(\mathbf{z}_k|\mathbf{x}_k) = \prod_n p(z_{k,n}|\mathbf{x}_k), \tag{6.13}$$

where $z_{k,n}$ is the nth measurement at time k. Assuming again the observation noise $\mathbf{w}_k$ as additive and Gaussian, the perception model of the single measure can be represented by a Gaussian distribution

$$p(z_{k,n}|\mathbf{x}_k) = \frac{1}{\sqrt{2\pi}\,\sigma_p}\mathrm{e}^{\frac{(z_{k,n}-\check{z}_{k,n}(\mathbf{x}_k))^2}{2\sigma_p^2}}, \tag{6.14}$$

where $\check{z}_{k,n}(\mathbf{x}_k)$ is the expected value of the nth measurement at time k as a function of the state (e.g., for a TOA measurement of the signal from the nth anchor node located in $\mathbf{x}_k^{\mathrm{anc}}$, $\check{z}_{k,n}(\mathbf{x}_k) = ||\mathbf{x}_k - \mathbf{x}_k^{\mathrm{anc}}||/c$).

One possible model for σ_p is the following: $\sigma_p^2 = c\,\sigma_0^2||\mathbf{x}_k - \mathbf{x}_k^{\mathrm{anc}}||^\beta$, where σ_0^2 is the variance of the ranging error at the reference distance of 1 m, and β is the path-loss exponent. Another possibility is to consider σ_p constant.

6.1.3 Generic Structure of a Bayesian Filter

Recalling the theory proposed in Section 6.1.1 and having in mind the contribution of the state-space model (6.8)–(6.9), a generic recursive Bayesian filter can be written as in Algorithm 6.1, where the contribution of the deterministic input $\mathbf{u}$ is included for completeness. $\hat{\mathbf{x}}_{k|k-1}$ is the predicted state, whereas $\hat{\mathbf{x}}_{k|k}$ is the optimum (in a Bayesian sense) estimate of the state $\mathbf{x}$ at the time instant k.

1: **for** $k = 1$ to ∞ **do**

2:     **Prediction**: estimate the predicted state

$$\hat{\mathbf{x}}_{k|k-1} = \mathbb{E}\{\mathbf{x}_k|\mathbf{z}_{1:k-1},\mathbf{u}_{1:k-1}\} = \int_{\chi} \mathbf{x}_k p(\mathbf{x}_k|\mathbf{z}_{1:k-1},\mathbf{u}_{1:k-1})d\mathbf{x}_k, \qquad (6.15)$$

    where:

$$p(\mathbf{x}_k|\mathbf{z}_{1:k-1},\mathbf{u}_{1:k-1}) = \int_{\chi} p(\mathbf{x}_k,\mathbf{x}_{k-1}|\mathbf{z}_{1:k-1},\mathbf{u}_{1:k-1})d\mathbf{x}_{k-1}$$

$$= \int_{\chi} p(\mathbf{x}_k|\mathbf{x}_{k-1},\mathbf{u}_{k-1})p(\mathbf{x}_{k-1}|\mathbf{z}_{1:k-1},\mathbf{u}_{1:k-1})d\mathbf{x}_{k-1}. \qquad (6.16)$$

3:     **Update**: update the state prediction with the new measurement and input at time k

$$\hat{\mathbf{x}}_{k|k} = \mathbb{E}\{\mathbf{x}_k|\mathbf{z}_{1:k},\mathbf{u}_{1:k}\} = \int_{\chi} \mathbf{x}_k p(\mathbf{x}_k|\mathbf{z}_{1:k},\mathbf{u}_{1:k})d\mathbf{x}_k, \qquad (6.17)$$

    where, applying Bayes' rule:

$$p(\mathbf{x}_k|\mathbf{z}_{1:k},\mathbf{u}_{1:k}) = p(\mathbf{x}_k|\mathbf{z}_k,\mathbf{u}_k,\mathbf{z}_{1:k-1},\mathbf{u}_{1:k-1}) \qquad (6.18)$$

$$= \frac{p(\mathbf{z}_k|\mathbf{x}_k,\mathbf{u}_k)p(\mathbf{x}_k|\mathbf{u}_k,\mathbf{z}_{1:k-1},\mathbf{u}_{1:k-1})}{p(\mathbf{z}_k|\mathbf{u}_k,\mathbf{z}_{1:k-1},\mathbf{u}_{1:k-1})} \qquad (6.19)$$

$$= \frac{p(\mathbf{z}_k|\mathbf{x}_k,\mathbf{u}_k)p(\mathbf{x}_k|\mathbf{z}_{1:k-1},\mathbf{u}_{1:k-1})}{\int_{\chi} p(\mathbf{z}_k|\mathbf{x}_k,\mathbf{u}_k)p(\mathbf{x}_k|\mathbf{z}_{1:k-1},\mathbf{u}_{1:k-1})d\mathbf{x}_k}. \qquad (6.20)$$

4: **end for**

Algorithm 6.1

Optimal recursive Bayesian filter structure

The state transition density $p(\mathbf{x}_k|\mathbf{x}_{k-1},\mathbf{u}_{k-1})$ is given by (6.8), whereas the posterior density (belief) at time $k-1$, $p(\mathbf{x}_{k-1}|\mathbf{z}_{1:k-1},\mathbf{u}_{1:k-1})$, is obtained in the earlier update step. Note that from (6.18) to (6.20), we have used the fact

$$p(\mathbf{x}_k|\mathbf{z}_{k-1},\mathbf{u}_{k-1},\mathbf{u}_k) = p(\mathbf{x}_k|\mathbf{z}_{k-1},\mathbf{u}_{k-1}),$$

which is equivalent to stating that the input $\mathbf{u}_k$ can be inferred using $\hat{\mathbf{x}}_{k|k-1}$. This allows the recursive relationship, since the a priori density $p(\mathbf{x}_k|\mathbf{z}_{k-1},\mathbf{u}_{k-1})$ is computed in the prediction step. The measurement likelihood function $p(\mathbf{z}_k|\mathbf{x}_k,\mathbf{u}_k)$ is given by (6.9).

Algorithm 6.1 provides an estimation framework that is optimal in the Bayesian sense and enjoys a recursive structure, which is a desirable property for online estimation. In the general case, however, this algorithm cannot be used in practice because the multidimensional integrals involved in (6.16) and (6.20) are intractable. For Markovian, nonlinear, and non-Gaussian state-space models, sequential Monte Carlo (SMC) methods [26, 27] provide a framework to obtain numerical approximations of such integrals by a stochastic sampling approach. This broad suitability comes at the expense of a high computational load that makes this solution difficult to embed in microprocessors or real-time applications.

The following sections describe several estimation algorithms applied to positioning and tracking problems. After a brief overview of the Kalman filter, we present some variants such as the extended Kalman filter (EKF), suitable to deal with low-order nonlinearities in the system model, the adaptive EKF, which includes online noise covariance estimations, and new approaches devoted to systems in which the noise is still Gaussian but their state-space equations are nonlinear, namely the square-root versions of the quadrature and cubature Kalman filters.

6.1.4 Kalman Filter and its Derivatives

The Kalman filter (KF) is an optimal (in MMSE sense) Bayesian recursive filter that estimates the state of a linear dynamic system from a series of noisy measurements when the following *assumptions* on (6.8)–(6.9) hold [74]:

- $\mathbf{v}_{k-1}$ and $\mathbf{w}_k$ are drawn from Gaussian densities of known parameters and are additive,
- $\mathbf{f}_{k-1}(\mathbf{x}_{k-1}, \mathbf{u}_{k-1}, \mathbf{v}_{k-1})$ is a known linear function of $(\mathbf{x}_{k-1}, \mathbf{u}_{k-1})$ and $\mathbf{v}_{k-1}$, and
- $\mathbf{h}_k(\mathbf{x}_k, \mathbf{w}_k)$ is a known linear function of $\mathbf{x}_k$ and $\mathbf{w}_k$.

The KF achieves optimal MMSE solution only under the above-mentioned highly constrained conditions. However, for most real-world systems, the assumptions are too tight. They may not hold in some applications where the dependence of measurements on states is nonlinear, or when noises cannot be considered normally distributed or zero-biased. In such situations, the MMSE estimator is intractable, and we have to resort to suboptimal Bayesian filters. Among the suboptimal filters, the EKF [21] has been widely used for several years. The main idea adopted in the EKF is to linearize the state transition and/or observation equations through a Taylor-series expansion around the mean of the relevant Gaussian RV and apply the linear KF to this linearized model.

EKF has been designed for many practical systems that have nonlinear state update and/or measurement equations [21]. In fact, EKF linearizes the observation model around the current mean of the state [74]. As a consequence, the performance of the EKF heavily depends on how the system dynamics and measurements are modeled.

If only the assumptions about the noises hold, the state-space model can be written as

$$\mathbf{x}_k = \mathbf{f}(\mathbf{x}_{k-1}, \mathbf{u}_{k-1}) + \mathbf{v}_k, \tag{6.21}$$

$$\mathbf{z}_k = \mathbf{h}(\mathbf{x}_k) + \mathbf{w}_k, \tag{6.22}$$

where $\mathbf{v}_k \sim \mathcal{N}(\mathbf{0}, \mathbf{Q}_k)$, $\mathbf{w}_k \sim \mathcal{N}(\mathbf{0}, \mathbf{R}_k)$, and $\mathbf{f}$ and $\mathbf{h}$ are nonlinear. Relying on this model, the EKF algorithm is described in Algorithm 6.2 (for derivation details, the reader may refer to Refs. [13, 21, 74]).

6.1.4.1 Basic Examples of EKF Models to Solve the Localization Problem

In this section, we show three examples of systems where the EKF can be fruitfully used to solve a basic localization problem. For simplicity, we consider a 2D geometry populated by nodes belonging to a WSN. Among them, there are both anchor nodes and "unknown" (in the positioning sense) nodes. The underlying wireless technology is UWB in the first example, ZigBee in the second example, and a mixture of the two in the third example.

1: **for** $k = 1$ to ∞ **do**

2:     **Prediction**:
    Estimate the predicted state

$$\hat{x}_{k|k-1} = \mathbf{F}_k \cdot \hat{x}_{k-1|k-1} + \mathbf{B}_k \cdot \mathbf{u}_k, \tag{6.23}$$

    where $\mathbf{B}_k$ represents the input matrix, $\mathbf{u}_k$ is the input to the system and $\mathbf{F}_k$ represents the linearized state transition matrix defined as

$$\mathbf{F}_k = \nabla_{\mathbf{x}} \mathbf{f}_k(\mathbf{x}, \mathbf{u}) \Big|_{\mathbf{x}=\check{\mathbf{x}}}, \tag{6.24}$$

    where $\nabla_{\mathbf{x}}$ is the Jacobian operator with respect to the vector $\mathbf{x}$ and $\check{\mathbf{x}}$ is the point where the linearization takes place.

3:     Estimate the predicted covariance matrix $\hat{\mathbf{P}}_{k|k-1}$ related to the current a priori state vector $\hat{x}_{k|k-1}$

$$\hat{\mathbf{P}}_{k|k-1} = \mathbf{F}_k \cdot \hat{\mathbf{P}}_{k-1|k-1} \cdot \mathbf{F}_k^T + \mathbf{Q}_k. \tag{6.25}$$

    Update:

4:     Compute the *innovation vector*, as the residual between the observed measurement $\mathbf{z}_k$ and the predicted measurement $\mathbf{h}\left(\hat{x}_{k|k-1}\right)$:

$$\tilde{\mathbf{y}}_k = \mathbf{z}_k - \mathbf{h}\left(\hat{x}_{k|k-1}\right). \tag{6.26}$$

5:     Compute the *Kalman gain*

$$\mathbf{K}_k = \hat{\mathbf{P}}_{k|k-1} \cdot \mathbf{H}_k^T \cdot \left(\mathbf{H}_k \cdot \hat{\mathbf{P}}_{k|k-1} \cdot \mathbf{H}_k^T + \mathbf{R}_k\right)^{-1}, \tag{6.27}$$

    where $\mathbf{H}_k$ is the linearized observation matrix and $\mathbf{R}_k$ represents the covariance matrix related to the observation vector. The observation matrix $\mathbf{H}_k$ is defined as

$$\mathbf{H}_k = \nabla_{\mathbf{x}} \mathbf{h}_k(\mathbf{x}). \tag{6.28}$$

6:     Compute the a posteriori state estimate

$$\hat{x}_{k|k} = \hat{x}_{k|k-1} + \mathbf{K}_k \cdot \tilde{\mathbf{y}}_k. \tag{6.29}$$

7:     Compute the a posteriori state covariance matrix

$$\hat{\mathbf{P}}_{k|k} = (\mathbf{I}_n - \mathbf{K}_k \cdot \mathbf{H}_k) \cdot \hat{\mathbf{P}}_{k|k-1}, \tag{6.30}$$

    where $\mathbf{I}_n$ represents the identity matrix of dimension n.

8: **end for**

Algorithm 6.2

Kalman filter structure

The state-space model that describes the unknown node motion is a position–velocity (PV) model and evidently has to be the same in the three cases, as well as the recursive structure of the EKF. On the contrary, the measurement model and its related equations are different in the three cases, as it is technology-dependent, and must be derived separately.

In these hypotheses, the vector state for each unknown node can be defined as

$$\mathbf{x}_k = [x_k, \, y_k, \, \dot{x}_k, \, \dot{y}_k]^T, \tag{6.31}$$

where $\dot{x}_k$ and $\dot{y}_k$ denote the speed of the target node along the x- and y-axes, respectively. Accordingly, the matrix $\mathbf{F}_k$ introduced in (6.23) can be written as follows:

$$\mathbf{F}_k = \begin{bmatrix} 1 & 0 & \Delta t_k & 0 \\ 0 & 1 & 0 & \Delta t_k \\ 0 & 0 & 1 & 0 \\ 0 & 0 & 0 & 1 \end{bmatrix}, \tag{6.32}$$

whereas the matrix $\mathbf{Q}$ is defined as follows [13]:

$$\mathbf{Q} = \begin{bmatrix} \frac{\Delta t_k^2}{2} I_2 \\ \Delta t_k I_2 \end{bmatrix} \cdot \begin{bmatrix} \sigma_{\ddot{x}}^2 & 0 \\ 0 & \sigma_{\ddot{y}}^2 \end{bmatrix} \cdot \begin{bmatrix} \frac{\Delta t_k^2}{2} I_2 \\ \Delta t_k I_2 \end{bmatrix}^T, \tag{6.33}$$

where $\sigma_{\ddot{x}}^2$ and $\sigma_{\ddot{y}}^2$ denote the variances of the acceleration noise along the x- and y-axes, respectively. The covariance matrix $\mathbf{P}_0$ related to the initial state vector $\mathbf{x}_0$ can be defined as:

$$\mathbf{P}_0 = \text{diag}\left(\sigma_{x_0}^2, \sigma_{y_0}^2, \sigma_{\dot{x}_0}^2, \sigma_{\dot{y}_0}^2\right), \tag{6.34}$$

where $\sigma_{x_0}^2, \sigma_{y_0}^2, \sigma_{\dot{x}_0}^2$, and $\sigma_{\dot{y}_0}^2$ represent the initial variances of the state vector components.

The technology-specific measurement models are defined in the following three examples.

Example 1: Observation Model for UWB-Distance Measurements

Indoor positioning systems can be based on UWB technology as UWB signals provide very accurate distance estimation [51]. Typically, UWB nodes perform Two-Way TOA $(\tilde{t}_{TW-TOA})$ [67] estimation. Consequently, the distance can be estimated as follows:

$$\tilde{d} = c \cdot \tilde{t}_{TOA} = c \cdot \frac{\tilde{t}_{TW-TOA}}{2},$$

where c is the speed of light. The EKF estimates the position of the mobile node by using the distance measurements between the mobile node and a set of anchor nodes [13]. Considering the two-dimensional case, the generic anchor node A_i has known coordinates $\mathbf{x}_{A_i} = [x_{A_i}, y_{A_i}]^T$ for $i = 1, \ldots, L$, where L represents the total number of UWB anchor nodes deployed in the environment. The observation vector $\mathbf{z}_{\text{dist},k}$ used by the EKF can be defined as:

$$\mathbf{z}_{\text{dist},k} = \begin{bmatrix} \tilde{d}_{A_1,k} & \tilde{d}_{A_2,k} & \cdots & \tilde{d}_{A_L,k} \end{bmatrix}, \tag{6.35}$$

where $\tilde{d}_{A_i,k}$ represents the estimated distance between the mobile node and the ith UWB anchor node at the current estimation time t_k. However, the vector of predicted measurements $\mathbf{h}_{\text{dist}}\left(\hat{\mathbf{x}}_{k|k-1}\right)$ (see (6.26)) is computed as the Euclidean distance between the predicted mobile node position and all anchor nodes at the estimation time t_k, expressed as follows:

$$\mathbf{h}_{\text{dist}}\left(\hat{\mathbf{x}}_{k|k-1}\right) = \begin{bmatrix} \text{dist}\left(\hat{\mathbf{x}}_{k|k-1}, \mathbf{x}_{A_1}\right) \\ \text{dist}\left(\hat{\mathbf{x}}_{k|k-1}, \mathbf{x}_{A_2}\right) \\ \vdots \\ \text{dist}\left(\hat{\mathbf{x}}_{k|k-1}, \mathbf{x}_{A_L}\right) \end{bmatrix}, \tag{6.36}$$

where dist$(\cdot)$ is the Euclidean distance operator. For the ith anchor node it is

$$\text{dist}\left(\hat{\mathbf{x}}_{k|k-1}, \mathbf{x}_{A_i}\right) = \sqrt{\left(\hat{x}_{k|k-1} - x_{A_i}\right)^2 + \left(\hat{y}_{k|k-1} - y_{A_i}\right)^2}.$$

The Jacobian $\mathbf{H}_{\text{dist},k}$ matrix of the predicted measurement vector $\mathbf{h}\left(\hat{\mathbf{x}}_k\right)$ around the a priori state vector $\hat{\mathbf{x}}_{k|k-1}$ is

$$\mathbf{H}_{\text{dist},k} = \begin{bmatrix} \dfrac{\hat{x}_{k|k-1} - x_{A_1}}{\text{dist}(\hat{\mathbf{x}}_{k|k-1}, \mathbf{x}_{A_1})} & \dfrac{\hat{y}_{k|k-1} - y_{A_1}}{\text{dist}(\hat{\mathbf{x}}_{k|k-1}, \mathbf{x}_{A_1})} & 0 & 0 \\ \vdots & \vdots & \vdots & \vdots \\ \dfrac{\hat{x}_{k|k-1} - x_{A_L}}{\text{dist}(\hat{\mathbf{x}}_{k|k-1}, \mathbf{x}_{A_L})} & \dfrac{\hat{y}_{k|k-1} - y_{A_L}}{\text{dist}(\hat{\mathbf{x}}_{k|k-1}, \mathbf{x}_{A_L})} & 0 & 0 \end{bmatrix}. \tag{6.37}$$

The covariance matrix $\mathbf{R}_{\text{dist},k}$ of the observation vector for the distance measurements can be defined as:

$$\mathbf{R}_{\text{dist},k} = \text{diag}\left(\sigma^2_{d_{A_1,k}} \quad \cdots \quad \sigma^2_{d_{A_L,k}}\right), \tag{6.38}$$

where $\sigma^2_{d_{A_i,k}}$ represents the variance of distance measurement for the ith anchor node.

Example 2: Observation Model for ZigBee-RSS Measurements

The RSS measurement capability is relatively cheap to implement in hardware. However, distances based on RSS are less accurate than TOA measurements [13]. The observation vector $\mathbf{z}_{\text{RSS},k}$ for RSS measurements at the estimation time t_k can be defined as:

$$\mathbf{z}_{\text{RSS},k} = \begin{bmatrix} \tilde{P}_{A_1,k} & \tilde{P}_{A_2,k} & \cdots & \tilde{P}_{A_M,k} \end{bmatrix}, \tag{6.39}$$

where $\tilde{P}_{A_i,k}$ represents the RSS measurement between the mobile node and the ith ZigBee anchor node. Typically, RSS measurements in WSN are modeled by using the log-normal shadowing path-loss model [13] that gives the power P in dB received at a distance d from the transmitter as follows:

$$P = P_0 - 10\alpha \log_{10}\left(d/d_0\right) + X_{\sigma_{\text{dB}}}, \tag{6.40}$$

where P_0 represents the received power at the close-in distance d_0 (typically 1 m for frequencies between 1 and 3 GHz), α represents the path-loss exponent (two for free space), and $X_{\sigma_{\text{dB}}} \sim \mathcal{N}\left(0, \sigma^2_{\text{dB}}\right)$ is an additive Gaussian random variable modeling the shadowing effect. The vector of the predicted measurements $\mathbf{h}_{\text{RSS}}\left(\hat{\mathbf{x}}_{k|k-1}\right)$ contains predicted RSS measurements between the predicted mobile node position and all anchor nodes at the estimation time t_k expressed as

$$\mathbf{h}_{\text{RSS}}\left(\hat{\mathbf{x}}_{k|k-1}\right) = \begin{bmatrix} P_{A_1}\left(\hat{\mathbf{x}}_{k|k-1}\right) \\ P_{A_2}\left(\hat{\mathbf{x}}_{k|k-1}\right) \\ \vdots \\ P_{A_M}\left(\hat{\mathbf{x}}_{k|k-1}\right) \end{bmatrix}, \tag{6.41}$$

where the received power from the ith ZigBee anchor node $P_{A_i}\left(\hat{\mathbf{x}}_{k|k-1}\right)$, expressed in dBm, can be defined as:

$$P_{A_i}\left(\hat{\mathbf{x}}_{k|k-1}\right) = P_0 - 10\alpha \log_{10} \frac{\sqrt{\left(\hat{x}_{k|k-1} - x_{A_i}\right)^2 + \left(\hat{y}_{k|k-1} - y_{A_i}\right)^2}}{d_0}. \tag{6.42}$$

For the PV model in the two-dimensional case, the Jacobian matrix $\mathbf{H}_{\text{RSS},k}$ is

$$\mathbf{H}_{\text{RSS},k} = -\frac{10\alpha}{\ln(10)} \begin{bmatrix} \frac{\hat{x}_{k|k-1} - x_{A_1}}{\text{dist}^2\left(\hat{\mathbf{x}}_{k|k-1}, \mathbf{x}_{A_1}\right)} & \frac{\hat{y}_{k|k-1} - y_{A_1}}{\text{dist}^2\left(\hat{\mathbf{x}}_{k|k-1}, \mathbf{x}_{A_1}\right)} & 0 & 0 \\ \vdots & \vdots & \vdots & \vdots \\ \frac{\hat{x}_{k|k-1} - x_{A_M}}{\text{dist}^2\left(\hat{\mathbf{x}}_{k|k-1}, \mathbf{x}_{A_M}\right)} & \frac{\hat{y}_{k|k-1} - y_{A_M}}{\text{dist}^2\left(\hat{\mathbf{x}}_{k|k-1}, \mathbf{x}_{A_M}\right)} & 0 & 0 \end{bmatrix}. \tag{6.43}$$

The covariance matrix $\mathbf{R}_{\text{RSS},k}$ of the observation vector for RSS measurements is also

$$\mathbf{R}_{\text{RSS},k} = \text{diag}\left(\sigma^2_{\text{dB}_{A_1},k} \quad \cdots \quad \sigma^2_{\text{dB}_{A_M},k}\right), \tag{6.44}$$

where $\sigma^2_{\text{dB}_{A_i},k}$ represents the initial variance of the shadowing for the ith ZigBee anchor node.

Example 3: Hybrid Observation Model for UWB-Distance and ZigBee-RSS Measurements

Higher positioning accuracy can be achieved by means of hybrid techniques which take advantage of smart combination of different types of measurements based on different technologies [1]. The hybrid technique considered in this example combines distance measurements based on UWB and RSS measurements based on ZigBee. In general, hybrid techniques, compared with positioning algorithms based only on a single type of range measurement, have an increased number of measurements in the observation vector as the total number of anchor nodes is increased. Consequently, both position estimation availability and position accuracy improve [78].

Let us consider L UWB and M ZigBee anchor nodes. Starting from (6.35) and (6.39), the hybrid observation vector $\mathbf{z}_k$ which feeds the hybrid EKF can be defined as follows:

$$\mathbf{z}_k = \begin{bmatrix} \mathbf{z}_{\text{dist},k} \\ \mathbf{z}_{\text{RSS},k} \end{bmatrix}. \tag{6.45}$$

The vector $\mathbf{h}\left(\hat{\mathbf{x}}_{k|k-1}\right)$ for the hybrid case can be defined using (6.36) and (6.41) as follows:

$$\mathbf{h}\left(\hat{\mathbf{x}}_{k|k-1}\right) = \begin{bmatrix} \mathbf{h}_{\text{dist}}\left(\hat{\mathbf{x}}_{k|k-1}\right) \\ \mathbf{h}_{\text{RSS}}\left(\hat{\mathbf{x}}_{k|k-1}\right) \end{bmatrix}. \tag{6.46}$$

Similarly, the hybrid Jacobian matrix $\mathbf{H}_k$ can be found using (6.37) and (6.43):

$$\mathbf{H}_k = \begin{bmatrix} \mathbf{H}_{\text{dist},k} \\ \mathbf{H}_{\text{RSS},k} \end{bmatrix}. \tag{6.47}$$

The hybrid covariance matrix $\mathbf{R}_k$ of the observation vector can be found from (6.38) and (6.44):

$$\mathbf{R}_k = \begin{bmatrix} \mathbf{R}_{\text{dist},k} & \mathbf{0}_{L \times M} \\ \mathbf{0}_{M \times L} & \mathbf{R}_{\text{RSS},k} \end{bmatrix}, \tag{6.48}$$

where $\mathbf{0}_{L \times M}$ and $\mathbf{0}_{M \times L}$ represent zero matrices of size $L \times M$ and $M \times L$, respectively.

6.1.4.2 *A More Complex Example: Extended Kalman Filter with Tracking of the NLOS Bias*

It is well known that NLOS positioning introduces a bias on TOA estimates, reducing the accuracy of positioning algorithms. In particular, because of the NLOS condition, the first arrival suffers from stronger attenuation than later arrivals and therefore wrong TOA information is obtained. In the multipath situation (be it LOS or NLOS), the late signal arrivals induce a displacement of the maximum of the impulse response, thus biasing timing estimates if algorithms which assume a simple frequency-flat channel are used. The resulting positioning system dealing with those estimates can only provide biased position estimation.

In order to reduce this kind of error, it is possible to define a more complete state-space model that includes the distance bias for each node as an additional parameter to be estimated [52]. Therefore, the objective of the associated EKF is to improve the positioning accuracy by tracking not only the position and speed of the unknown mobile node but also the bias because of the NLOS propagation. This approach is called extended Kalman filter with bias tracking (EKFBT) and is briefly described below.

The state vector is defined as $\mathbf{x}_k^{(\text{BT})} = \begin{bmatrix} \mathbf{x}_k^T & \mathbf{b}_k^T \end{bmatrix}^T$, where the vector $\mathbf{x}_k$ is that defined in (6.31) and the components of the vector $\mathbf{b}_k$ are the distance measurement bias at each node.

The state transition matrix $\mathbf{F}_k^{(\text{BT})}$ is

$$\mathbf{F}_k^{(\text{BT})} = \begin{bmatrix} \mathbf{F}_k & \mathbf{0} \\ \mathbf{0} & \mathbf{I}_M \end{bmatrix}, \tag{6.49}$$

where M is the number of reference points (anchor nodes).

The model noise vector is $\mathbf{v}_k = \begin{bmatrix} \mathbf{0}_2, & \mathbf{v}_s, & \mathbf{v}_b \end{bmatrix}^T$, where $\mathbf{v}_s$ and $\mathbf{v}_b$ are the (uncorrelated) speed and bias noise vectors with covariance matrices $\mathbf{Q}_{s,k}$ and $\mathbf{Q}_{b,k}$, respectively.

Given that the bias is additive with respect to the distance measurements, the observation equation for the EKFBT can be written as

$$\mathbf{z}_k = \mathbf{h}(\mathbf{x}_k) + \mathbf{b}_k + \mathbf{w}_k, \tag{6.50}$$

so that the TOA-based measurement matrix $\mathbf{H}_k^{(\text{BT})}$ turns out to be

$$\mathbf{H}_k^{(\text{BT})} = \begin{bmatrix} \mathbf{H}_{\text{dist},k} & \mathbf{I}_M \end{bmatrix}, \tag{6.51}$$

where $\mathbf{H}_{\text{dist},k}$ has been defined in (6.37).

6.1.4.3 *Adaptive EKF*

This section focuses on an *adaptive extended Kalman filter* (AEKF) technique proposed in Ref. [13] for tracking mobile targets moving at low dynamics by using RSS measurements from fixed anchor nodes.

When a KF is used, usually a sensitivity analysis is performed in order to find the numerical parameters that best fit the model to the actual target trajectories. Such analysis is strongly case-dependent since these parameters depend on factors that change in different scenarios such as target dynamics, sampling frequency, and noise variance. AEKFs attempt to automatically change relevant model parameters according to the conditions where the system is implemented and readapt them if the environment changes.

The idea behind this adaptive approach is to use the EKF to estimate some parameters of the log-normal shadowing path loss model (6.40) used for RSS measurements. Typically such parameters are found after a scene analysis campaign and a posterior linear regression on a semilogarithmic scale, where P_0 is the intercept at distance d_0, and the path-loss exponent α is the slope (see Refs. [53, 61] for additional details). The shadowing variance σ_{dB}^2 is extracted as the variance of the difference between actual measurements and the model.

Such off-line scene analysis can in principle be avoided if the propagation model parameters are estimated by the EKF itself. To do this, it is possible to include the path-loss exponent α and the mean received power P_0 at a given close-in distance into the state vector

$$\mathbf{x}_k^{(Ad)} = \left[\mathbf{x}_k^T, \alpha, P_0 \right]^T, \tag{6.52}$$

where $\mathbf{x}_k$ can be any state vector defined according to, for example, the P, PV, or PVA (Position, Velocity, or Acceleration) state models [13]. For instance, vector $\mathbf{x}_k$ for the PV model is that defined in (6.31).

Since parameters α and P_0 are rather static, their state transition function is the identity matrix, then

$$\mathbf{F}_k^{(Ad)} = \begin{bmatrix} \mathbf{F}_k & \mathbf{0}_{4 \times 2} \\ \mathbf{0}_{2 \times 4} & \mathbf{I}_2 \end{bmatrix}, \tag{6.53}$$

whereas the process noise covariance matrix becomes

$$\mathbf{Q}_k^{(Ad)} = \begin{bmatrix} \mathbf{Q}_k & \mathbf{0}_{4 \times 2} \\ \mathbf{0}_{2 \times 4} & \mathrm{diag}(\sigma_\alpha^2, \sigma_{P_0}^2) \end{bmatrix}, \tag{6.54}$$

where σ_α^2 and $\sigma_{P_0}^2$ are the noise variance related to the variable state α and P_0, respectively.

The observation is a vector including RSS measurements between target node and the M anchor nodes, as in (6.41). The corresponding Jacobian $\mathbf{H}_k^{(Ad)}$ (computed around the a priori state $\mathbf{x}_{k|k-1}^{(Ad)}$) is derived from the log-normal model (6.40) as

$$\mathbf{H}_k^{(Ad)} = \begin{bmatrix} & -10\log_{10}\left(\mathrm{dist}\left(\mathbf{x}_{k|k-1}, \mathbf{x}_{A_1}\right)/d_0\right) & 1 \\ \mathbf{H}_{RSS,k} & \vdots & \vdots \\ & -10\log_{10}\left(\mathrm{dist}\left(\mathbf{x}_{k|k-1}, \mathbf{x}_{A_M}\right)/d_0\right) & 1 \end{bmatrix}, \tag{6.55}$$

where $\mathbf{H}_{RSS,k}$ is that defined in (6.43). The measurement noise is Gaussian and white, as in (6.44).

Sage and Husa first presented in 1969 [59] the idea to estimate one or both measurement and process noise covariance matrices from the residual innovations. The main advantage of this algorithm is the capability to estimate online the noise covariances at each KF iteration, thus avoiding the need to perform statistical analysis on historical data. In the following, a slightly modified Sage–Husa adaptive algorithm that assumes diagonal covariance matrices (i.e., assuming independent state and measurement noise vector components) is presented.

The measurement noise covariance matrix to be used on the next iteration $\hat{\mathbf{R}}_{\text{RSS},k+1|k+1}$ is estimated as a convex combination of the current estimate $\hat{\mathbf{R}}_{\text{RSS},k|k}$ and the square of the innovations $\tilde{\mathbf{y}}_k$ and the a posteriori state estimation covariance matrix $\mathbf{P}_{k|k}$ projected into the measurements. For faster convergence, the learning rate λ_k is adapted like in an exponential moving average truncated to a window size w, allowing detection of further changes in the noise:

$$\hat{\mathbf{R}}_{\text{RSS},k+1|k+1} = (1 - \lambda_k)\,\hat{\mathbf{R}}_{\text{RSS},k|k} + \lambda_k \hat{\mathbf{R}}_{\text{RSS},k+1|k}, \tag{6.56}$$

$$\hat{\mathbf{R}}_{\text{RSS},k+1|k} = \text{diag}\left(\tilde{\mathbf{y}}_k\right)^2 + \mathbf{H}_k \mathbf{P}_{k|k} \mathbf{H}_k^T,$$

$$\lambda_k = \left(1 - (1/w)\right) \Big/ \left(1 - (1 - (1/w))^k\right).$$

Similarly, for the process noise covariance:

$$\hat{\mathbf{Q}}_{k+1|k+1}^{(\text{Ad})} = (1 - \lambda_k)\,\hat{\mathbf{Q}}_{k|k}^{(\text{Ad})} + \lambda_k \hat{\mathbf{Q}}_{k+1|k}^{(\text{Ad})}, \tag{6.57}$$

$$\hat{\mathbf{Q}}_{k+1|k}^{(\text{Ad})} = \text{diag}\left(\hat{\mathbf{x}}_{k|k} - \hat{\mathbf{x}}_{k|k-1}\right)^2 - \left(\mathbf{P}_{k|k} - \mathbf{P}_{k|k-1} + \hat{\mathbf{Q}}_{k|k}^{(\text{Ad})}\right).$$

Experimental Results

To test the performance of the earlier developed models, we present an experimental campaign executed in dynamic conditions. Eighteen wireless sensor nodes (Texas Instruments Chipcon 2430) are used as anchor nodes, placed on an indoor area of 15 m × 7 m. Simultaneous RSS measurements are performed each 0.25 s between a mobile node and the anchors, forwarded to a gateway (TI SmartRF) and stored in a database in order to perform off-line analysis and compare performance of the different tracking algorithms proposed. The target node is placed on board a LEGO Mindstorm NTX robot programmed to follow at constant speed (0.15 m/s) the path depicted in Fig. 6.2(b) by a solid line.

Figure 6.2 presents performance of the adaptive tracking techniques. The filters are designed with the PV state model and received signal strength (RSS) observation model discussed in Section 6.1.4.1, Example 1.

The baseline for comparison is the nonadaptive EKF introduced in Section 6.1.4.1 with measurement error defined according to the variance of the measurements from the scene analysis and parameters of process noise set according to the variance of the accelerations from a sensitivity analysis of the mobile node trajectory. This filter achieves an RMS radial accuracy of 0.62 m (not plotted in Fig. 6.2).

When the matrix $\mathbf{R}_{\text{RSS},k}$ is estimated adaptively as described earlier, no significant accuracy improvement can be observed (the radial RMSE decreases to 0.57 m) although the adaptive technique is able to estimate correctly the measurement noise variance. However, it can be observed that, estimating only the matrix $\mathbf{Q}_k^{(\text{Ad})}$, the tracking performance degrades to a worse accuracy than the baseline,

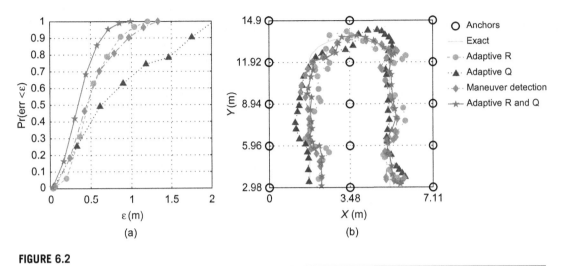

FIGURE 6.2

Adaptive EKF: Performance analysis. (a) Radial cumulative error distribution. (b) Tracking performance.

with a radial RMSE of 1.01 m. These results show that the introduction of online noise covariance estimation techniques keeps the tracking capabilities acceptable while avoiding lengthy off-line sensitivity analysis needed to correctly tune the filter.

The best performance, with a radial RMSE of 0.42 m, is achieved by a simultaneous adaptive estimation of both process noise and measurement noise covariance matrices.

6.1.4.4 *Advanced KF Architectures*

The kind of problems addressed in this chapter fall into the category of *probabilistic inference,* that is, the problem of optimally estimating the hidden variables of a system (states or parameters) given noisy observations. As shown in Section 6.1.1, the optimal solution is given by Bayesian estimation, which involves the use of the posterior pdf of the states. However, the difficulty in dealing with the analytical solution of the posterior and of the related moment integrals leads to the necessity of finding approximate solutions. EKF is one such solution. Sequential Monte Carlo methods, in particular the particle filters introduced in the next section, are another family of such approximated approaches. In the Gaussian hypothesis (i.e., both process and observation noises additive, Gaussian and mutually uncorrelated), other approximated solutions are represented by the unscented Kalman filter (UKF) [40] and by the sigma-point Kalman filter (SPKF) [70, 72].

SPKF is based on the application of the sigma-point approach to the recursive MMSE estimation of a state, where the sigma-point approach basically implies three steps: (1) Computation of a number of so-called sigma-points (of the order of $2N_x + 1$, where N_x is the state dimension). (2) Transformation of each sigma-point using the (nonlinear) process model **f**. (3) Reconstruction of the posterior statistics from propagated sigma-points, using a weighted sample mean and covariance [70]. The key benefits recognized for SPKFs are their accuracy (outperforming EKF), their computational complexity of the same order of EKF, and the fact that they do not require analytical derivatives.

Two examples of SPKFs are presented in this chapter (Section 6.3.3), in the context of the direct position estimation problem: The square-root quadrature Kalman filter (SQKF) and the multiple quadrature KF.

6.1.5 Particle Filters

Particle filters realize Bayes filter updates according to a sampling procedure, often referred to as sequential importance sampling. The key advantage of particle filters is their ability to represent arbitrary probability densities. Moreover, particle filters are very efficient because they automatically focus their resources (particles) on regions of the state space with high probability. However, because the worst case complexity of these methods grows exponentially in the dimensions of the state space, one has to be careful when applying particle filters to high-dimensional estimation problems.

The basic idea for the evaluation of the integrals involved in the prediction and update steps (6.15)–(6.17) is to approximate an expectation integral of the form

$$I(g) = \mathbb{E}_{p(\mathbf{x})}\{g(\mathbf{x})\} = \int_{\mathbf{X}} g(\mathbf{x})p(\mathbf{x})\, d\mathbf{x} \tag{6.58}$$

by the following average

$$I_{N_s}(g) = \frac{1}{N_s}\sum_{i=1}^{N_s} g\left(\mathbf{x}^{(i)}\right), \tag{6.59}$$

where $\left\{\mathbf{x}^{(i)}\right\}_{i=1}^{N_s}$ are independent samples drawn from the distribution $p(\mathbf{x})$. However, $p(\mathbf{x})$ could be a complicated posterior distribution from which it is difficult to draw samples in an efficient way, especially in high-dimensional, nonstandard cases. A possible solution framework is known as Markov Chain Monte Carlo (MCMC) methods [31], but its iterative structure is not well suited to recursive estimation problems. More sophisticated methods that allow recursions are represented by the family of the particle filters (PFs).

In particle filtering, the belief function $Bel(\mathbf{x}_k)$ is parameterized through a set of weighted samples,

$$\{< \mathbf{x}_k^{(i)}, \omega_k^{(i)} > | i = 1, \ldots, N_s\}, \tag{6.60}$$

where N_s is the number of particles, and $\omega_k^{(i)} \geq 0 \; \forall i, k$ are the weights normalized as $\sum_{i=1}^{N_s} \omega_k^{(i)} = 1$. The quality of the approximation in (6.60) depends on the number of particles N_s. In order to decrease the complexity, the sampling can be more dense in the regions where it is more probable that the object is located.

If we could generate N_s independent and identically distributed random samples $\{\mathbf{X}_{0:k}^{(i)}, i = 1, \ldots, N_s\}$ according to the posterior $p(\mathbf{X}_{0:k}|\mathbf{Z}_{1:k})$, we would estimate this distribution as

$$\hat{p}(\mathbf{X}_{0:k}|\mathbf{Z}_{1:k}) = \frac{1}{N_s}\sum_{i=1}^{N_s} \delta\left(\mathbf{X}_{0:k} - \mathbf{X}_{0:k}^{(i)}\right), \tag{6.61}$$

where $\delta(\cdot)$ is the Dirac delta function, and obtain the estimate

$$\hat{I}(g_k) = \int g_k(\mathbf{X}_{0:k})\hat{p}(\mathbf{X}_{0:k}|\mathbf{Z}_{1:k})d\mathbf{X}_{0:k} = \frac{1}{N_s}\sum_{i=1}^{N_s} g_k\left(\mathbf{X}_{0:k}^{(i)}\right), \tag{6.62}$$

which is proven to be unbiased and to converge to $I(g_k)$ under weak assumptions as N_s increases [26]. Unfortunately, it is usually impossible to efficiently draw samples from the posterior $p(\mathbf{X}_{0:k}|\mathbf{Z}_{1:k})$ because it is multivariate and far from being Gaussian.

This problem can be overcome by means of the *importance sampling* concept. Basically, it involves the approximation of the a posteriori probability density function (posterior) by a set of N_s weighted random samples,

$$p(\mathbf{x}_k|\mathbf{z}_{1:k}) \approx \sum_{i=1}^{N_s} \omega_k^{(i)} \delta\left(\mathbf{x}_k - \mathbf{x}_k^{(i)}\right). \tag{6.63}$$

Since the $p(\mathbf{x}_k|\mathbf{z}_{1:k})$ is not known, a so-called proposal distribution is defined, which is convenient to sample instead of $p(\mathbf{x}_k|\mathbf{z}_{1:k})$. Thus, the posterior is approximated by a set of weighted random samples (called *particles*) taken from an arbitrary *proposal density function* (or *importance density function*) $\pi(\mathbf{X}_{0:k}|\mathbf{Z}_{1:k})$. The proposal distribution is chosen so that it can be factorized:

$$\pi(\mathbf{x}_k|\mathbf{z}_{1:k}) = \pi(\mathbf{x}_k|\mathbf{x}_{k-1},\mathbf{z}_{1:k}) \cdot \pi(\mathbf{x}_{k-1}|\mathbf{z}_{1:k-1}). \tag{6.64}$$

Then, in the evaluation of $I(g_k)$ we can write

$$I(g_k) = \frac{\int g_k(\mathbf{X}_{0:k}) \omega(\mathbf{X}_{0:k}) \pi(\mathbf{X}_{0:k}|\mathbf{Z}_{1:k}) d\mathbf{X}_{0:k}}{\int \omega(\mathbf{X}_{0:k}) \pi(\mathbf{X}_{0:k})}, \tag{6.65}$$

where the term $\omega(\mathbf{X}_{0:k})$ is referred to as *importance weight* and is defined as

$$\omega(\mathbf{X}_{0:k}) = \frac{p(\mathbf{X}_{0:k}|\mathbf{Z}_{1:k})}{\pi(\mathbf{X}_{0:k}|\mathbf{Z}_{1:k})}. \tag{6.66}$$

Then, after generating N_s independent and identically distributed particles $\left\{\mathbf{X}_{0:k}^{(i)} ; i = 1,\ldots,N_s\right\}$ according to $\pi(\mathbf{X}_{0:k}|\mathbf{Z}_{1:k})$, we can compute the estimate of $I(g_k)$ as

$$\hat{I}(g_k) = \frac{\frac{1}{N_s}\sum_{i=1}^{N_s} g_k\left(\mathbf{X}_{0:k}^{(i)}\right) \omega\left(\mathbf{X}_{0:k}^{(i)}\right)}{\frac{1}{N_s}\sum_{i'=1}^{N_s} \omega\left(\mathbf{X}_{0:k}^{(i')}\right)} = \sum_{i=1}^{N_s} g_k\left(\mathbf{X}_{0:k}^{(i)}\right) \tilde{\omega}_k^{(i)}, \tag{6.67}$$

where $\tilde{\omega}_k^{(i)}$ are the normalized importance weights defined by

$$\tilde{\omega}_k^{(i)} = \frac{\omega\left(\mathbf{X}_{0:k}^{(i)}\right)}{\sum_{i'=1}^{N_s} \omega\left(\mathbf{X}_{0:k}^{(i')}\right)}. \tag{6.68}$$

However, this solution is not recursive, which is a very desirable feature for online processing of data. Each time new data $\mathbf{z}_{k+1}$ become available, all the importance weights over the entire state sequence need to be recomputed, with the computational complexity increasing with time.

One way to use the importance sampling solution while maintaining recursiveness is by means of the *sequential importance sampling* (SIS) concept. Resorting to the factorization of the proposal function (6.64) and iterating, we can get samples from

$$\mathbf{x}_k^{(i)} \sim \pi(\mathbf{X}_{0:k}|\mathbf{Z}_{1:k}) = \pi(\mathbf{x}_0) \prod_{m=1}^{k} \pi(\mathbf{x}_m|\mathbf{X}_{0:m-1},\mathbf{Z}_{1:m}), \tag{6.69}$$

with associated importance weights $\tilde{\omega}_k^{(i)}$ that can be computed recursively as

$$\tilde{\omega}_k^{(i)} \propto \tilde{\omega}_{k-1}^{(i)} \frac{p\left(\mathbf{z}_k \middle| \mathbf{x}_k^{(i)}\right) p\left(\mathbf{x}_k^{(i)} \middle| \mathbf{x}_{k-1}^{(i)}\right)}{\pi\left(\mathbf{x}_k^{(i)} \middle| \mathbf{X}_{0:k-1}^{(i)}, \mathbf{Z}_{1:k}\right)}. \tag{6.70}$$

Indeed, after the normalization, the estimation of $I(g_k)$ becomes

$$\hat{I}(g_k) = \sum_{i=1}^{N_s} g_k\left(\mathbf{x}_k^{(i)}\right) \tilde{\omega}_k^{(i)}. \tag{6.71}$$

The structure (6.70)–(6.71) is considerably simpler than the one expressed in (6.67)–(6.68) because the former does not need to modify past trajectories $\mathbf{X}_{0:k-1}^{(i)}$ at time k and saves a significant amount of memory and computational load.

The choice of $\pi(\cdot)$ is a critical issue in any SMC method design. In addition to being a distribution from which it is easy to generate samples, the importance function must have a support that includes the true posterior, which means that the set closure of arguments of function $\pi(\cdot)$ for which the value is not zero must includes the one of $p(\mathbf{X}_{0:k}|\mathbf{Z}_{1:k})$. The posterior approximation is then

$$\hat{p}(\mathbf{x}_k|\mathbf{Z}_{1:k}) = \sum_{i=1}^{N_s} \omega_k^{(i)} \delta\left(\mathbf{x}_k - \mathbf{x}_k^{(i)}\right). \tag{6.72}$$

This approximation converges almost surely to the true posterior as $N_s \to \infty$ under weak assumptions according to the Strong Law of Large Numbers. Notice that *almost surely* convergence is equivalent to convergence *with probability one*. These assumptions hold if the support of the chosen importance density ($\bar{\pi}$) includes the support of the posterior pdf ($\bar{p}$),

$$\bar{\pi} = \{\mathbf{x}_k \in \chi \,|\, \pi(\mathbf{x}_k|\mathbf{Z}_k) > 0\}$$
$$\bar{p} = \{\mathbf{x}_k \in \chi \,|\, p(\mathbf{x}_k|\mathbf{Z}_k) > 0\}$$
$$\text{and} \quad \bar{p} \subseteq \bar{\pi}. \tag{6.73}$$

However, for finite N_s, it is well known that when k increases, the weights $\tilde{\omega}_k^{(i)}$ tend to be zero for all particles except for one and, thus, the algorithm fails to represent adequately the posterior distribution. In other words, the unconditional variance of the importance weights increases over time [41]. This is known as the *degeneration phenomenon* and constitutes an important drawback for the SIS method. In order to avoid this, we can introduce a selection step usually known as *resampling*. A classical selection step consists of a resampling with replacement of N_s new particles from the generated set of particles according to its importance weight. Other more sophisticated schemes include multinomial, residual, or stratified resampling [25]. It is worthwhile mentioning that the resampling step constitutes a bottleneck in any parallel implementation because it is inherently sequential (it needs all the particles) and also cannot be executed concurrently (pipelined) with other operations. Some schemes that alleviate these constraints can be found in Ref. [46].

6.1.5.1 *Sequential Importance Resampling*
Sequential importance resampling (SIR) is an implementation of sequential importance sampling with a resampling step introduced to prevent the degeneration problem explained above. For the

choice of the importance density function $\pi\left(\mathbf{X}_{0:k}^{(i)}\middle|\mathbf{Z}_{1:k}\right)$, it is possible to use the transition density $p\left(\mathbf{x}_k\middle|\mathbf{x}_{k-1}^{(i)},\mathbf{z}_k\right)$, which can be defined by a motion equation. It has been shown that this importance function minimizes the variance of the importance weights conditioned upon the simulated trajectory $\mathbf{X}_{0:k}^{(i)}$ and the observations up to time k, $\mathbf{Z}_{1:k}$. Then, from (6.70), we see that the importance weights can be computed directly from the likelihood function $\omega_k^{(i)} = \omega_{k-1}^{(i)}p\left(\mathbf{z}_k\middle|\mathbf{x}_k^{(i)}\right)$. SIR, although requiring the ability to sample from $p\left(\mathbf{x}_k\middle|\mathbf{x}_{k-1}^{(i)},\mathbf{z}_k\right)$ and to evaluate $p\left(\mathbf{z}_k\middle|\mathbf{x}_k^{(i)}\right)$, which may not have an analytical solution in the general case, can be employed in the Gaussian state-space model with nonlinear transition equation. This is equivalent to taking Eqs (6.21) and (6.22) and considering the noise distributions $p_{\mathbf{v}_k}$ and $p_{\mathbf{w}_k}$ as mutually independent and Gaussian. For the resampling step, we make use of the effective sample size N_{eff} introduced in Ref. [41]. This is defined as

$$N_{\text{eff}} = \frac{N_s}{1 + \text{Var}_{\pi(\cdot|\mathbf{Z}_{1:k})}\{\omega(\mathbf{X}_{0:k})\}}, \tag{6.74}$$

which cannot be evaluated exactly but can be easily estimated by $\hat{N}_{\text{eff}} = \left(\sum_{i=1}^{N_s}\left(\omega_k^{(i)}\right)^2\right)^{-1}$. The resampling step will be applied whenever $\hat{N}_{\text{eff}}$ decreases below a given threshold N_{thres}, which is usually chosen to be $N_{\text{thres}} = 2N_s/3$. For the sake of clarity, Fig. 6.3 shows the general flow chart of the SIR procedure, whereas the method is outlined in Algorithm 6.3.

6.1.5.2 *Auxiliary Particle Filter*

The auxiliary particle filter (APF) [56] is a variation of the standard particle filter that tries to improve the efficiency of such algorithms. A fundamental problem of the SIR approach is that it is difficult to improve its simulation performance because of the expense of evaluating

$$p(\mathbf{x}_k|\mathbf{Z}_{1:k}) \propto p(\mathbf{z}_k|\mathbf{x}_k)\sum_{i=1}^{N_s}\omega_{k-1}^{(i)}p\left(\mathbf{x}_k\middle|\mathbf{x}_{k-1}^{(i)}\right). \tag{6.75}$$

The reason is that, when there is an outlier (i.e., a state occurring in the tail of the distribution), the weights $\omega^{(i)}$ will be very unevenly distributed and thus the required number of particles N_s being close to $p(\mathbf{x}_k|\mathbf{Z}_{1:k})$ will be very large. This effect can also be seen from Eq. (6.74), which makes evident that the SIR method will become very imprecise when $\omega^{(i)}$ becomes very variable. This happens when the likelihood is highly peaked compared with the prior, or when $p(\mathbf{z}_k|\mathbf{x}_k)$ is highly sensitive to $\mathbf{x}_k$. In order to deal with that problem, let us consider an auxiliary discrete random variable γ that indicates the particle from which the state $\mathbf{x}_k$ comes. Then, the joint density of the state and the new variable can be written as

$$p(\mathbf{x}_k,\gamma|\mathbf{Z}_{1:k}) \propto p(\mathbf{z}_k|\mathbf{x}_k)\omega_{k-1}^{(\gamma)}p\left(\mathbf{x}_k\middle|\mathbf{x}_{k-1}^{(\gamma)}\right). \tag{6.76}$$

Note that, if we marginalize index γ, we get again the distribution expressed in (6.75). Applying the importance sampling approach over (6.76) with the proposal function

$$\pi\left(\mathbf{x}_k,\gamma\middle|\mathbf{z}_k,\mathbf{x}_{k-1}^{(\gamma)}\right) \propto p\left(\mathbf{x}_k\middle|\mathbf{x}_{k-1}^{(\gamma)}\right)\omega_{k-1}^{(\gamma)}p\left(\mathbf{z}_k\middle|\hat{\boldsymbol{\mu}}_k^{(\gamma)}\right), \tag{6.77}$$

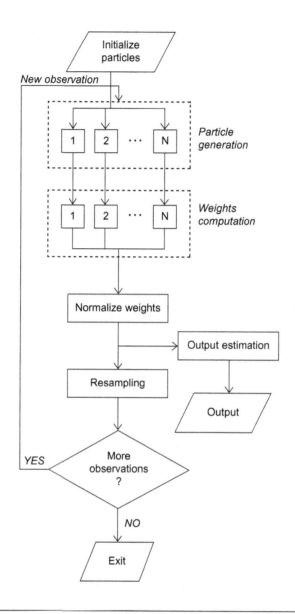

FIGURE 6.3

Flow chart of the SIR algorithm.

where $\hat{\boldsymbol{\mu}}_k^{(\gamma)}$ is a prediction of the state at instant k from the γth particle at instant $k-1$, leads to an implementation in which

$$\gamma^{(i)} \backsim \pi\left(\gamma \,|\, \mathbf{z}_k, \mathbf{x}_{k-1}^{(\gamma)}\right) \propto p\left(\mathbf{z}_k \,|\, \hat{\boldsymbol{\mu}}_k^{(\gamma)}\right) \omega_{k-1}^{(\gamma)}, \tag{6.78}$$

$$\mathbf{x}_k^{(i)} \backsim p\left(\mathbf{x}_k \,\Big|\, \mathbf{x}_{k-1}^{(\gamma^{(i)})}\right), \tag{6.79}$$

Require: $p_{\mathbf{x}_0}, p_{\mathbf{v}_k}, p_{\mathbf{w}_k}$, observations $\mathbf{Z}_{1:\mathcal{K}}$ and $\mathbf{U}_{1:\mathcal{K}}$.
Ensure: $\hat{\mathbf{x}}_k$ approximates $\mathbf{x}_k$ as $N_s \to \infty$.

1: Initialization: set $k = 0$, $\left\{\omega_0^{(i)}\right\}_{i=1}^{N_s} = \frac{1}{N_s}$ and generate N_s samples $\left\{\mathbf{x}_0^{(i)}\right\}_{i=1}^{N_s}$ from $p_{\mathbf{x}_0}$.

2: **for** $k = 1$ to $\mathcal{K}$ **do**

3:   **for** $i = 1$ to N_s **do**

4:     Sample $\tilde{\mathbf{x}}_k^{(i)} \sim p(\mathbf{x}_k | \mathbf{x}_{k-1}^{(i)}) \Leftrightarrow$ Generate $\mathbf{v}_{k-1} \sim p_{\mathbf{v}_{k-1}}$ and simulate
      $\tilde{\mathbf{x}}_k^{(i)} = \mathbf{F}\mathbf{x}_{k-1}^{(i)} + \mathbf{B}\mathbf{u}_{k-1} + \mathbf{G}\mathbf{v}_{k-1}$.

5:     Evaluate importance weight $\tilde{\omega}_k^{(i)} = \omega_{k-1}^{(i)} p\left(\mathbf{z}_k \middle| \tilde{\mathbf{x}}_k^{(i)}\right) = \omega_{k-1}^{(i)} p_{\mathbf{w}_k}\left(\mathbf{z}_k - \mathbf{h}\left(\tilde{\mathbf{x}}_k^{(i)}\right)\right)$.

6:   **end for**

7:   Normalize weights $\omega_k^{(i)} = \frac{\tilde{\omega}_k^{(i)}}{\sum_{i'=1}^{N_s} \tilde{\omega}_k^{(i')}}$.

8:   Compute $\hat{N}_{\text{eff}} = \frac{1}{\sum_{i=1}^{N_s} \left(\omega_k^{(i)}\right)^2}$

9:   **if** $\hat{N}_{\text{eff}} \geq N_{\text{threshold}}$ **then**

10:    Set $\mathbf{x}_k^{(i)} = \tilde{\mathbf{x}}_k^{(i)}$ $\forall i$.

11: **else**

12:    Resample with replacement N_s particles $\left\{\mathbf{x}_k^{(i)}\right\}_{i=1}^{N_s}$ from the set $\left\{\tilde{\mathbf{x}}_k^{(i)}\right\}_{i=1}^{N_s}$ according to the importance weights $\omega_k^{(i)}$.

13:    Reset weights to $\omega_k^{(i)} = \frac{1}{N_s}$ $\forall i$.

14: **end if**

15: Compute $\hat{\mathbf{x}}_k = \sum_{i=1}^{N_s} \omega_k^{(i)} \mathbf{x}_k^{(i)}$.

16: **end for**

Algorithm 6.3

Sequential importance resampling (SIR) algorithm

and, applying Eqs (6.77)–(6.70),

$$\omega_k^{(i)} \propto \frac{p\left(\mathbf{z}_k \middle| \mathbf{x}_k^{(i)}\right)}{p\left(\mathbf{z}_k \middle| \hat{\boldsymbol{\mu}}_k^{(\gamma^{(i)})}\right)}. \tag{6.80}$$

An example of APF is described in Algorithm 6.4. As can be seen in step 9, this method tries to reduce the variability of the importance weights by resampling the particles at time $k - 1$ according to a probability close to $p(\mathbf{x}_{k-1}|\mathbf{Z}_{1:k})$, that is, taking advantage of the new observation $\mathbf{z}_k$ to improve the particle stock at time k. This additional information makes the algorithm more efficient than the SIR implementation.

6.1.5.3 Cost-Reference Particle Filtering

Cost-reference particle filtering (CRPF) is another methodology to attain asymptotically an optimal estimation of the state of a nonlinear and/or non-Gaussian discrete-time dynamical system in a recursive way. First introduced in Ref. [48], this family of methods tries to overcome some limitations of general particle filtering algorithms. Namely, the need for a tractable and realistic probabilistic model

---

Require: $p_{\mathbf{x}_0}, p_{\mathbf{v}_k}, p_{\mathbf{w}_k}$, observations $\mathbf{Z}_{1:\mathcal{K}}$ and $\mathbf{U}_{1:\mathcal{K}}$.
Ensure: $\hat{\mathbf{x}}_k$ approximates $\mathbf{x}_k$ as $N_s \to \infty$.

1: Initialization: set $k = 0$, $\left\{\omega_0^{(i)}\right\}_{i=1}^{N_s} = \frac{1}{N_s}$ and generate N_s samples $\left\{\mathbf{x}_0^{(i)}\right\}_{i=1}^{N_s}$ from $p_{\mathbf{x}_0}$.

2: **for** $k = 1$ to $\mathcal{K}$ **do**

3:     **for** $i = 1$ to N_s **do**

4:         Prediction of the state: $\boldsymbol{\mu}_k^{(i)} = \mathbf{F}\mathbf{x}_{k-1}^{(i)} + \mathbf{B}\mathbf{u}_{k-1}$.

5:         Prediction of the likelihood: $\mathcal{P}_k^{(i)} = p\left(\mathbf{z}_k \middle| \boldsymbol{\mu}_k^{(i)}\right) = p_{\mathbf{w}_k}\left(\mathbf{z}_k - \mathbf{h}\left(\boldsymbol{\mu}_k^{(i)}\right)\right)$.

6:         Compute first stage weights: $\tilde{\lambda}_k^{(i)} = p\left(\mathbf{z}_k \middle| \boldsymbol{\mu}_k^{(i)}\right)\omega_{k-1}^{(i)}$.

7:     **end for**

8:     Normalize first stage weights: $\lambda_k^{(i)} = \frac{\tilde{\lambda}_k^{(i)}}{\sum_{i'=1}^{N_s} \tilde{\lambda}_k^{(i')}}$.

9:     Resample with replacement N_s particles $\left\{\tilde{\mathbf{x}}_{k-1}^{(i)}\right\}_{i=1}^{N_s}$ from the set $\left\{\mathbf{x}_{k-1}^{(i)}\right\}_{i=1}^{N_s}$ according to the first stage importance weights $\lambda_k^{(i)}$.

10:    **for** $i = 1$ to N_s **do**

11:        Generate $\mathbf{v}_{k-1} \sim p_{\mathbf{v}_{k-1}}$ and compute $\mathbf{x}_k^{(i)} = \mathbf{F}\tilde{\mathbf{x}}_{k-1}^{(i)} + \mathbf{B}\mathbf{u}_{k-1} + \mathbf{G}\mathbf{v}_{k-1}$.

12:        Computation of the likelihood: $\mathcal{L}_k^{(i)} = p_{\mathbf{w}_k}\left(\mathbf{z}_k - \mathbf{h}\left(\mathbf{x}_k^{(i)}\right)\right)$.

13:        Compute second stage weights $\tilde{\omega}_k^{(i)} = \frac{\mathcal{L}_k^{(i)}}{\mathcal{P}_k^{(i)}}$.

14:    **end for**

15:    Normalize second stage weights $\omega_k^{(i)} = \frac{\tilde{\omega}_k^{(i)}}{\sum_{i'=1}^{N_s} \tilde{\omega}_k^{(i')}}$.

16:    Compute $\hat{\mathbf{x}}_k = \sum_{i=1}^{N_s} \omega_k^{(i)}\mathbf{x}_k^{(i)}$.

17: **end for**

Algorithm 6.4

Auxiliary particle filter (APF) algorithm

(including, at least, the a priori pdf of the state, $p_{\mathbf{x}_0}$, the conditional density of the transition, $p(\mathbf{x}_k|\mathbf{x}_{k-1})$, and the likelihood function $p(\mathbf{z}_k|\mathbf{x}_k)$), the computational load (drawing samples from the state prior probability distribution, the evaluation of the likelihood function), and the bottleneck that the resampling step represents for parallel computation (as it involves the joint processing of all particles) are the major drawbacks of particle filtering. In order to overcome such problems, the CRPF methods perform dynamic optimization of an arbitrary cost function not necessarily tied to the statistics of the state and the observation processes instead of relying on a probabilistic model of the dynamic system. By a proper selection of this cost function, we can design and implement algorithms in a quite simple manner, regardless of the availability of the noise densities $p_{\mathbf{v}_k}$ and $p_{\mathbf{w}_k}$. In addition, CRPF algorithms have the potential of being independent of the noise distributions.

The procedure can be summarized as follows. Assume a lower-bounded real cost function

$$\mathcal{C} : \mathbb{R}^{N_x(k+1)} \times \mathbb{R}^{N_z k} \to \left[\mathcal{C}_k^{\text{opt}}, \infty\right), \tag{6.81}$$

where

$$\mathcal{C}_k^{\text{opt}} \triangleq \inf_{\mathbf{X}_{0:k}} \mathcal{C}(\mathbf{X}_{0:k}, \mathbf{Z}_{1:k}), \quad |\mathcal{C}_k^{\text{opt}}| < \infty, \tag{6.82}$$

is the optimal cost at time k. The role of function $\mathcal{C}$ is to give a quantitative value to a trial state sequence $\mathbf{X}_{0:k}$ given the actual sequence of observations $\mathbf{Z}_{1:k}$. Assume that function $\mathcal{C}$ has a recursive structure of the form

$$\mathcal{C}(\mathbf{X}_{0:k}, \mathbf{Z}_{1:k}) = \lambda \mathcal{C}(\mathbf{X}_{0:k-1}, \mathbf{Z}_{1:k-1}) + \Delta \mathcal{C}(\mathbf{x}_k, \mathbf{z}_k), \tag{6.83}$$

where $0 \leq \lambda \leq 1$ is a memory factor, and

$$\Delta \mathcal{C} : \mathbb{R}^{N_x} \times \mathbb{R}^{N_z} \to \left[\Delta \mathcal{C}_k^{\text{opt}}, \infty \right) \tag{6.84}$$

is called the incremental cost function. Again, the optimal incremental cost at time k is defined as

$$\Delta \mathcal{C}_k^{\text{opt}} \triangleq \inf_{\mathbf{x}_k} \mathcal{C}(\mathbf{x}_k, \mathbf{z}_k). \tag{6.85}$$

In navigation or tracking applications, it seems natural to apply the Euclidean distance as the choice for the cost function. Although other cost functions could be applied, it should be kept in mind that $\Delta \mathcal{C}$ has to be simple to evaluate. For example, Ref. [47] suggests using $\Delta \mathcal{C}(\mathbf{x}_k, \mathbf{z}_k) = \| \mathbf{z}_k - \mathbf{h}(\mathbf{x}_k) \|$.

Let us assume the availability of a set of N_s particles and their associated costs up to time $k-1$, that is, $\Omega_{k-1} = \left\{ \mathbf{X}_{0:k-1}^{(i)}, \mathcal{C}\left(\mathbf{X}_{0:k-1}^{(i)}, \mathbf{Z}_{1:k-1}^{(i)} \right) \right\}_{i=1}^{N_s}$. We now define a risk function

$$\mathcal{R} : \left[\mathcal{C}^{\text{opt}}\left(\mathbf{X}_{0:k-1}^{(i)}, \mathbf{Z}_{1:k-1}^{(i)} \right), \infty \right) \times \mathbb{R}^{N_x} \times \mathbb{R}^{N_z} \to \mathbb{R} \tag{6.86}$$

that measures the adequacy of a candidate state at time $k - 1$ to be propagated up to time k given the new observation $\mathbf{z}_k$. We will choose the risk function as a prediction of the cost at time k, that is,

$$\mathcal{R}\left(\mathcal{C}\left(\mathbf{X}_{0:k-1}^{(i)}, \mathbf{Z}_{1:k-1}^{(i)} \right), \mathbf{x}_{k-1}, \mathbf{z}_t \right) = \lambda \mathcal{C}\left(\mathbf{X}_{0:k-1}^{(i)}, \mathbf{Z}_{1:k-1}^{(i)} \right) + \Delta \mathcal{C}\left(\mathbf{F} \mathbf{x}_{k-1}^{(i)} + \mathbf{B} \mathbf{u}_k, \mathbf{z}_k \right). \tag{6.87}$$

Then, the CRPF consists of an initialization of N_s particles within a region $\mathcal{E}_0$ (drawn from an arbitrary pdf with support $\mathcal{E}_0$), a recursive loop that performs a selection of particles according to their risk and propagation, and finally an estimation of the state as a sum of particles weighted by its cost. This method is outlined in Algorithm 6.5.

6.2 HYBRID TERRESTRIAL LOCALIZATION BASED ON TOA + TDOA + AOA MEASUREMENTS

The EKF as in Section 6.1 allows us to track the position and speed of a terminal in motion, yielding an accurate location prediction algorithm. Tracking of the bias due to NLOS is also possible by increasing the dimension of the state vector with TOA bias as shown in Section 6.1.4.2.

In this section, we discuss a method to blend different kinds of measurements in a hybrid fully-terrestrial positioning system, exploiting the EKFBT algorithm introduced in Section 6.1.4.2.

We assume for simplicity a 2D geometry, where a mobile object needs to be located using wireless terrestrial measurements. A set of M reference points (or anchor nodes or access points) is available

---

Require: Observations $\mathbf{Z}_{1:\mathcal{K}}$ and $\mathbf{U}_{1:\mathcal{K}}$, memory factor λ.
Ensure: $\hat{\mathbf{x}}_k$ approximates $\mathbf{x}_k$ as $N_s \to \infty$.

1: Define region $\mathcal{E}_0$.
2: Draw N_s particles $\mathbf{x}_0^{(i)} \sim U(\mathcal{E}_0)$.
3: Set the particle costs to a constant $\mathcal{C}_0^{(i)} = c$.
4: **for** $k = 1$ to $\mathcal{K}$ **do**
5:      **for** $i = 1$ to N_s **do**
6:          Compute the risk $\mathcal{R}_k^{(i)} = \lambda \mathcal{C}_{k-1}^{(i)} + \left\| \mathbf{z}_k - \mathbf{F}\mathbf{x}_{k-1}^{(i)} - \mathbf{B}\mathbf{u}_k \right\|$.
7:      **end for**
8:      Obtain $\tilde{\Omega}_k = \left\{ \tilde{\mathbf{x}}_{k-1}^{(i)}, \tilde{\mathcal{C}}_{k-1}^{(i)} \right\}_{i=1}^{N_s}$ by resampling particles $\mathbf{x}_{k-1}^{(i)}$ according to their risk $\mathcal{R}_k^{(i)}$.
9:      Choose a radius ρ.
10:     **for** $i = 1$ to N_s **do**
11:        Define the region $\mathcal{E}_\rho\left(\tilde{\mathbf{x}}_{k-1}^{(i)}\right)$ centered at $\tilde{\mathbf{x}}_{k-1}^{(i)}$ with radius ρ.
12:        Draw $\mathbf{x}_k^{(i)} \sim U\left(\mathcal{E}_\rho\left(\tilde{\mathbf{x}}_{k-1}^{(i)}\right)\right)$.
13:        Set $\mathcal{C}_k^{(i)} = \lambda \tilde{\mathcal{C}}_{k-1}^{(i)} + \left\| \mathbf{z}_k - \mathbf{h}\left(\mathbf{x}_k^{(i)}\right) \right\|$.
14:     **end for**
15:     Compute $\omega\left(\mathcal{C}_k^{(i)}\right) \propto \dfrac{1}{\left(\mathcal{C}_k^{(i)} - \min_{n \in \{1,\dots,N_s\}} \left\{\mathcal{C}_k^{(n)}\right\} + \frac{1}{N_s}\right)^3}, \forall i$.
16:     Perform the estimation $\hat{\mathbf{x}}_k = \sum_{i=1}^{N_s} \omega\left(\mathcal{C}_k^{(i)}\right) \mathbf{x}_k^{(i)}$.
17: **end for**

Algorithm 6.5

Cost-reference particle filtering (CRPF) algorithm

---

at known locations. We also assume the mobile device to be equipped with a number of wireless interfaces, providing TOA, TDOA, and AOA measurements so that the measurement vector obtained from the different reference nodes at the time instant k is written as

$$\mathbf{z}_k = \begin{bmatrix} \mathbf{z}_{TDOA,k}^T & \mathbf{z}_{TOA,k}^T & \mathbf{z}_{AOA,k}^T \end{bmatrix}^T, \tag{6.88}$$

while the corresponding measurement noise is

$$\mathbf{w}_k = \begin{bmatrix} \mathbf{w}_{TDOA,k}^T & \mathbf{w}_{TOA,k}^T & \mathbf{w}_{AOA,k}^T \end{bmatrix}^T, \tag{6.89}$$

with covariance matrix (assumed time-invariant)

$$\mathbf{R}^{(hyb)} = \begin{bmatrix} \mathbf{R}_{TDOA} & \mathbf{0} & \mathbf{0} \\ \mathbf{0} & \mathbf{R}_{TOA} & \mathbf{0} \\ \mathbf{0} & \mathbf{0} & \mathbf{R}_{AOA} \end{bmatrix}. \tag{6.90}$$

Here, we consider that the obtained measures are uncorrelated and that

$$\mathbf{R}_{TOA} = \mathrm{diag}\left[\sigma_{TOA,1}^2 \quad \sigma_{TOA,2}^2 \quad \cdots \quad \sigma_{TOA,M}^2\right] \tag{6.91}$$

$$\mathbf{R}_{AOA} = \mathrm{diag}\left[\sigma_{AOA,1}^2 \quad \sigma_{AOA,2}^2 \quad \cdots \quad \sigma_{AOA,M}^2\right] \tag{6.92}$$

$$\mathbf{R}_{TDOA} = \mathbf{D}\mathbf{R}_{TOA}\mathbf{D}^T, \tag{6.93}$$

where $\mathbf{D}$ is the $(M-1) \times M$ matrix that defines the difference of times in the TDOA method [69]:

$$\mathbf{D} - \begin{bmatrix} 1 & -1 & 0 & \cdots & 0 & 0 \\ 0 & 1 & -1 & 0 & \cdots & 0 \\ \vdots & \vdots & \vdots & & \vdots & \vdots \\ 0 & 0 & 0 & \cdots & 1 & -1 \end{bmatrix}. \tag{6.94}$$

The nonlinear measurement relation $\mathbf{h}\left(\mathbf{x}_k^{(\mathrm{BT})}\right)$ has to be linearized in order to obtain the hybrid observation matrix $\mathbf{H}_k^{(\mathrm{hyb})}$. For the TOA measurements, the linearized observation equation can be written as in Section 6.1.4.2

$$\mathbf{h}\left(\mathbf{x}_k^{(\mathrm{BT})}\right) \approx \underbrace{\left[\mathbf{H}_{\mathrm{dist},k} \quad \mathbf{I}_M\right]}_{\mathbf{H}_{TOA,k}}\mathbf{x}_k^{(\mathrm{BT})}, \tag{6.95}$$

where $\mathbf{H}_{\mathrm{dist},k}$ has been written in Section 6.1.4.1, Example 1, and the approximation ($\approx$) accounts for the linearization operation. For the AOA measurements, defining as $\theta_0^{(m)}$ the angle of arrival of the signal from the mth reference point, $m = 1,\ldots M$, the observation matrix is written as

$$\mathbf{H}_{AOA,k} = \begin{bmatrix} \dfrac{-\sin\theta_0^{(1)}}{\mathrm{dist}(\mathbf{x}_{k|k-1},\mathbf{x}_{A_1})} & \dfrac{\cos\theta_0^{(1)}}{\mathrm{dist}(\mathbf{x}_{k|k-1},\mathbf{x}_{A_1})} & \vdots & \vdots \\ \vdots & \vdots & \mathbf{0}_{M\times2} & \mathbf{0}_{M\times M} \\ \dfrac{-\sin\theta_0^{(M)}}{\mathrm{dist}(\mathbf{x}_{k|k-1},\mathbf{x}_{A_M})} & \dfrac{\cos\theta_0^{(M)}}{\mathrm{dist}(\mathbf{x}_{k|k-1},\mathbf{x}_{A_M})} & \vdots & \vdots \end{bmatrix}, \tag{6.96}$$

where

$$\theta_0^{(m)} = \arctan\frac{(y - y_{A_m})}{(x - x_{A_m})} \tag{6.97}$$

in a four-quadrant arctangent codomain. Finally, for TDOA measurements, it is possible to write

$$\mathbf{H}_{TDOA,k} = \left[\boldsymbol{\Delta}\mathbf{H}_{\mathrm{dist},k} \quad \boldsymbol{\Delta}\right]. \tag{6.98}$$

Therefore, the matrix $\mathbf{H}_k^{(\mathrm{hyb})}$ can be expressed as a concatenation of linearization matrices for each type of measurement, that is,

$$\mathbf{H}_k^{(\mathrm{hyb})} = \begin{bmatrix} \mathbf{H}_{TDOA,k} \\ \mathbf{H}_{TOA,k} \\ \mathbf{H}_{AOA,k} \end{bmatrix}. \tag{6.99}$$

At this point, from the earlier state and measurement equations, the EKF prediction-update equations can be formulated as usual.

The proposed approach has been evaluated in a realistic universal mobile telecommunications system (UMTS) scenario. The measurement equipment consists of a 35-dBm-power UE transmitter with single dipole antenna of 4-dB gain, and four independent receivers connected to an antenna array of four elements with half wavelength separation between them and 15-dB gain. The bandwidth is 5 MHz at 1800 MHz. The baseband stage consists of a set of DSPs performing channel parameter estimation and mobile position computation. The transmitted signal is a pilot sequence of 256 chips shaped with a root-raised cosine pulse, sampled at two samples per chip. Additionally, a GPS receiver is located in the mobile terminal, so a baseline position is stored during the recording time and can be compared off-line with the estimated position. The number of channel estimates is around 200 per second and per receiving antenna. The dedicated physical channel (DPCH) that has been considered is strictly power controlled in every slot both in up-link (UL) and down-link (DL). It may be assumed that the objective E_b/N_0 is fixed by the serving base station (BS). Equal noise and interference powers are considered in all BSs, whereas the signal power is different for each BS, determining a different signal-to-noise-plus-interference ratio (SNIR) for each link. Positioning using time-of-arrival (TOA) and AOA measures is tested in this first experiment. It is assumed that the TOA is computed as one half of the round-trip time in the DPCH, and the AOA is computed at the BS equipped with an antenna array. In Fig. 6.4, positioning RMSE and the standard deviation are shown. These values are averaged over three different routes, emulating two BSs. In all cases, NLOS is found more than 50% of the time. The use of EKF with bias tracking of the TOA exhibits significant gains in positioning accuracy.

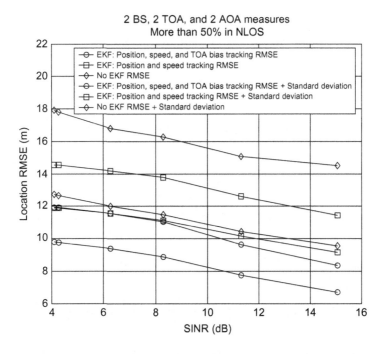

FIGURE 6.4

RMSE and standard deviation of the positioning error from TOA and AOA measures with two BSs.

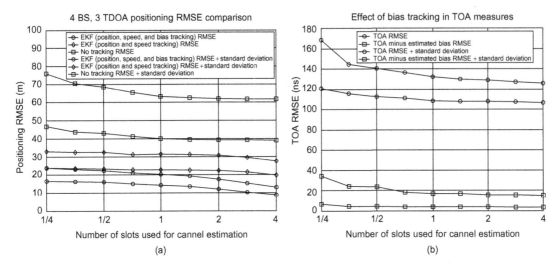

FIGURE 6.5

(a) RMSE and standard deviation of the positioning error from TDOA measures with four BSs. (b) RMSE of the TOA, in ns (from Ref. [52]).

Next, the common pilot channel (CPICH) in DL is considered. In the CPICH, no power control is used. Positioning using TDOA measures is tested in this case. Four different routes are combined in a multiple BS emulation. Three BSs are in NLOS more than 90% of the time with a mean bias of the TOA of 60 m and only one BS is in LOS more than 50% of the time. These conditions are typical of an urban environment. The results of this test are represented in Fig. 6.5(a) for different number of used slots for channel estimation (thus translated to different conditions of SNR). It may be observed that a reduction from 45 to 15 m of positioning error is achieved when using the EKF with bias tracking. Also, the variance is highly reduced from a standard deviation of 30 m to less than 10 m. The factor which mostly impacts this gain is the excellent tracking of the bias. Figure 6.5(b) shows the mean square in the TOA estimates when the bias is compensated through Kalman tracking.

6.3 HYBRID LOCALIZATION BASED ON GNSS AND INERTIAL SYSTEMS

It has already been discussed how the performance of GNSS systems severely degrades in critical scenarios such as urban canyons or indoor situations when there is no line of sight between the mobile device and the satellites. In these cases, the weak receiving power, the multipath effect, and interferences make a stand-alone GNSS receiver nearly useless. In order to alleviate this problem, the mobile device can include a low-cost inertial measurement unit (for instance, redundant accelerometer sets based on MEMS technology), which provides complementary information.

In contrast, an alternative and innovative approach is direct position estimation (DPE) as was already described in Section 4.4.1. DPE is a Bayesian technique that focuses on the estimation (directly from received and sampled IF signals from the GNSS satellites) of position coordinates, which are indeed the parameters of interest to the end user. Thus, the avoidance of intermediate estimation steps helps to partially overcome some limitations of current approaches, such as the degradation in position accuracy because of multipath and severe channel fading conditions. DPE has recently been shown to give remarkable improvements in the overall performance [15, 16].

Both linearized KF problem and DPE can be tailored to account for additional information provided by an inertial navigation system (INS), so as to cope with the problem of weak signals in critical environments. This section focuses on the algorithms that allow us to integrate GNSS with INS, considering both the classic EKF-based architectures and the Bayesian DPE one.

Notice that there is a subtle difference in using prior information in the DPE framework and its conventional counterpart. Although the latter operates at the *observable* level, DPE directly manipulates the received IF sampled signal. This has important consequences from a signal processing point of view since, in the conventional approach, prior information is introduced after pseudoranges are estimated, possibly affected by its inherent errors, for example multipath biases. In contrast, DPE presents an appealing framework for the inclusion of prior information, as addressed in the literature [7, 37]. Jointly considering prior information at the sampled IF signal level, some sources of accuracy degradation can be conveniently mitigated.

6.3.1 **Inertial Measurement Units and Inertial Navigation**

The history of inertial navigation started between the first and second world wars and developed principally for military purposes, namely for the guidance of rockets, spacecraft, guided missiles, and then civil aircraft.

Inertial navigation is based on two families of *inertial sensors*: *accelerometers* and *gyroscopes*. An accelerometer measures the acceleration that the sensor is subject to, along a predefined direction. Then, the time integral of the measured acceleration provides an estimate of the instantaneous speed of the body it is mounted on, provided that the initial speed has been correctly set. A second integration yields the distance traveled by the body along the predefined direction.

A gyroscope is a sensor able to measure the rotational motion, by measuring the spinning rate (angular rate) the sensor is subject to, around a predefined axis. Thus, the angular orientation of the body can be calculated by integrating the angular rate measurements, provided that an initial orientation of the sensor axis with respect to a reference is given.

Three orthogonal accelerometers rigidly mounted on a body provide measurements of acceleration, velocity, and position of the body itself in a three-dimensional frame. However, as long as the accelerometers are rigidly attached to the body in movement, their reference frame is the *body frame* (i.e., a frame integral with the body and centered in the center of the accelerometer system), which is usually insufficient for providing the body movement with respect to an external reference frame. In fact, the rotation (angular displacement) of the body with respect to the external reference frame, that is, the *body attitude*, must be computed from the instantaneous angular orientation of the body obtained from three gyroscopes rigidly mounted on the body along the orthogonal accelerometer axes. Thereby, the attitude information is used to resolve the accelerometer measurements into the reference frame. The system of equations used to compute the body trajectory (i.e., the instantaneous position, velocity, and attitude) in the selected reference frame from the inertial sensors measurements is called *mechanization*.

A system where the inertial sensors are mounted directly on the vehicle and move integrally with it is known as a *strapdown* system. Otherwise, the inertial sensors can be placed on a platform stabilized

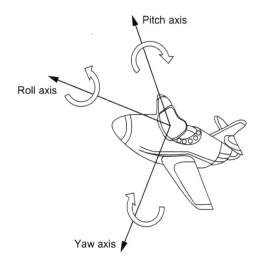

FIGURE 6.6

The body frame for an aircraft.

in space and then isolated from the body rotation (*gimballed* system). Strapdown systems are far more common than gimballed ones because of the simplicity of their mechanical realization.

A set of inertial sensors (accelerometers and gyroscopes, and sometimes a magnetometer and/or an altimeter) forms an *inertial measurement unit* (IMU). The IMU coupled with a computational unit to resolve the system mechanization is an *inertial navigation system* (INS). A comprehensive treatment of strapdown inertial navigation systems can be found in Refs. [28, 35, 68].

Reference Frames

Inertial navigation can be executed with respect to different Cartesian coordinate reference frames, orthogonal and right-handed. For the navigation in the vicinity of the Earth, four reference frames are commonly used:

Body frame (*b*-frame) It is aligned with the axes of the vehicles where the INS is rigidly mounted (Fig. 6.6). Vehicle axes are usually indicated as *roll* (x_b-axis, along the vehicle's head), *pitch* (y_b-axis), and *yaw* (z_b-axis, directed toward the Earth surface for a normally still vehicle).

Inertial frame (*i*-frame) It originates from the Earth's center; it has nonrotating axes with respect to the fixed stars and the z_i-axis coincides with Earth's polar axis.

Earth frame (*e*-frame) It originates from the Earth's center; it has nonrotating axes with respect to the Earth, the z_e-axis coincides with Earth's polar axis, and the x_e-axis is the intersection of the equatorial plane with the plane of the Greenwich meridian. The *e*-frame rotates along the z_i-axis with respect to the *i*-frame, with a rate equal to the Earth turn rate, Ω.

Navigation frame or north-east-down (NED) frame (*n*-frame) It is a local frame, originating on a certain point P over the Earth surface with axes aligned respectively with the local North, East, and Down directions. The Down axis coincides with the Earth's radius in P and is directed toward the Earth's center. The n-frame rotates with respect to the e-frame with a turn rate determined by the movement of the origin, P, with respect to the Earth.

The description of the body movement through mechanization depends on which reference frame is selected to execute the tracking of the trajectory.

6.3.1.1 *Principles of Strapdown Mechanization*

In strapdown systems, the accelerometers provide measurements in the body frame. In particular, they measure the so-called specific forces (i.e., accelerations) along their three axes. Specific forces in the b-frame are indicated as $\mathbf{f}^b$, where the superscript b indicates the reference frame in which the quantity is expressed. In order to express the accelerations in a different reference frame, a proper coordinate rotation operation has to be implemented. This can be done by defining a *coordinate transformation matrix* from the b-frame to another coordinate frame. If, for example, the inertial frame is chosen, coordinate rotation may be achieved by premultiplying the vector $\mathbf{f}^b$ by the *direction cosine matrix* (DCM), $\mathbf{C}_b^i$,

$$\mathbf{f}^i = \mathbf{C}_b^i \mathbf{f}^b, \tag{6.100}$$

where $\mathbf{C}_b^i$ depends on the attitude angles derived from the gyroscope measurements. It is worth noticing that $\mathbf{f}^b$ also contains the specific force opposed to the gravity ($-\mathbf{g}$), applied by the surface where the body moves. Therefore, the acceleration of the body in the vector position $\mathbf{r}$ in the inertial frame can be written by means of the *navigation equation*

$$\mathbf{a}^i = \left.\frac{d^2\mathbf{r}}{dt^2}\right|_i = \mathbf{f}^i + \mathbf{g}^i, \tag{6.101}$$

where the gravitational component $\mathbf{g}^i$ is added to $\mathbf{f}^i$ so that the sum $\mathbf{f}^i + \mathbf{g}^i$ is equal to $\mathbf{0}$ when the body is still ($\mathbf{a}^i = \mathbf{0}$; in this case, the accelerometers measure $\mathbf{f}^i = -\mathbf{g}^i$). The term $\mathbf{r}^i$ is the position vector observed by an observer in the i-frame.

However, the gyroscope output is a vector of three angular rates,

$$\boldsymbol{\omega}_b = \left[\omega_{bx},\ \omega_{by},\ \omega_{bz}\right]^T, \tag{6.102}$$

from which the body attitude $\boldsymbol{\psi}_b$ can be formally obtained through time integration. In particular, the direction cosine matrix (DCM) may be calculated from $\boldsymbol{\omega}_b$ using the relationship

$$\dot{\mathbf{C}}_b^i = \mathbf{C}_b^i \boldsymbol{\Omega}_b, \tag{6.103}$$

where $\dot{\mathbf{C}}_b^i$ is the time derivative of $\mathbf{C}_b^i$, and $\boldsymbol{\Omega}_b$ is the skew-symmetric matrix derived from $\boldsymbol{\omega}_b$.

When navigating in the vicinity of the Earth, the body position and velocity are influenced by the movement of the Earth itself (a "still" body on the Earth's surface in fact moves in the i-frame following the Earth's surface motion), but the information of interest is usually the body position and velocity *with respect to the Earth*, expressed in the selected reference frame. In fact, when navigating with respect to a rotating frame, such as the Earth, additional apparent forces have to be taken into account, which depend on the motion of the frame. In this case, the Coriolis theorem is introduced to express the velocity of a body with respect to the Earth, the *ground speed* $\mathbf{v}_e$,

$$\mathbf{v}_e = \left.\frac{d\mathbf{r}}{dt}\right|_e = \mathbf{v}_i - \boldsymbol{\omega}_E^i \times \mathbf{r} \quad \text{(Coriolis theorem)}, \tag{6.104}$$

Table 6.1 Mechanization of the Inertial Sensor Measurements in the *i*- and *e*-Frames

	Inertial Frame Mechanization	Earth Frame Mechanization
$\dot{\mathbf{r}}$	$\dot{\mathbf{r}}^i = \mathbf{v}_e^i$	$\dot{\mathbf{r}}^e = \mathbf{v}_e^e$
$\dot{\mathbf{v}}_e$	$\dot{\mathbf{v}}_e^i = \mathbf{f}^i - \boldsymbol{\omega}_E^i \times \mathbf{v}_e^i + \underbrace{\mathbf{g}^i - \boldsymbol{\omega}_E^i \times \left[\boldsymbol{\omega}_E^i \times \mathbf{r}\right]}_{\mathbf{g}_\ell^i}$	$\dot{\mathbf{v}}_e^e = \mathbf{f}^e - 2\boldsymbol{\omega}_E^e \times \mathbf{v}_e^e + \underbrace{\mathbf{g}^e - \boldsymbol{\omega}_E^e \times \left[\boldsymbol{\omega}_E^e \times \mathbf{r}\right]}_{\mathbf{g}_\ell^e}$
$\mathbf{f}$	$\mathbf{f}^i = \mathbf{C}_b^i\,\mathbf{f}^b$	$\mathbf{f}^e = \mathbf{C}_b^e\,\mathbf{f}^b$
$\dot{\mathbf{C}}$	$\dot{\mathbf{C}}_b^i = \mathbf{C}_b^i\boldsymbol{\Omega}_b$	$\dot{\mathbf{C}}_b^e = \mathbf{C}_b^e\boldsymbol{\Omega}_e$
$\boldsymbol{\omega}$	$\boldsymbol{\omega}_b^i = \boldsymbol{\omega}_b$	$\boldsymbol{\omega}_b^e = \boldsymbol{\omega}_b - \mathbf{C}_b^{e^T}\boldsymbol{\omega}_E^e$

Notes:

$\mathbf{g}_\ell \triangleq \mathbf{g}^i - \boldsymbol{\omega}_E^i \times \left[\boldsymbol{\omega}_E^i \times \mathbf{r}\right]$ is known as local gravity vector.

$\boldsymbol{\omega}_E^i \times \left[\boldsymbol{\omega}_E^i \times \mathbf{r}\right]$ represents the vector of the centripetal acceleration.

$\boldsymbol{\Omega}_s$ is the skew-symmetric form of the vector $\boldsymbol{\omega}_b^s$.

$\boldsymbol{\omega}_b^i$ is the body turn rate with respect to the inertial frame, as measured by the gyroscopes.

$\boldsymbol{\omega}_b^e$ is the body turn rate expressed with respect to the Earth frame.

where $\mathbf{v}_i$ is the inertial body velocity from which the velocity of the Earth's surface in the body position $(\mathbf{v}_E^i(\mathbf{r}) \triangleq \boldsymbol{\omega}_E^i \times \mathbf{r})$ is subtracted. Here, $\boldsymbol{\omega}_E^i = [0\ 0\ \Omega]^T$ is the Earth's turn rate with respect to the *i*-frame, and the notation $\times$ indicates vector cross product.

As anticipated, mechanization consists of expressing the body trajectory (position $\mathbf{r}$ and velocity $\mathbf{v}_e$) in a selected reference frame, "purged" from the component because of the Earth's motion. Taking into account the apparent forces expressed by the Coriolis theorem, mechanization equations appear different from one reference frame to another. Table 6.1 reports the results of mechanization in the *i*-and *e*-frames. Proof of the formulas shown can be found in Ref. [68].

Inertial Sensors

INSs are prone to errors due to estimate drift over time. The reason is the fact that body position and velocity are obtained from double and single integration, respectively, of biased acceleration measurements. Similarly, angular displacements used to compute frame rotations are obtained from turn rate measurements. Therefore, INS performance strongly depends on the quality of the sensors, in terms of *bias* on accelerometers and gyroscopes and *noise*.

6.3.2 Classic Integration of a GNSS Receiver with Inertial Sensors

The goal of integrating INSs with GNSS is to take advantage of the complementary characteristics of the two systems. INSs experience relatively low-noise solutions, but they tend to drift over time due to biased sensor measures; therefore, trajectory errors are potentially unbounded. The rate of the solutions output by an INS is usually higher than that offered by a GNSS receiver. However, a GNSS receiver produces a solution affected by a small bias depending on the quality of the receiver, but with a large error variance because of propagation channel noise; furthermore, in particular conditions the satellite signals could be unavailable for the receiver (e.g., in urban canyons, tunnels, under dense foliage, under water, etc.), thus causing an outage of the solution. The final goal of integrating the two systems is to improve their performance in those conditions when one system alone would fail to work, or would work with poor performance.

Three conceptual approaches can be identified for integrating INS-based positioning and GNSS-based positioning (*system hybridization*): (1) *loose integration,* (2) *tight integration,* and (3) *ultra-tight integration.* They differ in the degree of integration of the two systems, that is, in the nature of the information that is used in the hybridization process, as well as in the architecture of the interactions between the two systems.

The integration mechanism is usually the same for all the approaches: A *complementary Kalman filter* [11, 73] (also known as error-state Kalman filter because it estimates the errors of the estimated states) is the most common "hybridization engine." In the complementary configuration, a KF estimates the difference between the current INS and GNSS outputs, that is, it estimates the error in the (primary) inertial estimates by using the GNSS as a second (redundant) reference [73]. This error estimate is then used to correct the inertial estimates.

The complementary KF employs *incremental states* defined as the difference between the *nominal* states value and their estimate [11]. The basic states $\mathbf{x}_k$ are incremental body position in the Earth frame $\Delta\mathbf{p}^e{}_k$, incremental body velocity $\Delta\mathbf{v}^e{}_k$, body attitude errors $\Delta\boldsymbol{\psi}^e{}_k$, accelerometer and gyroscope biases in the body frame, $\mathbf{b}^b_{ak}, \mathbf{b}^b_{gk}$, respectively:

$$\mathbf{x}_k = \left[\Delta\mathbf{p}^{eT}_k, \ \Delta\mathbf{v}^{eT}_k, \ \Delta\boldsymbol{\psi}^{eT}_k, \ \mathbf{b}^{bT}_{ak}, \ \mathbf{b}^{bT}_{gk} \right]^T. \tag{6.105}$$

Other additional states may also be used [43, 55].

Without entering into the derivation details, which may be found in Refs. [11, 55], the discrete-time state-space model governing the above states can be written as follows:

$$\mathbf{x}_{k+1} = \mathbf{F}_k\mathbf{x}_k + \mathbf{G}_k\mathbf{v}_k, \tag{6.106}$$

where $\mathbf{v}_k$ is the process noise vector on accelerometers and gyroscopes (see below). Taking into account the mechanization equations resolved in the *e*-frame, the time-dependent state transition matrix for the model (6.106) can be written as [55]

$$\mathbf{F}_k = \begin{bmatrix} \mathbf{I}_3 & T_s\mathbf{I}_3 & \mathbf{0} & \mathbf{0} & \mathbf{0} \\ \mathbf{N}^e & \mathbf{I}_3 - 2T_s\mathbf{\Omega}^e_E & -T_s\mathbf{\Phi}_k & T_s\mathbf{C}^e_{bk} & \mathbf{0} \\ \mathbf{0} & \mathbf{0} & \mathbf{I}_3 - T_s\mathbf{\Omega}^e_E & \mathbf{0} & -T_s\mathbf{C}^e_{bk} \\ \mathbf{0} & \mathbf{0} & \mathbf{0} & \mathbf{I}_3 + T_s\mathbf{D}_a & \mathbf{0} \\ \mathbf{0} & \mathbf{0} & \mathbf{0} & \mathbf{0} & \mathbf{I}_3 + T_s\mathbf{D}_g \end{bmatrix} \in \mathbb{R}^{15\times 15}, \tag{6.107}$$

where T_s is the sampling interval, $\mathbf{N}^e$ is the *tensor of gravity gradients,* including the gravitational and centripetal acceleration components, $\mathbf{\Omega}^e_E$ is the skew-symmetric matrix describing the Earth turn rate as derived from the vector ω^e_E (it is assumed constant in time), $\mathbf{\Phi}_k$ is the skew-symmetric matrix of the specific force measurements $\mathbf{f}^b_k$, $\mathbf{D}_a$ and $\mathbf{D}_g$ are diagonal matrices defining the accelerometer and gyroscope bias model, respectively (a Gauss–Markov model is commonly chosen) [55].

Under the Gauss–Markov hypothesis for the IMU biases, the model noise vector is

$$\mathbf{v}_k = \left[\mathbf{v}^{(a),T}_k, \ \mathbf{v}^{(g),T}_k, \ \mathbf{v}^{(aa),T}_k, \ \mathbf{v}^{(gg),T}_k, \ \right]^T \tag{6.108}$$

whereas the noise-state transition matrix $\mathbf{G}_k$ is

$$\mathbf{G}_k = \begin{bmatrix} \mathbf{0} & \mathbf{0} & \mathbf{0} & \mathbf{0} \\ T_s \mathbf{C}_{bk}^e & \mathbf{0} & \mathbf{0} & \mathbf{0} \\ \mathbf{0} & -T_s \mathbf{C}_{bk}^e & \mathbf{0} & \mathbf{0} \\ \mathbf{0} & \mathbf{0} & T_s \mathbf{I}_3 & \mathbf{0} \\ \mathbf{0} & \mathbf{0} & \mathbf{0} & T_s \mathbf{I}_3 \end{bmatrix} \in \mathbb{R}^{15 \times 12}. \tag{6.109}$$

It is evident that a different frame mechanization, a different structure of the state vector, or a different IMU noise model would yield a different formulation of both the state transition matrix and of the noise-state transition matrix. However, the observation model depends on the kind of integration the two systems are subject to, as presented hereafter.

6.3.2.1 *Loose Integration*

In a loosely integrated architecture, the GNSS receiver and the INS run as independent navigation systems. Their navigation solutions (body position and velocity) are coupled in order to obtain a better estimate of the body trajectory.

The conceptual block diagram of a loosely integrated architecture is shown in Fig. 6.7. The system measurements entering the KF are the body positions and velocities as computed, respectively, by the INS and the GNSS receiver. Their difference is taken so as to define the *observation vector* of the complementary KF. Note that this vector can be computed only when both trajectory solutions are available, but typically the solution rate of the INS is higher than that of the GNSS receiver. Furthermore, in the case of GNSS outage, due to environmental conditions (e.g., indoor navigation) the GNSS solution could be unavailable for relatively long periods of time.

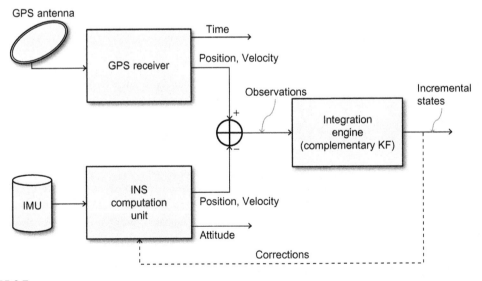

FIGURE 6.7

Block diagram of a GNSS–INS loosely integrated hybrid architecture.

The *observation equation* for such an architecture can be written as

$$\mathbf{z}_k^{lo} = \mathbf{H}^{lo}\mathbf{x}_k + \mathbf{w}_k^{lo},$$ (6.110)

where

$\mathbf{z}_k^{lo} = \boldsymbol{\pi}_k^{sat} - \boldsymbol{\pi}_k^{ins}$ is the observation vector;

$\boldsymbol{\pi}_k^{sat} = \left[\mathbf{p}_k^{sat T} \ \mathbf{v}_k^{sat T}\right]^T$ is the vector of the body position and velocity estimated by the GNSS receiver at the time instant k;

$\boldsymbol{\pi}_k^{ins} = \left[\mathbf{p}_k^{e T} \ \mathbf{v}_k^{e T}\right]^T$ is the vector of the body position and velocity estimated by the INS at the time instant k;

$\mathbf{H}^{lo}$ is the *observation matrix* for the loose architecture, written as

$$\mathbf{H}^{lo} = \begin{bmatrix} \mathbf{I}_3 & \mathbf{0}_3 & \mathbf{0}_3 & \mathbf{0}_3 & \mathbf{0}_{3\times 6} \\ \mathbf{0}_3 & \mathbf{I}_3 & \mathbf{0}_3 & \mathbf{0}_3 & \mathbf{0}_{3\times 6} \end{bmatrix};$$ (6.111)

$\mathbf{w}_k^{lo}$ is the observation noise.

Notice that the loose integration scheme, whose system is constructed as in (6.106) and (6.110), is inherently linear. Thus, assuming Gaussianity, this problem can be optimally handled by the KF.

6.3.2.2 *Tight Integration*

In a tightly coupled architecture, the two systems interact at the "pseudorange level," thus implementing an architecture that promises to be more robust than the loose one. In tight integration, GNSS pseudorange information enters directly in the update stage of the integration filter. This way the information available to bound the INS errors is in some sense proportional to the number of satellites in view. However, the integration is theoretically possible even with just one satellite in view (recall that, on the contrary, the loosely coupled architecture requires at least four satellites in view to obtain an estimate of the body position and velocity from the GNSS receiver).

The concept of a tightly integrated architecture is shown in Fig. 6.8. The system measurements entering the integration KF are the set of pseudorange and pseudorange rates measured by the GNSS receiver, as well as a set of "predicted" pseudorange and pseudorange rates computed on the basis of the inertially-estimated body position and velocity [38, 55]. The difference between the two sets of measurements is taken so as to determine the observations of the complementary KF. Thus, the *observation vector* for the tight integration, $\mathbf{z}_k^{ti}$, is defined as

$$\mathbf{z}_k^{ti} = \boldsymbol{\zeta}_k^{sat} - \boldsymbol{\check{\zeta}}_k,$$ (6.112)

where

$\boldsymbol{\zeta}_k^{sat} = \left[\boldsymbol{\rho}_k^T \ \dot{\boldsymbol{\rho}}_k^T\right]^T$ is the vector of the precorrected GNSS-measured pseudoranges $\boldsymbol{\rho}_k$ and pseudorange rates $\dot{\boldsymbol{\rho}}_k$ at the time instant k;

$\boldsymbol{\check{\zeta}}_k$ is the predicted pseudorange and pseudorange rate measurements computed from the last estimate of the body trajectory, indicated by $\boldsymbol{\check{\pi}}_k$.

Thereby, the *observation equation* for the tight architecture can be written as

$$\mathbf{z}_k^{ti} = \mathbf{H}_k^{ti}\mathbf{x}_k + \mathbf{w}_k^{ti},$$ (6.113)

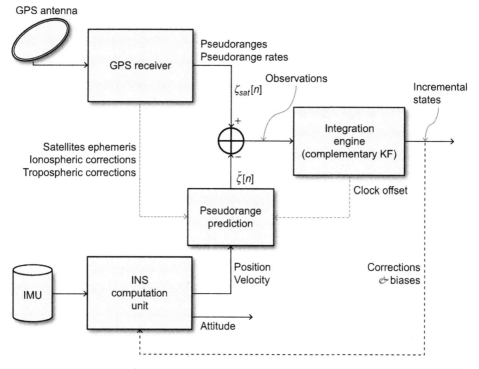

FIGURE 6.8

Block diagram for GNSS–INS tight coupling.

where $\mathbf{w}_k^{ti}$ is the *observation noise* and $\mathbf{H}_k^{ti}$ is the *observation matrix* relating the pseudorange measurements to the current system states.

Although the relationship between receiver position and satellite-to-receiver pseudoranges is not linear, a local *problem linearization* is possible [49], yielding

$$\mathbf{H}_k^{ti} \triangleq \begin{bmatrix} \mathbf{U}_k & \mathbf{0} & \mathbf{0} & \mathbf{0} & \mathbf{0} \\ \mathbf{0} & \mathbf{U}_k & \mathbf{0} & \mathbf{0} & \mathbf{0} \end{bmatrix}, \in \mathbb{R}^{2N_{\mathrm{sat}} \times 17} \tag{6.114}$$

$$\mathbf{U}_k \triangleq \begin{bmatrix} -\mathbf{u}_{1\,k}^T & 1 \\ -\mathbf{u}_{2\,k}^T & 1 \\ \vdots & \vdots \\ -\mathbf{u}_{N_{\mathrm{sat}}\,k}^T & 1 \end{bmatrix}, \tag{6.115}$$

$$\mathbf{u}_{nk} \triangleq \frac{\mathbf{p}_{nk} - \breve{\mathbf{p}}(t)}{\|\mathbf{p}_{nk} - \breve{\mathbf{p}}_k\|}, \forall\, n = 1,\dots,N_{\mathrm{sat}}, \tag{6.116}$$

where N_{sat} is the number of satellites in view at time instant k, $\mathbf{p}_{nk}$ is the position of the nth satellite, and $\breve{\mathbf{p}}_k$ is an estimate of the current body position. Furthermore, as regards matrix $\mathbf{U}_k$, the predicted pseudorange measurements $\check{\boldsymbol{\zeta}}_k$ can be computed as

$$\check{\boldsymbol{\zeta}}_k = \begin{bmatrix} \mathbf{U}_k & \mathbf{0} \\ \mathbf{0} & \mathbf{U}_k \end{bmatrix} \check{\boldsymbol{\pi}}_k, \tag{6.117}$$

where $\check{\boldsymbol{\pi}}_k \triangleq \begin{bmatrix} \check{\mathbf{p}}_k^T & \check{\mathbf{v}}_k^T \end{bmatrix}^T$.

Ultra-Tight Integration

Ultra-tight integration is a derivation of tight integration, where it is the in-phase and quadrature signals produced by the correlators of the GNSS tracking loops that are employed as observations. As for the earlier architectures, the GNSS information is conceptually used as a refinement of the INS information: The GNSS data are exploited to counteract the intrinsic derivation of the INS solution, correcting the INS trajectory. The way they are used to compute and implement the correction is the peculiarity of the ultra-tight hybridization scheme.

In an ultra-tightly integrated architecture, GNSS and INS may interact at the "loop-filter level," where Doppler frequency and code phase are directly updated by the *hybrid navigation filter*, driven by both the GNSS receiver data and the corresponding INS estimates [54]. This approach turns out to be more robust in the case of outages of the satellite signals and can better track the dynamics of the incoming signals, thanks to the presence of the INS.

The states of the channel filter, which also enter the navigation filter, are the signal amplitude, the code phase tracking error (in units of chips), the phase tracking error (in units of rad), the frequency tracking error (in units of rad/s), and the frequency rate error of the NCO (in units of rad/s^2). The system measurements entering the Kalman filter are the early, late, and prompt outputs of the I (in-phase) and Q (quadrature) correlators for each channel. The resulting evolving system is nonlinear; thus, it must be handled resorting to suboptimal filters. Some architectures proposed in the literature operate with the differential correlation error [6], similarly to the two earlier integration schemes. Other proposals use the output of the tracking filter loops as the measurement model itself [8].

The main drawback of the ultra-tightly integrated receiver is that, because all satellite tracking channels are intimately related, any error in one channel can potentially affect other channels adversely [54].

6.3.3 **Bayesian Direct Position Estimation with Inertial Information**

To introduce the Bayesian direct position estimation (DPE) approach for a GNSS receiver architecture, it is necessary to rely on the state-space model introduced earlier in (6.8)–(6.9). Many alternatives exist for the choice of the state equation, as discussed in Ref. [14]. Depending on whether the signals are measurable or not, they may be components of either the state vector $\mathbf{x}_k$ or the input signal $\mathbf{u}_k$. In particular, we address the use of IMU's data: Since the inertial data are measurable, they are included in the input signal. Although measures and inertial data are not expected to be delivered at the same rate, for the sake of simplicity, we will assume that $\mathbf{u}_k$ and $\mathbf{z}_k$ are aligned in time.

6.3.3.1 *Measurement Equation*

A global navigation satellite system (GNSS) antenna receives measurements which are considered to be a superposition of plane waves corrupted by noise and, possibly, interferences and multipath. An antenna receives M scaled, time-delayed, and Doppler-shifted signals with known structure. Each one corresponds to the LOS signal of one of the M visible satellites. The received complex baseband signal can be modeled as

$$y(t) = \sum_{m=1}^{M} \alpha_m q_m(t - \tau_m) \exp\{j2\pi f_{d_m} t\} + \eta(t), \tag{6.118}$$

where $q_m(t)$ is the transmitted complex baseband (low-rate) navigation signal spread by the pseudoran-dom code of the mth satellite, considered known, and α_m is its complex amplitude, τ_m is its time-delay, f_{d_m} the Doppler deviation, and $\eta(t)$ represents zero-mean additive noise and other unmodeled terms. Equivalently, we can express the signal model in its compact form:

$$y(t) = \boldsymbol{\alpha}^T \mathbf{d}(t; \boldsymbol{\tau}, \mathbf{f}_d) + \eta(t), \tag{6.119}$$

where $\boldsymbol{\alpha} = [\alpha_1, \ldots, \alpha_M]^T$; $\boldsymbol{\tau}, \mathbf{f}_d \in \mathbb{R}^{M \times 1}$ are column vectors containing time-delays and Doppler shifts of each satellite; and

$$\mathbf{d}(t, \boldsymbol{\tau}, \mathbf{f}_d) = \left[d_1(t; \tau_1, f_{d_1}), \ldots, d_M(t; \tau_M, f_{d_M}) \right]^T \in \mathbb{C}^{M \times 1}, \tag{6.120}$$

where the mth component is the delayed Doppler-shifted narrowband signal envelope defined by

$$d_m(t; \tau_m, f_{d_m}) = q_m(t - \tau_m) \exp\{j2\pi f_{d_m} t\}.$$

The signal model presented so far is valid for the conventional approach, where the incoming signals are parameterized by synchronization parameters, that is, $\boldsymbol{\tau}$ and $\mathbf{f}_d$. DPE's approach is based on a simple fact: Synchronization parameters of each satellite can be expressed as functions of the same common parameters (including user position) [15]. After inspecting GNSS observables, one can easily identify that

$$\boldsymbol{\tau} \triangleq \boldsymbol{\tau}(\mathbf{x}), \tag{6.121}$$

$$\mathbf{f}_d \triangleq \mathbf{f}_d(\mathbf{x}), \tag{6.122}$$

with $\mathbf{x} \in \mathbb{R}^{N_x}$ being a vector gathering all considered motion parameters, whose simplest configuration is $\mathbf{x} = \left[\mathbf{p}^T, \mathbf{v}^T, b\right]^T$, that is, receiver's position, velocity, and clock bias, respectively. However, DPE is a quite general approach, and $\mathbf{x}$ can include a number of additional parameters [7, 37]. Specifically, the mth satellite's synchronization parameters can be written in terms of the elements of $\mathbf{x}$ as:

$$c\tau_m = \| \mathbf{p}_m - \mathbf{p} \| + c(b - b_m) \tag{6.123}$$

$$f_{d_m} = -(\boldsymbol{v}_i - \boldsymbol{v})^T \frac{\mathbf{p}_m - \mathbf{p}}{\| \mathbf{p}_m - \mathbf{p} \|} \frac{f_c}{c}, \tag{6.124}$$

with $\mathbf{p}_m, \boldsymbol{v}_m$, and b_m being the known position, velocity, and clock bias of the mth satellite, respectively; f_c is the carrier frequency; c represents the speed of light; and $\| \cdot \|$ is the Euclidean norm. Notice that, in (6.123), τ_m is not the actual TOA, but the estimated time-delay w.r.t. the receiver's clock.

Hereinafter, we consider the K-snapshot signal model, parameterized by $\mathbf{x}_k$ as introduced in Section 4.4.1. We use $k \in \mathbb{N}$ to denote the kth record of K samples and the states associated with that record. Therefore, we assume that states are piecewise constant during each observation interval k, although allowing evolution in the long term among different observation windows. The model is then given by

$$\mathbf{y}_k = \boldsymbol{\alpha}_k^T \mathbf{D}(\mathbf{x}_k) + \boldsymbol{\eta}_k, \tag{6.125}$$

where

- $\mathbf{y}_k \in \mathbb{C}^{1 \times K}$ is the observed signal vector, digitized at a suitable sampling rate $f_s = 1/T_s$,
- $\mathbf{D}(\mathbf{x}_k) = [\mathbf{d}(kKT_s; \mathbf{x}_k), \dots, \mathbf{d}((k+1)KT_s - T_s; \mathbf{x}_k)] \in \mathbb{C}^{M \times K}$ is known as the basis-function matrix, and
- $\boldsymbol{\eta}_k \in \mathbb{C}^{1 \times K}$ represents K snapshots of zero-mean additive Gaussian noise with piecewise constant variance σ_n^2 during the observation interval.

Using (6.125) as our measurement model, we would face the so-called curse of dimensionality effect. That is to say, the number of samples to be processed is extremely large. To overcome this, we propose to process samples at the output of the correlator, that is, after correlating the received signal $\mathbf{y}_k$ with a local replica of the ranging signal. Then, L_c samples ($L_c \ll K$) of the correlator output are recorded

$$\mathbf{z}_k = \left[z_k^{(0)}, z_k^{(1)}, \dots, z_k^{(L_c - 1)} \right]^T, \tag{6.126}$$

where

$$z_k^{(i)} = \frac{1}{K} \mathbf{y}_k \left(\hat{\boldsymbol{\alpha}}_{k|k-1}^T \mathbf{D}(\boldsymbol{\Psi}_i) \right)^H, \tag{6.127}$$

and $\boldsymbol{\Psi} = \{\boldsymbol{\Psi}_i\}_{i=0}^{L_c - 1}$ is the set of $\mathbf{x}$-points where the correlator in (6.127) is evaluated at. Notice that, being at the position domain, $\boldsymbol{\Psi}$ plays the role of early-prompt-late samples in conventional delay lock loop (DLL) architectures.

The selection of such $\mathbf{x}$-points is a design choice. Here, we propose the following scheme, which ensures that we are exploring $\mathbf{X}$ using orthogonal vectors centered at a prompt point, denoted as $\mathbf{x}_p$. Then, the correlation points are selected such that they are distributed on the surface of a ball centered at that estimate (mimicking early-late samples in one-dimensional delay tracking [22]). This ball is defined as

$$\mathcal{B}(\mathbf{x}_p, \boldsymbol{\Delta}_x) = \left\{ \tilde{\mathbf{x}} \in \mathbf{X} : \; (\tilde{\mathbf{x}} - \mathbf{x}_p)^T \boldsymbol{\Delta}_x^{-2} (\tilde{\mathbf{x}} - \mathbf{x}_p) = 1 \right\}, \tag{6.128}$$

where $\boldsymbol{\Delta}_x$ is a diagonal $N_x \times N_x$ matrix whose entries provide the distance of each dimension to $\mathbf{x}_p$. For instance, we can set $\mathbf{x}_p = \hat{\mathbf{x}}_{k-1}$, that is, the earlier state estimate. Thus, the deterministically selected set of points are

$$\boldsymbol{\Psi}_0 = \hat{\mathbf{x}}_{k-1}$$
$$\boldsymbol{\Psi}_i = \hat{\mathbf{x}}_{k-1} + \left[\boldsymbol{\Delta}_{x,1} \right]_i \quad i = 1, \dots, N_x$$
$$\boldsymbol{\Psi}_i = \hat{\mathbf{x}}_{k-1} - \left[\boldsymbol{\Delta}_{x,1} \right]_i \quad i = N_x + 1, \dots, 2N_x, \tag{6.129}$$

with $[\boldsymbol{\Delta}]_i$ denoting the ith column of matrix $\boldsymbol{\Delta}$.

This scheme involves the computation of $L_c = 2N_x + 1$ correlation samples following (6.126) and (6.127). $\boldsymbol{\Psi}$ can be extended to include points defined in other balls, that is, with relative distances to $\hat{\mathbf{x}}_{k-1}$ given by $\boldsymbol{\Delta}_{x,2}$:

$$\boldsymbol{\Psi}_i = \hat{\mathbf{x}}_{k-1} + \left[\boldsymbol{\Delta}_{x,2} \right]_i \quad i = 2N_x + 1, \dots, 3N_x$$
$$\boldsymbol{\Psi}_i = \hat{\mathbf{x}}_{k-1} - \left[\boldsymbol{\Delta}_{x,2} \right]_i \quad i = 3N_x + 1, \dots, 4N_x, \tag{6.130}$$

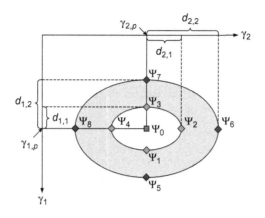

FIGURE 6.9

Selection of **x**-points with two balls in the two-dimensional case (here $\gamma_1, \gamma_2 = x_1, x_2$) (from Ref. [80]).

resulting in $L_c = 4N_x + 1$ correlation samples. This extension of the correlator set of **x**-points has its analogy in double delta correlation operation [44], where two early and two late samples are used. We can generalize the point generation to the use of M_b balls; in that case, we are using $L_c = M_b 2N_x + 1$ correlation samples.

For the sake of clarity, Fig. 6.9 shows a representation of the **x**-points selection process for the case of $M_b = 2$ balls and a two-dimensional state space such that $\mathbf{x} = [x_1, x_2]^T$ and $\mathbf{x}_p = [x_{1,p}, x_{2,p}]^T$. Distance matrices are defined as $\boldsymbol{\Delta}_{x,1} = \mathrm{diag}\{d_{1,1}, d_{1,2}\}$ and $\boldsymbol{\Delta}_{x,2} = \mathrm{diag}\{d_{2,1}, d_{2,2}\}$. This setup results in the generation of $L_c = 9$ correlation points.

In conclusion, the ith entry on our measurement equation turns out to be

$$z_k^{(i)} = \frac{1}{K} \boldsymbol{\alpha}_k^T \mathbf{D}(\mathbf{x}_k) \mathbf{D}^H(\boldsymbol{\Psi}_i) \hat{\boldsymbol{\alpha}}_{k|k-1}^* + w_k^{(i)}, \tag{6.131}$$

where $w_k^{(i)}$ is the ith entry of the measurement noise vector $\mathbf{w}_k$, distributed according to $\mathbf{w}_k \sim \mathcal{N}(\mathbf{0}, \mathbf{R})$. This formulation of the problem, working after jointly despreading received signals, reduces the number of samples to process and consequently the computational load.

6.3.3.2 *State Equation*

In this section, we describe the time evolution of the motion parameters of the GNSS receiver [37]. In addition, in order to perform the integration, we would like to relate this model to the biased accelerations provided by an inertial measurement unit (IMU).

From the differential equations $\dot{\mathbf{p}}(t) = \mathbf{v}(t)$, $\dot{\mathbf{v}}(t) = \mathbf{a}(t)$ and Newton's law (relating forces to acceleration), we know that under a constant acceleration regime, the position evolves as $\mathbf{p}(t) = \mathbf{p}_0 + \mathbf{v}_0 t + \mathbf{a}_0 \frac{t^2}{2}$ in a continuous-time model. We consider the discrete-time model by introducing the *states update period* $\mathcal{T}_s$. The value of $\mathcal{T}_s$ has to be chosen according to the time elapsed between two consecutive time instants k and $k+1$. Agreeing with the definitions in (6.125), we can identify that $\mathcal{T}_s = KT_s$ in our case. We can broadly express the linear discrete-time state model as

$$\mathbf{x}_k = \mathbf{F}\mathbf{x}_{k-1} + \mathbf{B}\mathbf{u}_{k-1} + \mathbf{G}\mathbf{v}_{k-1}, \tag{6.132}$$

where $\mathbf{F}$ is the transitional matrix, $\mathbf{B}$ is the matrix that relates inputs with states and $\mathbf{G}$ links with the noise term. The noise term $\mathbf{v}_k$ is regarded as zero-mean and Gaussian distributed with covariance matrix $\mathbf{Q}$.

In our setup, acceleration is the measured input signal (as provided by the INS), and the state vector is composed of the position, velocity, acceleration bias, and receiver's clock bias. Thus, our state equation is of the form:

$$
\underbrace{\begin{pmatrix} \mathbf{p}_k \\ v_k \\ \delta\mathbf{a}_k \\ b_k \end{pmatrix}}_{\mathbf{x}_k} = \underbrace{\begin{pmatrix} \mathbf{F}_1 & \mathbf{0}_{9\times1} \\ \mathbf{0}_{1\times9} & 1 \end{pmatrix} \begin{pmatrix} \mathbf{p}_{k-1} \\ v_{k-1} \\ \delta\mathbf{a}_{k-1} \\ b_{k-1} \end{pmatrix}}_{\mathbf{Fx}_{k-1}}
\tag{6.133}
$$
$$
+ \underbrace{\begin{pmatrix} \frac{\mathcal{T}_s^2}{2}\cdot\mathbf{I}_3 \\ \mathcal{T}_s\cdot\mathbf{I}_3 \\ \mathbf{0}_3 \\ \mathbf{0}_{1\times3} \end{pmatrix} \mathbf{a}_{k-1}}_{\mathbf{Bu}_{k-1}} + \underbrace{\begin{pmatrix} \frac{\mathcal{T}_s^3}{6}\cdot\mathbf{I}_3 \\ \frac{\mathcal{T}_s^2}{2}\cdot\mathbf{I}_3 \\ \mathcal{T}_s\cdot\mathbf{I}_3 \\ \mathbf{0}_{1\times3} \end{pmatrix} \mathbf{v}_{k-1}}_{\mathbf{Gv}_{k-1}},
$$

with

$$
\mathbf{F}_1 = \begin{pmatrix} \mathbf{I}_3 & \mathcal{T}_s\cdot\mathbf{I}_3 & \frac{\mathcal{T}_s^2}{2}\cdot\mathbf{I}_3 \\ \mathbf{0}_3 & \mathbf{I}_3 & \mathcal{T}_s\cdot\mathbf{I}_3 \\ \mathbf{0}_3 & \mathbf{0}_3 & \mathbf{I}_3 \end{pmatrix},
\tag{6.134}
$$

where we have included the bias of the IMU in the vector state. Notice that, for the sake of simplicity, we assumed that $\mathcal{T}_i = \mathcal{T}_s$. In general, the latter does not hold and (6.133) must be modified accordingly if the assumption is not considered.

6.3.3.3 *Conceptual DPE Solution and Algorithms*

From a Bayesian standpoint, the posterior distribution $p(\mathbf{x}_{0:k}|\mathbf{z}_{1:k},\mathbf{u}_{1:k})$ provides all necessary information about the state of the system $\mathbf{x}_{0:k}$, given all measurements $\mathbf{z}_{1:k}$, all inputs $\mathbf{u}_{1:k}$, and the prior $p(\mathbf{x}_{0:k})$. Nevertheless, we are interested in the marginal distribution $p(\mathbf{x}_k|\mathbf{z}_{1:k},\mathbf{u}_{1:k})$, which can be computed sequentially in the two known stages: prediction (6.4) and update (6.7).

As we already know, unfortunately there are only few cases where the integrals involved in prediction and update can be solved analytically [58]; that is, the case of linear/Gaussian models where the Kalman filter yields to the optimal solution. However, we saw that our measurement model in (6.131) is nonlinear in the state variables and, thus, we have to resort to suboptimal filtering alternatives. In the following two subsections, we describe two approaches that can handle such nonlinearities: The standard particle filter and the *square-root derivative-free Kalman filters*, recently developed. After evaluation and comparison of their computational complexity, the results of simulative tests of the two approaches are discussed in Section 6.3.3.4.

Particle Filter for DPE

Among all the possible variants, we consider for DPE the sequential importance resampling (SIR) algorithm presented in Section 6.1.5.1 as a benchmark.

Square-Root Derivative-Free Kalman Filters

The broad suitability of PFs comes at the expense of a high computational load that makes this solution difficult to embed in real-time applications. A key assumption to reduce the computational load is to consider that both process noise $\mathbf{v}_k$ and measurement noise $\mathbf{w}_k$ are independent random additive processes with Gaussian distribution. This leads us to view the state transition density $p(\mathbf{x}_k|\mathbf{x}_{k-1},\mathbf{u}_{k-1})$ and the measurement likelihood function $p(\mathbf{z}_k|\mathbf{x}_k,\mathbf{u}_k)$ as Gaussians, which in turn reverts to a Gaussian representation of the posterior density $p(\mathbf{x}_k|\mathbf{z}_{1:k},\mathbf{u}_{1:k})$. As shown in Ref. [39], for an arbitrary nonlinearity, we can derive a recursive algorithm by equating the Bayesian formulae with respect to the first and second moments of the distributions, that is, means and covariances of the conditional densities involved. Since all the appearing multidimensional integrals are of the form

$$I = \int_{\mathbf{X}} \text{nonlinear function} \cdot \text{Gaussian density} \cdot d\mathbf{x}, \tag{6.135}$$

(see (6.58)) the problem of Bayesian estimation under the Gaussian assumption reduces to a numerical evaluation of such integrals. Possible ways to approximate the definite integral of a function such as (6.135) are the Gauss–Hermite quadrature rules [32] or the cubature rules [63]. These approaches are based on a weighted sum of function values at specified (i.e., deterministic) points within the domain of integration, as opposite to the stochastic sampling performed by particle filtering methods.

A further refinement of these schemes comes from the fact that when we propagate the covariance matrix through a nonlinear function, the filter should preserve the properties of a covariance matrix, namely, its symmetry, and positive definiteness. In practice, however, because of lack of arithmetic precision, numerical errors may lead to a loss of these properties. To circumvent this problem, a square-root filter is introduced to propagate the square root of the covariance matrix instead of the covariance itself. Even more, it avoids the inversion of the updated covariance matrix, entailing additional computational saving. Although this idea is not new [57], it has recently been applied to the UKF [71] (although it does not guarantee positive definiteness of the covariance matrix) and, more successfully, to the SQKF [3] and the square-root cubature Kalman filter (SCKF) [4]. In the case of the quadrature rule, it requires a fixed number of points $L = D^{N_x}$, where D is the number of points per dimension. In our particular setup, we have $N_x = 10$, which implies a huge computational load even for $D = 3$ and thus making its real-time implementation unfeasible. A complexity analysis will be performed at the end of this section.

The resulting procedure is a square-root, derivative-free scheme sketched in Algorithm 6.6. In steps 7 and 16, $\mathbf{S} = \text{Tria}(\mathbf{A})$ denotes a general triangularization algorithm (for instance, the QR decomposition), where $\mathbf{A} \in \mathbb{R}^{p \times q}$, $p < q$, and $\mathbf{S}$ is a lower triangular matrix. For computational reasons, we prefer to keep the square root as a triangular matrix of the dimension $p \times p$. This can be achieved by the thin QR decomposition [33, § 5.2, Theorem 5.2.2], which has a computational complexity of $\mathcal{O}\left(qp^2\right)$ flops. When the matrix $\mathbf{A}^T$ is decomposed into an orthogonal matrix $\mathbf{Q} \in \mathbb{R}^{q \times p}$ and an upper triangular matrix $\mathbf{R} \in \mathbb{R}^{p \times p}$ such that $\mathbf{A}^T = \mathbf{Q}\mathbf{R}$, the covariance matrix $\mathbf{\Sigma}$ becomes

$$\mathbf{\Sigma} = \mathbf{A}\mathbf{A}^T = \mathbf{R}^T\mathbf{Q}^T\mathbf{Q}\mathbf{R} = \mathbf{R}^T\mathbf{R} = \mathbf{S}\mathbf{S}^T. \tag{6.136}$$

Require: $\mathbf{z}_{1:K}$, $\mathbf{u}_{0:K}$, $\hat{\mathbf{x}}_{0|0}$, $\mathbf{P}_0 = \mathbf{S}_{x,0|0}\mathbf{S}_{x,0|0}^T$, $\mathbf{Q} = \mathbf{S}_v\mathbf{S}_v^T$, $\mathbf{R} = \mathbf{S}_w\mathbf{S}_w^H$.

Initialization:

1: Define sigma points and weights $\{\boldsymbol{\xi}_i, \omega_i\}_{i=1,\dots,L}$ by using Algorithm 6.7 or any other rule.

Tracking:

2: **for** $k = 1$ to ∞ **do**

3:    **Time update**:

4:    Evaluate the sigma points:
$$\mathbf{x}_{i,k-1|k-1} = \mathbf{S}_{x,k-1|k-1}\boldsymbol{\xi}_i + \hat{\mathbf{x}}_{k-1|k-1}, \ i = 1,\dots,L.$$

5:    Evaluate the propagated sigma points using equation (6.133):
$$\tilde{\mathbf{x}}_{i,k|k-1} = \mathbf{f}(\mathbf{x}_{i,k-1|k-1}, \mathbf{u}_{k-1}).$$

6:    Estimate the predicted state:
$$\hat{\mathbf{x}}_{k|k-1} = \frac{1}{2N_x}\sum_{i=1}^{L}\tilde{\mathbf{x}}_{i,k|k-1}.$$

7:    Estimate the square-root factor of the predicted error covariance:
$$\mathbf{S}_{x,k|k-1} = \mathrm{Tria}\left(\left[\ \tilde{\mathcal{Z}}_{k|k-1}\ \vdots\ \mathbf{S}_v\ \right]\right), \text{ where:}$$
$$\tilde{\mathcal{Z}}_{k|k-1} = \frac{1}{2N_x}\left[\tilde{\mathbf{x}}_{1,k|k-1} - \hat{\mathbf{x}}_{k|k-1}\ \cdots\ \tilde{\mathbf{x}}_{L,k|k-1} - \hat{\mathbf{x}}_{k|k-1}\right].$$

8:    **Measurement update**:

9:    Evaluate the sigma points:
$$\mathbf{x}_{i,k|k-1} = \mathbf{S}_{x,k|k-1}\boldsymbol{\xi}_i + \hat{\mathbf{x}}_{k|k-1}, \ i = 1,\dots,L.$$

10:    Evaluate the propagated sigma points using equation (6.131):
$$\tilde{\mathbf{z}}_{i,k|k-1} = \mathbf{h}(\mathbf{x}_{i,k|k-1}).$$

11:    Estimate the predicted measurement:
$$\hat{\mathbf{z}}_{k|k-1} = \frac{1}{2N_x}\sum_{i=1}^{L}\tilde{\mathbf{z}}_{i,k|k-1}.$$

12:    Estimate the square-root of the innovation covariance matrix:
$$\mathbf{S}_{z,k|k-1} = \mathrm{Tria}\left(\left[\ \mathcal{Z}_{k|k-1}\ \vdots\ \mathbf{S}_n\ \right]\right), \text{ where:}$$
$$\mathcal{Z}_{k|k-1} = \frac{1}{2N_x}\left[\tilde{\mathbf{z}}_{1,k|k-1} - \hat{\mathbf{z}}_{k|k-1}\ \cdots\ \tilde{\mathbf{z}}_{L,k|k-1} - \hat{\mathbf{z}}_{k|k-1}\right].$$

13:    Estimate the cross-covariance matrix
$$\boldsymbol{\Sigma}_{xz,k|k-1} = \mathcal{Y}_{k|k-1}\mathcal{Z}_{k|k-1}^T, \text{ where:}$$
$$\mathcal{Y}_{k|k-1} = \frac{1}{2N_x}\left[\mathbf{x}_{1,k|k-1} - \hat{\mathbf{x}}_{k|k-1}\ \cdots\ \mathbf{x}_{L,k|k-1} - \hat{\mathbf{x}}_{k|k-1}\right].$$

14:    Estimate the Kalman gain
$$\mathbf{K}_k = \left(\boldsymbol{\Sigma}_{xz,k|k-1}/\mathbf{S}_{z,k|k-1}^T\right)/\mathbf{S}_{z,k|k-1}.$$

15:    Estimate the updated state
$$\hat{\mathbf{x}}_{k|k} = \hat{\mathbf{x}}_{k|k-1} + \mathbf{K}_k\left(\mathbf{z}_k - \hat{\mathbf{z}}_{k|k-1}\right).$$

16:    Estimate the square-root factor of the corresponding error covariance:
$$\mathbf{S}_{x,k|k} = \mathrm{Tria}\left(\left[\ \mathcal{Y}_{k|k-1} - \mathbf{K}_k\mathcal{Z}_{k|k-1}\ \vdots\ \mathbf{K}_k\mathbf{S}_n\ \right]\right).$$

17: **end for**

Algorithm 6.6

Square-root, derivative-free nonlinear Kalman filter

1: Set $L = 2N_x$.

2: Set the cubature points $\boldsymbol{\xi}_i = \sqrt{N_x} \left[\mathbf{I}_{N_x \times N_x} \mid -\mathbf{I}_{N_x \times N_x} \right]_i$, where $[\cdot]_{i=1,\dots,L}$ indicates the ith column.

3: Set the cubature weights $\omega_i = \frac{1}{2N_x}, i = 1,\dots,L$.

Algorithm 6.7

Generation of sigma points and weights for third-degree spherical–radial cubature rule

---

This triangular structure also allows backward/forward substitution instead of matrix inversion (step 14), resulting in additional computational saving ($\mathcal{O}\left(p^2\right)$ instead of $\mathcal{O}\left(p^3\right)$).

Computational Complexity of the Algorithms for DPE

Algorithms proposed in the earlier section should run in embedded digital processors and should be executed in real time. For that reason, it is important to analyze them to determine the amount of resources (such as time and storage) necessary to execute them. In this section, we will be concerned only with asymptotic time complexity. The time complexity of an algorithm can be viewed as the number of basic operations it performs. We will assume that multiplication of matrices of size n_1 by n_2 and n_2 by n_3 costs $\mathcal{O}(n_1 n_2 n_3)$. The meaning of this notation is the following: A function $p(n)$ is $\mathcal{O}(g(n))$ if and only if there exist a real, positive constant C and a positive integer n_0 such that $p(n) \leq Cg(n) \; \forall n \geq n_0$. Regarding the cost of the evaluation of function $\mathbf{h}(\cdot)$, defined in Eq. (6.131), we have only considered the cost of the matrix operations, without taking into account the generation of matrix $\mathbf{D}$ that can be done by means of a look-up table or by any other method.

Table 6.2 analyzes the complexity of the SIR algorithm for a single iteration. We assume that the process and measurement noises are Gaussian, and the computational cost of drawing one multidimensional sample from a Gaussian distribution is $\mathcal{O}(n^2)$, where n is the dimension of the sample [66]. We also assume that the evaluation of a multivariate Gaussian is $\mathcal{O}(n^3)$. In our particular setup, we have that $N_x = 10$, $N_z = 2M_b N_x + 1$, K depends on the duration of a spreading code and the sampling frequency, whereas the number of particles N_s, the number of tracked satellites M, and the number of

Table 6.2 Complexity of the SIR Algorithm (Algorithm 6.3) Applied to the DPE State-Space Model, (6.131) and (6.132)

Computation	Operation	Size	Cost
$\tilde{\mathbf{x}}_k^{(i)} \sim p(\tilde{\mathbf{x}}_k \mid \mathbf{x}_{k-1}^{(i)}, \mathbf{u}_{k-1})$	N_s evaluations of $\mathbf{f}(\cdot)$ N_s drawn from $p_{\mathbf{v}}(\cdot)$	$N_s \times 3 \times N_x \times N_x$	$\mathcal{O}(N_s N_x^2) N_x$ $\mathcal{O}(N_s N_x^2)$
$\tilde{\omega}_k^{(i)} = p(\mathbf{z}_k \mid \tilde{\mathbf{x}}_k^{(i)})$	N_s evaluations of $\mathbf{h}(\cdot)$ N_s evaluations of $p_{\mathbf{w}}(\cdot)$	$N_s \times M \times K \times M$ N_z	$\mathcal{O}(N_s K M^2)$ $\mathcal{O}(N_s N_z^3)$
$\omega_k^{(i)} = \frac{\tilde{\omega}_k^{(i)}}{\sum_{m=1}^{N_s} \tilde{\omega}_k^{(m)}}$	N_s sums, 1 division	N_s	$\mathcal{O}(N_s)$
Resampling	See Ref. [10]	N_s	$\mathcal{O}(N_s)$
$\hat{\mathbf{x}}_k = \sum_{i=1}^{N_s} \omega_k^{(i)} \mathbf{x}_k^{(i)}$	Scalar–vector product and sum	$N_s \times N_x$	$\mathcal{O}(N_s N_x)$

Table 6.3 Complexity of Algorithm 6.6

Computation	Operation	Size	Cost
$\mathbf{S}_{x,k-1|k-1}\boldsymbol{\xi}_i + \hat{\mathbf{x}}_{k-1|k-1}$	L matrix–vector products	$L \times N_x \times N_x$	$\mathcal{O}(LN_x^2)$
$\mathbf{f}(\mathbf{x}_{i,k-1|k-1}, \mathbf{u}_{k-1})$	L evaluations of $\mathbf{f}(\cdot)$	$L \times 3 \times N_x \times N_x$	$\mathcal{O}(3LN_x^2)$
$\frac{1}{2N_x}\sum_{i=1}^{L}\tilde{\mathbf{x}}_{i,k|k-1}$	Scalar–vector product and sum	$L \times N_x$	$\mathcal{O}(LN_x)$
$\mathrm{Tria}\left(\left[\tilde{\mathcal{Y}}_{k|k-1} \,\vdots\, \mathbf{S}_v\right]\right)$	Thin QR	$N_x \times (L + N_x)$	$\mathcal{O}\left((L + N_x)N_x^2\right)$
$\mathbf{S}_{x,k|k-1}\boldsymbol{\xi}_i + \hat{\mathbf{x}}_{k|k-1}$	L matrix–vector products	$L \times N_x \times N_x$	$\mathcal{O}(LN_x^2)$
$\mathbf{h}(\mathbf{x}_{i,k|k-1})$	L evaluations of $\mathbf{h}(\cdot)$	$L \times M \times K \times M$	$\mathcal{O}(LKM^2)$
$\frac{1}{2N_x}\sum_{i=1}^{L}\tilde{\mathbf{z}}_{i,k|k-1}$	Scalar–vector product and sum	$L \times N_z$	$\mathcal{O}(LN_z)$
$\mathrm{Tria}\left(\left[\mathcal{Z}_{k|k-1} \,\vdots\, \mathbf{S}_n\right]\right)$	Thin QR	$N_z \times (L + N_z)$	$\mathcal{O}\left((L + N_z)N_z^2\right)$
$\mathcal{Y}_{k|k-1}\mathcal{Z}_{k|k-1}^T$	Matrix–matrix product	$N_x \times L \times N_z$	$\mathcal{O}(N_x L N_z)$
$\left(\boldsymbol{\Sigma}_{xz,k|k-1}/\mathbf{S}_{z,k|k-1}^T\right)/\mathbf{S}_{z,k|k-1}$	Two backward substitutions	$N_x \times N_z$	$\mathcal{O}(N_z^2)$
$\hat{\mathbf{x}}_{k|k-1} + \mathbf{K}_k\left(\mathbf{z}_k - \hat{\mathbf{z}}_{k|k-1}\right)$	Matrix–vector product	$N_x \times N_z$	$\mathcal{O}(N_x N_z)$
$\mathrm{Tria}\left(\left[\mathcal{Y}_{k|k-1} - \mathbf{K}_k\mathcal{Z}_{k|k-1} \,\vdots\, \mathbf{K}_k\mathbf{S}_n\right]\right)$	Thin QR	$N_x \times (L + N_z)$	$\mathcal{O}\left((L + N_z)N_x^2\right)$
	Matrix–matrix product $\mathbf{K}_k\mathcal{Z}_{k|k-1}$	$N_x \times N_z \times L$	$\mathcal{O}(LN_x N_z)$
	Matrix–matrix product $\mathbf{K}_k\mathbf{S}_n$	$N_x \times N_z \times N_z$	$\mathcal{O}\left(N_x N_z^2\right)$

balls M_b that define the x-points are design parameters. Inspecting Table 6.2, we see that the bottleneck is the computation of the particle weights. In general, we can state that the complexity of the algorithm grows linearly with N.

Table 6.3 shows the complexity analysis of the SCKF also for a single iteration. In this case, the number of cubature points is not a design parameter but is fixed to $L = 2N_x$. We can see that the algorithm is $\mathcal{O}(N_z^3)$, but in this case, the bottleneck is the evaluation of the measurement equation $\mathbf{h}(\cdot)$.

Regarding the implementation, basic linear algebra subprograms (BLAS) is a de facto application programming interface standard for software libraries to perform basic linear algebra operations such as vector and matrix multiplication [9, 42]. They are used to build larger packages, such as the linear algebra package (LAPACK) [2], and both provide the key foundations for performing mathematical computations in other usual computing tools such as MATLAB. Heavily used in high-performance computing, optimized implementations of the BLAS interface have been developed by hardware vendors such as by Intel as well as by other authors (e.g., ATLAS [75] is a portable self-optimizing BLAS). There are also highly optimized machine code versions that can be executed by digital signal processors or FPGA devices [79]. Thus, the operations required for DPE tracking can be implemented by calls to the LAPACK or BLAS libraries, which contain routines for matrix–vector operations, matrix multiplication, and allow for the efficient implementation of the thin QR decomposition and forward/backward substitution required in Algorithm 6.6.

Table 6.4 Parameters of the Generated Signal

Parameter	Value	Units
r_c	1.023	MHz
f_s	$40/7 \cdot r_c$	MHz
B	1.1	MHz
K/f_s	10^{-3}	s
M	7	–
PRN ♯	9–12–17–18–26–28–29	–
Elev. cutoff	$10°$	degrees

Table 6.5 Azimuth and Elevation Values (in Degrees) of the Visible Satellites

PRN ♯	9	12	17	18	26	28	29
Azimuth, θ	288.9	215.2	87.9	295.4	123.5	46.1	130.6
Elevation, ϕ	46.9	24.5	29.1	32.1	71.5	24.4	60.7

6.3.3.4 *Performance Comparison of Simulation Results*

The performance of the presented Bayesian filters under the DPE framework can be studied by computer simulation. The main characteristics of the recreated environment are summarized in Table 6.4. We consider a static receiver using the civilian GPS C/A navigation signal. The receiver has a precorrelation filter with a 1.1 MHz cutoff frequency, and the signal is digitized at an intermediate frequency of 4.308 MHz with a sampling rate $f_s = 5.714$ MHz. This frequency plan results in a baseband replica of the signal of interest. The observation time is 1 ms. The recreated scenario corresponds to a realistic constellation geometry, with an elevation mask of $10°$, $M = 7$ visible satellites, and with the nominal C/N_0 being 45-dB-Hz. Table 6.5 shows PRN code numbers, azimuth, and elevations of the simulated satellites.

For the sake of simplicity, and without loss of generality, we consider that $\mathbf{x}$ is composed of the three unknown receiver coordinates, thus $N_x = 3$. We study the effect of having different numbers of correlation samples by playing with the number of balls used to generate the outputs, as described in (6.128)–(6.130). Specifically, we consider two setups: (1) one ball with the same distance in each coordinate direction of 50 m and (2) two equidistant balls with radii 50 and 100 m. This represents the use of $L_c = 7$ or $L_c = 13$ correlation samples. Each setup is used to test the SCKF and SIR particle filtering algorithms. Two numbers of particles were used for the latter, one coinciding with the number of generated SCKF sigma points, that is, $N_s = 2N_x$, and another with $N_s = 100$ particles.

Figures 6.10(a) and 6.10(b) show the RMSE of the position estimate in meters after 5 s of operation (data is processed each millisecond). The RMSE is computed as the square root of the sum of squared coordinate errors, averaged over 50 Monte Carlo runs. The results show that using more correlation samples improve the accuracy of the algorithms. We can also claim from the results that, under the same number of generated particles/sigma points, the SCKF outperforms an SIR PF approach. Nevertheless, increasing the number of particles results in improvements in the SIR PFs algorithm at the expense of a higher computational cost.

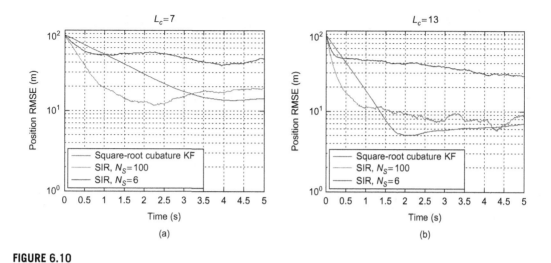

FIGURE 6.10

DPE: Position RMSE with $L_c = 7$ and $L_c = 13$ correlation samples (from Ref. [80]).

6.3.3.5 *Multiple Quadrature Kalman Filtering*

From the earlier sections, it should be clear that in Bayesian inference the computational complexity associated with the estimation of the mean and covariance of the necessary distributions can be high, depending on the dimensionality of the state-space model. Actually, most algorithms are prone to fail in high-dimensional systems, what is typically referred to as *curse of dimensionality*. For instance, PFs are known to require a large sample pool to characterize the filtering distribution when N_x increases [20]. To circumvent this problem, some solutions were reported in the literature in the context of particle filtering. First, a marginalization of linear states was proposed in Ref. [60], where the core idea was to use a KF to optimally deal with these states, while reducing the dimension of the state space that the PF has to explore. This was termed the marginalized PF. Then, a multiple PF was studied in Refs. [12, 24], where the idea was to divide the state space into subsets and run a PF for each partition.

In this section, we borrow the idea of the multiple PF and apply it to the dimension reduction of the recently introduced square-root quadrature Kalman filter (SQKF) [3, 64], a kind of sigma-point Kalman filter (SPKF) [45]. SPKFs are a family of derivative-free Gaussian filters, which are based on a weighted sum of function evaluations at L specified (i.e., deterministic) points within the domain of integration, as opposed to the stochastic sampling performed by particle filtering methods. Different SPKF-like algorithms involve different sigma points and weights. As said earlier, we focus on the SQKF which resorts to the Gauss–Hermite quadrature rule [32]. A single parameter is required for the use of the rule, which is the number of sigma points per dimension α. Indeed, this parameter can be used to adjust the algorithm, with the quadrature rules being optimal for nonlinearities of order $2\alpha - 1$.

The remarkable performance exhibited by the SQKF makes it an appealing tool for practitioners. However, its main drawback is the exponential increase of the number of generated sigma points with the dimension of the state space: $L = \alpha^{N_x}$. Thus, there is a need to reduce the dimensionality of the problems.

In the sequel, we consider that the state space can be split into S subspaces, possibly with different dimensions. Therefore, although the measurement equation remains unaltered, the state equation can be equivalently expressed as[2]

$$
\begin{pmatrix}
\mathbf{x}_k^{(1)} \\
\mathbf{x}_k^{(2)} \\
\vdots \\
\mathbf{x}_k^{(S)}
\end{pmatrix}
=
\begin{pmatrix}
\mathbf{f}_{k-1}^{(1)}\left(\mathbf{x}_{k-1}^{(1)}, \mathbf{x}_{k-1}^{(-1)}\right) \\
\mathbf{f}_{k-1}^{(2)}\left(\mathbf{x}_{k-1}^{(2)}, \mathbf{x}_{k-1}^{(-2)}\right) \\
\vdots \\
\mathbf{f}_{k-1}^{(S)}\left(\mathbf{x}_{k-1}^{(S)}, \mathbf{x}_{k-1}^{(-S)}\right)
\end{pmatrix}
+
\begin{pmatrix}
\mathbf{v}_{k-1}^{(1)} \\
\mathbf{v}_{k-1}^{(2)} \\
\vdots \\
\mathbf{v}_{k-1}^{(S)}
\end{pmatrix},
\tag{6.137}
$$

where each function $\mathbf{f}_{k-1}^{(s)}(\cdot)$, with $s \in \{1, \dots, S\}$, can be different, and where we defined the dimension of each subspace $N_x^{(s)} = \dim\{\mathbf{x}_k^{(s)}\}$ such that $\sum_{s=1}^{S} N_x^{(s)} = N_x$. The sth process noise is distributed as $\mathbf{v}_{k-1}^{(s)} \sim \mathcal{N}\left(\mathbf{0}, \mathbf{Q}_{k-1}^{(s)}\right)$. Notice that the process noise vector in (6.137) is equivalent to $\mathbf{v}_{k-1}$ in (6.9) only in the case where the cross-covariances among subspaces are zero. The next section deals with the implementation of parallel SQKFs to track each subspace.

Multiple SQKF Algorithm

As we have mentioned, the key idea is to divide the state space into S subspaces and apply an SQKF to each partition. The Gaussian filter in charge of the sth subspace is mainly interested in the following marginal predictive and filtering distributions:

$$
p(\mathbf{x}_k^{(s)} \mid \mathbf{z}_{1:k-1}) = \mathcal{N}\left(\hat{\mathbf{x}}_{k|k-1}^{(s)}, \mathbf{P}_{k|k-1}^{(s)}\right)
\tag{6.138}
$$

$$
p(\mathbf{x}_k^{(s)} \mid \mathbf{z}_{1:k}) = \mathcal{N}\left(\hat{\mathbf{x}}_{k|k}^{(s)}, \mathbf{P}_{k|k}^{(s)}\right).
\tag{6.139}
$$

The equations involved in the prediction and update steps can be found in Table 6.6. In that table, we omit the dependence with time of process and measurement functions for the sake of clarity. Inspecting the resulting equations, we realize that a number of integrals must be solved by the filter. These integrals can be seen as expectations of some known (nonlinear) function of the states over a distribution, which can be either the predictive or the filtering distribution. Under the Gaussian assumption, these distributions are also Gaussian as given by (6.138) and (6.139). Although the integrals could be solved by Monte Carlo integration [23], the assumption done allows us to resort to the Gauss–Hermite quadrature rules. Those rules were seen to be a powerful tool to approximate integrals of the kind we mentioned [32], constituting the core concept of SQKF.

The operation of the multiple SQKF is shown in Algorithm 6.8. Our new algorithm is constructed as a bank of S parallel SQKFs, then initially we have to specify the state-space partitioning. Notice that step 8 (predictions) could be done in parallel for each subspace, and the same holds for step 12 (updates). Each filter computes sigma points and weights resorting to the Gauss–Hermite quadrature rules [5, 29]. The corresponding deterministic points are indeed a function of the dimension of the subspace and the number of points per dimension α. Hereinafter, the latter is considered equal for all filters.

---

[2] $\mathbf{x}^{(s)}$ denotes the sth element (possibly a vector) in a vector $\mathbf{x}$ and $\mathbf{x}^{(-s)}$ is the vector of all elements in $\mathbf{x}$ except $\mathbf{x}^{(s)}$.

Table 6.6 Prediction and Update Equations under the Gaussian Assumption

Prediction

$$\hat{\mathbf{x}}_{k|k-1}^{(s)} = \int \mathbf{f}(\mathbf{x}_{k-1}^{(s)}, \mathbf{x}_{k-1}^{(-s)}) p(\mathbf{x}_{k-1}^{(s)} \mid \mathbf{z}_{1:k-1}) d\mathbf{x}_{k-1}^{(s)}$$

$$\mathbf{P}_{k|k-1}^{(s)} = \int \mathbf{f}(\mathbf{x}_{k-1}^{(s)}, \mathbf{x}_{k-1}^{(-s)}) \mathbf{f}^T(\mathbf{x}_{k-1}^{(s)}, \mathbf{x}_{k-1}^{(-s)}) p(\mathbf{x}_{k-1}^{(s)} | \mathbf{z}_{1:k-1}) d\mathbf{x}_{k-1}^{(s)} - \hat{\mathbf{x}}_{k|k-1}^{(s)} \left(\hat{\mathbf{x}}_{k|k-1}^{(s)} \right)^T + \mathbf{Q}_{k-1}^{(s)}$$

Update

$$\hat{\mathbf{x}}_{k|k}^{(s)} = \hat{\mathbf{x}}_{k|k-1}^{(s)} + \mathbf{K}_k^{(s)} \left(\mathbf{z}_k - \hat{\mathbf{z}}_{k|k-1}^{(s)} \right)$$

$$\mathbf{P}_{k|k}^{(s)} = \mathbf{P}_{k|k-1}^{(s)} - \mathbf{K}_k^{(s)} \boldsymbol{\Sigma}_{zz,k|k-1}^{(s)} \left(\mathbf{K}_k^{(s)} \right)^T$$

$$\mathbf{K}_k^{(s)} = \boldsymbol{\Sigma}_{xz,k|k-1}^{(s)} \left(\boldsymbol{\Sigma}_{zz,k|k-1}^{(s)} \right)^{-1}$$

$$\boldsymbol{\Sigma}_{xz,k|k-1}^{(s)} = \int \mathbf{x}_k^{(s)} \mathbf{h}^T(\mathbf{x}_k^{(s)}, \mathbf{x}_k^{(-s)}) p(\mathbf{x}_k^{(s)} | \mathbf{z}_{1:k-1}) d\mathbf{x}_k^{(s)} - \hat{\mathbf{x}}_{k|k-1}^{(s)} \left(\hat{\mathbf{z}}_{k|k-1}^{(s)} \right)^T$$

$$\boldsymbol{\Sigma}_{zz,k|k-1}^{(s)} = \int \mathbf{h}(\mathbf{x}_k^{(s)}, \mathbf{x}_k^{(-s)}) \mathbf{h}^T(\mathbf{x}_k^{(s)}, \mathbf{x}_k^{(-s)}) p(\mathbf{x}_k^{(s)} | \mathbf{z}_{1:k-1}) d\mathbf{x}_k^{(s)} - \hat{\mathbf{z}}_{k|k-1}^{(s)} \left(\hat{\mathbf{z}}_{k|k-1}^{(s)} \right)^T + \mathbf{R}_k$$

$$\hat{\mathbf{z}}_{k|k-1}^{(s)} = \int \mathbf{h}(\mathbf{x}_k^{(s)}, \mathbf{x}_k^{(-s)}) p(\mathbf{x}_k^{(s)} | \mathbf{z}_{1:k-1}) d\mathbf{x}_k^{(s)}$$

Require: α, $\hat{\mathbf{x}}_{0|0}$, $\boldsymbol{\Sigma}_{x,0|0}$, $\mathbf{Q}_k$, $\mathbf{R}_k$, $\{N_x^{(s)}\}_{s=1,\ldots,S}$

Initialization

1: **for** $s = 1$ to S **do**
2:     Define sigma points and weights for the sth filter $\{\boldsymbol{\xi}_i^{(s)}, \omega_i^{(s)}\}_{i=1,\ldots,L_s}$ using quadrature rules.
3:     Set $\mathbf{W}^{(s)} = \text{diag}(\sqrt{\omega_i^{(s)}})$.
4: **end for**
5: Set $k \Leftarrow 1$.
6: Set $\mathbf{S}_{x,k-1|k-1}$ as the square-root factor of $\boldsymbol{\Sigma}_{x,k-1|k-1}$.

SQKF Prediction

7: **for** $s = 1$ to S **do**
8:     Compute $\hat{\mathbf{x}}_{k|k-1}^{(s)}$ and $\mathbf{S}_{x,k|k-1}^{(s)}$ following Algorithm 6.9.
9: **end for**
10: Construct $\hat{\mathbf{x}}_{k|k-1}$ with $\hat{\mathbf{x}}_{k|k-1}^{(1)}, \ldots, \hat{\mathbf{x}}_{k|k-1}^{(S)}$.

SQKF Update

11: **for** $s = 1$ to S **do**
12:     Compute $\hat{\mathbf{x}}_{k|k}^{(s)}$ and $\mathbf{S}_{x,k|k}^{(s)}$ following Algorithm 6.10.
13: **end for**
14: Construct $\hat{\mathbf{x}}_{k|k}$ with $\hat{\mathbf{x}}_{k|k}^{(1)}, \ldots, \hat{\mathbf{x}}_{k|k}^{(S)}$.
15: Set $k \Leftarrow k + 1$ and go step 7.

Algorithm 6.8

Multiple SQKF

Require: $\hat{\mathbf{x}}_{k-1|k-1}, \mathbf{S}^{(s)}_{x,k-1|k-1}, \mathbf{Q}^{(s)}_{k-1}$.
1: Evaluate the sigma points:
$$\mathbf{x}^{(s)}_{i,k-1|k-1} = \mathbf{S}^{(s)}_{x,k-1|k-1} \boldsymbol{\xi}^{(s)}_i + \hat{\mathbf{x}}^{(s)}_{k-1|k-1}, \; i = 1,\dots,L_s.$$
2: Evaluate the propagated sigma points:
$$\tilde{\mathbf{x}}^{(s)}_{i,k|k-1} = \mathbf{f}_{k-1}\left(\mathbf{x}^{(s)}_{i,k-1|k-1}, \hat{\mathbf{x}}^{(-s)}_{k-1|k-1}\right).$$
3: Estimate the predicted state:
$$\hat{\mathbf{x}}^{(s)}_{k|k-1} = \sum_{i=1}^{L_s} \omega^{(s)}_i \tilde{\mathbf{x}}^{(s)}_{i,k|k-1}.$$
4: Estimate the square-root factor of the predicted error covariance:
$$\mathbf{S}^{(s)}_{x,k|k-1} = \text{Tria}\left(\left[\tilde{\mathcal{X}}^{(s)}_{k|k-1} \; \vdots \; \mathbf{S}^{(s)}_{Q_{k-1}} \right]\right), \text{ where:}$$

$\mathbf{S}^{(s)}_{Q_{k-1}}$ is a square-root factor of $\mathbf{Q}^{(s)}_{k-1}$ such that
$\mathbf{Q}^{(s)}_{k-1} = \mathbf{S}^{(s)}_{Q_{k-1}} \left(\mathbf{S}^{(s)}_{Q_{k-1}}\right)^T$, and
$\tilde{\mathcal{X}}^{(s)}_{k|k-1} = \left[\tilde{\mathbf{x}}^{(s)}_{1,k|k-1} - \hat{\mathbf{x}}^{(s)}_{k|k-1}, \dots, \tilde{\mathbf{x}}^{(s)}_{L_s,k|k-1} - \hat{\mathbf{x}}^{(s)}_{k|k-1}\right] \mathbf{W}^{(s)}.$

Algorithm 6.9

Square-root quadrature Kalman filter running over the sth subspace (prediction)

Notice that SQKF computes the square-root factor of the estimation covariances, $\mathbf{S}^{(s)}_{x,k|k-1}$ and $\mathbf{S}^{(s)}_{x,k|k}$. The distributions in (6.138) and (6.139) can be readily obtained using $\mathbf{P}^{(s)}_{k|k-1} = \mathbf{S}^{(s)}_{x,k|k-1}\left(\mathbf{S}^{(s)}_{x,k|k-1}\right)^T$ and $\mathbf{P}^{(s)}_{k|k} = \mathbf{S}^{(s)}_{x,k|k}\left(\mathbf{S}^{(s)}_{x,k|k}\right)^T$.

At k, when a new measurement becomes available, the multiple filter proceeds in the usual manner: Prediction and update of the filtering distributions. The pseudo-code of these steps can be consulted in Algorithms 6.9 and 6.10, respectively. At each step, the sth filter is aware of the estimates delivered by the rest of the filters. The connection among filters composing the multiple SQKF is not evident. As one can see in Table 6.6, the sth filter is connected to the rest through the knowledge of $\mathbf{x}^{(-s)}_{k-1}$ and $\mathbf{x}^{(-s)}_k$, in prediction and update steps, respectively. Following Refs. [12, 24], here we propose to set

$$\mathbf{x}^{(-s)}_{k-1} = \hat{\mathbf{x}}^{(-s)}_{k-1|k-1} \quad \text{and} \quad \mathbf{x}^{(-s)}_k = \hat{\mathbf{x}}^{(-s)}_{k|k-1}, \tag{6.140}$$

with these values being available at the required step.

Although the number of points used by an SQKF is $L = \alpha^{N_x}$, the total number of points generated by the multiple SQKF is $L_M = \sum_{s=1}^{S} \alpha^{N^{(s)}_x}$. The computational complexity of such filters is indeed dependent on the number of quadrature points; thus, a reduction on that value has an impact on the implementation cost. Our interest now is to show that the multiple architecture provides such a reduction.

Proposition 1. Let $L = \alpha^{N_x}$ and $L_M = \sum_{s=1}^{S} \alpha^{N^{(s)}_x}$ with $N_x \in \mathbb{Z}$, $N^{(s)}_x \in \{1,\dots,N_x\} \subset \mathbb{Z}$, and guaranteeing that $\sum_{s=1}^{S} N^{(s)}_x = N_x$. Then,

$$L > L_M \quad \text{if} \quad \alpha > S^{\frac{1}{S+1}} \tag{6.141}$$

for $S \geq 2$ and $\alpha > 2$.

Require: $\mathbf{z}_k, \hat{\mathbf{x}}_{k|k-1}, \mathbf{S}^{(s)}_{x,k|k-1}, \mathbf{R}_k$.

1: Evaluate the sigma points:
$$\mathbf{x}^{(s)}_{i,k|k-1} = \mathbf{S}^{(s)}_{x,k|k-1}\boldsymbol{\xi}^{(s)}_i + \hat{\mathbf{x}}^{(s)}_{k|k-1}, \; i = 1,\ldots,L_s.$$

2: Evaluate the propagated sigma points:
$$\tilde{\mathbf{z}}^{(s)}_{i,k|k-1} = \mathbf{h}_k(\mathbf{x}^{(s)}_{i,k|k-1}, \hat{\mathbf{x}}^{(-s)}_{k|k-1}).$$

3: Estimate the predicted measurement:
$$\hat{\mathbf{z}}^{(s)}_{k|k-1} = \textstyle\sum_{i=1}^{L_s} \omega^{(s)}_i \tilde{\mathbf{z}}^{(s)}_{i,k|k-1}.$$

4: Estimate the square-root of the innovation covariance matrix:
$$\mathbf{S}^{(s)}_{z,k|k-1} = \mathrm{Tria}\left(\left[\,\mathcal{Z}^{(s)}_{k|k-1} \;\middle|\; \mathbf{S}_{\mathbf{R}_k}\,\right]\right), \text{ where:}$$

$\mathbf{S}_{\mathbf{R}_k}$ denotes a square-root factor of $\mathbf{R}_k$ such that
$\mathbf{R}_k = \mathbf{S}_{\mathbf{R}_k}(\mathbf{S}_{\mathbf{R}_k})^T$, and
$$\mathcal{Z}^{(s)}_{k|k-1} = \left[\tilde{\mathbf{z}}^{(s)}_{1,k|k-1} - \hat{\mathbf{z}}^{(s)}_{k|k-1}, \ldots, \tilde{\mathbf{z}}^{(s)}_{L_s,k|k-1} - \hat{\mathbf{z}}^{(s)}_{k|k-1}\right]\mathbf{W}^{(s)}.$$

5: Estimate the cross-covariance matrix
$$\boldsymbol{\Sigma}^{(s)}_{xz,k|k-1} = \mathcal{X}^{(s)}_{k|k-1}\left(\mathcal{Z}^{(s)}_{k|k-1}\right)^T, \text{ where:}$$

$$\mathcal{X}^{(s)}_{k|k-1} = \left[\mathbf{x}^{(s)}_{1,k|k-1} - \hat{\mathbf{x}}^{(s)}_{k|k-1}, \ldots, \mathbf{x}^{(s)}_{L_s,k|k-1} - \hat{\mathbf{x}}^{(s)}_{k|k-1}\right]\mathbf{W}^{(s)}.$$

6: Estimate the Kalman gain
$$\mathbf{K}^{(s)}_k = \left(\boldsymbol{\Sigma}^{(s)}_{xz,k|k-1} \middle/ \left(\mathbf{S}^{(s)}_{z,k|k-1}\right)^T\right)\middle/ \mathbf{S}^{(s)}_{z,k|k-1}.$$

7: Estimate the updated state
$$\hat{\mathbf{x}}^{(s)}_{k|k} = \hat{\mathbf{x}}^{(s)}_{k|k-1} + \mathbf{K}^{(s)}_k\left(\mathbf{z}_k - \hat{\mathbf{z}}^{(s)}_{k|k-1}\right).$$

8: Estimate the square-root factor of the corresponding error covariance:
$$\mathbf{S}^{(s)}_{x,k|k} = \mathrm{Tria}\left(\left[\,\mathcal{X}^{(s)}_{k|k-1} - \mathbf{K}^{(s)}_k\mathcal{Z}^{(s)}_{k|k-1} \;\middle|\; \mathbf{K}^{(s)}_k\mathbf{S}_{\mathbf{R}_k}\,\right]\right).$$

Algorithm 6.10

Square-root quadrature Kalman filter running over the sth subspace (update)

Proof. Let us operate with the difference between L and L_M:

$$\Delta = L - L_M \tag{6.142}$$

$$= \alpha^{\sum_{s=1}^{S} N^{(s)}_x} - \sum_{s=1}^{S}\alpha^{N^{(s)}_x} \tag{6.143}$$

$$= \prod_{s=1}^{S}\alpha^{N^{(s)}_x} - \sum_{s=1}^{S}\alpha^{N^{(s)}_x}, \tag{6.144}$$

we want to see if $\Delta > 0$ and under which conditions. We resort to the procedure used in the proof of Weierstrass's inequality [50], which tackles a similar problem.

If we define $a_s = \alpha^{N_x^{(s)}}$, we can express the function of interest as

$$f(a_1,\ldots,a_S) = \prod_{s=1}^{S} a_s - \sum_{s=1}^{S} a_s. \tag{6.145}$$

Our aim is to minimize (6.145) and evaluate its lower bound. Suppose now that all but one variable are fixed, and then we obtain a function of a single variable

$$f(a_s) = (\kappa_1 - 1)a_s - \kappa_2, \tag{6.146}$$

where

$$\kappa_1 = \prod_{\substack{s'=1 \\ s'\neq s}}^{S} a_{s'} \quad \text{and} \quad \kappa_2 = \sum_{\substack{s'=1 \\ s'\neq s}}^{S} a_{s'}. \tag{6.147}$$

Function (6.146) is linear on a_s, whose slope is $\kappa_1 - 1$. We can bound

$$\alpha^{S-1} \le \kappa_1 \le \alpha^{S(S-1)} \tag{6.148}$$

and observe that $\alpha^{S-1} > 1$ if $S \ge 2$ and $\alpha > 2$. Therefore, the slope cannot be negative under these assumptions, and the minimum of (6.146) is found at $a_s = \alpha$. Extending the result for the rest of the variables in (6.145) yields

$$f(a_1,\ldots,a_S) \ge f(\alpha,\ldots,\alpha) \tag{6.149}$$

$$= \alpha^S - S\alpha, \tag{6.150}$$

from which we can obtain $\alpha > S^{\frac{1}{S+1}}$, a sufficient condition which concludes the proof. $\qquad\square$

Notice that the condition on α is always attained in our setup. For $S > 0$, we can easily show that $S^{\frac{1}{S+1}} < 2$ and, since $\alpha > 2$, we have that $\alpha > S^{\frac{1}{S+1}}$ always holds. Therefore, we proved that the multiple SQKF is always reducing the number of generated quadrature points. We assess its performance by computer simulations in the next section compared with a standard SQKF.

Simulated Performance

We consider the academic example proposed in Ref. [4], where state dimension can be arbitrarily set. States evolve according to $\mathbf{x}_k \sim \mathcal{N}(0.8 \cdot \mathbf{x}_{k-1}, \mathbf{I})$, and measurements are obtained as

$$z_k = \left(\sqrt{1 + \mathbf{x}_k^T \mathbf{x}_k}\right)^q + w_k, \tag{6.151}$$

where $q = 5$ is a parameter used to tune the nonlinearity of the function and $w_k \sim \mathcal{N}(0,2)$.

With this setup, in Fig. 6.11 we compare the RMSE of four different nonlinear filters based on the SQKF with $\alpha = 3$ and $N_x = 6$. Namely, (1) the conventional SQKF that operates over the complete state-space system ($L = \alpha^{N_x} = 729$ sigma points); (2) the multiple SQKF using a single subset ($S = 1$) of the state space ($L_M = 729$ sigma points); (3) $S = N_x/3$ filters running in parallel to track the state space in triples ($L_M = 2\alpha^3 = 54$ sigma points); (4) $S = N_x/2$ filters running in parallel to track the state

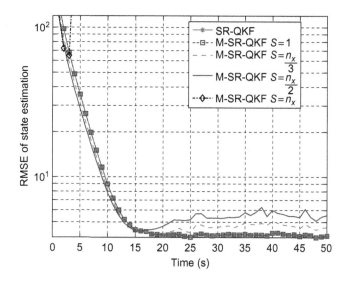

FIGURE 6.11

RMSE of several implementations of the multiple SQKF after averaging 200 realizations.

space in pairs ($L_M = 3\alpha^2 = 27$ sigma points); (5) and each of the six states tracked by its own filter ($L_M = 6\alpha = 18$ sigma points and $S = N_x$).

We can see that the multiple SQKF with $S = 1$ partition is equivalent to the conventional SQKF; it is plotted in order to validate the implementation. As simulations show, a remarkable reduction of the computational cost can be achieved by employing the multiple architecture with low (depending on the state-space division) accuracy degradation.

6.4 HYBRID LOCALIZATION BASED ON GNSS AND PEER-TO-PEER TERRESTRIAL SIGNALING

This section focuses on hybrid GNSS+terrestrial algorithms suitable for a peer-to-peer (P2P) scenario, where "peers" are mobile devices (e.g., cars and other vehicles, or even pedestrians) equipped with both a GNSS receiver and a wireless communications interface.

It is well known that GNSSs provide highly reliable positioning when at least four satellites are visible at the node receiver antenna. This is generally guaranteed in open sky scenarios. On the contrary, in urban canyons, under dense foliage and indoors, the line of sight (LOS) between satellites and receiver's antenna is often obstructed; consequently, GNSS-based localization heavily degrades or completely fails. Therefore, in such environments, hybrid GNSS+terrestrial localization methods, which use both pseudorange and terrestrial ranging measurements, have recently been proposed in the literature to improve both positioning availability and accuracy [34, 65]. However, the proposed schemes have been designed for noncooperative scenarios. In fact, they estimate one unknown peer by leveraging on terrestrial range measurements from fixed anchors. On the contrary, in this section, we present a hybrid algorithm which estimates the position of several unknown peers using a cooperative approach, where unknown peers help each other to improve position accuracy.

The algorithm is able to provide seamless position estimation in the case of GNSS-only, terrestrial-only, and GNSS+terrestrial available signals.

6.4.1 Hybrid Distributed Weighted Multidimensional Scaling

We introduce the *hybrid dwMDS* (hdwMDS) algorithm which extends the earlier distributed weighted multidimensional scaling (dwMDS) algorithm proposed in Ref. [18] for terrestrial positioning only. The hdwMDS algorithm combines measurements from neighboring peers and satellites and provides an estimation of peers positions in a "distributed" and "cooperative" fashion, thus keeping the same features of the original algorithm.

Notations

For notation purposes, we consider a network of $N = N_u + N_a + N_s$ devices in a D-dimensional space (in our case $D = 3$), where:

- N_u is the number of *unknown peers* which have either no knowledge or some imperfect a priori knowledge of their coordinates,
- N_a is the number of *anchor peers* which have perfect a priori knowledge of their coordinates,
- N_s is the number of satellites which, as anchor peers, have perfect a priori knowledge of their coordinates.

Let $\{x_i\}_{i=1}^N$, $x_i = \begin{bmatrix} x_i & y_i & z_i \end{bmatrix}^T \in \mathbb{R}^D$, be the actual vector coordinates of devices. Let $\{x_i^e\}_{i=1}^N$, $x_i^e = \begin{bmatrix} x_i & y_i & z_i & b_i \end{bmatrix}^T \in \mathbb{R}^{D+1}$, be the same vector extended to the P2P scenario to include the bias component $b_i = c \cdot \delta t_i$ expressed in meters. δt_i is the bias of peer i with respect to the satellite constellation time. For satellites and anchor peers, the bias component is assumed to be zero; hence, it does not need to be estimated. Let $\mathbf{X} \in \mathbb{R}^{D \times N}$ be the matrix of actual peer and satellite coordinates defined as follows:

$$\mathbf{X} = [\underbrace{x_1, \ldots, x_{N_u}}_{\text{unknown peers}}, \underbrace{x_{N_u+1}, \ldots, x_{N_u+N_a}}_{\text{anchor peers}}, \underbrace{x_{N_u+N_a+1}, \ldots, x_N}_{\text{satellites}}].$$

Similarly, matrix $\mathbf{X}^e \in \mathbb{R}^{(D+1) \times N}$ denotes the extension of matrix $\mathbf{X}$ composed of the set $\{x_i^e\}_{i=1}^N$ and arranged similarly to $\mathbf{X}$. Imperfect a priori knowledge about *unknown peers* ($i \leq N_u$) is encoded by parameters r_i and $\bar{x}_i$, meaning that with accuracy r_i, x_i is believed to lie around $\bar{x}_i$. If no such knowledge is available, $r_i = 0$.

The localization problem addressed by this algorithm is the estimation of the *unknown peer* coordinates and bias components, $\{x_i^e\}_{i=1}^{N_u}$, given the coordinates of the anchor peers, $\{x_i\}_{i=N_u+1}^{N_u+N_a}$, the coordinates of satellites, $\{x_i\}_{i=N_u+N_a+1}^N$, the imperfect a priori knowledge of unknown peers, $\{(r_i, \bar{x}_i)\}_{i=1}^{N_u}$, range measurements, $\{\delta_{ij}^{(t)}\}$, and pseudorange measurements, $\{\pi_{ij}^{(t)}\}$, taken over time $t = 1 \ldots K$. Range measurement, $\delta_{ij}^{(t)}$, is a noisy measurement of the Euclidean distance between peers i and j defined as

$$d_{ij}(\mathbf{X}) = d(x_i, x_j) = \|x_i - x_j\| = \sqrt{(x_i - x_j)^T (x_i - x_j)}.$$

Pseudorange measurement, $\pi_{ij}^{(t)}$, is a noisy measurement of the exact pseudorange between peer i and satellite j defined as

$$\rho_{ij} = d_{ij}(\mathbf{X}) + b_i, \tag{6.152}$$

where b_i is the bias of the unknown peer i.

Range and Pseudorange Models

The generic range measurement performed by peer i from peer j can be modeled as

$$\delta_{ij}^{(t)} = d_{ij}(\mathbf{X}) + w_r, \tag{6.153}$$

where w_r is the peer-to-peer range measurement noise which can be, for instance, Gaussian distributed, $w_r \sim \mathcal{N}(0, \sigma_r^2)$. Similarly, the generic pseudorange measurement performed by peer i from satellite j can be modeled as

$$\pi_{ij}^{(t)} = \rho_{ij} + w_\rho, \tag{6.154}$$

where w_ρ is the pseudorange measurement noise, which is usually assumed to be Gaussian distributed, $w_\rho \sim \mathcal{N}(0, \sigma_\rho^2)$.

Let us also define the "pseudo-Euclidean" distance $d'(\cdot)$ as follows:

$$d'_{ij}(\mathbf{X}^e) = d_{ij}(\mathbf{X}) + |b_i|. \tag{6.155}$$

Since the bias b_i associated with a generic peer i can be either positive or negative, the actual pseudorange ρ_{ij} coincides with the "pseudo-Euclidean" distance $d'_{ij}(\mathbf{X}^e)$ if and only if the bias b_i is positive. Because we are interested in using the pseudo-Euclidean distance $d'_{ij}(\mathbf{X}^e)$ as exact value of the pseudorange $\rho_{ij}^{(t)}$, we introduce a new bias $b'_i = b_i + B$, where B is a known positive constant large enough such that the new bias b'_i is positive for every unknown peer $1 \le i \le N_u$. Therefore, if the hdwMDS algorithm uses as input the modified set of pseudorange measurements $\left\{ \pi_{ij}^{(t)'} \right\}$, where each element is given by $\pi_{ij}^{(t)'} = \pi_{ij}^{(t)} + B$, the following equality holds:

$$\pi_{ij}^{(t)'} = d'_{ij}(\mathbf{X}^{e'}) + w_\rho, \tag{6.156}$$

where $\mathbf{X}^{e'}$ is a matrix equal to $\mathbf{X}^e$ except for the first N_u columns related to the unknown peers defined as $\mathbf{x}_i^{e'} = \begin{bmatrix} x_i & y_i & z_i & b'_i \end{bmatrix}^T$.

The hdwMDS Cost Function

The hdwMDS algorithm estimates the position of unknown peers that minimizes the following cost function (also known as the STRESS function [19]):

$$S = 2 \sum_{i=1}^{N_u} \sum_{j=i+1}^{N_u+N_a} \sum_{t=1}^{K} \omega_{ij}^{(t)} \left(\delta_{ij}^{(t)} - d_{ij}(\mathbf{X}) \right)^2$$

$$+ 2 \sum_{i=1}^{N_u} \sum_{j=N_u+N_a+1}^{N} \sum_{t=1}^{K} \omega_{ij}^{(t)} \left(\pi_{ij}^{(t)'} - d'_{ij}(\mathbf{X}^{e'}) \right)^2 + \sum_{i=1}^{N_u} r_i \|\mathbf{x}_i - \bar{\mathbf{x}}_i\|^2. \tag{6.157}$$

The weight parameter $\omega_{ij}^{(t)}$ $(t = 1,\ldots,K)$ associated with the range or pseudorange measurement performed at time t by i from j reflects the accuracy of the range or pseudorange measurement, such that less accurate measurements are down-weighted in the overall cost function [18]. For example, if range and pseudorange measurements are affected by an additive Gaussian distributed noise, one might select $\omega_{ij} = 1/\sigma_\rho^2$ and $\omega_{ij} = 1/\sigma_r^2$, respectively. If a measurement between i and j is not available, or its accuracy is zero, then $\omega_{ij}^{(t)} = 0$. It is assumed that $\omega_{ii}^{(t)} = 0$ and $\omega_{ij}^{(t)} = \omega_{ji}^{(t)}$, that is, the weights are symmetric.

After simple manipulations, the global cost function can be rewritten as $S = \sum_{i=1}^{N_u} S_i + \alpha$, where local cost functions S_i are defined for each unknown-location peer (i.e. $1 \le i \le N_u$) as

$$S_i = \sum_{\substack{j=1 \\ j \ne i}}^{N_u} \omega_{ij}\left(\bar{\delta}_{ij} - d_{ij}(\mathbf{X})\right)^2 + \sum_{j=N_u+1}^{N_u+N_a} 2\bar{\omega}_{ij}\left(\bar{\delta}_{ij} - d_{ij}(\mathbf{X})\right)^2$$

$$+ \sum_{j=N_u+N_a+1}^{N} 2\bar{\omega}_{ij}\left(\bar{\pi}'_{ij} - d'_{ij}(\mathbf{X}^{e'})\right)^2 + r_i\|\mathbf{x}_i - \bar{\mathbf{x}}_i\|^2, \tag{6.158}$$

where α is a constant independent of the peer locations $\mathbf{X}$. In (6.158), the K weights, range, and pseudorange measurements between i and j are summarized by single components as $\bar{\omega}_{ij} = \sum_{t=1}^{K} \omega_{ij}^{(t)}$, $\bar{\delta}_{ij} = \sum_{t=1}^{K} \omega_{ij}^{(t)}\delta_{ij}^{(t)}/\bar{\omega}_{ij}$ and $\bar{\pi}'_{ij} = \sum_{t=1}^{K} \omega_{ij}^{(t)}\pi_{ij}^{(t)'}/\bar{\omega}_{ij}$, respectively.

Minimization of the hdwMDS Cost Function

As S_i in (6.158) depends only on the measurements available at the unknown peer i and on the positions of neighboring peers and visible satellites (i.e., devices for which $\omega_{ij}^{(t)} > 0$ for some t), it can be viewed as the local cost function at peer i.

By assuming that each unknown peer receives position estimates from neighboring peers, $S_i(\mathbf{x}_i^{e'})$ can be minimized iteratively and locally by each peer using quadratic majorizing functions as in SMACOF (Scaling by MAjorizing a COmplicated Function [36]).

A majorizing function $T_i(\mathbf{x}_i^{e'},\mathbf{y}_i^{e'})$ of $S_i(\mathbf{x}_i^{e'})$ is a function $T_i : \mathbb{R}^{D+1} \times \mathbb{R}^{D+1} \to \mathbb{R}$ that satisfies (i) $S_i(\mathbf{x}_i^{e'}) \le T_i(\mathbf{x}_i^{e'},\mathbf{y}_i^{e'})$ for all $\mathbf{y}_i^{e'}$, and (ii) $S_i(\mathbf{x}_i^{e'}) = T_i(\mathbf{x}_i^{e'},\mathbf{x}_i^{e'})$. This function can then be used to implement an iterative minimization scheme. Starting from an initial condition $\mathbf{x}_i^{e'(0)}$, the function $T_i(\mathbf{x}_i^{e'},\mathbf{x}_i^{e'(0)})$ is minimized as a function of $\mathbf{x}_i^{e'}$. The newly found minimum, $\mathbf{x}_i^{e'(1)}$, can then be used to define a new majorizing function $T_i(\mathbf{x}_i^{e'},\mathbf{x}_i^{e'(1)})$ to be minimized, and so forth until convergence is reached (see Ref. [36] for details). In order to obtain an effective iterative and distributed process, a simple majorizing function that can be minimized analytically, e.g., a quadratic function, has to be defined. Following Ref. [36], S_i can be rewritten as:

$$S_i(\mathbf{x}^{e'}) = \eta_\delta^2 + \eta^2(\mathbf{X}^{e'}) - 2\rho(\mathbf{X}^{e'}), \tag{6.159}$$

where

$$\eta_\delta^2 = \sum_{\substack{j=1 \\ j\neq i}}^{N_u} \overline{\omega}_{ij}\,\overline{\delta}_{ij}^2 + \sum_{j=N_u+1}^{N_u+N_a} 2\,\overline{\omega}_{ij}\,\overline{\delta}_{ij}^2 + \sum_{j=N_u+N_a+1}^{N} 2\,\overline{\omega}_{ij}\,\overline{\pi}_{ij}^{\prime 2}, \tag{6.160}$$

$$\eta^2(\mathbf{X}^{e'}) = \sum_{\substack{j=1 \\ j\neq i}}^{N_u} \overline{\omega}_{ij} d_{ij}^2(\mathbf{X}) + \sum_{j=N_u+1}^{N_u+N_a} 2\,\overline{\omega}_{ij} d_{ij}^2(\mathbf{X}) + \sum_{j=N_u+N_a+1}^{N} 2\,\overline{\omega}_{ij} d_{ij}^{\prime 2}(\mathbf{X}^{e'}) + r_i\|\mathbf{x}_i - \overline{\mathbf{x}}_i\|^2, \tag{6.161}$$

$$\rho(\mathbf{X}^{e'}) = \sum_{\substack{j=1 \\ j\neq i}}^{N_u} \overline{\omega}_{ij}\overline{\delta}_{ij} d_{ij}(\mathbf{X}) + \sum_{j=N_u+1}^{N_u+N_a} 2\,\overline{\omega}_{ij}\overline{\delta}_{ij} d_{ij}(\mathbf{X}) + \sum_{j=N_u+N_a+1}^{N} 2\,\overline{\omega}_{ij}\overline{\pi}_{ij} d_{ij}'(\mathbf{X}^{e'}). \tag{6.162}$$

Note that the term $\eta^2(\mathbf{X}^{e'})$ depends on the square of the pseudo-Euclidean distance given by $d_{ij}^{\prime 2}(\mathbf{X}^{e'}) = d_{ij}^2(\mathbf{X}) + b_i^{\prime 2} + 2\,d_{ij}(\mathbf{X})b_i'$. In order to simplify the calculation needed to minimize the cost function (6.159) and thus to obtain a feasible distributed algorithm, instead of $\eta^2(\mathbf{X}^{e'})$, we use $\tilde{\eta}^2(\mathbf{X}^{e'})$ which does not take into account the double product contribution $2d_{ij}(\mathbf{X})b_i'$ in the term $d_{ij}^{\prime 2}(\mathbf{X}^{e'})$. Simulation results prove that the above approximation does not imply any negative effect on the minimization of the cost function.

In conclusion, term η_δ^2 does not depend on $\mathbf{x}_i^{e'}$, term $\tilde{\eta}^2(\mathbf{X}^{e'})$ is quadratic in $\mathbf{x}_i$, and only term $\rho(\mathbf{X}^{e'})$ depends on $\mathbf{x}_i$ through a more complicated (sum of square roots) function.

In the majorizing step, we exploit the Cauchy–Schwarz inequality rewritten as follows:

$$d_{ij}(\mathbf{X}) = \frac{d_{ij}(\mathbf{X})d_{ij}(\mathbf{Y})}{d_{ij}(\mathbf{Y})} \geq \frac{(\mathbf{x}_i - \mathbf{x}_j)^T(\mathbf{y}_i - \mathbf{y}_j)}{d_{ij}(\mathbf{Y})}. \tag{6.163}$$

By applying in equation (6.162) the above inequality (6.163) to both the Euclidean distance $d_{ij}(\mathbf{X})$ and the pseudo-Euclidean one $d'(\mathbf{X}^{e'})$, it can be easily seen that S_i is majorized by $T_i(\mathbf{x}_i^{e'}, \mathbf{y}_i^{e'})$:

$$T_i(\mathbf{x}_i^{e'}, \mathbf{y}_i^{e'}) = \eta_\delta^2 + \tilde{\eta}^2(\mathbf{X}^{e'}) - 2\,\rho(\mathbf{X}^{e'}, \mathbf{Y}^{e'}), \tag{6.164}$$

where

$$\rho(\mathbf{X}^{e'}, \mathbf{Y}^{e'}) = \sum_{\substack{j=1 \\ j\neq i}}^{N_u} \overline{\omega}_{ij} \frac{\overline{\delta}_{ij}}{d_{ij}(\mathbf{Y})}(\mathbf{x}_i - \mathbf{x}_j)^T(\mathbf{y}_i - \mathbf{y}_j) + \sum_{j=N_u+1}^{N_u+N_a} 2\,\overline{\omega}_{ij} \frac{\overline{\delta}_{ij}}{d_{ij}(\mathbf{Y})}(\mathbf{x}_i - \mathbf{x}_j)^T(\mathbf{y}_i - \mathbf{y}_j) \tag{6.165}$$

$$+ \sum_{j=N_u+N_a+1}^{N} 2\,\overline{\omega}_{ij} \frac{\overline{\pi}_{ij}'}{d_{ij}'(\mathbf{Y}^{e'})}(\mathbf{x}_i^{e'} - \mathbf{x}_j^e)^T(\mathbf{y}_i^{e'} - \mathbf{y}_j^e).$$

Note that terms $\mathbf{x}_j^e$ and $\mathbf{y}_j^e$ do not have the superscript '′' since they refer to satellites which have a null bias.

Minimizing S_i through a majorizing algorithm consists of finding the minimum of T_i:

$$\frac{\partial T_i(x_i^{e\prime}, y_i^{e\prime})}{\partial x_i^{e\prime}} = 0 \iff \begin{cases} \frac{\partial T_i(x_i^{e\prime}, y_i^{e\prime})}{\partial x_i} = 0 \\ \frac{\partial T_i(x_i^{e\prime}, y_i^{e\prime})}{\partial b_{(x)i}'} = 0 \end{cases}, \tag{6.166}$$

where $b_{(x)i}'$ is the bias component of the vector $x_i^{e\prime}$. An expression of the gradient $\partial T_i(x_i^{e\prime}, y_i^{e\prime})/\partial x_i^{e\prime}$ is reported in Ref. [62].

We denote by $\mathbf{X}^{(k)}$ the matrix whose columns contain the position estimates for all devices at iteration k, and by $\mathbf{X}^{e\prime(k)}$ its extension, which also contains the bias estimation at iteration k. By using the gradient expression defined in Ref. [62], it is possible to update the bias at peer i as

$$b_i'^{(k+1)} = \left(\sum_{j=N_u+N_a+1}^{N} \overline{\omega}_{ij} \right)^{-1} \cdot \sum_{j=N_u+N_a+1}^{N} \overline{\omega}_{ij} \overline{\pi}_{ij}'/d_{ij}'(\mathbf{X}^{e\prime(k)}) b_i'^{(k)}. \tag{6.167}$$

In order to speed up convergence, in (6.168) we substituted $b_i'^{(k)}$ with $(\overline{\pi}_{ij}' - d_{ij}(\mathbf{X}^{(k)}))$, thus obtaining the following new bias update formula:

$$b_i'^{(k+1)} = \left(\sum_{j=N_u+N_a+1}^{N} \overline{\omega}_{ij} \right)^{-1} \cdot \sum_{j=N_u+N_a+1}^{N} \overline{\omega}_{ij} \overline{\pi}_{ij}'/d_{ij}'(\mathbf{X}^{e\prime(k)})(\overline{\pi}_{ij}' - d_{ij}(\mathbf{X}^{(k)})). \tag{6.168}$$

Similarly, it is possible to update the corresponding position as

$$x_i^{(k+1)} = a_i \left(r_i \overline{x}_i + \mathbf{X}^{(k)} c_i^{(k)} \right), \tag{6.169}$$

where

$$a_i^{-1} = \sum_{\substack{j=1 \\ j\neq i}}^{N_u} \overline{\omega}_{ij} + \sum_{j=N_u+1}^{N_u+N_a} 2\overline{\omega}_{ij} + \sum_{j=N_u+N_a+1}^{N} 2\overline{\omega}_{ij} + r_i, \tag{6.170}$$

and $c_i^{(k)} = [c_{i1}, \ldots, c_{iN}]^T$ is a vector column whose definition is reported in Ref. [62]. As the weights $\omega_{ij}^{(t)}$ can be zero when peer i is not connected with neighboring peers j, peer i is updated according to its neighborhood.

Algorithm

The proposed hdwMDS algorithm is summarized in Algorithm 6.11. The global cost function S is computed at each iteration, and the execution of the algorithm is stopped when the difference between the earlier value of S and the one at the current iteration falls under a prefixed threshold ϵ. Several approaches exist to calculate S. For instance, as reported in Algorithm 6.11, at each iteration each unknown peer calculates its own local cost function S_i and then sends the result to a unique specific peer in charge of calculating the global cost function S. This peer then notifies to all unknown peers that the algorithm execution should be interrupted since convergence is reached. Alternatively, in order to save energy and reduce latency, each unknown peer can stop the local algorithm execution when the convergence criterion is met by its own S_i.

Require: $\left\{\delta_{ij}^{(t)}\right\}, \left\{\pi_{ij}^{(t)}\right\}, \left\{\omega_{ij}^{(t)}\right\}, N_a, N_s, \{r_i\}, \{\bar{x}_i\}, \epsilon$, initial condition $\mathbf{X}^{e\prime(0)}$

1: **initialize:** $k = 0, S^{(0)}, \left\{\pi_{ij}^{(t)\prime}\right\} = \left\{\pi_{ij}^{(t)}\right\} + B$, compute a_i from (6.170)

2: **repeat**

3:     $k \leftarrow k + 1$

4:     **all unknown peers in parallel** $(1 \leq i \leq N_u)$

5:         compute $b_i^{\prime(k)}$ from (6.168)

6:         compute $x_i^{(k)} = a_i \left(r_i \bar{x}_i + \mathbf{X}^{(k-1)} c_i^{(k-1)} \right)$

7:         compute $S_i^{(k)}$ from (6.159)

8:         communicate $x_i^{e\prime(k)}$ and $S_i^{(k)}$ to a specific peer

9:     **end parallel**

10: **until** $S^{(k-1)} - S^{(k)} < \epsilon$

Algorithm 6.11

Hybrid dwMDS

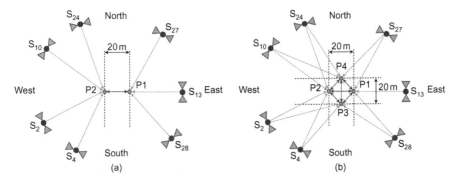

FIGURE 6.12

(a) P2P scenario with two cooperating peers. (b) P2P scenario with four cooperating peers.

Simulation Results

Performance of the hdwMDS can be evaluated through simulation focusing on two simple but challenging scenarios depicted in Fig. 6.12. The first one is composed of two unknown peers and the second of four fully connected unknown peers. Actual coordinates of peers and satellites are listed in Table 6.7, expressed in meters according to the ECEF reference system.

In particular, the four peers are geographically located in Torino, Italy, at the ISMB premises. They have different heights, and the maximum distance between each other is 20 m. Satellites are taken from the GPS constellation considering a specific instant of time as seen from the ISMB premises. Peer P2 is the only one able to see four satellites, whereas the other three peers are able to see only three satellites. Consequently, only peer P2 is able to localize itself, whereas the other three peers are not able to localize themselves without peer-to-peer information. Pseudorange measurements are corrupted by additive Gaussian noise with standard deviation σ_ρ assuming at each different simulation

Table 6.7 Actual Peer and Satellite Positions Expressed in the WGS-84 ECEF Reference System

Id	X [m]	Y [m]	Z [m]	b [m]
P1	4472421.095	601452.586	4492698.629	84708.471
P2	4472420.907	601432.788	4492695.798	−713236.849
P3	4472428.024	601443.631	4492690.144	−1054343.544
P4	4472413.979	601441.743	4492704.283	84704.492
S_2	20479525.971	−13805259.387	9180247.642	0
S_4	25968694.141	−5438581.602	−3544129.914	0
S_{10}	9783415.408	−11324267.214	21824890.846	0
S_{13}	7903408.541	17567293.861	18212109.389	0
S_{24}	−5022069.932	−14586093.187	21822066.500	0
S_{27}	17103426.847	7808290.115	19504730.641	0
S_{28}	22022105.206	11634350.179	−9387065.701	0

one of the following four values $\{3,4,5,6\}$ m. Range measurements between peers are corrupted by additive Gaussian noise with standard deviation $\sigma_r = 20$ cm. The fifth column of Table 6.7 also reports the bias of the four peers. Bias values have been calculated assuming that for each peer the actual pseudorange associated with the visible satellite with the shortest distance from the peer itself is equal to the average pseudorange value, corresponding to the propagation time $\delta t = 70$ ms.

Figure 6.13 shows the performance of the hdwMDS compared with the Cooperative Least Squares (CLS) implemented according to the iterative descent algorithm proposed in Ref. [76] and extended to the hybrid GNSS and terrestrial ranging scenario. Performance shown in Fig. 6.13 refers to the scenario depicted in Fig. 6.12(a) with only two cooperating peers (P1 and P2) as a function of the time slot (at each time slot a new set of range and pseudorange measurements are available). Initial peer positions are randomly set to be on average 25,000 m away from the actual ones. As can be observed, the CLS does not properly converge, though more than 200 consecutive time slots are simulated.

As can be seen from Fig. 6.13, after three time slots, the hdwMDS algorithm converges. For all simulations, the following settings have been used, $r_i = \frac{1}{3}$ and $\omega_{ij} = 1$ (for both range and pseudorange measurements), resulting in an average number of iterations equal to 15.

Table 6.8 shows the performance for the two-peer scenario in both cooperative and noncooperative cases. The performance is evaluated considering 200 time slots. The first five position estimations are removed from the statistical computation because they are affected by large errors. As can be observed from Table 6.8, P2 shows better performance than P1. This is because of the fact that P2 sees four satellites whereas P1 only sees three. Note that peer P1 can be localized only in the cooperative case. In addition, peer P2 in the cooperative case has better performance with respect to the noncooperative case, thanks to the cooperation with peer P1. Table 6.9 shows the performance for the four-peer cooperative scenario depicted in Fig. 6.12(b). Again, peer P2 shows better performance than the other peers. Note that peers P1, P3, and P4 can be localized even if they see three only satellites thanks to the P2P cooperation.

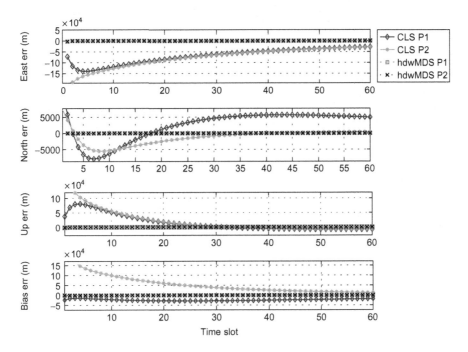

FIGURE 6.13

Location performance related to the two-peer scenario (composed of P1 and P2), $\sigma_\rho = 5$ m, $\sigma_r = 20$ cm.

Table 6.8 Positioning Performance Results in the Two-Peer Scenario, for the Cooperative and the Noncooperative Scheme

	Two-Peer Cooperative Scenario						Two-Peer Noncooperative Scenario					
	Peer P1 RMS [m]			Peer P2 RMS [m]			Peer P1 RMS [m]			Peer P2 RMS [m]		
σ_ρ [m]	H_{err}	V_{err}	b_{err}	H_{err}	V_{err}	b_{err}	H_{err}	V_{err}	b_{err}	H_{err}	V_{err}	b_{err}
3	3.64	2.76	2.42	3.16	2.65	2.07	NA	NA	NA	3.94	3.19	2.21
4	4.85	3.61	3.02	4.05	3.47	2.91	NA	NA	NA	5.28	4.78	3.06
5	5.79	4.54	3.58	5.03	4.79	3.83	NA	NA	NA	6.82	6.19	4.02
6	7.57	5.75	4.87	6.52	5.74	4.74	NA	NA	NA	8.24	7.79	4.99

Finally, Fig. 6.14 shows the performance of peer P2 as a function of the pseudorange standard deviation σ_ρ in three different cases. P2 shows better results with the four-peer cooperative scenario and worst performance in the noncooperative scenario, which confirms the beneficial effect of cooperation among peers.

Table 6.9 Positioning Performance Results in the Four-Peer Scenario

σ_ρ [m]	Four-Peer Cooperative Scenario											
	Peer P1 RMS [m]			Peer P2 RMS [m]			Peer P3 RMS [m]			Peer P4 RMS [m]		
	H_{err}	V_{err}	b_{err}	H_{err}	V_{err}	b_{err}	H_{err}	V_{err}	b_{err}	H_{err}	V_{err}	b_{err}
3	2.09	2.02	1.86	1.97	1.89	1.78	1.95	1.69	1.88	2.23	2.39	2.44
4	2.90	2.64	2.54	2.75	2.32	2.30	2.83	2.36	2.62	3.03	2.99	3.00
5	4.00	3.90	3.54	3.46	4.22	3.12	3.82	4.23	3.61	3.91	4.07	4.24
6	5.28	4.87	4.57	4.54	5.31	3.91	5.23	5.64	4.46	5.19	4.96	5.24

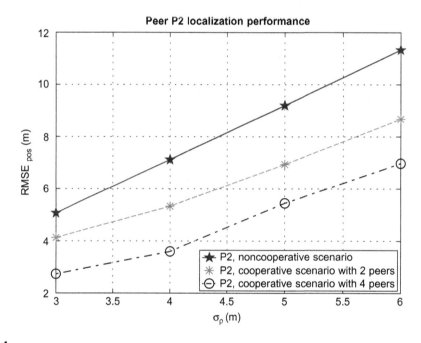

FIGURE 6.14

Positioning performance associated with peer P2, in terms of RMS error.

References

[1] T. Alhmiedat, S. Yang, A survey: Localization and tracking mobile targets through wireless sensors network, in: Proceedings of the 8th Annual PostGraduate Symposium on the Convergence of Telecommunications, Networking and Broadcasting, John Moores University, Liverpool, June 2007.

[2] E. Anderson, Z. Bai, C. Bischof, L.S. Blackford, J. Demmel, J. Dongarra, et al., LAPACK Users' Guide, third ed., Society for Industrial and Applied Mathematics, Philadelphia, PA, 1999.

[3] I. Arasaratnam, S. Haykin, Square-root quadrature Kalman filtering, IEEE Trans. Signal Process. 56 (6) (2008) 2589–2593.

[4] I. Arasaratnam, S. Haykin, Cubature Kalman filters, IEEE Trans. Automatic Control 54 (6) (2009) 1254–1269.

[5] I. Arasaratnam, S. Haykin, R.J. Elliot, Discrete-time nonlinear filtering algorithms using Gauss–Hermite quadrature, Proc. IEEE 95 (5) (2007) 953–977.

[6] R. Babu, J. Wang, Ultra-Tight GPS/INS/PL Integration: A System Concept and Performance Analysis. Technical report, School of Surveying and Spatial Information Systems, The University of New South Wales, Sydney, Australia, 2004.

[7] Y. Bar-Shalom, X.R. Li, T. Kirubarajan, Estimation with Applications to Tracking and Navigation: Theory Algorithms and Software, John Wiley & Sons, 2001.

[8] D. Bernal, P. Closas, J.A. Fernández-Rubio, Particle filtering algorithm for ultra-tight GNSS/INS integration, in: Proceedings of the ION GNSS 2008, The Institute of Navigation, Savannah, GA, 2008, pp. 2137–2144.

[9] L.S. Blackford, J. Demmel, J. Dongarra, I. Duff, S. Hammarling, G. Henry, et al., An updated set of basic linear algebra subprograms (BLAS), ACM Trans. Math. Soft. 28 (2) (2002) 135–151.

[10] M. Bolić, P. Djurić, S. Hong, Resampling algorithms for particle filters: A computational complexity perspective, EURASIP J. Appl. Signal Process. 15 (2004) 2267–2277.

[11] R.G. Brown, P.Y.C. Hwang, Introduction to Random Signals and Applied Kalman Filtering, second ed., John Wiley & Sons, Inc., 1992.

[12] M.F. Bugallo, T. Lu, P.M. Djurić, Target tracking by multiple particle filtering, in: Proceedings of IEEE Aerospace Conference, Big Sky, MT, 2007, pp. 1–7.

[13] M. Caceres, F. Sottile, M.A. Spirito, Adaptive location tracking by Kalman filter in wireless sensor networks, in: IEEE International Conference on Wireless and Mobile Computing, Networking and Communications (WiMob 2009), Marrakech, Morocco, 2009, pp. 123–128.

[14] P. Closas, Bayesian signal processing techniques for GNSS receivers: from multipath mitigation to positioning, PhD thesis, Department of Signal Theory and Communications, Universitat Politècnica de Catalunya (UPC), Barcelona, Spain, 2009.

[15] P. Closas, C. Fernández-Prades, J.A. Fernández-Rubio, Maximum likelihood estimation of position in GNSS, IEEE Signal Process. Lett. 14 (5) (2007) 359–362.

[16] P. Closas, C. Fernández-Prades, J.A. Fernández-Rubio, Cramér–Rao bound analysis of positioning approaches in GNSS receivers, IEEE Trans. Signal Process. 57 (10) (2009) 3775–3786.

[17] A. Conti, D. Dardari, L. Guerra, L. Mucchi, M.Z. Win, Experimental characterization of diversity tracking systems, IEEE Syst. J. (submitted).

[18] J.A. Costa, N. Patwari, A.O. Hero III, Distributed weighted-multidimensional scaling for node localization in sensor networks, ACM Trans. Sens. Netw. 2 (1) (2006) 39–64.

[19] T. Cox, M. Cox, Multidimensional Scaling, Chapman & Hall, London, 2001.

[20] F. Daum, J. Huang, Curse of dimensionality and particle filters, in: Proceedings of IEEE Aerospace Conference, vol. 4, Big Sky, MT, 2003, pp. 1979–1993.

[21] M. Di Rocco, F. Pascucci, Sensor network localisation using distributed extended Kalman filter, in: Proceedings of the IEEE/ASME Int. Conf. on Advanced Intelligent Mechatronics, ETH, Zurich, Sept. 2007, pp. 1–6.

[22] A.J.V. Dierendonck, P. Fenton, T. Ford, Theory and performance of narrow correlator spacing in a GPS receiver, Navigation J. Inst. Navig. 39 (3) (1992) 265–283.

[23] P.M. Djurić, J.H. Kotecha, J. Zhang, Y. Huang, T. Ghirmai, M.F. Bugallo, et al., Particle filtering, IEEE Signal Process. Mag. 20 (5) (2003) 19–38.

[24] P.M. Djurić, T. Lu, M.F. Bugallo, Multiple particle filtering, in: Proceedings of IEEE International Conference on Acoustics, Speech and Signal Processing, ICASSP 2007, Honolulu, HI, 2007.

[25] R. Douc, O. Cappé, E. Moulines, Comparison of resampling schemes for particle filtering, in: Proceedings of the 4th International Symposium on Image and Signal Processing and Analysis, ISPA'05, Zagreb, Croatia, 2005, pp. 64–69.

[26] A. Doucet, N. Gordon, V. Krishnamurthy, Particle filters for state estimation of jump Markov linear systems, IEEE Trans. Signal Process. 49 (2001) 613–624.

[27] A. Doucet, X. Wang, Monte Carlo methods for signal processing: A review in the statistical signal processing context, IEEE Signal Process. Mag. 22 (6) (2005) 152–170.

[28] J.A. Farrell, M. Barth, The Global Positioning System & Inertial Navigation, McGraw-Hill, 1999.

[29] C. Fernández-Prades, J. Vilà-Valls, Bayesian nonlinear filtering using quadrature and cubature rules applied to sensor data fusion for positioning, in: Proceedings of the IEEE International Conference on Communications, ICC, Cape Town (South Africa), 2010.

[30] V. Fox, J. Hightower, L. Liao, D. Schulz, G. Borriello, Bayesian filtering for location estimation, IEEE Pervasive Comput. 2 (3) (2003) 24–33.

[31] W. Gilks, S. Richardson, D. Spiegelhalter (Eds.), Markov Chain Monte Carlo in Practice: Interdisciplinary Statistics, CRC Interdisciplinary Statistics Series, Chapman & Hall, 1996.

[32] G. Golub, V. Pereyra, The differentiation of pseudo-inverses and nonlinear least squares problems whose variables separate, SIAM J. Numer. Anal. 10 (1973) 413–432.

[33] G.H. Golub, C.F. van Loan, Matrix Computations, third ed., The John Hopkins University Press, 1996.

[34] J. González, J.L. Blanco, C. Galindo, A. Ortiz-de Galisteo, J.A. Fernández-Madrigal, F.A. Moreno, et al., Combination of UWB and GPS for indoor–outdoor vehicle localization, in: 9th International Symposium on Signal Processing and Its Application (ISSPA), Sharjah, UAE, Feb. 2007.

[35] M.S. Grewal, L.R. Weill, A.P. Andrews, Global Positioning Systems, Inertial Navigation and Integration, John Wiley & Sons, 2001.

[36] P. Groenen, The Majorization Approach to Multidimensional Scaling: Some Problems and Extensions, DSWO Press, 1993.

[37] F. Gustafsson, F. Gunnarsson, N. Bergman, U. Forssell, J. Jansson, R. Karlsson, et al., Particle filters for positioning, navigation and tracking. IEEE Trans. Signal Process. 50 (2) (2002) 425–437.

[38] N. Hjortsmarker, Experimental system for validating GPS/INS integration algorithms, Master's thesis, Department of Computer Science and Electrical Engineering, Luleå University of Technology, 2005.

[39] K. Ito, K. Xiong, Gaussian filters for nonlinear filtering problems, IEEE Trans. Automatic Control 45 (5) (2000) 910–927.

[40] S.J. Julier, J.K. Uhlmann, A new extension of the Kalman filter to nonlinear systems, in: Proceedings of AeroSense: The 11th Int. Symp. on Aerospace/Defence Sensing, Simulation and Controls, Orlando, FL, April 1997.

[41] A. Kong, J. Liu, W. Wong, Sequential imputations and Bayesian missing data problems, J. Am. Stat. Assoc. 89 (425) (1994) 278–288.

[42] C.L. Lawson, R.J. Hanson, D. Kincaid, F.T. Krogh, Basic linear algebra subprograms for Fortran usage, ACM Trans. Math. Soft. 5 (3) (1979) 308–323.

[43] W. Lijun, Z. Huichang, Y. Xiaoniu, The modeling and simulation for GPS/INS integrated navigation system, in: Proceedings of the Int. Conf. on Microwave and Millimeter Wave Technology, Beijing, China, 2008.

[44] G. McGraw, M. Braash, GNSS multipath mitigation using gated and high resolution correlator concepts, in: Proceedings of the ION GPS/GNSS, 1999, pp. 333–342.

[45] R.V.D. Merwe, E. Wan, Sigma-point Kalman filters for probabilistic inference in dynamic state-space models, in: Proceedings of the Workshop on Advances in Machine Learning, Montreal, Canada, 2003.

[46] J. Míguez, Analysis of parallelizable resampling algorithms for particle filtering, Signal Process. 87 (12) (2007) 3155–3174.

[47] J. Míguez, Analysis of selection methods for cost-reference particle filtering with applications to maneuvering target tracking and dynamic optimization, Elsevier Digital Signal Process. 17 (4) (2007) 787–807.

[48] J. Míguez, M. Bugallo, P. Djurić, A new class of particle filters for random dynamical systems with unknown statistics, EURASIP J. Appl. Signal Process. 2004 (1) (2004) 2278–2294.

[49] P. Misra, P. Enge, Global Positioning System – Signals, Measurements, and Performance, first ed., Ganga-Jamuna Press, 2001.

[50] D.S. Mitrinović, P.M. Vasić, Analytic Inequalities, Springer-Verlag, New York, 1970.

[51] A.F. Molisch, J.R. Foerster, M. Pendergrass, Channel models for ultrawideband personal area networks, in: IEEE Wireless Communications, vol. 10, No. 6, 2003, pp. 14–21.

[52] M. Najar, J.M. Huerta, J. Vidal, J.A. Castro, Mobile location with bias tracking in non-line-of-sight, in: Proceedings of IEEE International Conference on Acoustics, Speech and Signal Processing, 2004.

[53] N. Patwari, A.O. Hero, M. Perkins, N.S. Correal, R.J. O'Dea, Relative location estimation in wireless sensor networks, IEEE Trans. Signal Process. 51 (8) (2003) 2137–2148.

[54] M. Petovello, C. O'Driscoll, G. Lachapelle, Weak signal carrier tracking using extended coherent integration with an ultra-tight GNSS/IMU receiver, in: European Navigation Conference 2008, Toulouse, France, 2008.

[55] M.G. Petovello. Real-Time Integration of a Tactical-Grade IMU and GPS for High-Accuracy Positioning and Navigation, PhD thesis, Department of Geomatics Engineering, University of Calgary, 2003.

[56] M.K. Pitt, N. Shephard, Filtering via simulation: Auxiliary particle filters, J. Am. Stat. Assoc. 94 (446) (1999) 590–599.

[57] J.E. Potter, R.G. Stern, Statistical filtering of space navigation measurements, in: Proceedings of the AIAA Guidance, Navigation, and Control Conference, Cambridge, MA, 1963.

[58] B. Ristic, S. Arulampalam, N. Gordon, Beyond the Kalman Filter: Particle Filters for Tracking Applications, Artech House, Boston, 2004.

[59] A.P. Sage, G.W. Husa, Algorithms for sequential adaptive estimation of prior statistics, in: Proceedings of IEEE Symposium on Adaptive Processes (8th) Decision and Control, vol. 8, 1969, pp. 6a1–6a10.

[60] T. Schön, F. Gustafsson, P. Nordlund, Marginalized particle filters for mixed linear/nonlinear state-space models, IEEE Trans. Signal Process. 53 (7) (2005) 2279–2289.

[61] F. Sottile, R. Giannantonio, M.A. Spirito, F.L. Bellifemine, Design, deployment and performance of a complete real-time zigbee localization system, in: Proceedings of the IEEE IFIP Wireless Days Conference, Dubai, United Arab Emirates, 2008, pp. 1–5.

[62] F. Sottile, M.A. Spirito, M.A. Caceres, J. Samson, Distributed-weighted multidimensional scaling for hybrid peer-to-peer localization, in: Proceedings of Ubiquitous Positioning, Indoor Navigation and Location-Based Service (UPINLBS) 2010, Kirkkonummi (Helsinki), Finland, 2010, pp. 1–6.

[63] A.H. Stroud, Approximate Calculation of Multiple Integrals, Prentice-Hall, Englewood Cliffs, NJ, 1971.

[64] W.I. Tam, D. Hatzinakos, An efficient radar tracking algorithm using multidimensional Gauss–Hermite quadratures, in: Proceedings of IEEE International Conference on Acoustics, Speech and Signal Processing, ICASSP 1997, Munich, 1997.

[65] K.M. Tan, C.L. Law, GPS and UWB integration for indoor positioning, in: 6th International Conference on Information, Communications & Signal Processing, Singapore, 2007, pp. 1–5.

[66] D.B. Thomas, W. Luk, Multivariate Gaussian random number generation targeting reconfigurable hardware, ACM Trans. Reconfigurable Technol. Syst. 1 (2) (2008) 12:1–12:29.

[67] N.J. Thomas, D.G.M. Cruickshank, D.I. Laurenson, A robust location estimator architecture with biased Kalman filtering of TOA data for wireless systems, in: IEEE 6th International Symposium on Spread Spectrum Tech. & Applications, 2000, pp. 296–300.

[68] D.H. Titterton, J.L. Weston, Strapdown Inertial Navigation Technology, Paul Zarchan Editor, 2004.

[69] D. Torrieri, Statistical theory of passive location systems, IEEE Trans. AES 20 (2) (1984) 183–198.

[70] R. van der Merwe, Sigma-Point Kalman Filters for Probabilistic Inference in Dynamic State-Space Models, PhD thesis, OGI School of Science & Engineering at Oregon Health & Science University, 2004.

[71] R. Van der Merwe, E.A. Wan, The square-root unscented Kalman filter for state and parameter estimation, in: Proceedings of ICASSP'01, vol. 6, Salt Lake City, UT, 2001, pp. 3461–3464.

[72] R. van der Merwe, E.A. Wan, Gaussian mixture sigma-point particle filters for sequential probabilistic inference in dynamic state-space models, in: Proceedings of the IEEE International Conference on Acoustics, Speech and Signal Processing, Hong Kong, 2003, pp. 701–704.

[73] G. Welch, G. Bishop, An introduction to the Kalman filter, Technical Report, SIGGRAPH 2001, Course 8, Los Angeles, CA. "www.cs.unc.edu/~tracker/media/pdf/SIGGRAPH2001_CoursePack_08.pdf", 2001 (accessed 01.10).

[74] G.F. Welch, G. Bishop, An introduction to the Kalman filter, Technical Report, University of North Carolina, Chapel Hill, NC, 1995.

[75] R.C. Whaley, J. Dongarra, Automatically tuned linear algebra software, in: Proceedings of the Ninth SIAM Conference on Parallel Processing Scientific Computing, San Antonio, TX, 22–24 March 1999.

[76] H. Wymeersch, J. Lien, M. Win, Cooperative localization in wireless networks, Proc. IEEE 97 (2) (2009) 427–450.

[77] J. Youssef, Contributions to navigation solutions based on ultra wideband radios, inertial measurement units, and the fusion of both modalities, PhD thesis, Technical Report, University of Grenoble, Grenoble, France, 2010.

[78] Y. Zhu, A. Shareef, Comparisons of three Kalman filter tracking algorithms in sensor networks, in: Int. Workshop on Networking, Architecture, and Storages, 2006, IWNAS '06, Shengyang, China, 2006, p. 2.

[79] L. Zhuo, V.K. Prasanna, High performance linear algebra operations on reconfigurable systems, in: Proceedings of the ACM/IEEE Conference on Supercomputing, Seattle, WA, 2005, pp. 2–13.

[80] P. Closas, C. Fernández-Prades, Bayesian nonlinear filters for direct position estimation, in: Proc. IEEE Aerospace Conf., Big Sky, MT, 6–13 March 2010.

Casting Signal Processing to Real-World Data

Pau Closas, Andrea Conti, Davide Dardari, Nicoló Decarli, Emanuela Falletti,
Carles Fernández-Prades, Mohammad Reza Gholami, Montse Nájar, Eva Lagunas,
Marco Pini, Mats Rydström, Francesco Sottile, Erik G. Ström

Previous chapters have introduced several signal processing techniques related to positioning and navigation, whose performance was mainly analyzed via theory and simulation. The successive step toward a full validation of a new algorithm relies on testing on real-world data, where all the effects and the impairments due to propagation conditions, implementation losses, nonidealities, and nonlinearities of the hardware devices etc. affect the observed signals.

This chapter presents the results of a testing and validation activity articulated in the following three steps:

1. the setup of a measurement campaign and the setup of a shared database
2. the usage of this database to test algorithms
3. the development of a software defined radio (SDR)-based architecture for a GNSS receiver, to be used for advanced laboratory tests on innovative GNSS-related signal processing algorithms.

A measurements database provides a useful framework for the comparison of various algorithms in a real-world scenario. In the literature, very few studies on positioning algorithms based on realistic environments [6, 11, 22] are available.

In addition, positioning systems are migrating toward *hybridization* where data coming from heterogeneous technologies are fused to improve localization accuracy and coverage. These reasons motivate the need of measurement campaigns focused on the collection, in controlled and calibrated scenarios, of different position-related metrics, for example RSS, TOA, accelerations, and so on, obtained with different devices and technologies (ZibBee and UWB sensors, inertial sensors, etc.). This allows creation of *databases* containing measurement data that can be used as a common baseline for successive activities of performance assessment and comparison of ranging techniques, localization, and tracking schemes.

The first part of this chapter presents the description of an extensive measurement campaign performed on different indoor network configurations of wireless sensors, with the purpose of collecting position-related measurement metrics suitable for successive uses, for example the test of distributed synchronization algorithms, of positioning and tracking algorithms, of cooperative positioning procedures, and so on.

The second part of this chapter is then devoted to show the examples of exploitation of the measurements database to test some of the terrestrial indoor positioning algorithms described in the previous chapters, such as terrestrial hybrid localization and tracking algorithms and LOS/NLOS detection and mitigation schemes. The practical approach shown here is particularly instructive since it allows a validation of the performance of previously discussed theoretical solutions when processing real-world data.

The third section introduces the concept of *software defined radio* (SDR) as an enabling technology to realize extremely flexible and reconfigurable software-based radio platforms. An SDR represents a fundamental experimental tool to quickly include innovative signal processing algorithms in a laboratory terminal. A paradigmatic example of SDR is then discussed, addressing two different prototypal architectures of a GNSS receiver, realized in the laboratory.

7.1 THE NEWCOM++ BOLOGNA TEST SITE

The measurement campaign took place at the University of Bologna, Cesena Campus (Italy) in May 2009, under a joint activity promoted by the Network of Excellence NEWCOM++. It involved the usage of three different technologies to collect measurements in the same indoor environment: ZigBee-based RSS measurements, UWB-based TOA measurements, and inertial sensor-based acceleration measurements. The measurement scenario was a typical office environment. Two scenarios were investigated: *static* and *dynamic*. In the dynamic scenario, ranging measurements were collected using a small robot following predefined paths at controlled speed and acceleration; dynamic measurements were integrated with inertial data [7].

After the postprocessing phase, the NEWCOM++ WPR.B database (WPR.B-DB) has been made available to the community on the ViCE-WiCom website.[1] A vectorial file, which can be imported in common CAD tools such as Autocad, containing the precise topology of the environment is provided within the database.

7.1.1 Hardware Setup

As already mentioned, three different kinds of measurements were considered in constructing the database: RSS, TOA, and inertial accelerations. The hardware setup used to obtain such measurements is described in the following.

Equipment for RSS Measurements

During the measurement campaign, a total of 22 ZigBee-based devices, model CC2430 made by Texas Instruments, were used to measure RSS values. Each of them is a single chip solution which integrates microprocessor and radio interface compliant with the IEEE 802.15.4 standard at 2.4 GHz. Hardware characteristics of the TI CC2430 devices are reported in Table 7.1 [19].

All but one ZigBee devices were programmed as *anchor* nodes while the remaining device was programmed as a *mobile* node. The RSS measurement procedure can be summarized in the following steps, also depicted in Fig. 7.1:

---

[1]http://www.vicewicom.eu/General-Area/Contents/WPR.B-Database.

Table 7.1 Features of the ZigBee TI CC2430 Devices	
ZigBee TI CC2430	
Processor	Intel 8051 @ 32 MHz
RAM	8 KB
Flash	128 KB
Radio chip	IEEE 802.15.4 CC2420
ADC	8 channels, 8–14 bits
TX frequency	2.4 GHz
PHY layer	IEEE 802.15.4

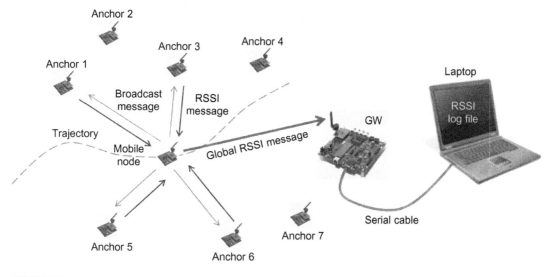

FIGURE 7.1

Procedure for the RSS measurements setup. RSSI, received signal strength indicator.

1. Every 50 milliseconds, the mobile node sends a one-hop broadcast message to the anchor nodes.
2. On reception of the broadcast message, anchor nodes perform the corresponding RSS measurements and reply immediately to the mobile node sending a message containing the RSS measurement and the anchor node identifier.
3. The mobile node collects all the RSS responses from the anchor nodes.
4. The mobile node sends a global RSS message to the gateway (GW) node connected to a laptop.
5. The GW stores the information in a log file after having inserted a time-stamp.

Equipment for TOA Measurements

Two UWB-based devices, model PulsOn220 made by Time Domain [20], were used to perform TOA ranging measurements in both static and dynamic scenarios, as shown in Fig. 7.2(a). The UWB

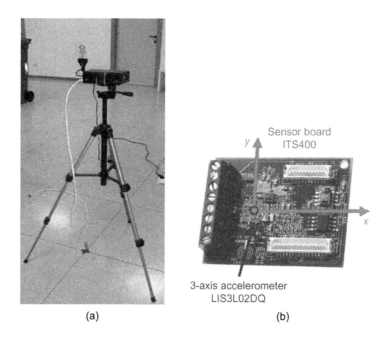

(a)   (b)

FIGURE 7.2

(a) PulsOn220 UWB device. (b) Sensor board ITS400 equipped with the three-axis accelerometer LIS3L02DQ.

Table 7.2 Features of the UWB PulsOn220 Devices	
UWB PulsOn220	
Center frequency (radiated)	Approximately 4.7 GHz
Bandwidth (10 dB radiated)	3.2 GHz
Pulse repetition frequency	9.6 MHz
EIRP	−12.8 dBm (FCC mask)
Power consumption	5.7 W
Dimensions	16.5 cm × 10.2 cm × 5.1 cm (housing w/o antenna)

node operates at a center frequency of 4.7 GHz and with a 10 dB bandwidth of 3.2 GHz. Further characteristics of the PulsOn220 devices are reported in Table 7.2 [20].

The two UWB devices were programmed to perform distance estimation using the following procedure:

1. Device 1 sends a UWB pulse and waits for a reply signal from device 2.
2. Device 2, which is in scanning mode, receives the signal and responds immediately.
3. After receiving the reply signal from device 2, device 1 makes an estimation of two-way TOA (TW-TOA).
4. Based on TW-TOA, ranging estimation can be obtained by getting an ensemble mean of several TW-TOA measurements.

Equipment for Inertial Measurements

For the dynamic measurement campaign, a wireless sensor device, model Imote2 made by Cross-bow, was used to acquire acceleration data of the mobile node. The Imote2 module can be connected with the ITS400 sensor board equipped with a three-axis accelerometer (resolution $\pm 2g$) named LIS3L02DQ and shown in Fig. 7.2(b). The hardware characteristics of the Imote2 device and ITS400 sensor are reported in Table 7.3 [4, 5].

The Imote2 device was programmed to produce one acceleration measurement every 2 milliseconds. Each sample is composed of three acceleration values, one for each axis. In particular, the Imote2 sends a message to the gateway node containing 10 acceleration samples every 20 milliseconds. The gateway node is connected through a USB cable to a laptop where the acceleration messages are stored in a log file.

7.1.2 Reference Scenarios

7.1.2.1 Static Scenario

A grid of 20 possible target positions (numbered 1–20 in Fig. 7.3(a)) defines the locations at which range (distance) measurements were taken at 113 cm height. Positions are indexed from 1 to 20. To collect static measurements, one UWB device, acting as a target node, was located in one position, while the other one, acting as anchor node, was located in a position such that the two devices were able to communicate and perform ranging measurements. Figure 7.3(b) shows the connectivity matrix between each pair of anchor–target positions. Given a specific position of the target node, the number of positions of the anchor node allowed to perform ranging measurements depends on the quality of the communication link: if UWB communication is not available between two nodes, then a null entry is present in the connectivity matrix. In order to test cooperative positioning algorithms, one thousand range measurements were also taken between each possible pair of positions in the grid. During the experiment, there were no moving objects between the two UWB devices. Reflections and attenuations

Table 7.3 Features of the Imote2 and ITS400 Devices Used for Inertial Measurements

Imote2	ITS400 sensor board
Marvell PXA271 XScale processor at 13 MHz	Compatible with the Imote2 processor/radio board
Marvell wireless MMX DSP coprocessor	Three-axis accelerometer (resolution $\pm 2g$), chip LIS3L02DQ [18]
256 KB SRAM, 32 MB FLASH, 32 MB SDRAM	Temperature sensor (resolution $\pm 1.5°C$)
Integrated 802.15.4 radio (CC2420)	Humidity sensor
Integrated 2.4 GHz antenna, optional external SMA connector	Light sensor
USB client with on-board mini-B	Four channels with 12-bit analog input
Standard I/O: $3 \times$ UART, $2 \times$ SPI, I2C, SDIO, GPIOs	

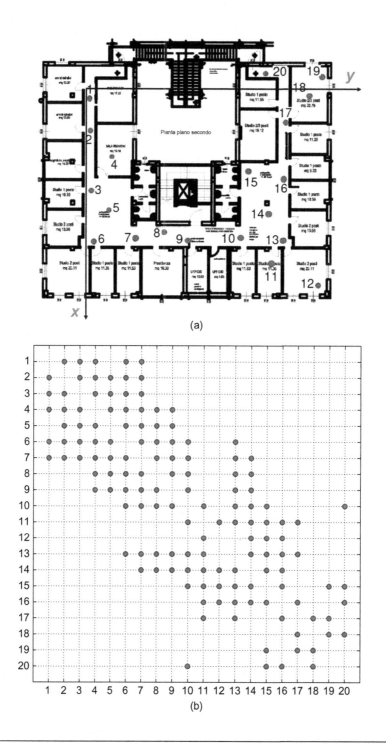

FIGURE 7.3

Deployment of the static scenario: (a) Location of the UWB static nodes. (b) Connectivity matrix: The markers show which pairs of nodes were connected during the measurement campaign and, consequently, for whom a ranging measurement exists.

of the UWB signal were mainly caused by walls, furniture, or by people that were standing behind the two devices. Some measurements were obtained in LOS, while others were in NLOS conditions.

By using UWB range measurements, different positioning algorithms can be tested to estimate the target position, also including cooperative procedures. Besides ranging measurements, the received *waveforms* for each configuration were also collected for future use.

7.1.2.2 *Dynamic Scenario*

Measurements in the dynamic scenario were performed using a mobile node that followed a predefined path. The mobile node is composed of a small Lego robot carrying the UWB device, the ZigBee node, and the inertial measurement mote. The final robot setup is depicted in Fig. 7.4. The robot follows a predefined path along two sections of the office corridor. It has two straight segments and a ninety degree curve. The path was established by means of a rail deployed on the office floor.

Two sets of anchor nodes have been used in the dynamic scenario, namely, 21 for ZigBee and six for UWB. The ZigBee anchor nodes were placed on the floor with an arbitrary distribution surrounding the robot path, while the UWB anchor nodes were placed at a height of 1.13 meters from the floor, and the related positions represent a subset of the UWB node positions defined for the static scenario.

The movement and the speed of the robot was controlled by means of a connection established between the the Lego robot and a laptop. A Matlab application was developed and used on the laptop in order to send messages to control the movement and the speed of the robot along the path. The robot speed control was programmed to produce two different behaviors: (1) a constant speed "mission" and (2) a variable speed "mission" with fourteen preset speed steps during the robot movement. The actual

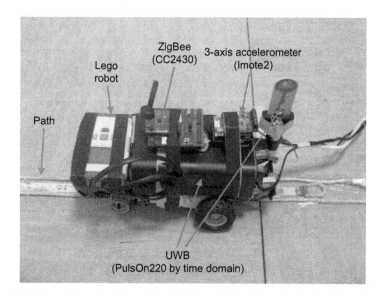

FIGURE 7.4

Dynamic scenario: Lego robot equipped with UWB, ZigBee, and inertial devices.

robot position was derived using a meter tape attached in the middle of the rail and a video-recording device.

Because of the availability of only two UWB nodes, only one UWB anchor node could be activated at a time. Consequently, the measurement was repeated six times, one for each UWB anchor node position. To synchronize the different measurements, the video recording was used to correct the start and stop time differences between trials.

During the movement of the robot UWB-based range, the ZigBee-based RSS and the three-axis acceleration were recorded. The anchor nodes position and the robot path for the dynamic scenario are reported in Fig. 7.5 (Fig. 7.5(a) for the ZigBee anchor nodes and Fig. 7.5(b) for the UWB ones). During the movement of the robot, the following measurements were performed:

- UWB-based range to anchor nodes every 500 ms.
- ZigBee-based RSS to anchor nodes every 50 ms.
- Three-axis acceleration every 2 ms.

A laptop computer placed on top of a trolley was used to control the robot movement and store the measurements collected by every mobile wireless device. For instance, the measured acceleration along the *x*-axis with respect to the robot reference system is shown in Fig. 7.6 in the case of variable speed trial.

Perturbations from the Environment

Taking into account the characteristics of the dynamic scenario, the perturbations from the environment can be grouped in the following categories:

- Office static environment (walls, office furniture, ceiling, and floor)
- Office dynamic environment (persons, doors)
- Measurement activity related (robot, trolley, and robot operators)
- Environmental interferences.

These perturbations affect both TOA and RSSI measurements, so that the database contains dynamic data that are useful to test the performance of different tracking algorithms in realistic situations.

7.2 APPLICATION OF SIGNAL PROCESSING ALGORITHMS EXPERIMENTAL DATA

The measurement database described so far has been used to test a few of the positioning algorithms presented in the previous chapters. Some of the most instructive results are presented in the following sections, related in particular to the hybridization algorithms discussed in Chapter 6.

7.2.1 Hybridization of Radio Measurements with Inertial Acceleration Corrections

The inertial accelerometer mounted on board of the robot was meant to provide inertial measurements to be fused with the radio ones to improve localization accuracy. However, since the adopted inertial sensor was a low-cost, low-performance device, the corresponding accelerometer measurements were not so accurate. In addition, the acceleration measurements were also affected by vibrations caused

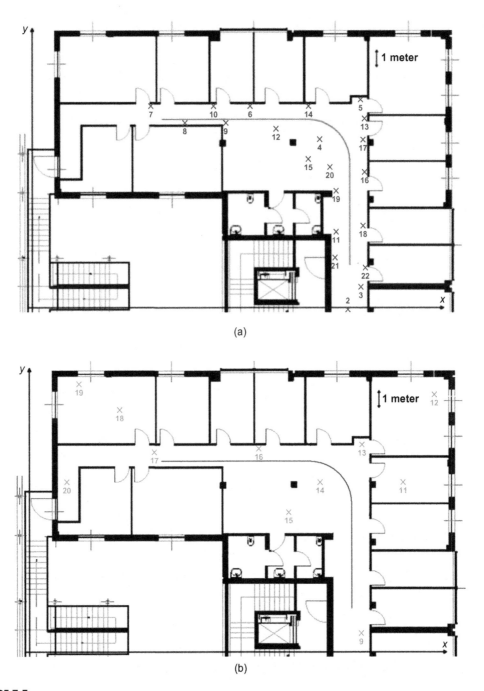

(a)

(b)

FIGURE 7.5

Deployment of the dynamic scenario: movement path and location of ZigBee (a) and UWB (b) anchor nodes.

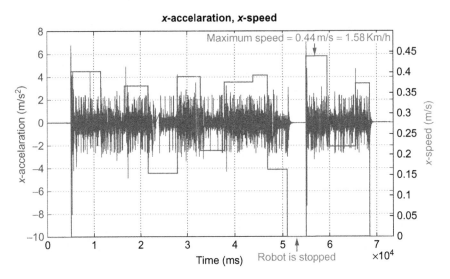

FIGURE 7.6

Dynamic scenario: Evolution of the x-acceleration measurements for a variable speed trial. The thick curve represents the average estimated speed of the robot.

by the movement of the robot. For these reasons, it is expected that such acceleration measurements cannot add very much to our observation model. Nonetheless, the acceleration data were used to *detect the motion* of the mobile node. When the mobile node is detected as "in motion" then the Bayesian recursion of the position-tracking filters is performed, whereas if the averaged inertial acceleration goes below a predefined threshold, then the mobile node is detected as "static" and the recursion is partially stopped: Only the update phase of the filters is executed to correct the previous predicted position. Even this simple motion detection technique is shown to be capable of improving the positioning accuracy.

7.2.2 EKF and SIR-PF for Hybrid Terrestrial Navigation

In this section, we present the performance assessment of the hybrid positioning approaches, introduced in Chapter 6. Specifically, both the extended Kalman filter (EKF) (Section 6.1.4.1) and the sequential importance resampling (SIR)-particle filter (PF) (Section 6.1.5.1) are analyzed.

Performance is evaluated in terms of the RMS position error:

$$\text{RMSE} = \sqrt{\frac{1}{K} \sum_{k=1}^{K} \left\| \mathbf{x}_k - \hat{\mathbf{x}}_k \right\|^2}. \tag{7.1}$$

The *outage probability* is also evaluated and reported.

7.2.2.1 *Performance of the SIR Particle Filter*

In our application of particle filtering, the SIR filter has been implemented with $N_s = 2000$ particles and with an interval between two successive measurements $\Delta t_k = 500$ ms [3]. The parameters of the mobility models are as follows: $\sigma_p = 0.26$ m (UWB), $\sigma_p = 0.5$ m (ZigBee), $\sigma_p = 0.5$ m (UWB+ZigBee), $\sigma_m = 0.07$ m (M1–M3 with constant speed), $\sigma_m = 0.056$ m (M1–M3 with variable speed), $\sigma_m = 0.08$ m (M2 with inertial data).

The impact of the mobility model in UWB positioning is investigated in Figs 7.7 and 7.8 where the position error CDFs obtained with the mobility models M1–M3 (Section 6.1.2) have been compared with the CDF obtained when the mobility model is not used. It is apparent that the mobility model better approximates the real movement of the object; the performance of the positioning system improves.

The performance is summarized in Table 7.4 for constant speed and in Table 7.5 for variable speed. The outage probability is reported for two different thresholds, $e_{th} = 1$ m and $e_{th} = 0.5$ m. Again, performance improves when mobility models incorporating more information about the actual mobile node speed are employed (i.e., M2 and M3), especially in the variable speed scenario.

Figure 7.9(a–c) presents the results for SIR-based positioning in the case of the UWB-only observation model, ZigBee-only observation model, and hybrid UWB–ZigBee with the M3 mobility model. The CDFs of the position error for the three observation models are also compared in Fig. 7.9(d). The variable speed scenario is considered. The corresponding outage probability and RMSE are reported in Table 7.6. The tracking accuracy of the UWB system is almost equal to that of the ZigBee system

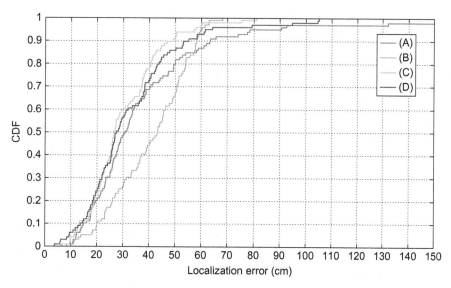

FIGURE 7.7

SIR particle filter. CDF of the position error for the UWB system; constant speed scenario; four mobility models considered: (A) no mobility model, (B) M1, (C) M2, and (D) M3 with $N_v = 18$.

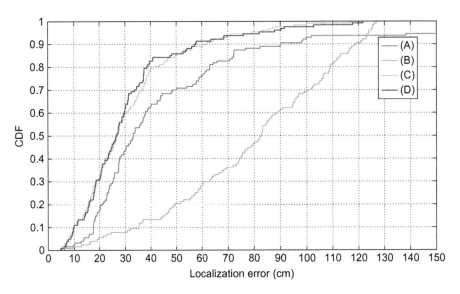

FIGURE 7.8

SIR particle filter. CDF of the position error for the UWB system; variable speed scenario; four mobility models considered: (A) no mobility model, (B) M1, (C) M2, and (D) M3 with $N_v = 18$.

Table 7.4 SIR-PF Positioning Outage Probability and RMSE when Only the UWB System Is Used for Positioning. M1, M2, M3 Mobility Models Are Considered with Constant Mobile Node Speed

Mobility Model	$\mathcal{P}_{out}$ ($e_{th} = 1$ m)	$\mathcal{P}_{out}$ ($e_{th} = 0.5$ m)	RMSE (cm)
No mobility	0.03	0.19	48
M1	≈ 0	0.33	43
M2	≈ 0	0.08	33
M3	0.02	0.14	37

Table 7.5 SIR-PF Positioning Outage Probability and RMSE when Only the UWB System Is Used for Positioning. M1, M2, M3 Mobility Models Are Considered with Variable Mobile Node Speed

Mobility Model	$\mathcal{P}_{out}$ ($e_{th} = 1$ m)	$\mathcal{P}_{out}$ ($e_{th} = 0.5$ m)	RMSE (cm)
No mobility	0.07	0.29	110
M1	0.31	0.79	85
M2	0.02	0.15	38
M3	0.02	0.14	38

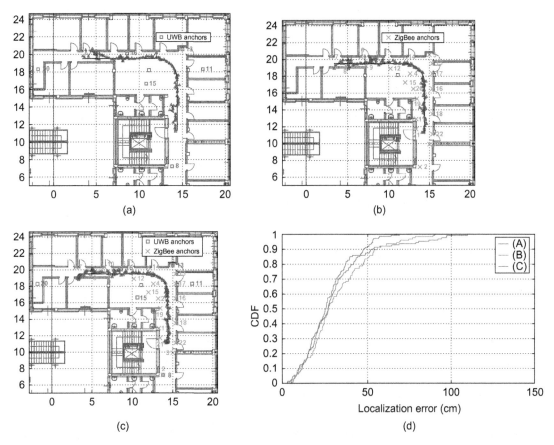

FIGURE 7.9

SIR-PF applied to the M3 mobility model with $N_{v=18}$. The plots show the true trajectory of the mobile object (solid line), the estimated positions of the mobile object (dots), and the position of the anchors (squares for UWB, crosses for ZigBee). (a) UWB-only model, (b) ZibBee-only model, (c) Hybrid UWB and ZigBee model. (d) CDF of the positioning error for the UWB-only (A), the ZigBee-only (B), and the hybrid UWB + ZigBee (C) models. Variable speed scenario.

Table 7.6 SIR-PF Positioning Outage Probability and RMSE when Mobility Model M3 Is Used. Comparison between Different Technologies with Variable Mobile Node Speed

Technology	$\mathcal{P}_{out}$ ($e_{th} = 1$ m)	$\mathcal{P}_{out}$ ($e_{th} = 0.5$ m)	RMSE (cm)
UWB	0.02	0.14	38
ZigBee	≈ 0	0.15	36
Hybrid	≈ 0	0.08	31

mainly because the number of ZigBee anchor nodes is larger than that of UWB anchors, and ZigBee anchors are in LOS condition in almost all locations. Data fusion, that is, the joint use of UWB and ZigBee sensors, represents a possible solution for improving the accuracy of positioning.

7.2.2.2 *Performance of the EKF*

As for the previous algorithms, the performance of the EKF presented in Chapter 6, Section 6.1.4.1 was evaluated considering UWB-only ranging measurements, ZigBee-only RSS measurements, and in the hybrid case.

The algorithm is evaluated for both dynamic constant-speed and dynamic variable-speed scenarios, with the performance shown in Table 7.7. The performance of the EKF algorithm is slightly better for constant-speed as compared to the variable-speed scenario. Moreover, the position estimation *accuracy* obtained with the EKF based on UWB-only ranging measurements is higher than the accuracy achieved by the EKF based on RSS-only measurements. Further improvement is obtained using the hybrid (ranging + RSS) EKF algorithm.

The positioning performance was also evaluated using the acceleration-based motion detection approach introduced in Section 7.2.1. The associated performance for constant- and variable-speed

Table 7.7 EKF Tracking Performance for Constant-Speed and Variable-Speed Scenarios (Without Inertial Aiding)

Performance Metric	Mobility Model	UWB (Ranging)	ZigBee (RSS)	Hybrid
Average positioning error (m)	Constant-speed	0.377	0.585	0.304
	Variable-Speed	0.415	0.549	0.332
Standard deviation positioning error (m)	Constant-speed	0.271	0.383	0.185
	Variable-Speed	0.240	0.419	0.169
RMS positioning error (m)	Constant-speed	0.464	0.691	0.356
	Variable-Speed	0.479	0.691	0.372

Table 7.8 EKF Tracking Performance for Constant-Speed and Variable-Speed Scenarios, with Acceleration-Driven Motion Detection

Performance Metric	Mobility Model	UWB (Ranging)	ZigBee (RSS)	Hybrid
Average positioning error (m)	Constant-Speed	0.288	0.426	0.225
	Variable-Speed	0.332	0.397	0.274
Standard deviation positioning error (m)	Constant-Speed	0.243	0.248	0.132
	Variable-Speed	0.296	0.291	0.194
RMS positioning error (m)	Constant-Speed	0.377	0.493	0.261
	Variable-Speed	0.445	0.493	0.336

scenarios also including acceleration-based motion detection is shown in Table 7.8. A comparison of Tables 7.7 and 7.8 allows us to conclude that motion detection based on acceleration measurements further improves the final positioning accuracy.

7.2.3 Coping with NLOS Measurements: A Comparison among EKF with Bias Tracking, Cubature PF, and Cost-Reference PF

Part of the UWB-ranging measurements in the database was obtained in NLOS conditions (because of the presence of walls and other furniture), so that there was the opportunity of testing and comparing tracking algorithms specifically designed for NLOS signals, namely the extended Kalman filter with bias tracking (EKFBT) (introduced in Chapter 6, Section 6.1.4.2), the cubature Kalman filter (CKF) (Chapter 6, Sections 6.3.3–6.3.3.5), and the cost-reference particle filter (CRPF) (Chapter 6, Section 6.1.5.3). Both Zigbee–RSS measurements (all LOS) and UWB-ranging measurements (LOS and NLOS) will be used in the following.

The dynamic state model of the mobile object (the robot) is written as usual as

$$\begin{bmatrix} x_k \\ y_k \end{bmatrix} = \begin{bmatrix} x_{k-1} \\ y_{k-1} \end{bmatrix} + \begin{bmatrix} v_{1,k} \\ v_{2,k} \end{bmatrix}, \tag{7.2}$$

where x_k and y_k are the robot coordinates at time k, while $\mathbf{v}_k = [v_{1,k} \quad v_{2,k}]^T$ is the process noise at time k. The process noise sequence is assumed to be a zero-mean Gaussian process with covariance $\mathbf{Q}$, that is, $\mathbf{v} \sim \mathcal{N}(\mathbf{0}, \mathbf{Q})$. The observations were collected through a set of N sensors with known locations as discussed above. For the case of ZigBee sensors, the widely used log-normal model was used to model the received power. The observation at time k for the ith sensor is

$$z_{i,k}(\text{dBm}) = P_0(\text{dBm}) - 10\alpha \log_{10}\left(\frac{||\mathbf{r}_k - \mathbf{r}_i||}{d_0}\right) + w_{i,k}, \quad i = 1,\ldots,N, \tag{7.3}$$

where $P_0 = -49.134$ dBm is the power transmitted by the ZigBee node mounted on the robot, $\alpha = 3.313$ is the path-loss exponent, $d_0 = 1$ m is the reference distance related to the log-normal model for RSS measurements, $\mathbf{r}_k$ is the position of the robot at time k, and $\mathbf{r}_i$ is the known position of the ith anchor node. $\mathbf{w}_k = [w_{1,k} \quad w_{2,k} \ldots w_{N,k}]^T$ is the measurement noise sequence which is again assumed to be zero-mean Gaussian with covariance $\mathbf{R} \in \mathbb{R}^{N \times N}$. The diagonal elements of $\mathbf{R}$ is σ_{dB}^2, that is, the variance of the shadowing, where in our case $\sigma_{\text{dB}} = 5.55$ dB.

For UWB sensors, the observation is directly in the form of noisy range measurements. The range measurement at time k for the ith sensor is

$$z_{i,k} = ||\mathbf{r}_k - \mathbf{r}_i|| + w_{i,k} \quad i = 1,\ldots,N. \tag{7.4}$$

7.2.3.1 Setup of the Observation Noise Covariance

There is no need to specify the covariances for state and measurement models in the case of CRPF. On the contrary, the other statistical reference filters need such a specification. It was found that a sufficiently small value of state covariance should be used for proper tracking.

To find the covariance for the measurement model, in EKF, CKF, and CRPF, the error of the measurements was calculated. For this purpose, the true trajectory of the robot and its velocity were calculated first, as we know how the robot was moved at each sampling instant. Knowing the true position at each sampling instant, the theoretical value of measurements was then calculated for each node.

For instance, with ZigBee nodes, the theoretical RSS for node i at time k can be calculated as:

$$(\text{RSS}_{\text{theoretical}})_{k,i} = P_0 - 10\alpha \log_{10}\left(\frac{d_{k,i}}{d_0}\right), \tag{7.5}$$

where $d_{k,i}$ is the distance between the anchor node i and the true position of robot at time k.

The error was then calculated by evaluating the difference of the real measurement and the theoretical one. This procedure was done off-line, as the setup we had used was controlled and reproducible. It was found that, due to NLOS and hardware-related problems,[2] the average error for the UWB was of the order of 2 meters. Hence, the measurement covariance was taken to be $2^2 = 4$. Similarly, the average error was found to be of the order of around 5 dB for the case of ZigBee.

For EKFBT, the covariance for the measurement model is updated each time a new measure is available. To do that, the standard deviation of each node is updated at each iteration, following

$$\sigma_n = \beta\sigma_{n-1} + (1 - \beta)\sigma_{\text{new}}, \tag{7.6}$$

where β is the update factor fixed at 0.5, σ_{n-1} is the variance at the previous step, and σ_{new} is the actualization.

7.2.3.2 *Tracking Algorithms*

Three filtering algorithms, namely EKF, CKF, and CRPF, were implemented and compared in different scenarios. The EKFBT algorithm was also implemented but it was not directly compared with the other ones because of the different initialization assumptions. The focus was to study the behavior of the filters in different circumstances.

For the experimental setup used, the magnitude of error in the measurements is very high. This means that the filter should rely less on the measurement and more on the state model. This validates the reason for taking small covariance for the state model. $\mathbf{Q}$ was chosen to be $0.1\mathbf{I}_2$ for EKF, CKF, and CRPF, and $0.001\mathbf{I}_2$ for EKFBT.

Figure 7.10 shows the trajectory of EKF, CKF, and CRPF for the case of constant speed with measurements taken using twenty-one ZigBee sensors. The EKFBT trajectory for the same circumstances is shown in Fig. 7.11(a).

The RMSEs of the estimated trajectories using EKF, CKF, CRPF, and EKFBT were calculated using the expression of the form:

$$\text{RMSE}_{\text{location},k} = \sqrt{\frac{1}{J}\sum_{j=1}^{J}\left[(x_{k,j} - \hat{x}_{k,j})^2 + (y_{k,j} - \hat{y}_{k,j})^2\right]}, \tag{7.7}$$

---

[2]Such hardware-dependent problems have been resolved in the second version of the database, currently available.

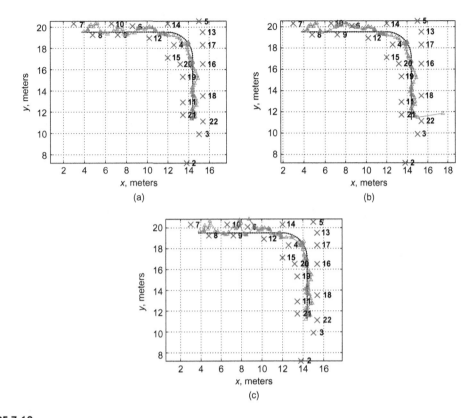

FIGURE 7.10

Trajectory of the robot moving with constant speed, estimated by different filters using ZigBee measurements: (a) EKF; (b) CKF; (c) CRPF. Crosses indicate ZigBee sensor locations, the solid curve is the true trajectory of the robot, and the curve with markers its corresponding estimate.

where J is the total number of independent filter runs,[3] (x, y) is the true location, and $(\hat{x}, \hat{y})$ is the estimated location; 100 Monte Carlo runs were simulated for this purpose.

Figure 7.12 shows the RMSE plots for the three filters in the case of constant speed and using (a) ZigBee sensor measurements and (b) UWB. With ZigBee measurements (Fig. 7.12(a)), the performance of EKF and CKF is very similar and that of CRPF is slightly better than the other two filters. In fact, all four filters seem to perform quite well with ZigBee measurements. This is due to the fact that there are many ZigBee sensors close to the trajectory for the given area of the experimental setup. Similarly, using UWB measurements (Fig. 7.12(b)), the performance of CRPF is appreciably better than the other filters. Also, the CKF is seen to perform better than EKF, which is exactly as expected.

---

[3]The average in (7.7) is over the same dataset but considering different random initialization points for the filters and also that particle filtering methods are stochastic methods, and thus their realizations might vary within independent realizations.

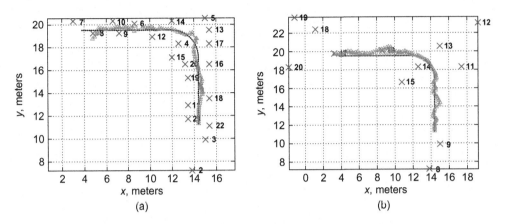

FIGURE 7.11

Trajectory of the robot moving with constant speed, estimated by the EKFBT using both ZigBee (a) and UWB (b) measurements. Crosses indicate sensor locations, the solid curve is the true trajectory of the robot, and the curve with markers its corresponding estimate.

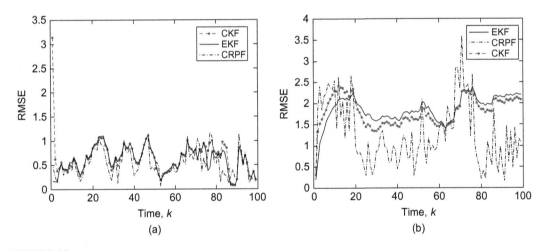

FIGURE 7.12

RMSE (in meters) of the estimated trajectories of the robot moving at constant speed. (a) ZigBee measurements. (b) UWB measurements.

Figure 7.13 shows the RMSE plots for the EKFBT in the case of constant speed and using ZigBee and UWB sensor measurements. Measurements with UWB are somewhat better than measurements with ZigBee, although only 12 UWB nodes are available compared to the 21 ZigBee-based RSS measurements. Even so, the RMSE is lower using UWB measurements.

The experiment for the case when the robot was moving with varying speed showed that the performance was similar to the case of constant speed, but with slightly degraded RMSE for all filters.

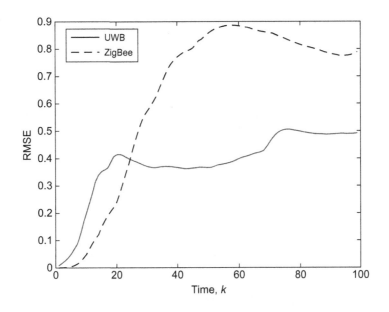

FIGURE 7.13

RMSE (in meters) of estimated trajectories using EKFBT for constant speed of the robot.

One of the major advantages of CRPF is that it utilizes a user-defined cost function rather than relying on the probabilistic model of the system. So, it is often possible to improve the performance of the filtering algorithm by making some changes in the cost function which is suitable to the application. Keeping this in mind, the cost function of the filter was modified by multiplying it by an appropriate weight matrix.

In the case of UWB measurements, the inverse of the rough range measurement obtained from the UWB sensors itself, after normalization, was taken as the weight. The assumption in doing so is that if the sensors are close to the target, then the path loss is less, and hence these close ranges are more reliable. On the other hand, if the range measurement is large, then the sensor is far from the moving target, and it will suffer more path loss. Hence, far sensors are less reliable, and thus smaller weights are assigned to them.

The same principle is applied to calculate the weight matrix for ZigBee measurements. In our experimental setup, the RSS was in the range from −90 dBm to −35 dBm. The weight is chosen such that it is inversely proportional to the absolute value of the RSS. This implies that the weight matrix for ZigBee can be calculated in exactly the same way as with the case of UWB measurements, but replacing the range measurement by RSS measurements.

There was a good improvement in the performance with the inclusion of the weight matrix for the case of UWB measurements, and some improvement for the ZigBee measurements as well. The performance of the EKF, CKF, and CRPF filters was further improved by removing the outliers in the measurements. With the knowledge of the expected range of measurements, a certain threshold was set above which the measurements were considered as outliers, thus neglected. Since the measurement

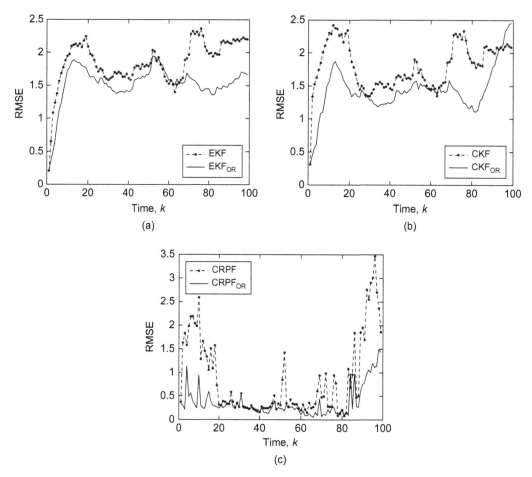

FIGURE 7.14

Comparison of RMSE (in meters) of estimated trajectories with and without the removal of outliers, for UWB measurements: (a) EKF; (b) CKF; (c) CRPF.

data obtained in our experiment were highly corrupted with errors, especially the ones close to the beginning and end of the run of the target, the removal of outliers resulted in a good performance improvement. This improvement is demonstrated in Fig. 7.14 through comparison of RMSE plots with and without the outlier measurements (subscript "OR" in the curves' labels denotes outlier removal). Figure 7.15 shows the estimated trajectory using CRPF for UWB measurements, using both weight matrix and the removal of outliers.

In Fig. 7.16, the estimated trajectory using EKFBT for both ZigBee and UWB measurements without removing outliers is shown. The covariance of the state model is the main difference with the EKFBT with constant speed. The variations in the speed can also be reflected in faster variations of the bias.

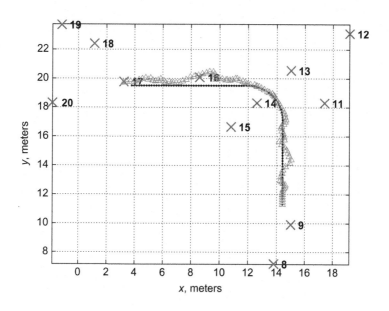

FIGURE 7.15

Estimated trajectory using CRPF with weight matrix and outliers removal, using UWB measurements. Crosses indicate UWB sensor locations, the solid curve is the true trajectory of the robot, and the curve with markers its estimate.

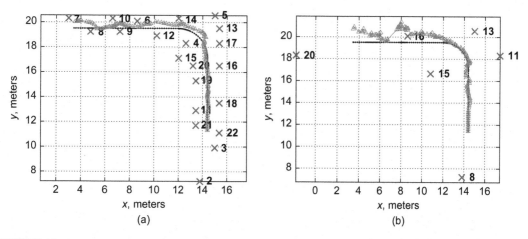

FIGURE 7.16

Estimated trajectory using EKFBT when the robot was moving with varying speed without the removal of outliers: (a) ZigBee; (b) UWB. Crosses indicate sensor locations, solid curve is the true trajectory of the robot, and the curve with markers its corresponding estimate.

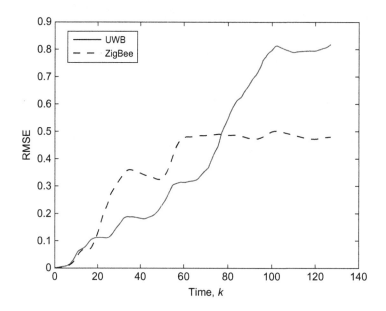

FIGURE 7.17

RMSE (in meters) of estimated trajectories using EKFBT for variable speed of the robot.

In Fig. 7.17, the RMSE is plotted for the EKFBT in the case of variable speed and using ZigBee and UWB sensor measurements. Again, measurements with UWB are somewhat better than measurements with ZigBee. In this case, only six UWB nodes have been considered.

7.2.3.3 *Conclusions*

From our tests, we found that, although the standard particle filter emerged as a promising solution to address non-Gaussian distributions, it requires a good knowledge of the actual distribution.

On the other hand, the CRPF seems to address the problem of uncertainty of the model distributions quite nicely as it does not make any assumption about the noise distribution and simply propagates the particles based on some user-defined ad hoc cost function. The CRPF turns out to be more robust than other filters and works well with any noise distribution. A minor change in the cost function, in the form of a weight matrix, was seen to give good improvement as shown in the robot tracking application.

7.2.4 **Experimental Results on LOS versus NLOS Propagation Condition Identification**

We remarked in Section 3.1.3.3 that most of the LOS/NLOS identification techniques proposed in the literature can be summarized according to the classic binary detection scheme (3.37), where the detection is performed by extracting a certain number N of features $\boldsymbol{\gamma} = \{\gamma_1, \gamma_2, \ldots, \gamma_N\}$ out of the received signal. In this section, a different approach for NLOS/LOS detection is illustrated and a performance comparison between several schemes is reported using experimental data described in Section 7.1.

Distribution-Based Identification Approach

A different method to jointly exploit the complete set of waveforms instead of taking a decision on the single waveform has been investigated in Ref. [8]. The idea is to provide an estimation of the probability distribution of the parameter of interest, and to compare it with the reference ones corresponding to LOS and NLOS propagation. The decision is taken in favor of the hypothesis for which the estimated distribution is at minimum "distance" to the reference one. The distance between distributions has to be defined according to a certain metric. Examples of such metrics are the Euclidean distance and the relative entropy or Kullback–Leibler distance. The decision criterion is then given by

$$\frac{D(\widehat{p}_\gamma \| p_\gamma^{(\mathrm{nlos})})}{D(\widehat{p}_\gamma \| p_\gamma^{(\mathrm{los})})} \underset{\mathcal{H}_1}{\overset{\mathcal{H}_0}{\gtrless}} \frac{p(\mathrm{NLOS})}{p(\mathrm{LOS})}, \tag{7.8}$$

where $\widehat{p}_\gamma$ denotes the estimated joint distribution, while $p_\gamma^{(\mathrm{los})}$ and $p_\gamma^{(\mathrm{nlos})}$ are the reference distributions of the two hypotheses. For $N = 1$ and equal prior probabilities for the two channel states, we have

$$D(\widehat{p}_\gamma \| p_\gamma^{(\mathrm{nlos})}) \underset{\mathcal{H}_1}{\overset{\mathcal{H}_0}{\gtrless}} D(\widehat{p}_\gamma \| p_\gamma^{(\mathrm{los})}). \tag{7.9}$$

The experimental results relating to this identification method presented in the next section are obtained using the (squared) Euclidean distance given by

$$D(p \| q)^2 = \int_{-\infty}^{+\infty} [p(x) - q(x)]^2 \mathrm{d}x. \tag{7.10}$$

Experimental Results

Waveforms used for testing the identification algorithms were collected during the measurements campaign presented in Section 7.1 and stored in the resulting database. The collected waveforms have a length of about 60 ns and are sampled at 24.2 GHz. In Fig. 7.18, typical waveforms from, respectively, LOS and NLOS conditions are presented.

The set of collected waveforms has been split into two disjoint sets: a *training set* and a *validation set*. The former is used for the computation of the reference probability distributions under LOS and NLOS hypotheses and the choice of the decision thresholds; the latter is used for testing the identification algorithms. In Fig. 7.19, an example of relative frequency distributions for the rms channel delay spread τ_{rms} in LOS and NLOS conditions is shown; from that, probability distributions are approximated. Each of the two sets contains 500 waveforms collected for each pair of nodes. In so doing, we have four disjointed sets: (1) the training set and (2) the validation set for nodes in LOS conditions and the same (3, 4) for nodes in NLOS conditions.

In the case of classical identification, the parameters τ_{rms} and the kurtosis $\mathcal{K}$ are first computed from the waveform under test; then a decision based on (3.37) is taken. The percentage of agreements is considered as an indicator of the quality of the identification method.

In the case of the distribution-based identification approach, the observation is taken on a certain number of waveforms belonging to the validation set related to the same pair of nodes instead of on the single waveform. Specifically, the decision is taken according to (7.9), having previously built the reference distributions through the training set of waveforms, and again the percentage of agreements is taken as an indicator of the quality. Waveforms have been filtered with a bandpass filter compliant with

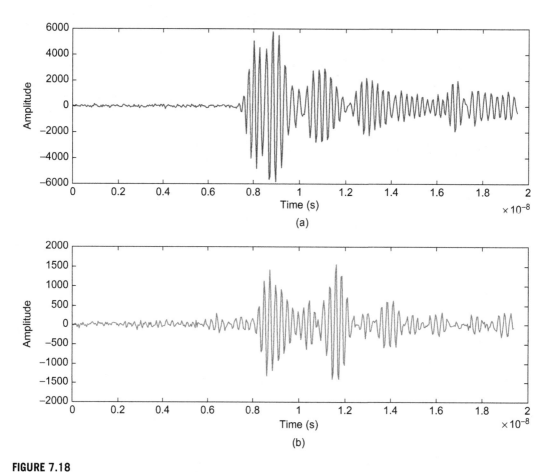

FIGURE 7.18

Example of LOS (a) and NLOS (b) waveforms (from Ref. [8]).

spectral emission of the devices used during measurements; subsequently they have been normalized to unit energy.

Table 7.9 shows the rate of correct and incorrect channel condition identification using both classic and distribution-based identification schemes. As can be noted, for the classic approach τ_{rms} and $\mathcal{K}$ features give similar results. The third column considers the joint distribution

$$\frac{p(\gamma_1, \gamma_2 | \text{LOS})}{p(\gamma_1, \gamma_2 | \text{NLOS})} = \frac{p(\gamma_1 | \text{LOS})}{p(\gamma_1 | \text{NLOS})} \frac{p(\gamma_2 | \text{LOS})}{p(\gamma_2 | \text{NLOS})} \overset{\mathcal{H}_0}{\underset{\mathcal{H}_1}{\gtrless}} \frac{p(\text{NLOS})}{p(\text{LOS})} \tag{7.11}$$

derived from (3.37) by approximating τ_{rms} and $\mathcal{K}$ as independent random variables (RVs), which leads to an improvement in the detection performance.

The second part of Table 7.9 refers to the distribution-based approach: For each couple of nodes, the distribution is computed using 100 waveforms of the validation set. This approach gives in general

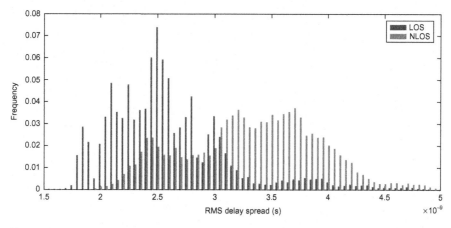

FIGURE 7.19

Example of relative frequency for RMS delay spread τ_{rms} in LOS and NLOS conditions derived from experimental data (from Ref. [8]).

Table 7.9 Parameters of LOS/NLOS Channel Identification Approaches (Classic and Distribution-Based Methods)

Decision	τ_{rms}	$\mathcal{K}$	Joint τ_{rms}–$\mathcal{K}$
Classic Approach			
$\mathcal{H}_0\|\mathcal{H}_0$	0.77	0.74	0.88
$\mathcal{H}_1\|\mathcal{H}_1$	0.79	0.77	0.79
Error rate	0.22	0.25	0.17
Distribution-Based Approach			
$\mathcal{H}_0\|\mathcal{H}_0$	0.81	0.96	0.93
$\mathcal{H}_1\|\mathcal{H}_1$	0.76	0.71	0.82
Error rate	0.22	0.17	0.13

better results, especially when using the kurtosis as parameter. Even in this case using the joint distribution instead of considering a single parameter improves the detection performance.

7.3 SOFTWARE-DEFINED RADIO: AN ENABLING TECHNOLOGY TO DEVELOP AND TEST ADVANCED POSITIONING TERMINALS

Mass-market GNSS receivers are based on application-specific integrated circuit (ASIC) technology, an approach with high development costs but extremely low cost per unit and relatively high performance. However, recent and forthcoming changes in the space segment (the modernization of global positioning system (GPS) and global orbiting navigation satellite system (GLONASS), and the advent of Galileo) are pushing designers to quickly develop new solutions for new receivers and applications. This requires more flexibility in the design and implementation processes and leads to more agile development tools. Although ASIC technology will remain pervasive for mass-market applications, other technologies such as field-programmable gate arrays (FPGAs) could be of interest for limited-market applications such as reference stations, geodesy and surveying, timing, control, and others. Finally, DSP and ordinary personal computers (PCs) constitute an invaluable platform for research and education. In this context, the SDR approach constitutes a trend that has received a lot of attention in the last decade.

Software radios are flexible all-purpose radios that can implement new and different standards or protocols through reprogramming [17]. They can reduce the cost of manufacturing and testing, while providing a quick way to upgrade the product to take advantage of newer signal processing techniques and new wireless applications.

Flexibility in programming, testing, and upgrading is obtained by replacing dedicated hardware components with programmable digital signal processors (DSPs) wherever possible. Flexibility and efficiency have in fact to be traded for processing speed [17]. The drawback of SDRs comes from the requirement of large computational power over programmable devices whose cost is nearly proportional to their computational capability and speed. However, though this proportionality was roughly exponential ten years ago [1], the corresponding gain in flexibility is increasing faster and faster. This evolved situation, clearly advantageous for SDR, is pictorially represented in Fig. 7.20.

From a practical point of view, there are substantial differences in the hardware resource allocation of a system implemented in dedicated hardware and the same system implemented in software radio: Software radio design requires additional flexibility and reusability, which impacts the partitioning of a process into its hardware and software components. This resource allocation is also closely related to the data flow of the system, since data have to be processed with minimal delays and overheads.

In light of these considerations, *software receiver* refers to a receiver that processes the received data stream without using dedicated hardware to implement specific functions, but using a generic programmable platform implementing suitable software modules. In so doing, the same platform can be reprogrammed to perform different functions (flexibility), at the expense of requirement of high computational power.

7.3.1 The Software-Defined Radio Concept

In software radios, the most challenging section where analog signal processing is replaced by digital signal processing is the so-called digital front-end (DFE) [12]. The digital front-end is the part of the transceiver implementing front-end functionalities like frequency conversion and channel filtering. According to the "SDR doctrine," the boundary between analog and digital sections of the front-end has to be pushed as much as possible towards the antenna, so as to minimize the presence of analog components. Thereby, the digital front-end turns out to be one of the most power and time-critical functionalities of the software radio terminal. This is due to the combination of large bandwidth and high dynamic range of the signals to be processed [12].

At the boundary between the analog and digital part of the radio front-end, the analog-to-digital converters (ADCs) need a large number of quantization bits in order to meet the high dynamic range requirements of wideband signals. Although advanced ADCs applying noise-shaping techniques

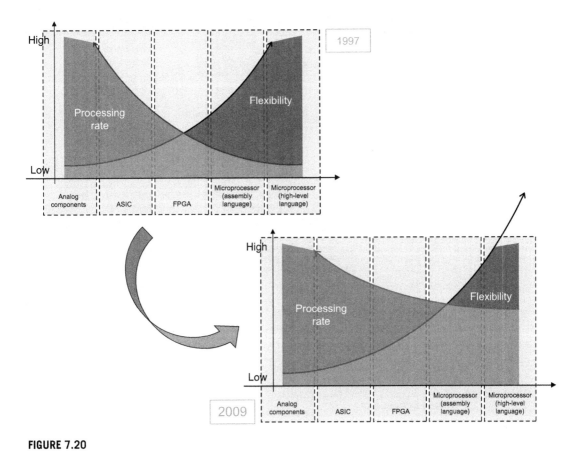

FIGURE 7.20

Performance trends of SDR architectures in the last ten years.

(e.g., "sigma-delta" converters) can provide extreme dynamic range at relatively low expense, the ADC cost and performance, as well as the data storage capacity and transfer rates necessary to handle huge amounts of digitized data, remain the bottleneck that generally prevents the realization of an "ideal" and low-cost SDR front-end. The most commonly adopted solution is to migrate from a *full-band digitization* approach, where the whole signal bandwidth entering the antenna stage is digitized at the radio frequency (RF), to a *partial-band digitization* scheme, where the maximum frequency carried by the signal to be digitized is reduced by means of analog conversion and filtering at an intermediate frequency (IF). The partial-band digitization concept results in architectures that employ IF (or nearly baseband) sampling, as shown in Fig. 7.21.

The digital front-end basically performs channelization and sample rate conversion. Channelization comprises all tasks necessary to select the channel of interest. In communications, this includes conversion to baseband, channel filtering, and possibly despreading. The great challenge when designing a digital front-end of a software radio terminal for communications is the exploitation of commonalities of the different standards of operation.

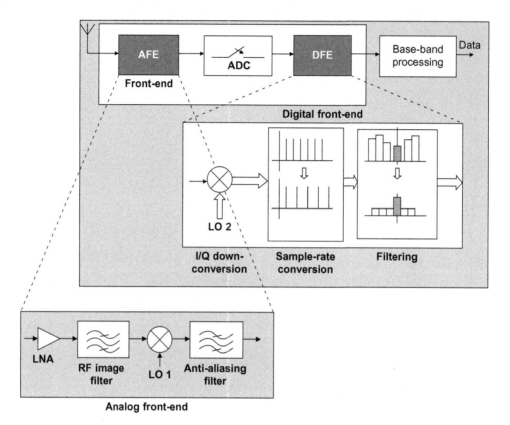

FIGURE 7.21

Conceptual scheme of a software radio receiver employing partial-band digitization and IF sampling.

7.3.2 SDR Technology in Localization

It is nowadays accepted that the SDR technology applied to localization receivers (specifically, GNSS receivers) produces significant benefits for prototyping new equipment and analyzing signal quality and performance, thanks to the replacement of some hardware components with more flexible and easier-to-test software-based signal processors. Developers are now attempting to extend SDR to commercial end-user products, including mobile devices incorporating GNSS functionality [21].

There is fervent activity in the design of novel architectures for multistandard and multimedia receivers. The software radio, with its high level of flexibility, represents the enabling technology for the commercial success of such architectures. As a step forward, future cognitive receivers are expected to autonomously reconfigure their own architecture to adapt to the environment they operate in, both in terms of communication signals and localization approaches.

In the field of localization, several, extremely different types of receiver architectures have been generally indicated as SDR. However, they can be classified in at least three main categories [21]:

- The category of the "postprocessing receivers" (that is, the most populated) includes the software tools developed to test new algorithms. An example is the Matlab-based software receiver [2]; another example, representing an "SDR front-end simulator" for GNSS signals, is the Matlab-based N-FUELS tools [10, 14]. They cannot operate in real time and are to be considered as realistic simulation tools.
- True SDR receivers are those presenting real-time processing capabilities, offered by DSPs and, in the very recent years, by conventional PCs. Today, specialized processors for embedded applications are also gaining popularity, partially replacing the use of DSPs. PC-based and embedded SDRs also differ substantially in the overall software design, even if they share common signal processing and navigation algorithms [21]. Embedded systems are generally size and power-limited (they address the market segment of the portable devices); therefore, their computational capability, as well as their target functions, is limited. On the other hand, the far higher computational capability offered by a PC-based solution enables the realization of complex, multifunctional SDR receivers. An example of PC-based GNSS receiver is the N-GENE software receiver [13].
- The last category is represented by FPGA-based implementations. FPGAs offer the option for allocating a number of processing resources operating in parallel, with full algorithmic and architectural reconfigurability. In addition, they present specific features for integration with other hardware and for (re)programmability in the field. For this reason, they can be seen as a sort of "hybrid," hardware/software SDR devices, conceptually different from other PC-based or embedded approaches. As has already been discussed, the most critical challenge of an SDR architecture is the coexistence of high computational efficiency featuring a wide software reconfigurability in a relatively small device with reduced power consumption (particularly in embedded systems). In this context, the joint use of FPGA platforms and 32-bit embedded microcontrollers might provide a good compromise between the two opposite requirements. FPGAs enable parallelization and reconfigurability of the processing resources, while microcontrollers provide the necessary flexibility and quick implementation times, for those procedures that need not be implemented on the FPGA.

7.3.2.1 *Two Examples of SDR Architecture for a GNSS Receiver*

To have a more practical idea of the different implementation architecture implies, we discuss two examples of SDR implementation of a GNSS receiver, the first one being a hybrid FPGA-based receiver, the second one falling in the category of PC-based architectures [16].

Let us consider the architecture outline of a generic GNSS receiver shown in Fig. 7.22. Following the SDR concept, the blocks contained in the dashed rectangle can be realized with at least two conceptually different architectures:

- The first implementation is a hybrid FPGA microprocessor architecture, implementing a hybrid hardware/software GPS/Galileo receiver able to receive the signals transmitted on the L1/E1 bands, as well as one among those broadcast on the GPS L2 and Galileo E5, E5a, E5b, or E6 bands.
- The second realization is a fully software GPS/Galileo receiver, belonging to the category of real-time, PC-based receivers, able to process the signal broadcast on GPS L1 and Galileo E1 bands. The processing and storage capacity available at the time of receiver development (2007–2009) on a standard portable laptop limits this architecture to processing E1/L1 signals only, while a higher computational efficiency allows a wider band allocation in the hybrid architecture.

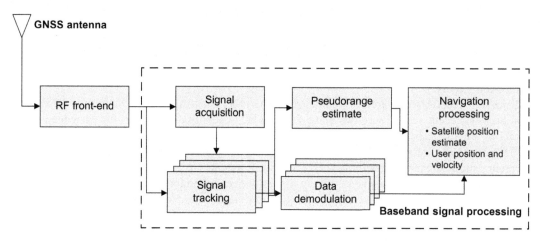

FIGURE 7.22

Functional block diagram of a GNSS receiver.

The conceptual architecture of the two solutions, which maps all the functionalities appearing in Fig. 7.22, is depicted in Fig. 7.23. Note that a multifrequency RF front-end serves both receivers.

Hybrid Receiver

The main aspect of the hybrid receiver based on reconfigurable hardware is the partitioning between hardware and software resources. In other words, the partitioning defines the receiver functions that will be implemented on the FPGA and those that will be programmed on the microcontroller.

The rate of the operations performed in real time is the main constraint that drives partitioning. Among the functional units of a generic GNSS receiver (signal acquisition, signal tracking, data demodulation, pseudorange computation, PVT), the most challenging is the signal acquisition and in general correlation with the ranging codes. For such functions, the operating rate is equal to the raw sample rate at the ADC output (i.e., of the order of tens of MHz, depending on the bandwidth). On the contrary, the operational rate can be lowered to a few kHz for all of the other functional blocks.

As a consequence, the partitioning for the DFE of the addressed GNSS receiver is shown in Fig. 7.24: For each channel, the correlation between the input signal and the local codes is implemented on the hardware platform, while the control signal of the tracking loops is generated by proper software routines through a discrimination function, after every integration period. Pseudorange computation, navigation data demodulation, and calculation of user's position are low-rate operations, and as such are implemented on the microcontroller.

Software Receiver

The software receiver addressed in this example is described in Refs. [13, 16]. The digitized data at the output of the front-end are sent to the PC through a standard USB 2.0 interface, enabling the transport of a maximum theoretical data flow of 480 Mbit/s. Using 8 bits to represent each sample, the corresponding maximum useful sampling frequency is 60 MHz. Obviously, this value is theoretical because it does not take into account the inherent overhead introduced by the USB protocol, nor the

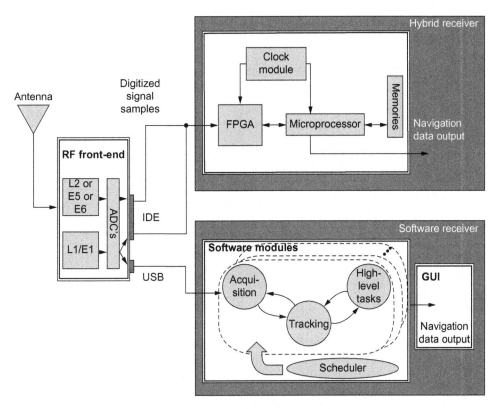

FIGURE 7.23

Conceptual architecture of the two SDR-based GNSS receivers.

hardware limitations of the low-cost devices currently available on the market. However, this value is higher with respect to the values necessary for reliably representing the Galileo and the GPS signals transmitted on L1 (typically less than 20 MHz).

The problem of real-time data management has a strong impact on the selection of the algorithms to be used in the design of the receiver functional blocks, as well as on their possible approximation and simplification. The optimal real-time data flow management also requires a dedicated "software partitioning" among functions that should be written in a high-level language (e.g., C language), and functions that may need the use of a lower-level language (e.g., Assembler language). The ANSI-C language is used to program all the functions that do not operate at the same rate as the input data, such as satellite search during the acquisition phase, update of the control signal of the tracking loops, data demodulation, pseudorange evaluation, and PVT computation. On the contrary, computation of the correlations between the received and the local signals, as well as generation of such local signals (both pseudorandom codes and sinusoidal signals), must operate with a high-rate data flow; these functions can take advantage of the parallelism offered by the "single instruction multiple data" (SIMD) facility of the Pentium processors, and therefore they are realized in machine language.

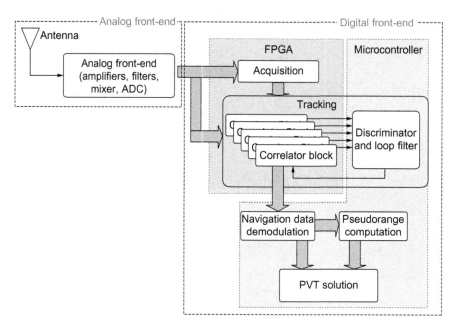

FIGURE 7.24

HW/SW partitioning for a hybrid FPGA/microcontroller-based GNSS receiver.

A high-level block diagram that describes the architecture of the fully software receiver is shown in Fig. 7.25, where the inherent tasks of the GNSS receiver (signal processing functions) are shown separated from the specifically user-related tasks. The core of the receiver is the *receiver core engine scheduler*, a scheduler of the activities that manages the parallelism between the channels and their relative operations, like for instance the acquisition or tracking state. The scheduler also handles visualization of the data on the graphical interface and takes care of both the demodulation of the navigation message and the calculation of the user's PVT.

7.3.2.2 *Exploiting SDR to Realize an Advanced Laboratory Prototype*

An example of a fully software GNSS receiver is represented by the laboratory prototype of a *peer-aided GNSS receiver* [9, 15].

The objective was to provide a proof-of-concept of an innovative hybrid cooperation technique (named *peer-based aiding to GNSS*) based on GNSS signals and terrestrial radio communications, derived from an assisted-GPS approach, but adapted to a scenario where "peer users" assist each other in their positioning task, in the absence of any AGPS-enabled base station. All the peers are equipped with a GNSS receiver and an end-to-end communication module and are organized in a partially connected network, without the necessity of any master node or cluster head. A portion of the peers is in *light-indoor* condition, where terminals have just partial visibility of the GNSS satellites and therefore their signal acquisition performance may be degraded. In this situation, the peer-based aiding approach

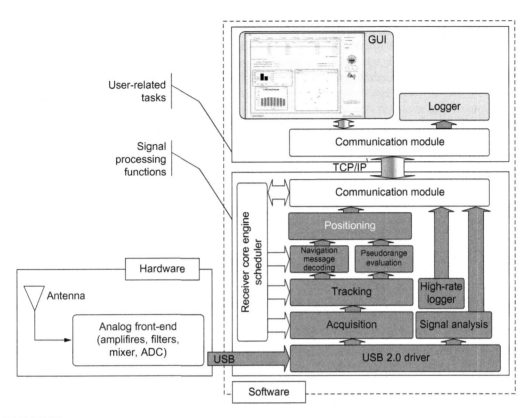

FIGURE 7.25

Block diagram of the fully software receiver. Dark-shaded blocks represent GNSS-specific signal processing functions, while light-shaded ones represent user-related tasks. Dark arrows indicate the signal data flow, while the white ones indicate the control/schedule data flow.

exploits the communication link between a degraded peer and one (or more) other peer with open-sky visibility to the satellites to exchange proper GNSS data that may assist the light-indoor terminal in its acquisition procedures.

In the peer-based cooperation architecture, each terminal in open-sky visibility acts as a kind of assisted GPS (AGPS) server, sending its GNSS positioning data to another user, which employs them to improve its own GNSS localization performance. In this case, the former terminal is the *aiding user*, while the latter is the *aided user*.

The aiding data transmitted by each aiding user are represented by the following: Parameters of the visible satellites (list of satellite vehicle numbers, associated C/N_0 ratio, almanac, ephemeris, UTC time), Doppler frequency, navigation data synchronization, signal quality control information. Because of the relative physical proximity among users (far closer than in the average terminal-to-BS case for the AGPS), the values of the Doppler frequency and code phase received from an aiding user are with high probability also valid to perform a raw satellite signal acquisition on the aided user side, who can consequently reduce the dimension of the search space in its acquisition phase, thus reducing its time-to-first-fix (TTFF) or increasing its sensitivity. Nonetheless, these aiding data, in particular those related to the code phase and data synchronization, can be effective only under some hypotheses, like the short distance between users and also the precise synchronization of their clocks. While the first is valid in local area networks, the second is not guaranteed with low-cost GPS receivers, which are generally equipped with commercial temperature-compensated crystal oscillators (TCXOs), without dedicated hardware to ensure a fine synchronization. Considering the rate of the open GPS C/A code, a synchronization with an offset lower than 1 μs (i.e., offset lower than a chip width) is required in order to guarantee acceptable performance of the peer-based aiding technique [9].

A novel acquisition algorithm based on the concepts of peer-based aiding has been presented in Ref. [15]. It includes a procedure to estimate the synchronization error between a couple of peers and uses it as an additional aiding data [9]. Through evaluation of the synchronization error, the aided terminal can adjust the time stamp related to each aiding data.

The fully SW receiver prototype described in the previous section has been used to validate the effectiveness of the mentioned algorithm through some tests in real indoor conditions.

The complete test scenario is depicted in Fig. 7.26, showing the Istituto Superiore Mario Boella (ISMB) premises in Torino, Italy. The PC on the left, equipped with the fully SW receiver programmed to support the peer-based aiding procedure, represents the aiding user: It is connected to an antenna on the laboratory rooftop, with full visibility of the GPS satellites. The second PC, again equipped with a copy of the fully SW receiver, was placed in the laboratory corridor, close to the window. In such a location, its patch antenna could acquire only a few GPS satellites, thus being unable to resolve

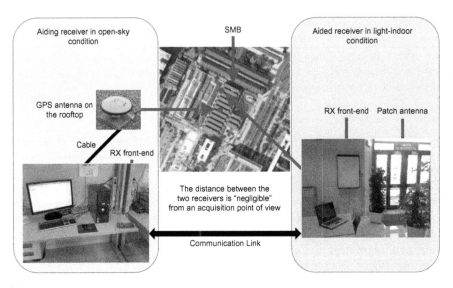

FIGURE 7.26

Setup of the laboratory test for the peer-based aided acquisition.

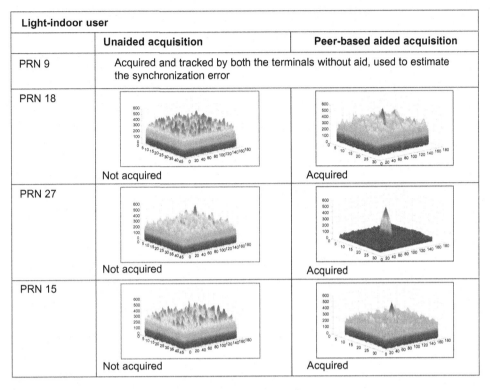

FIGURE 7.27

Acquisition in light-indoor conditions: Search spaces computed without aiding (left column) and employing the peer-based aiding procedure (right column).

its position. The two receivers were connected through the Ethernet LAN network that allowed the exchange of synchronization messages (although in the final version of the demonstrator the wired LAN connection was replaced by a wireless one).

During the test, the receiver in light-indoor condition could autonomously track only one satellite, identified as PRN 9. This PRN was also tracked by the aiding user; therefore, it was used to estimate the clock offset among the two PCs. Based on the signal processing of PRN 9, the aiding user created the aiding message. After several tests, it was possible to conclude that about 80% of the estimations of the GPS bit transition had an error less than 12 μs, starting from a synchronization offset of several ms. With such an estimate, it was possible to perform long coherent integrations, avoiding data bit transitions and thus enhancing the signal acquisition sensitivity.

Through the use of the aiding message in the acquisition procedure, the second user could acquire more satellites and consequently resolve its position. Figure 7.27 shows the acquisition search spaces computed by the light-indoor terminal without (left) and with (right) the aiding algorithm, revealing the presence of correlation peaks that were impossible to find without aiding. Thanks to the aiding data, the frequency range of the search space was limited to 250 Hz around the likely value, subdivided in 10 frequency bins, 25 Hz each. All the correlation peaks rose above the noise floor around the center of the search space, proving that the estimates of the frequency shift and code delay in the aiding message were also reasonably accurate for the aided terminal.

References

[1] D.M. Akos, A Software-Radio Approach to Global Navigation Satellite System Receiver Design, Ph. D. dissertation, Russ College of Engineering and Technology, Ohio University, 1997.

[2] K. Borre, D.M. Akos, N. Bertelsen, P. Rinder, S.H. Jensen, A Software Defined GPS and Galileo Receiver: A Single-Frequency Approach, Birkhauser, Boston, 2007.

[3] A. Conti, D. Dardari, M. Guerra, L. Mucchi, M.Z. Win, Experimental characterization of diversity tracking systems, IEEE Syst. J. (submitted).

[4] Crossbow, IMOTE2 datasheet. http://bullseye.xbow.com:81/Products/Product_pdf_files/Wireless_pdf/Imote2_Datasheet.pdf (accessed: June 2011).

[5] Crossbow, ITS400 datasheet. http://bullseye.xbow.com:81/Products/Product_pdf_files/Wireless_pdf/TTS400 Datasheet.pdf (accessed: June 2011).

[6] D. Dardari, A. Conti, J. Lien, M.Z. Win, The effect of cooperation on localization systems using UWB experimental data, EURASIP J. Adv. Signal Process. (2008) (Special Issue on Cooperative Localization in Wireless Ad Hoc and Sensor Networks).

[7] D. Dardari, F. Sottile (Eds.), WPR.B database: Annex of progress report II on advanced localization and positioning techniques: Data fusion and applications. Deliverable DB.3 Annex, 216715 Newcom++ NoE, WPR.B, 2009.

[8] N. Decarli, D. Dardari, S. Gezici, A.A. D'Amico, LOS/NLOS detection for UWB signals: A comparative study using experimental data, in: 5th IEEE International Symposium on Wireless Pervasive Computing (ISWPC 2010), IEEE Press, Piscataway, NJ, 2010, pp. 169–173.

[9] S. Digenti, M. Nicola, L.L. Presti, M. Pini, Technique for the estimation of PC clock offset in a GNSS-aided network of collaborative users, in: 5th Esa Workshop on Satellite Navigation Technologies (Navitec 2010), Noordwijk, The Netherlands, IEEE, 2010.

[10] E. Falletti, D. Margaria, B. Motella, A complete educational library of GNSS signals and analysis functions for navigation studies, Coordinates V (8) (2009) 30–34.

[11] M.R. Gholami, E.G. Strom, F. Sottile, D. Dardari, A. Conti, S. Gezici, et al., Static positioning using UWB range measurements, in: P. Cunningham, M. Cunningham (Eds.), Future Network and Mobile Summit 2010 Conference Proceedings, IIMC International Information Management Corporation, Florence, Italy, 2010, pp. 1–10.

[12] T. Hentschel, M. Henker, G. Fettweis, The digital front-end of software radio terminals, IEEE Pers. Commun. 6 (4) (August) (1999) 40–46.

[13] A. Molino, M. Nicola, M. Pini, M. Fantino, N-Gene GNSS software receiver for acquisition and tracking algorithms validation, in: 17th European Signal Processing Conference (EUSIPCO), Glasgow, Scotland, 2009.

[14] NavSas group, N-FUELS website. http://www.navsas.ismb.it/ns/index.php?option=com_content&task=view&id=53&Itemid=78 (accessed: June 2011).

[15] M. Panizza, C. Sacchi, J. Varela-Miguez, S. Morosi, L. Vettori, S. Digenti, et al., Feasibility study of an SDR-based reconfigurable terminal for emergency applications, in: 2011 IEEE Aerospace Conference, Big Sky, Montana, IEEE, 2011.

[16] M. Pini, M. Fantino, P. Mulassano, L.L. Presti, L. Bragagnini, The IRGAL project: Innovation and research on Galileo in Piedmont, in: ION-GNSS 2008 Conference, Savannah, GA, Sept. 2008, pp. 1514–1519.

[17] S. Srikanteswara, J.H. Reed, P. Athanas, R. Boyle, A soft radio architecture for reconfigurable platforms, IEEE Commun. Mag. 38 (2) (February) (2000) 140–147.

[18] STMicroelectronics, LIS3L02DQ datasheet. http://www.datasheetcatalog.org/datasheet2/1/036ete1wq40k5k gkoi91389ap83y.pdf (accessed: June 2011).

[19] Texas Instruments, CC2430 datasheet. http://focus.ti.com/lit/ds/symlink/cc2430.pdf (accessed: June 2011).

[20] Time Domain®, PulsOn220 website. http://www.timedomain.com/datasheets/P220aSK.php (accessed: June 2011).

[21] J.H. Won, T. Pany, G.W. Hein, GNSS software-defined radio, Inside GNSS 1 (5) (July/August) (2006) 48–56.

[22] J. Youssef, Contributions to navigation solutions based on ultra wideband radios, inertial measurement units, and the fusion of both modalities, Technical report, University of Grenoble, Grenoble, France, PhD Thesis, 2010.

Index

Printed in the United States
By Bookmasters